CORROSION RESISTANCE

Edited by **Hong Shih**

Corrosion Resistance
Edited by Hong Shih

Published by InTech
Janeza Trdine 9, 51000 Rijeka, Croatia

Publishing Process Manager Marija Radja

Technical Editor Teodora Smiljanic

Cover Designer InTech Design Team

Additional hard copies can be obtained from orders@intechopen.com

Corrosion Resistance, Edited by Hong Shih
p. cm.
ISBN 978-953-51-0467-4

For EU product safety concerns:
IN TECH d.o.o., Prolaz Marije Krucifikse Kozulić 3, 51000 Rijeka, Croatia,
info@intechopen.com or visit our website at intechopen.com.

Meet the editor

Dr. Hong Shih, Senior Technology Director at Lam Research Corporation, received his Ph.D. in Metallurgy from the Pennsylvania State University in 1986. He was a Post Doctor and a Research Professor at University of Southern California. He holds B.S. in Chemistry from Beijing University in 1970 and M.S. in Electrochemistry from Academic Sinica in 1981. He has been working at Applied Materials (1994-2002) and at Lam Research Corporation (2002-). He has 120 publications in the referred journals and books, 160 technical consulting reports,200 semiconductor equipment specifications and best known methods, 1,300 technical reports, and 190 USA and International Patents granted/published. His pioneer work in the study of materials under high density plasma opened a new scientific and engineering field. Dr. Shih is a well-known scientist on electrochemical impedance spectroscopy, corrosion monitoring and control, electrochemical techniques and expert in plasma resistance materials, anodized aluminum, surface coatings, precision wet cleaning, defect, particle, and metal contamination reduction on plasma etching equipment served in semiconductor wafer fabrication.

Contents

Preface

The book titled "Corrosion Resistance" has a significant meaning because the terminology "Corrosion Resistance" is one of the most important indications of materials under corrosion study and characterization.

The book has covered the state-of-the-art technologies, development, and research progress of corrosion studies in a wide range of research and application fields. The authors have contributed their chapters, each on their special field, on corrosion characterization and corrosion resistance. The applications of corrosion resistance materials will also bring great values to reader's work at different fields. In addition to traditional corrosion study, the book also contains chapters dealing with energy, fuel cell, daily life materials, corrosion study in green materials, and in semiconductor industry.

It has been an awarding process for me to learn from all the authors throughout the editing process. I would like to take this opportunity to express appreciation from my heart to all the authors for your contributions to this book. Your dedication and your enriched expertise will add value to this book. Your contribution will definitely be rewarded by the readers. I am very glad to see that the book "Corrosion Resistance" will become one of the critical books in Corrosion Sciences & Engineering.

Finally, I would like to express many thanks to InTech Open Access Publisher, particularly to Mrs. Marija Radja, Editor Relations Consultant, for her tremendous work. Without her help and dedication, it would be impossible to have this book published so smoothly.

Hong Shih, Ph.D.
Etch Products Group, Lam Research Corporation Fremont,
California,
USA

1

A Systematic Study and Characterization of Advanced Corrosion Resistance Materials and Their Applications for Plasma Etching Processes in Semiconductor Silicon Wafer Fabrication

Hong Shih
Etch Products Group, Lam Research Corporation, Fremont, California, USA

1. Introduction

Corrosion resistance is a quantitative measure of materials under study in a special corrosion environment. With a continuous development in semiconductor IC industry on silicon wafer fabrication and the rapid shrinkage of silicon wafer feature size as of to 32nm, 25nm and even smaller, the requirement on **corrosion resistance chamber materials** under high density plasma becomes extremely critical and difficult. Therefore, the study, characterization and new development of **corrosion resistance chamber materials** have been a critical task for technologists in semiconductor IC industry. Without the correct selection of **corrosion resistance chamber materials**, it is impossible for semiconductor IC industry to achieve current technology levels. Among steps of semiconductor wafer fabrication, plasma dry etching is the most difficult and comprehensive step which has a very high standard for the selection of **corrosion resistance chamber materials**.

Different from the traditional corrosion study, materials under high density plasma during dry etching processes should meet a comprehensive requirement. First of all, chamber materials must demonstrate a very **high corrosion/erosion resistance** under high density plasma during etching processes as well as in the defined wet chemicals. Since different etching processes use different reactive gases and chamber conditions, **chamber materials selected** have to vary in order to meet the variations of etch processes and chamber conditions. Secondly, chamber materials should have low particles and defects during etching processes because the particles and defects generated from chamber materials will fall on the silicon wafer, serve as the killer defects, and cause the loss of wafer production yield. Thirdly, chamber materials should avoid metal contamination issues on silicon wafer. The high metal contamination generated from chamber materials such as Na, K, Fe, Ni, Cr, Cu et al will electrically shorten the dies on a silicon wafer and directly impact wafer production yield. In addition to the above requirements for **advance corrosion resistance chamber materials**, chamber etching process stability and transparence, chamber impedance matching and stability, thermal and dielectric properties, capable of surface texturing, microstructure, wet cleaning compatibility, resistance to in-situ waferless plasma dry cleaning (WAC), RF coupling/grounding efficiency, adhesion of etch by-products and polymer, bonding strength of surface coatings, fundamental mechanical properties, manufacture ability and reproducibility, and cost of the materials have to be considered as a whole. After reviewing the overall requirements of chamber materials, one can see that it is not an easy task to find a suitable chamber material for semiconductor IC wafer fabrication which can meet all the above requirements. A comprehensive study has to be performed in order to find and to determine **the best chamber materials** among the existing materials in the world for a special etching application. Due to the complexity, the qualification processes of a new advanced corrosion resistance material for plasma etching processes are not only very time-consuming, but also very expensive. The fundamentals and applications of plasma dry etching and the applications on equipment of semiconductor silicon wafer fabrication have been described and studied extensively [1-20].

Let's take some examples. During metal etch processes (etching aluminum line), Cl_2 and BCl_3 are the main reactive gases to etch aluminum. Ar, N_2, CF_4, CHF_3, C_2H_4, or O_2 are also used during etching and WAC processes. Therefore, **the selected chamber materials** have to demonstrate **high corrosion (and erosion) resistance** to these gases under the high density plasma. For silicon etch processes, SF_6, NF_3, HBr and HCl are the main reactive gases used to etch silicon. Other gases may also be used in the etching and WAC processes. **The selected chamber materials** should have a **high corrosion resistance** to both F-based gases and HBr corrosion. In particular, the corrosion of HBr mixed with a very tiny amount of water on the heat effected zone of stainless steel has been an issue for a long time. For dielectric etching processes, C_xF_x based reactive gases are usually used with a high applied power in order to etch oxide. Chamber materials selected have to show high corrosion and erosion resistance at a relatively high temperature and high power. For special etch processes such as metal hard mask etch, MRAM etch, high K etch and Bevel etch, special process gases and chamber conditions are applied. Therefore, the requirements to **corrosion resistance chamber materials** may be different. Since some plasma etching processes even etch noble metals such as Pt, Ru and Ir, one has to find chamber materials which can survive in these aggressive plasma etching conditions. Therefore, chamber materials which are submitted to sputtering, chemical etching, ion-enhanced etching, as well as ion-enhanced inhibitor etching have to be studied and characterized thoroughly for each special etching applications. There is no any material which can meet all plasma etching applications. In summary, some of the key requiements of chamber materials is listed below [21-39]:

- Low erosion rate under vigorous plasma bombardment.
- Low chemical reaction rate under many chemistries such as
- Cl_2/BCl_3-containing plasma,
- Fluorine-containing plasma,
- $HBr/HCl/Cl_2$-containing plasma,
- Oxygen-containing plasma.
- Low transition metal transport to the workpiece.
- Low or zero particle contamination from surfaces.
- Strong interface bonding of surface coatings for long part lifetime.
- Excellent and repeatable dielectric properties for RF energy coupling.
- Pore-free ceramic materials and low porosity surface coating to avoid undercut corrosion and to eliminate substrate attack.
- Excellent adhesion of etch by-products and polymers.
- Excellent corrosion resistance in wet chemistry cleaning.

- Cost effective in manufacturing.
- Excellent repeatability from part to part and wafer to wafer.

The relationship among chamber materials used in semiconductor etching equipment, etching, wet cleaning, sputtering, and etch by-products is shown in Fig. 1[21, 22, 35, 36].

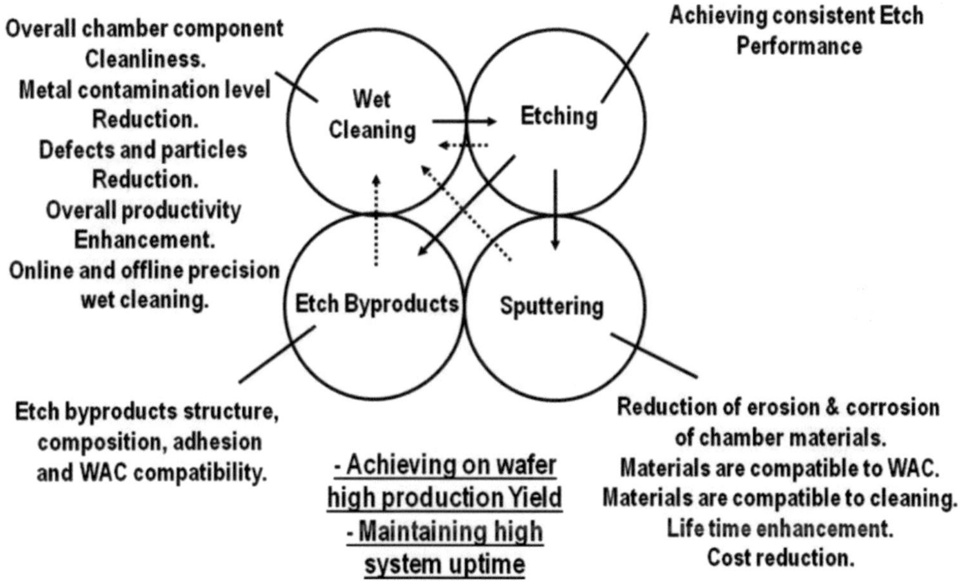

Fig. 1. The relationship of chamber materials, etching processes, precision wet cleaning and etch byproducts in a plasma etching chamber [21, 22, 35, 36].

For etching process requirement, a metal etch film stack and common issues are shown in Fig. 2

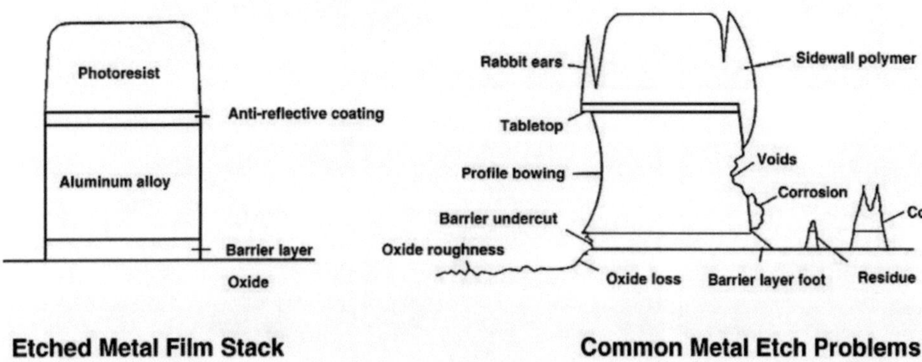

Fig. 2. Aluminum metal film stack and common issues in etching processes [21, 22].

The killer defects which are generated during metal etching processes fall on metal lines and cause the loss of production yield in wafer fabrication. The killer defects may either come from chamber materials or etch by-products [21, 22, 23, 25, 27].

Fig. 3. Killer defects generated in aluminum metal etch processes.

The corrosion/erosion patterns of chamber materials showed three different patterns under plasma. Fig. 4 shows the three different patterns [21, 22, 28, 35, 36].

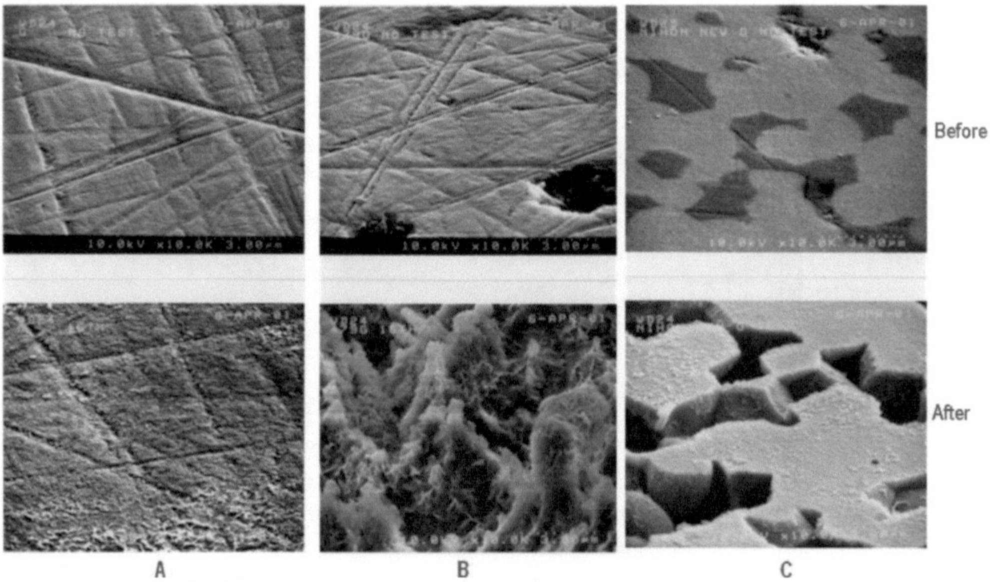

Fig. 4. Corrosion/erosion patterns of chamber materials under plasma etching (pictures are at 10,000x magnification). Model A indicates a uniform corrosion/erosion which can either be higher or low; Model B shows the attack at grains of materials; and Model C shows the attack at grain boundaries of materials.

In pattern A, chamber materials can be etched/sputtered by plasma uniformly. The etch rate can be very low or very high. The etch rate depends on the plasma chemistry, process recipe, and materials. For example, high purity Y_2O_3 has showed very high plasma resistance in both Metal and Silicon etch processes. A uniform corrosion/erosion pattern is observed [21, 22, 25, 30]. For anodized aluminum, a very high corrosion/erosion rate is observed under BCl_3-containing plasma during metal etch processes. In fact, an anodized aluminum film with a 75 μm in thickness (hot deionized water sealed) can only hold up to 1,800 wafers in some etch process recipes in production. This became a severe problem on the lifetime of anodized aluminum in aluminum etch processes. For Silicon etch processes, the lifetime of anodized aluminum has no issue because there is no obvious attack of reactive gases to anodized aluminum in Silicon etch processes. The only concern is the formation of AlOF on anodized aluminum surface when SF_6 and NF_3 are used in the etching processes. The formed AlOF can either have chamber particle issue or cause etch process shift due to the surface impedance change on anodized aluminum surface. The wet cleaning to fully remove AlOF film on anodized aluminum surface is very critical to achieve a consistent and reliable etching performance on wafer fabrication. Fig. 5 shows an anodized aluminum metal etch chamber after 1,800 wafer fabrication in production. The anodized aluminum is fully removed under Cl_2/BCl_3 high density plasma [21, 22, 25, 30]. The major attack of anodized aluminum is due to the chemical reaction between BCl_3 and Al_2O_3 under the high density plasma. The reaction rate of the attack to anodized aluminum highly depends on the gas concentration of BCl_3 and the plasma density. Chamber erosion test indicates that Cl_2 has little attack to anodized aluminum [21, 22, 25, 30].

$$BCl_3 + Al_2O_3 = B_2O_3 + AlCl_3$$

Fig. 5. Anodized aluminum is fully removed under Cl_2/BCl_3 plasma after only 1,800 wafers in production (about 60 RF hours). The special attacking pattern depends on the local plasma density and gas concentration.

The high density plasma reaction rate of BCl_3 with anodized aluminum or high purity alumina at different flow is shown in Fig. 6. The high reaction rate occurs on chamber top window due to both high density plasma and gas flow. On the chamber wall, the reaction rate of BCl_3 with Al_2O_3 is almost a liner relationship, but the reaction rate is much lower than that on the chamber top window. It also indicates that without BCl_3 flow, the reaction rate of Cl_2 plasma has almost no attack to anodized aluminum or to high purity alumina. In the plasma reaction rate study, the total flow is fixed as of 205 sccm. The Argon gas flow is fixed at 40 sccm. The test starts at 165 sccm Cl_2 flow and zero flow of BCl_3, then 155 sccm Cl_2 flow and 10 sccm BCl_3 flow, until the final flow of Cl_2 is 85 sccm and BCl_3 flow is 80 sccm. The test coupons are either on chamber top window or on the chamber wall. Nine different types of anodized aluminum and high purity alumina are used in the test [21, 22, 25, 30]. The reaction rate is in the unit of mils per RF hour.

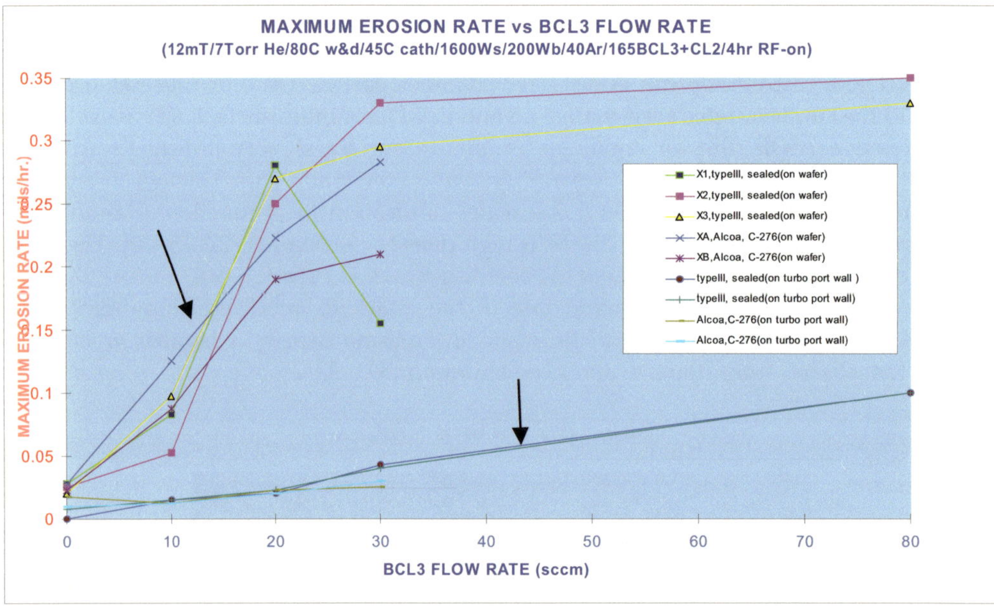

Fig. 6. The maximum reaction rate of Al_2O_3 at different BCl_3 gas flow under high density plasma [21, 22, 25, 30].

In pattern B, chamber materials suffered the attack of grains under plasma. CVD SiC grains can be attacked by Cl_2-containing plasma and SiC material cannot be used in aluminum etch processes as a chamber material. Grains of high purity ceramic (99.5% or higher alumina) can also be attached by BCl_3 in metal etch processes and the glass phases such as SiO_2, CaO, and MgO remain. It is obvious that BCl_3 can attack anodized aluminum and alumina under high density plasma. For high purity AlN, AlN grains are attacked by fluorine-containing plasma such as SF_6 and NF_3, the grain boundaries remain.

In pattern C, chamber materials are attacked at grain boundaries only. A typical example is high purity alumina (99.5% or higher in Al_2O_3), glass phases such as SiO_2, MgO, and CaO can react with fluorine-containing gases. In this case, grains of alumina remain. The formation of AlOF may occur on alumina surface. Fig. 7 shows a ceramic ESC surface which is covered by a layer of AlOF after exposure to plasma in silicon etch processes [35, 36].

A 33% atomic% of F is detected on electrostatic chuck ceramic surface (high purity alumina) indicating the formation of AlOF on high purity alumina surface under fluorine-containing plasma. The chemical treatment to remove AlOF using TMAH (tetramethylammonia hydroxide) is also demonstrated in Fig. 8 [35, 36, 40].

Fig. 7. A uniform AlOF film (rainbow color) covers the ceramic surface of a used electrostatic chuck after silicon etch processes.

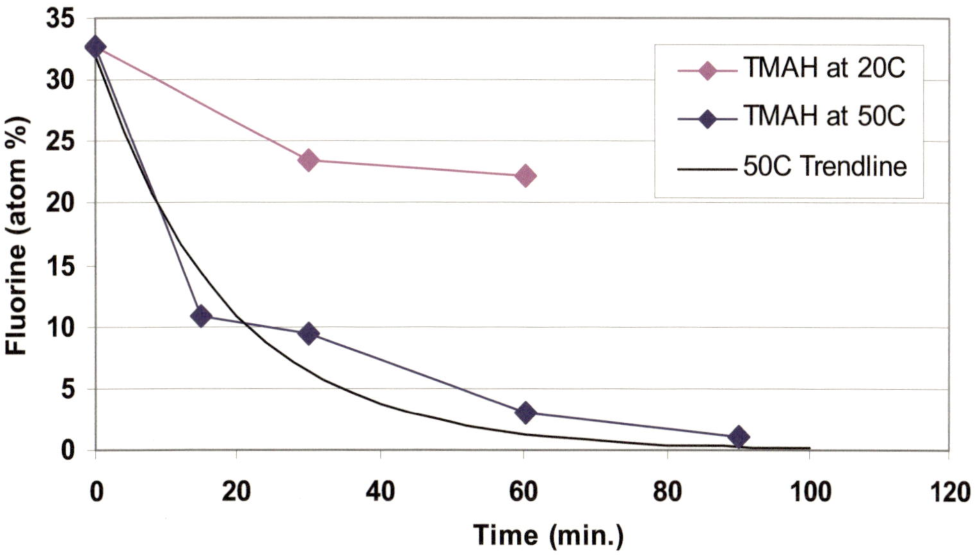

Fig. 8. The ceramic surface of a used electrostatic chuck contains about 33 atom% fluorine on the surface film with a rainbow color.

Since the limitation of the space of this chapter, anodized aluminum, boron carbide, and Y_2O_3 as chamber materials will be demonstrated.

2. Experimental and discussion

Since the limitation of the space of this chapter, anodized aluminum and boron carbide coating as chamber materials will be demonstrated and discussed in details.

All the ceramic and CVD test coupons (except anodized aluminum coupons) are polished to mirror surface finish with the average surface roughness less than 1.0 μ-in in Ra. The anodized aluminum coupons are anodized with the surface roughness less than 32 μ-in (as-received). The thermal spray coatings keep the as-coated surface. Ceramic and CVD coated coupons weigh the pre-test weight. Anodized aluminum and thermal spray coating coupons were measured to obtain the average coating layer thickness before test. All the test coupons were soaked in IPA for 5 minutes, wiped by both IPA and acetone, rinsed by deionized water (DIW) for 1 minute and baked at 110°C for 30 minutes. A special thermal conductive tape was used to mount the test coupons on locations in the etching chamber. Three locations were selected to mount the test coupons. Test coupons are mounted on chamber top window, chamber wall, and electrostatic chuck surface, respectively. Etching systems used in this study include Applied Materials 200mm and 300mm etching tools and Lam 2300 etching tools. During materials characterization on chamber wall, on chamber top window and on electrostatic chuck, a dummy aluminum wafer was used to cover the electrostatic ceramic surface. The etching process recipe keeps running for three minutes, followed by a cooling down process for about two minutes, than repeat the etching recipe. The minimum process time (RF hours) cycled is 120 RF hours and the longest process time cycled is 200 RF hours.

A typical process recipe under a 200mm etch tool is shown below [21, 22, 25, 30]:

Step 1. Plasma Etching & Dechuck Steps
- 12mT/85Cl_2/80BCl_3/40Ar/1600W_s/200W_b/45°C cathode/80°C wall & top window/7 Torr He flow/180 seconds.
- 100Ar/TFO/500W_s/100W_b/5sec.

Step 2. Cooling Down Step
- 12mT/200Ar/45°C cathode/80°C wall & top window/120 seconds.

Repeat process recipe (step 1 and step 2) until the accumulated RF hours achieve 120 RF hours or 200 RF hours, respectively.

After plasma etching processes, all the test coupons were removed from the chamber. A post wet cleaning was carried out to remove polymer, etch by-products, and other contaminants. All the coupons were then DIW rinsed and baked at 110°C for 30 minutes. Post weight measurements were carried out to obtain the average thickness loss per RF hour. For anodized aluminum and spray coating coupons, post thickness measurements were carried out in order to obtain the coating thickness loss per RF hour.

All the coupons are studied by SEM before and after plasma etching process. The corrosion/erosion rates of different test coupons are recorded and compared as mils per RF hour. Test coupons on etch chamber top window, chamber wall and on electrostatic chuck surfaces are shown in Fig. 9.

Fig. 9. Test coupons in etching chamber are mounted on chamber top window (left) and on chamber wall (right) and on the dummy aluminum wafer on an electrostatic chuck surface (right, white surface).

Fig. 10 shows the test results of various materials obtained from worldwide suppliers. The letters of A, B, C, D et al represent the suppliers and their materials. Agreements were signed for not allowing to release the names of the worldwide suppliers and their materials. The plasma etching rate is in the unit of mils (1 mil = 25.4 μm). It is obvious that either YAG (solid solution of Al_2O_3 and Y_2O_3) and solid Y_2O_3 can reduce the plasma etching rate at the order of 40-50 times in comparison with the previously used chamber materials such as high purity alumina. That is the reason why Y_2O_3 has been as one of the leading chamber materials in plasma etching tools in the past 10 years for the leading semiconductor etching equipment companies.

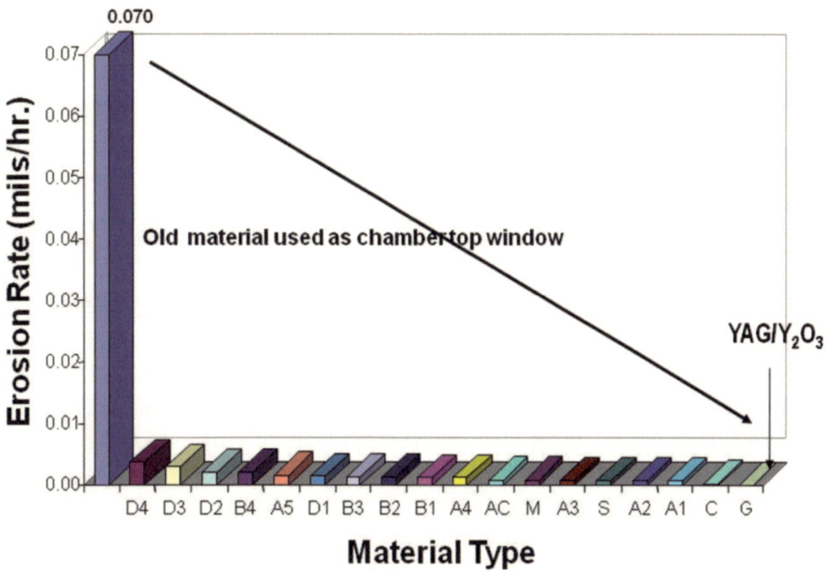

Fig. 10. Test results of new and old chamber materials in plasma etching on chamber top window. The etch rate reduction of new chamber materials can reduce the etching rate by 40 to 50 times.

For test results on chamber wall, the etching rate of anodized aluminum from various suppliers w/wo hot DI water seal is between 0.050 to 0.070 mils / RF hour. For boron carbide coating through a thermal spray method, the etching rate is below 0.001 mils/RF hour. For sintered or hot pressed boron carbide, the etching rate is between 0.0001 to 0.0007 mils/RF hours. It is also obvious that the plasma etching resistance of boron carbide can improve the plasma resistance by 50 times or higher. In fact, boron carbide coated chamber has been using at worldwide wafer fabrication customer sites since 1998. 50 to 100 times chamber life improvement has been demonstrated since 1998 up to today [21, 22, 25, 30, 41].

In order to select the best configuration of surface coatings such as B_4C (boron carbide), three configurations are considered. Configuration 1 is the coating of B_4C on bare aluminum surface. Configuration 2 is the B_4C coating on anodized aluminum surface. Configuration 3 is the B_4C coating on anodized aluminum surface and then HL126 sealant is used to seal the pores in the spray coating layer. HL126 contains methacrylate esters and it can fill very tiny pores. The metal contamination levels of HL126 is pretty low. All metal levels are below 1 ppm except the sodium level at 57 ppm. Permabond HL126 is a high strength and low viscosity anaerobic threadlocker. Its properties are listed in the attached table below:

PHYSICAL PROPERTIES OF THE UNCURED ADHESIVE *	
Properties	
Base Resin,	Methacrylate Esters
Solid, %	100
Color	Green
Viscosity, cP, 25°C (77°F)	20
Consistency	Liquid
Gap Filling, in	0.005
Specific Gravity	1.09
Flash Point, °C (°F)	>110(230)
Shelf Life stored at or below 27°C (80°F), months	12
Electric Properties	
Dielectric strength, MV/m	11
Electric Resistance, ohm-cm	10^{15}
Performance properties of the cured sealant	
Operating temperature °C (°F)	150 (300)

Table 1. Properties of HL126 sealant

The corrosion resistance of boron carbide coated coupons after plasma etching is tested by HCl bubble test method which was first proposed by Shih in 1992 and was used as a standard technique in anodization study for IC industry in 1994 [42]. The fundamental concept of the defined HCl bubble test method can be explained as follows. The dilute HCl solution can go through the pores and micro-cracks on coating and anodized aluminum layer to react with bare aluminum under the coating or under the anodized aluminum. When HCl reacts with aluminum alloy, hydrogen bubbles will generate. Streams of hydrogen bubbles can be observed and the time to start the continuous hydrogen bubbles can be recorded and compared for different coating configuration and different types of anodized aluminum before and after plasma etching processes. Shih [43, 44] has set up the method at two major semiconductor equipment companies since 1994 and the method has been widely accepted by worldwide anodization suppliers. The method is simple, low cost and fast in comparison with ASTM standard salt spray test method [45, 46]. The test results show that boron carbide coating on anodized aluminum and sealed with HL126 provide the best corrosion resistance among the four configurations as shown in Fig. 11 [25].

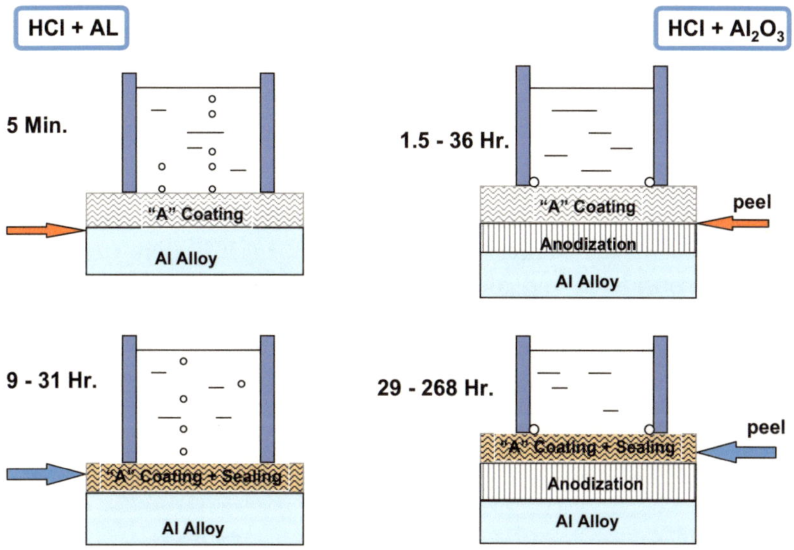

Fig. 11. After plasma etching for 200 RF hours, Boron carbide coated anodized aluminum sealed with HL126 sealant provides the best corrosion in all configuration.

The wet cleaning compatibility of four configurations is also tested by soaking the large size B_4C coated rings in saturated $AlCl_3$ solution at pH=0 for 90 minutes, then put the rings in an environmental chamber to monitor the time when boron carbide coating starts to peel off. The test sequence is shown in Fig. 12 [25].

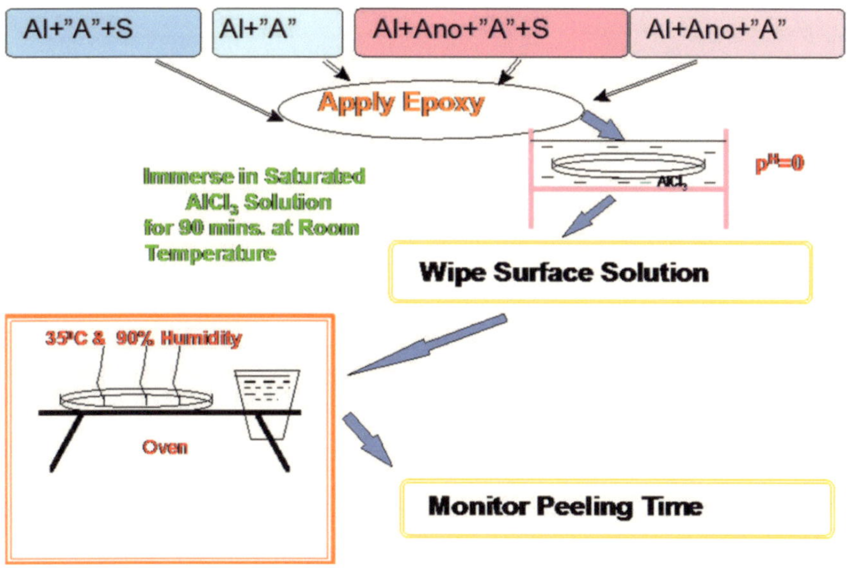

Fig. 12. Wet cleaning compatibility test of four configurations of boron carbide coated rings.

For coating on bare aluminum alloy, the entire coating layer peeled off during immersion in the saturated $AlCl_3$ solution at pH=0.0. The coating on anodized aluminum can hold 45 hours in the environmental chamber and the coating layer peeled off completely at 47 hours. Both coating on bare aluminum alloy and on anodized aluminum with the use of HL126 sealant can hold up to 114 hours in the environmental chamber without any failure. At 114 hours, the environmental chamber test was stopped. From the test results of HCl bubble test and wet cleaning compatibility test, coating on anodized aluminum with the use of HL 126 sealant can provide the best corrosion resistance. This configuration is selected as the final configuration as the new chamber wall material.

In order to qualify boron carbide coating as a new chamber material, many aspects have to be considered. One of the concerns is the impact to ICF (ion current flux). Three configurations are considered and compared in the etching chamber. The ICF of anodized aluminum chamber is used as the baseline. Boron carbide coatings on bare aluminum or on anodized aluminum are studied through ICF measurements. The results showed that the three configurations have the compatible ICF. The results of ICF measurements are shown in Fig. 13 [21, 22, 25, 30].

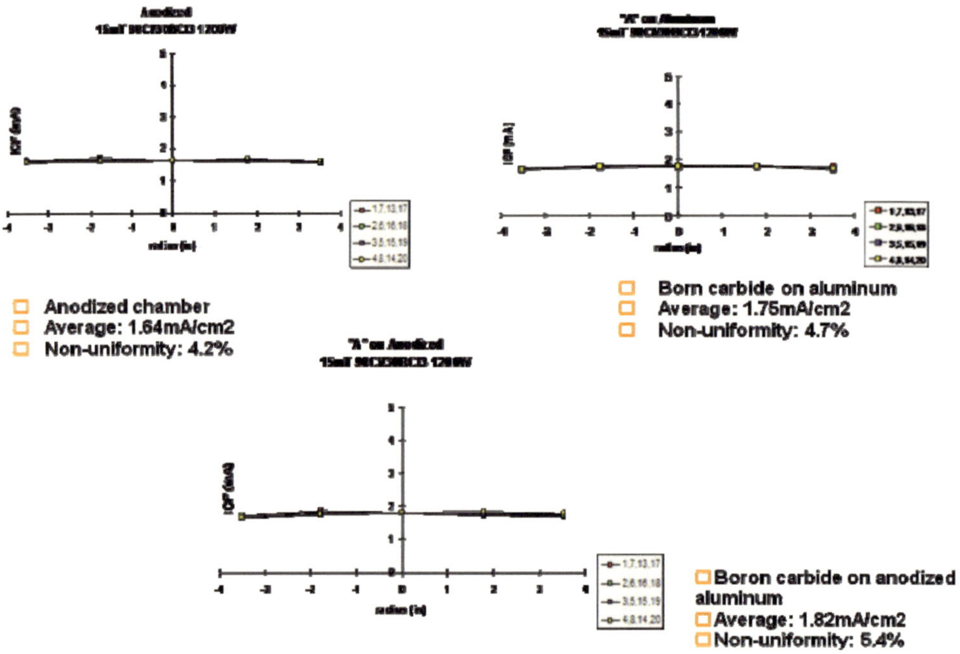

Fig. 13. ICF measurements on the wafer during the use of three configuration chambers.

Another concern is the potential damage to gate oxide. The leakage current measurements on the gate oxide show that born carbide coating does not introduce damage to gate oxide.

The measurements of leakage current of gate oxide are shown in Fig. 14 [21, 22, 25, 30].

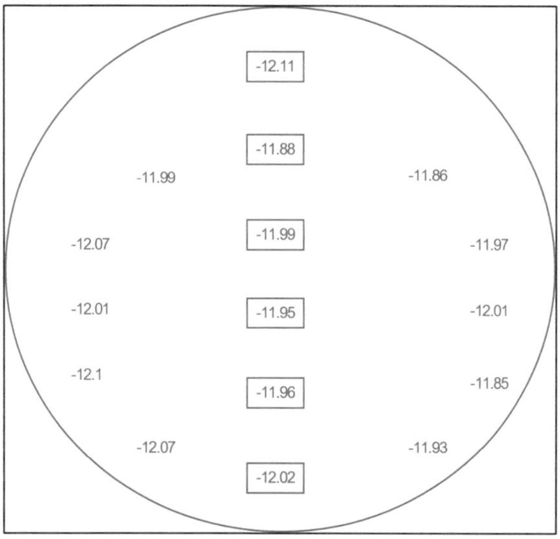

Fig. 14. The leakage current of gate oxide in log scale indicates that there is no damage to gate oxide when a born carbide coated chamber is used.

The metal contamination using a boron carbide coated chamber has shown meeting the specification of metal contaminations such as Ca, Co, Cr, Cu, Fe, K, Mg, Mn, Mo, Na, Ni, Ti, and Zn in 1,000, 2,000, and 3,000 wafer marathons, respectively.

There is no metal contamination introduced when a high purity boron carbide coating is introduced as the new chamber wall coating.

The monitoring data of on-wafer aluminum etch rate, etch rate non-uniformity, defect and particle performance, and thickness measurements of boron carbide coating before and after plasma etching processes are shown in the following figures and tables. In Fig. 15, the particle data during a 3,000 wafer marathon are provided and compared with the specification requested by customers. It is obvious that new B_4C coated chamber wall can meet the requirement of particles. In this study, particles at and larger than 0.2 μm are recorded. The B_4C coated chamber wall can also provide excellent aluminum etch rate and etch rate non-uniformity through the entire 3,000 wafer marathon as shown in Fig.16 [47].

The boron carbide coated chamber is also qualified through a 2,000 wafer marathon for etching of 0.15 μm feature size. Excellent aluminum etching performance is demonstrated as shown in Fig. 17 [48].

On a 300mm etch tool, boron carbide coated chamber was also used in a 1,000 wafer marathon. The boron carbide coated chamber meets all the requirements including aluminum etch rate and etch rate non-uniformity, etch profiles, defects and particles, metal contamination [49]. The particle performance at 0.12 μm or larger is the critical requirement. It is obvious that the boron carbide coated chamber can meet the requirement. The up limit of particle allowance at 0.12 μm or larger is defined as 50 adders/per wafer.

After plasma etching O_2/Cl_2 for 120 RF hours, the thickness of pre and post boron carbide coating on anodized aluminum is measured and the data are listed in Table 2

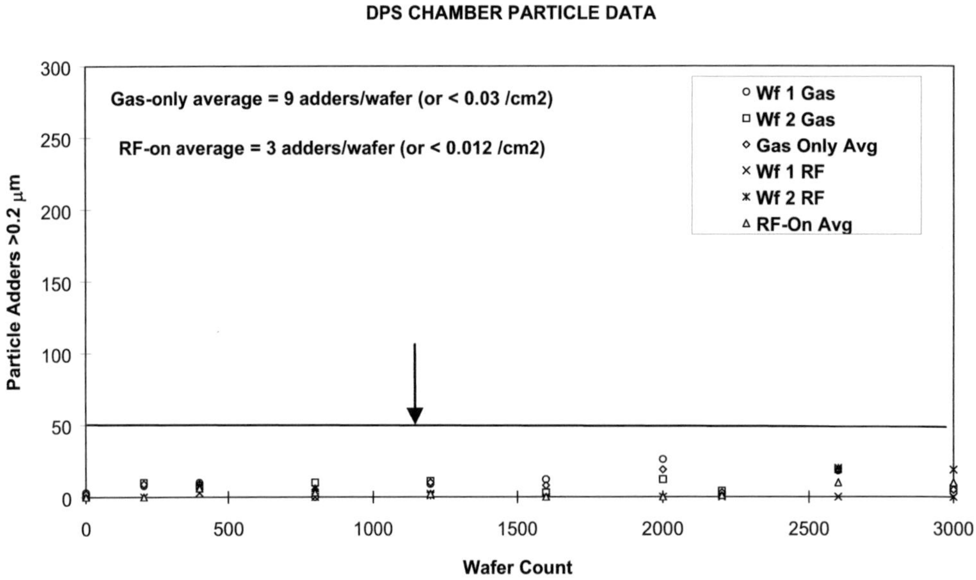

Fig. 15. Gas-only and RF-on particles during a 3,000 wafer marathon. The up limit of allowance of defect and particles is defined as 50 adders/ per wafer.

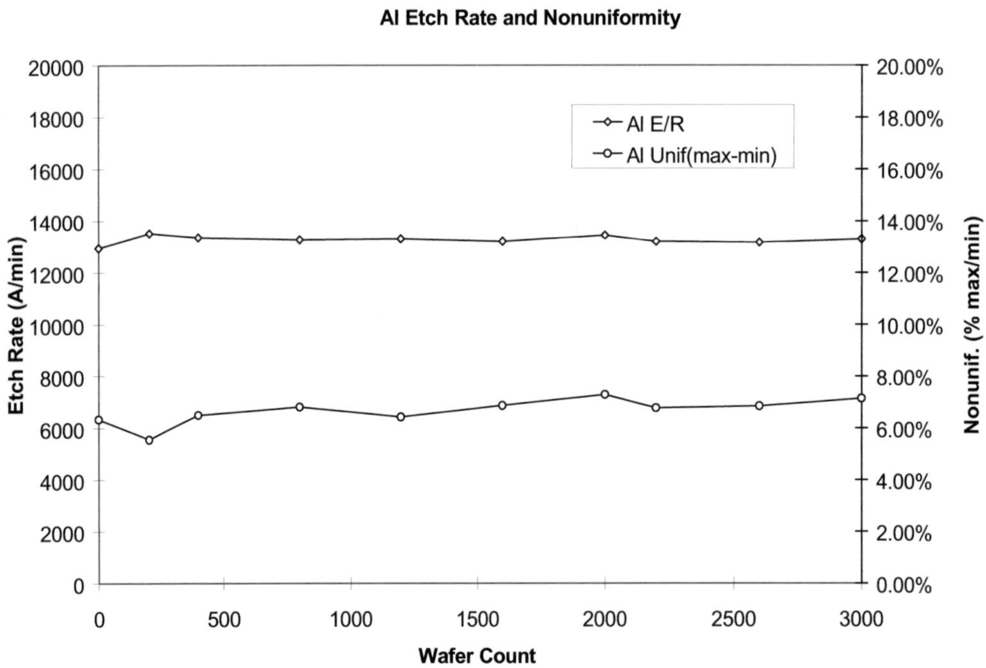

Fig. 16. Al etch rate and etch rate non-uniformity during a 3,000 wafer marathon.

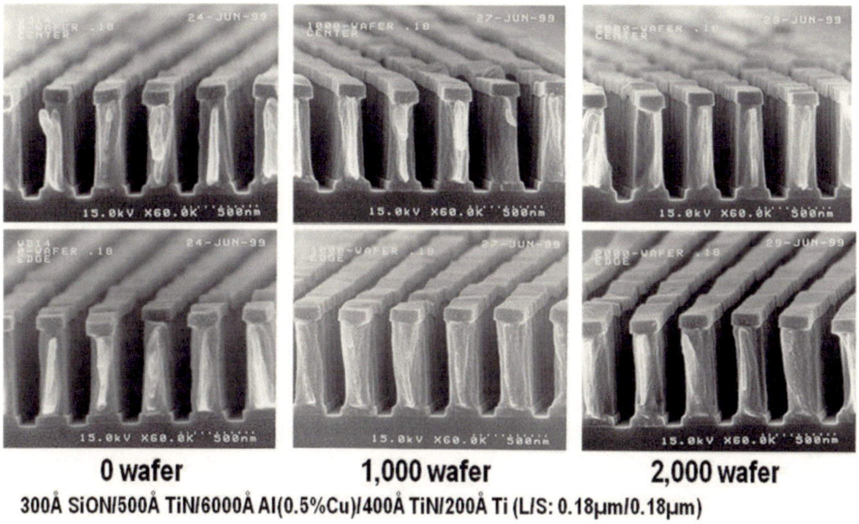

Fig. 17. The boron carbide coated chamber shows an excellent aluminum etch performance on a feature size as of 0.15 μm through a 2,000 wafer marathon.

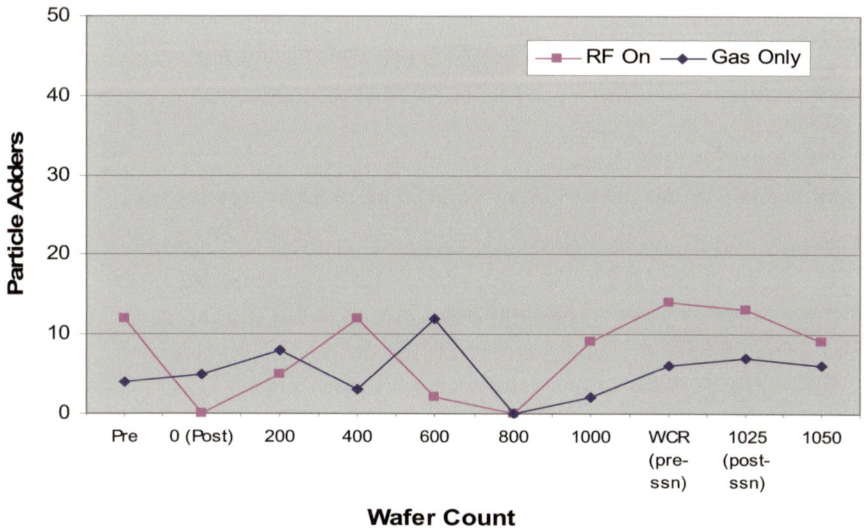

Fig. 18. Particle adders as of ≥ 0.12 μm size in a 1,000 wafer marathon on a 300mm etch tool. The boron carbide coating on anodized aluminum with HL126 sealant is used as the chamber coating to replace anodized aluminum.

Location	Pre (mils)	STD(mils)	Post (mils)	STD(mils)	Delta
A (10 data)	17.60	0.413	17.98	0.236	+0.38
B (10 data)	17.55	0.363	17.52	0.225	- 0.03
C (10 data)	18.76	0.287	18.56	0.201	- 0.20
30 data	**17.97 (average)**		**18.02 (average)**		**+0.05**

Table 2. The overall coating thickness before and after plasma etching under O_2/Cl_2 plasma for 120 RF hours.

It is obvious that there is little coating thickness loss after 120 RF hours under O_2/Cl_2 plasma. The main purpose of O_2/Cl_2 plasma is to test the performance of HL126 sealant under O_2/Cl_2 plasma condition.

After the detail study through a thorough process qualification, the new boron carbide coated chamber wall is used to replace the previously anodized aluminum surface. The new ceramic material such as YAG or Y_2O_3 is used to replace original high purity alumina. This configuration was introduced to semiconductor wafer fabrication for evaluation. Excellent etch performance, enhanced defect and particle reduction, and 50 to 100 times chamber lifetime improvement are reported. The production yield of the wafer fabrication also improved about 7% in production at the customer site (see Fig.19) [41]. The following data provide some of the information. The sequence of the data collection is as follows:

Baseline configuration using the old chamber hard ware submitted to gas-only and RF-on particle measurements without seasoning. After 1st RF-on particle measurement, five oxide wafers were used for seasoning the chamber, then RF-on particles were measured again. Two PR wafers were used to seasoning the chamber before final RF-on particle measurement. The test data are shown in Table 3 [41].

Condition	Gas-only w/o seasoning	RF-on(1) w/o seasoning	RF-on(2) 5 ox seasoning	RF-on(3) 2 PR seasoning
Old chamber configuration	10	34	53	48
New chamber Configuration	2	5	6	5

*: Unit in particle counts/wafer and particle adders at 0.2 μm or larger are recorded.

Table 3. Gas-only and RF-on particles of old and new chamber configurations

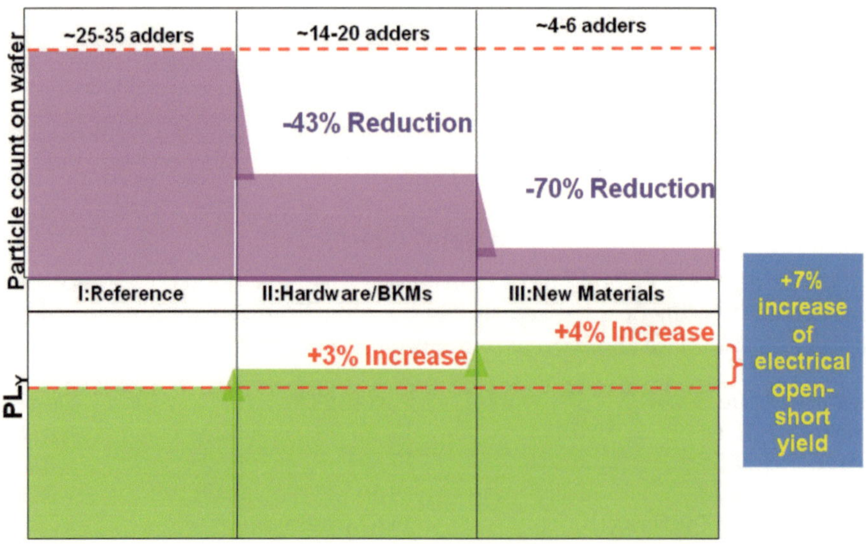

Fig. 19. Production yield improvement at wafer fabrication when new chamber material, hardware and best-known method are implemented.

About 7% production yield is reported in comparison with the old chamber configuration. The lifetime of chamber of chamber wall and chamber top window can improve about 50 times [41].

The new boron carbide coating has been introducing to worldwide wafer fabrication for over 10 years with over 1,000 chambers introduced to wafer fabrication in IC industry. The chamber lifetime has demonstrated to improve from the worse case as of 60 RF hours (1,800 wafers) under BCl_3/Cl_2 etching plasma to over 4,000 RF hours or longer in semiconductor wafer fabrication in the world. It also demonstrates that the chamber materials play a critical role in semiconductor etching equipment, particularly, for the cost reduction. A short comparison of anodized aluminum and born carbide coating is highlighted in Table 4 [21, 22, 25, 30].

Items	Anodized aluminum	Boron Carbide Coating
Maximum etch rate	0.07 mils/RF hour	0.001 mils/RF hour
Mininum lifetime	1,800 wafers	120,000 wafers
Particle performance	normal*	better
Metal contamination	normal	better
Micro-cracks	yes	no, but with coating pores
Coating bonding	very high	less than anodized Al
Surface roughness	normal	higher
Polymer adhesion	normal	better
Wet cleaning recovery	normal	normal
Process performance	normal	normal
Production yield	normal	better
Gate oxide damage	no	no
Water adsorption	normal	normal
Micro-hardness (100g)	360-450	3,000
Localized attack	yes, through cracks	no, with HL126 sealant.
Etch process window	normal	normal
Risk of undercut corrosion	no	no, with HL126 sealant
HCl bubble time (5wt% HCl solution)	≤ 10 minutes (non hot DIW seal) 30 minutes to 24 hours (after hot DIW seal)	> 50 hours, with HL126 sealant
Effect of base aluminum alloys to coating quality	yes, large impact	no

Table 4. Comparison of Anodized Aluminum and Boron Carbide Coating

Anodized aluminum has been using as the major etching tools surface coatings since 1980. It still received a lot of applications in plasma etching tools because of its low cost, easy to manufacture, easy to make large or small sizes of the parts, wide applications, easy to refurbish, and achieving good quality control at different suppliers in the world. Therefore, the study of anodized aluminum has always been a major task for the major semiconductor etching tool manufacturers. For high purity Y_2O_3 thermal spray coating, it has been qualified and applied as one of the major chamber components in plasma etching tools in the past 10 years. It is still one of the major materials as coating or as a solid sintered material which is used in plasma etching tools. At Lam Research Corporation, great attentions have been paid in the improvements and the new development of anodized aluminum and Y_2O_3 coatings.

These studied have been highlighted in John and Hong's presentation [28, 31-33, 35-39] and the publications of the new anodized aluminum study with Mansfeld et al [38, 39].

The study and characterization of anodized aluminum and the methodology are shown below. But the techniques are not limited to the techniques listed below:

- Admittance measurements on anodized aluminum to check sealing quality.
- Micro-hardness on surface and through the anodized aluminum layer.
- SEM cross section and Eddy current meter for anodized layer thickness.
- X-ray diffraction for the phase analysis of anodized aluminum.
- SEM cross section of micro-structure and secondary phases observation.
- TEM analysis to estimate the barrier layer thickness.
- EDX analysis for element analysis on the surface or through the layer.
- ICPMS analysis to obtain the surface cleanliness before and after cleaning.
- GD-OES analysis of the depth profile of elements in anodized layer.
- HCl bubble test to obtain the acidic corrosion resistance of anodized layer.
- Dielectric voltage breakdown of the anodic layer.
- Color and color uniformity of the anodic film.
- Electrochemical impedance spectroscopy to obtain the overall impedance.
- Surface roughness and coating thickness and their variations.
- Taber abrasion test to obtain wear resistance of the anodic film.
- Microhardness on surface and through coating cross section.
- Coating weight.
- Erosion/corrosion rate under high density plasma with different chemistries.
- Raw aluminum alloys analysis through different manufacturing processes.
- Intermetallic inclusions and their chemical composition analysis.
- Thermal property of anodic film after thermal cycling at different temperatures.

Although there are so many techniques used in the anodized aluminum study, there are only key techniques which are selected as a routine quality monitoring of worldwide anodization suppliers. The basic techniques are surface roughness, thickness of anodic film, color and color uniformity, dielectric voltage breakdown, acidic corrosion resistance through HCl bubble test, electrochemical impedance in 3.5wt% NaCl solution, surface micro-hardness, SEM cross section to observe the anodic layer micro-cracks, and admittance under 3.5wt% K_2SO_4 solution at 1000 Hz. For the surface cleanliness of anodization, ICPMS analysis of post precision wet cleaning has been used as a standard technique for metal contamination control. Since the requirements to anodized aluminum quality, corrosion resistance, and surface cleanliness for plasma etching tools are much strict and higher than the traditional industry applications, improvements of corrosion resistance and surface cleanliness are always the tasks. Lam Research has defined the surface cleanliness and the corrosion resistance of anodized aluminum specification for a standard type III and advanced anodized aluminum [28, 31-33, 35-39, 44].

The reaction mechanism of aluminum oxidation is summarized by Macdonald [50] as a reasonable model. The oxides grow as bilayer structures with an inner layer due to movement of oxygen vacancies from metal/film interface and an outer layer due to the movement of cations outward from the film/environmental interface. The vacancy concentrations vary exponentially with distance. The cathode consumes electrons by evoloving hydrogen and reducing oxygen. Barrier layer grows into metal phase via reaction.

Outer film grows via precipitation of Al^{3+} due to hydrolysis. The fundamental reactions for anodized aluminum systems are shown as follows:

Metal	film	Environment
(1) $m + V_M^{x'} = M_M + V_m + Xe^-$		(3) $M_M = M^{x+} + V_m^{x'}$
(2) $m = M_M + (x/2)\, V_o'' + Xe^-$		(4) $V_o'' + H_2O = O_o + 2H$

The principal crystallographic defects are (1) vacancies: $V_o''$and $V_M^{x'}$ for $MO_{x/2}$; (2) interstitials: O_i^{2-} and M_i^{x+}. In fact, oxide films can be described as exponentially-doped semiconductor junctions. The fundamentals and process optimization of anodized aluminum have been studied thoroughly by Brace, Thompson, Wood, Mansfeld, and recent years by Shih through the comprehensive studies of anodization of different aluminum alloys, different anodization processes, and different manufacturing processes [51 - 60]. The interface model of anodized aluminum with hot DIW seal has been described by Mansfeld, Kendig, Shih and others [61- 72] as shown in Fig.20.

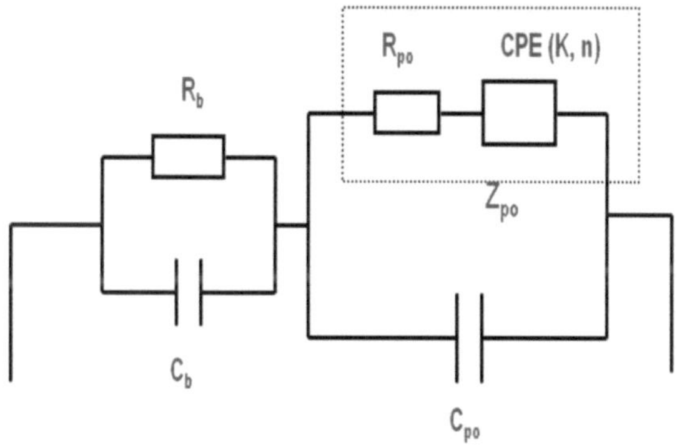

Fig. 20. The typical interface model of anodized aluminum with a hot DI water seal.

$$Z(\omega) = R_s + R_b/\{1+(j\omega C_b R_b)^{\alpha 2}\}+(R_{po}+CPE)/\{1+(j\omega C_{po}(R_{po}+CPE))^{\alpha 1}\}$$

$$\text{where} \quad C_b = \varepsilon^o \varepsilon_b A/D_b;\ C_{po} = \varepsilon^o \varepsilon_{po} A/D_{po} \text{ and } CPE = k(j\omega)^n$$

$\varepsilon^o = 8.854 \times 10^{-14}$ F/cm and is the permittivity of free space.

In Fig.20, C_b and R_b are barrier layer capacitance and resistance, respectively. R_{po} and CPE are the total impedance of the porous layer defined as Z_{po} which equals to R_{po} + CPE. C_{po} is the capacitance of the porous layer. CPE represents the constant phase element (CPE). A two-time constant interface model and suitable values of R_b and Z_{po} indicate a good quality of anodized aluminum. Z_{po} values highly depend on the quality control of hot DI water sealing process and it is very important for the improvement of the corrosion resistance of anodized aluminum [73-82]. R_b values depend on the voltage applied during anodization as well as the overall process control during anodization. A uniform and thick barrier layer helps to improve the dielectric voltage breakdown of the anodized aluminum. Mansfeld and Shih [63-69] developed a software package specially for the analysis of electrochemical impedance spectroscopy (EIS) data of anodized aluminum and the software has been widely applied for EIS data analysis. The EIS data of the new anodized aluminum developed and qualified at Lam Research Corporation show that the anodized aluminum has no corrosion in 3.5wt% NaCl (similar to seawater) for 365 days as shown in Fig. 21 [28, 38].

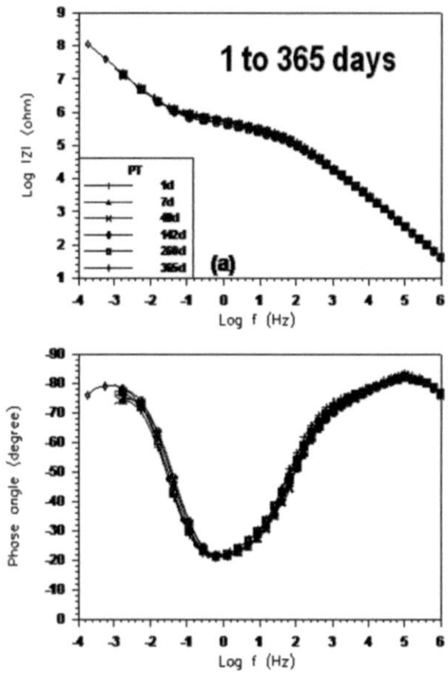

Fig. 21. EIS data of anodized aluminum in 3.5wt% NaCl solution for 365 days.

The overall impedance and HCl bubble test results are shown in Table 5. EIS data of three test coupons after immersion in 365 days in 3.5wt% NaCl solution are analyzed using the software written by Shih and Mansfeld called "ANODAL" [63-69].

Coupon ID	Z_{po} (ohm-cm^2)	R_b (ohm-cm^2)	HCl bubble time in hours
365D005	1.462×10^7	1.257×10^{10}	> 24
365D107	1.568×10^7	1.641×10^{10}	> 24
365D073	1.396×10^7	1.363×10^{10}	> 24

Table 5. The overall impedance and HCl Bubble Time of Test Coupons After Immersion in 3.5wt% NaCl solution for 365 days (coupons were prepared in three different batches of anodization processes) [78].

The Bode-plots of the three EIS data after 365 day's immersion in 3.5wt% NaCl solution is shown in Fig. 22. The anodized aluminum shows an excellent corrosion resistance and high quality of process control.

The complete EIS data analysis of the three test anodized aluminum samples is listed in Table 6 below. It is obvious that a consistent and an excellent corrosion resistance on both porous layer and barrier layer have been demonstrated. It is very important to improve the overall corrosion resistance of anodized aluminum through a well-controlled hot DI water sealing process. The parameters of hot DI water sealing tank water purity level, temperature range, sealing time, hot DI water pH value, and the pre-cleaning of the anodized aluminum before loading to the hot DI water tank will impact the quality of the sealing quality. The anodized anodization as a chamber coating for semiconductor IC industry moved from previously used non-sealed type III anodization or other types of non-sealed anodization to a well-controlled hot DI water sealed anodization for over 15 years because the hot DI water sealed anodized aluminum has demonstrated much better overall corrosion resistance in plasma etching chamber.

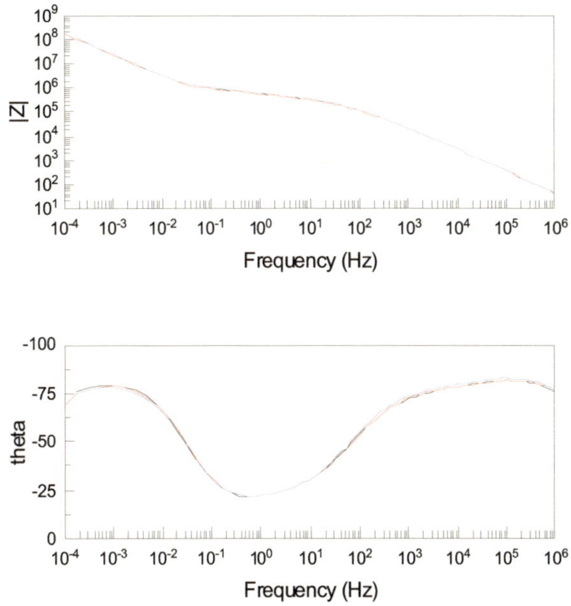

Fig. 22. EIS Bode-plots of advanced anodized aluminum coupons in 3.5wt% NaCl solution for 365 days. Black - 365D005; Red - 365D107; Blue - 365D073.

Parameters	D005	D107	D073
C_{po} (nF)	9.15	9.20	9.33
C_b (μF)	8.45	9.19	8.97
Z_{po} (Mohm)	0.731	0.784	0.698
R_b (Mohm)	628.4	820.5	681.5
n	0.1814	0.1733	0.1733
α_1	0.887	0.900	0.887
α_2	0.958	0.930	0.940
A (area in cm^2)	20.0	20.0	20.0
Chi-sq	2.785×10^{-3}	3.761×10^{-3}	2.775×10^{-3}
Z_{po} (ohm-cm^2)	1.462×10^{7}	1.568×10^{7}	1.396×10^{7}
R_b (ohm-cm^2)	1.257×10^{10}	1.641×10^{10}	1.363×10^{10}

Table 6. Detailed of EIS Data Analysis of test Samples D005, D107 and D073 after 365 Day's Immersion in 3.5wt% NaCl Solution

In Table 6, α is called frequency dispersion which is related to surface inhomogeneties with different dimensions [83]. Chi-sq is the fitting error between the experimental data and fitted data at each frequency. The detailed calculation is shown as below [84] and is the sum of the fitting error at each frequency multiplying 100 and dividing the total data points.

$$\text{Chi-sq} = (100/N)\ \Sigma\ \{[\ |Z_{exp}(f_i) - Z_{fit}(f_i)|]/Z_{exp}(f_i)\}$$

TEM is also used to obtain the barrier layer thickness of different types of anodized aluminum [80]. In Fig. 23, a standard type III anodization achieves about 50nm thickness of the barrier layer. The thickness of a new anodization can be as thick as 100nm due to the higher voltage applied in the anodization process. The thicker barrier layer can provide a higher barrier layer resistance during the EIS study as shown in Table 5 and Table 6. By combining both an excellent hot DI water seal processing to obtain an excellent corrosion resistance of the porous layer and a thicker barrier layer of the anodic film, the anodic film can hold 365 days in seawater without corrosion.

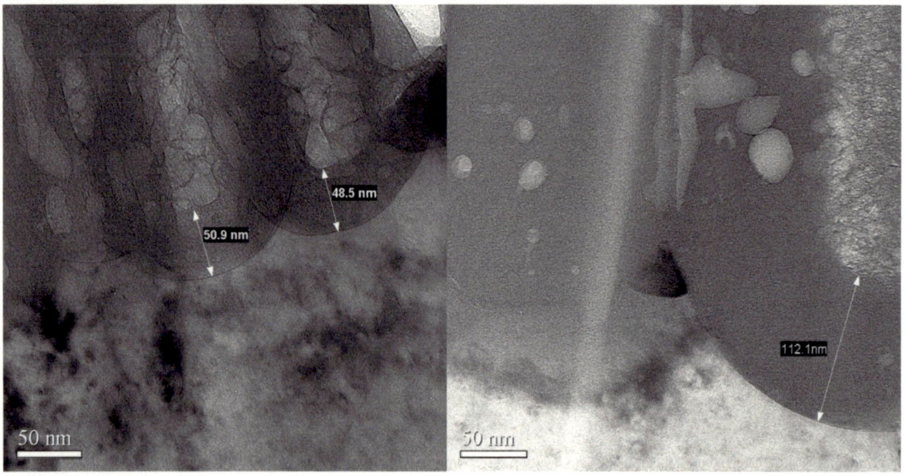

Fig. 23. TEM pictures of a standard type III hard anodization (left) and a mixed acid anodization (right) [80]

For acidic corrosion resistance of anodized aluminum, HCl bubble test is an easy and very effective method to obtain the corrosion resistance. From Fig. 24 below, one can see the good and poor anodized aluminum under the solution of 5wt% HCl solution (28, 31-33, 44, 78]. On the left of Fig.24, there is no any hydrogen bubble generated under the attack of a strong acid within 2 hours immersion. It indicates a high quality of anodized aluminum. On the right of Fig. 24, anodized aluminum generates a lot of hydrogen bubbles in 5wt% HCl solution only after 10 minutes immersion in the acid. It indicates a poor anodized aluminum.

Fig. 24. Acidic corrosion resistance of two test coupons of anodized aluminum. On the left, anodized aluminum does not show any acidic corrosion in two hours and on the right, anodized aluminum shows severe acidic corrosion after only 10 minutes immersion in the acid.

The HCl bubble test can be processed at any position of etching chamber before and after etching process. In Fig. 25, one process chamber is studied on its corrosion resistance in 5.0wt% HCl solution [76]. This method has received a wide application for the corrosion resistance study of anodized aluminum.

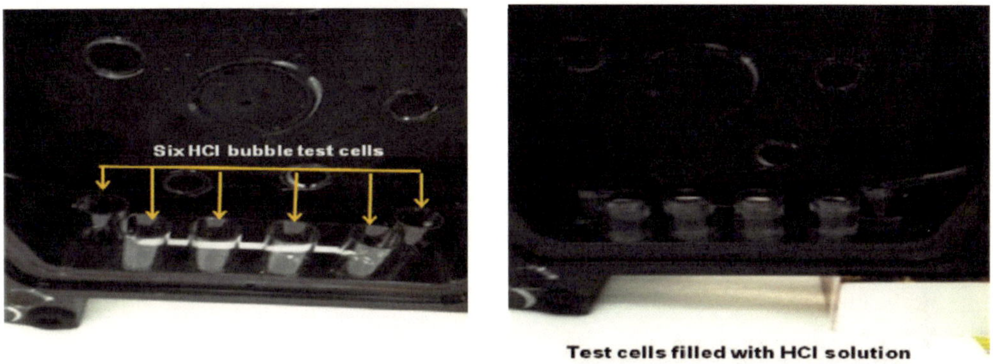

Fig. 25. HCl bubble test on a used process chamber after 12,000 wafers processing. Six locations are selected to run the HCl bubble test [76].

A systematic study of anodized aluminum made of Al6061-T6 11" thick block was carried out. It is obvious that HCl bubble test can be studied at different thickness positions to compare the differences of corrosion resistance [77]. The detail positions of different tests are shown in Fig. 26. Eleven different test methods are applied to the study.

HCl bubble test can be carried at different thickness to evaluate the corrosion resistance at different thickness in a thick aluminum block [77].

The thermal properties of anodized aluminum have also been studied. One of the typical studies was published in the work with Mansfeld [39]. A lot of studied have been carried out at Lam through the years [39, 73-82]. All these studies show that anodized aluminum film can be degraded through the use at a relatively high temperature. Both porous layer and barrier layer can be impacted depending on the operation temperature. Radii before anodizing on aluminum parts have to be controlled and the micro-cracks at corners and edges depend on the type of anodized aluminum and final thickness of anodic layer.

High purity Y_2O_3 has advantages in comparison with anodized aluminum and ceramic such as high purity alumina in many aspects. First of all, it can reduce the plasma etching rate for both metal etch and silicon etch by a factor of 40 times. It can bring cost saving in etch tools for semiconductor wafer fabrication. It can also reduce metal contamination too. The comparison of anodized aluminum and thermal spray coating of high purity Y_2O_3 are summarized by John and Shih [28, 31]. The advantages of Y_2O_3 coated anodized aluminum are as follows:

- Particle and defect reduction due to the elimination of aluminum fluoride.
- Metal contamination reduction due to its lower transition metal content.
- Better resistance to dielectric breakdown due to a thicker coating.
- Chamber material lifetime improvement due to much lower etch rate under plasma.
- Cost reduction due to extensive chamber materials lifetime.

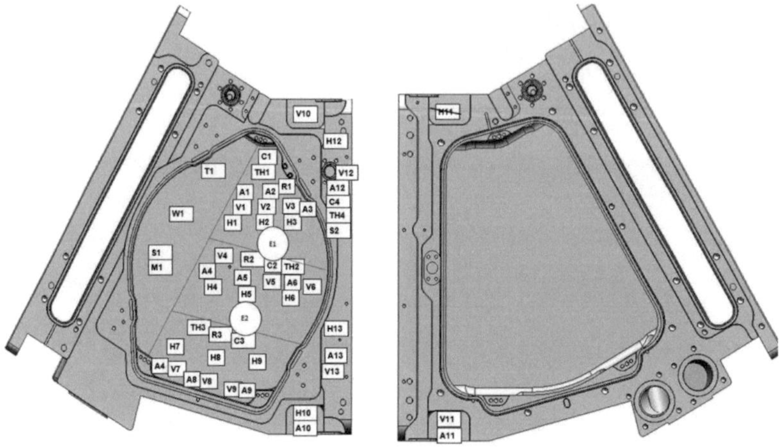

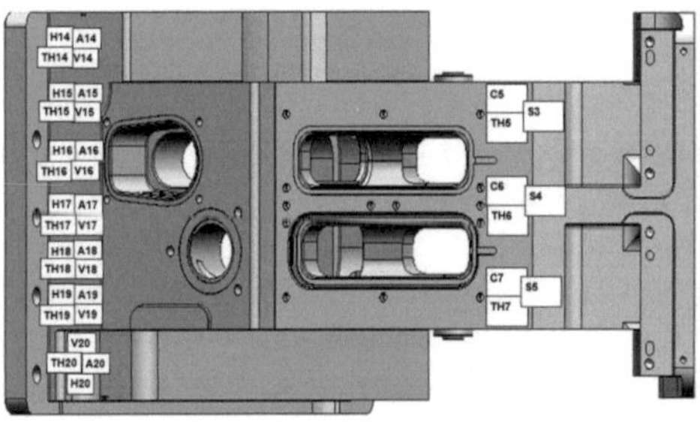

Legend		Total test locations
TH	Thickness	7
C	Color	7
R	Ra roughness	3
A	Admittance	20
H	HCl bubble test	13
V	Breakdown voltage	20
E	EIS	2
T	Taber abrasion	1
W	Coating weight	1
M	Microhardness	1
S	SEM/EDX	5

Fig. 26. A systematic study the anodized aluminum of Al6061-T6 11" thick block.

Although thermal spray and sintered Y_2O_3 has been widely using as one of the chamber materials in wafer fabrication, the study of this material as well as its coating has never been stopped because of the challenges. These studies for semiconductor IC wafer fabrication contain the following studies, but not limit to these techniques [28, 78].

- Impurity levels of raw powder.
- Particle size distribution of the raw powder.
- Impurity levels of the raw powder.
- Environmental control of coating processes.
- Optimization of coating process to reduce porosity and to eliminate non-melted particles.
- Coating thickness, roughness, color and the uniformities.
- Bonding strength and bending strength through pulling and bending tests.
- Dielectric voltage breakdown.
- Acidic corrosion resistance under 5wt% HCl solution.
- Micro-hardness.
- Overall admittance (coating/anodize/aluminum).
- ICPMS analysis for surface cleanliness.
- Surface particles after precision wet cleaning.
- Wet chemicals compatibility study.
- Porosity estimation through SEM cross section analysis.
- Overall impedance and interface model of coating through EIS study.
- Plasma resistance under both BCl_3/Cl_2 and SF_6/HBr gases.

In order to study Y_2O_3 coating on anodized aluminum, the following electrochemical cell configuration is used to study the overall impedance and interface model of the coated samples or parts as shown in Fig. 27 [28, 78].

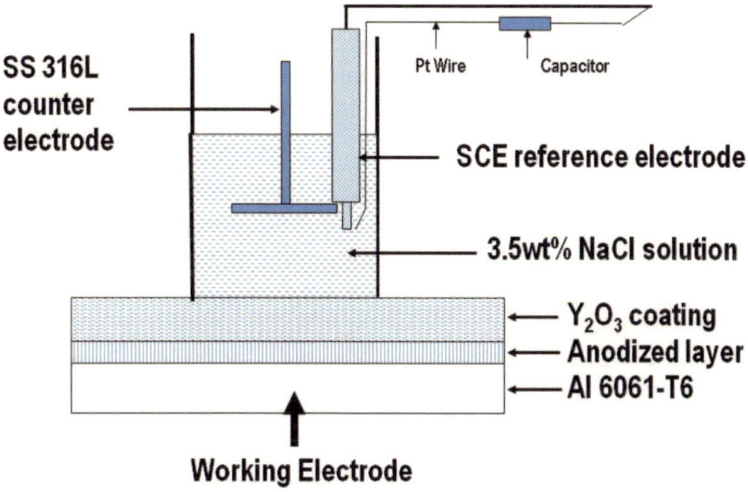

Fig. 27. Electrochemical cell configuration during EIS study of Y_2O_3 coated anodized aluminum in 3.5wt% NaCl solution [78].

An interface model of Y_2O_3 coated anodize aluminum shows a three-time constant interface model indicating a Y_2O_3 coated layer, the porous layer of anodized aluminum, and the barrier layer of anodized aluminum as shown in Fig. 28 [28, 78]

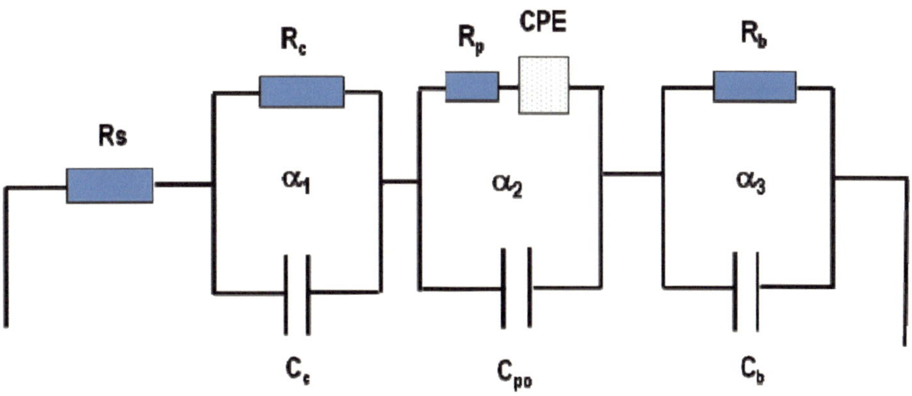

Fig. 28. The proposed interface model of Y_2O_3 coated anodized aluminum in EIS study

Assuming that C_c, C_{po} and C_b are capacitances which represent the capacitances of Y_2O_3 coating layer, porous layer of anodized aluminum and barrier layer of anodized aluminum, respectively. The interface parameters can be obtained and the coating quality can be monitored. The interface model can be described as the following equation [28, 78].

$$Z(\omega) = R_b/\{1+(j\omega C_b R_b)^{\alpha 3}\} + (R_p + CPE)/\{1+(j\omega C_{po}(R_p+CPE))^{\alpha 2}\} + R_c/\{1+(j\omega C_c R_c)^{\alpha 1}\} + R_s$$

Where Z is the total impedance, R_b is the barrier layer resistance of anodization, R_p is the porous layer resistance of porous layer of anodized aluminum, CPE is the constant phase element of the porous layer, R_c is the coating resistance, and R_s is the solution resistance.

Soaking three spraycoated Y_2O_3 on anodized aluminum in 3.5wt% NaCl solution for 7 days, the EIS data are shown in Fig.29. The EIS data indicate that samples coated at different time have the similar overall impedance and the quality control of coating process is consistent. The complete analysis of the EIS data using a three-time constant interface model is shown below (Table 7).

Parameters	Sample 1	Sample 2	Sample 3
R_s (ohm)	2.0	2.0	2.0
R_c (ohm)	40	40	40
R_p (ohm)	1.2×10^5	1.1×10^5	1.2×10^5
R_b (ohm)	5.0×10^7	5.0×10^7	5.1×10^7
K (ohm)	8.0×10^4	6.6×10^4	7.5×10^4
A (area in cm^2)	20.0	20.0	20.0
Z_c (ohm-cm^2)	800	800	800
Z_{po} (ohm-cm^2)	4.0×10^6	3.5×10^6	3.9×10^6
Z_b (ohm-cm^2)	1.0×10^9	1.0×10^9	1.0×10^9

Table 7. Interface parameters of three spraycoated Y_2O_3 on anodized aluminum

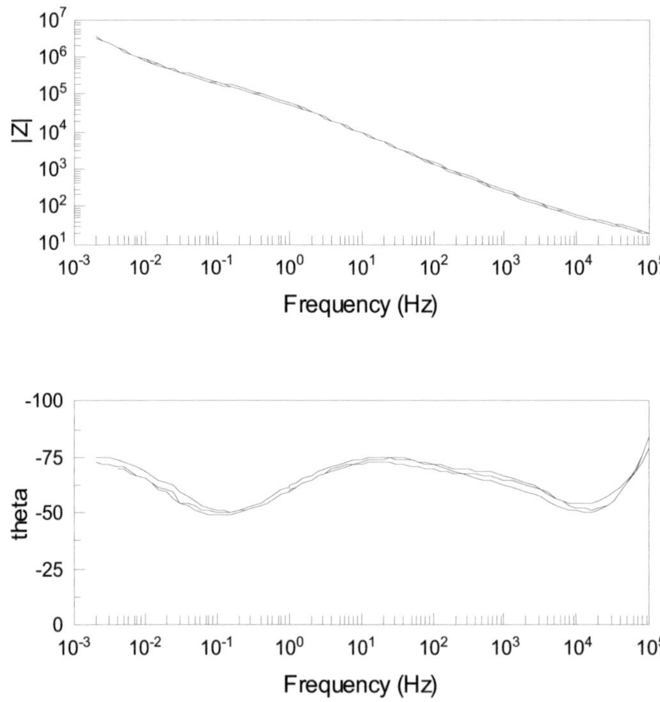

Fig. 29. EIS data of three spraycoated Y_2O_3 on anodized aluminum in 3.5wt% NaCl solution for 7 days. Excellent coating quality control is observed through the EIS study [28, 78].

3. Conclusions

- The revolutionary chamber materials study under high density plasma has opened a new scientific field in the characterization of materials. Meeting the comprehensive requirements of plasma etching tools in semiconductor wafer fabrication with the technology node shrinkage is not an easy task. The efforts and methodology developed through these studies have built up the foundation in the advanced materials characterization, development and application.
- A systematic study of chamber materials for plasma etching tools has been demonstrated. The materials studied include anodized aluminum, boron carbide coating, Y_2O_3 coating and sintered Y_2O_3 as well as the solid solution of aluminum oxide and yttrium oxide (YAG).
- In addition to the discussion of systematic study of anodized aluminum, an advanced anodized aluminum developed has shown no corrosion in seawater environment for 365 days.
- Thermal spray coating of boron carbide has been studied thoroughly. It has been introducing to worldwide semiconductor wafer fabrication since 1998. The coating has demonstrated over 50 times even 100 times lifetime improvement in production of semiconductor wafer fabrication and has been one of major chamber materials up to today.
- Y_2O_3 coating has also been studied thoroughly through the years since 2002. It has demonstrated longer chamber lifetime under plasma for both metal etch and silicon etch applications. It has been introduced to worldwide wafer fabrication since 2002 with a great cost saving and improved etch performance. As of today, it is still the major chamber coating of the current plasma etching tools.
- Lam Research Corporation has been putting great efforts and support for the new chamber materials development. It becomes more critical for semiconductor plasma etching equipment companies to develop new and advanced materials for current and next generation plasma etching feature size applications.

4. Acknowledgement

The author would like to express great thanks to Professor H. W. Pickering and Professor D. D. Macdonald for their guidance during author's Ph.D. study at the Pennsylvania State University between 1981 and 1986. Great thanks to Professor F. B. Mansfeld for author's post doctor and research work at University of Southern California between 1986 and 1990. Thanks to Dr. M. W. Kendig and Professor W. J. Lorenz. Many thanks to Dr. Richard Gottscho, Dr. John Daugherty, and Dr. Vahid Vahedi at Lam Research Corporation. Both Lam Research Corporation and Applied Materials provided the author the opportunity to carry on the systematic study of advanced materials under high density plasma on plasma etching tools. The author would like to express thanks to many individuals during the study of chamber materials through the past 18 years. There have been hundreds of individuals who gave support and encouragement to the author. Due to the limitation of space, the author cannot list all the individuals. Some of the individuals are Dr. Duane Outka, Dr. Tuochuan Huang, Chris Chang, Patrick Barber, Declan Hayes, Dr. Shun Wu, Dr. Harmeet Singh, Dr. Yan Fang, Dr. Armen Avoyan, Dr. Siwen Li, Shenjian Liu, Dr. Qian Fu, Dr. Steve Lin, Nianci Han, Dr. Peter Loewenhardt, Dr. Diana Ma, Mike Morita, Tom Stevenson, Sivakami Ramanathan, John Mike Kerns, Dean Larson, Alan Ronne, Hilary Haruff, David

Song, Dr. You Wang, Dr. Yan Chun, Dr. Qi Li, Dr. Allan Zhao, Joe Sommers, Dr. Griff O'Neil, Dr. Raphael Casaes, Dr. T. W. Kim, Dr. Lin Xu, Dr. Catherine Zhou, Dr. Yijun Du, Dr. Daxing Ren, Ho Fang, May He, Josh Cormier, Fangli Hao, Dr. Steve Mak, Dr. Gerald Yin, Dr. Xikun Wang, Dr. Danny Lu, Yoshi Tanase, Dr. Samantha Tan, Dr. Fuhe Li, Shang-I Chou, Tim. Su, John Holland, Peter Holland, Jason Augustine, Dr. Y. L. Huang, Lucy Chen, Jie Yuan, Hui Chen, Li Xu, Charlie Botti, Han Sellege, Dr. Arthur W. Brace, and many others.

5. List of related patents

[1] Hong Shih and M. S. Mekhjian, "Method and Apparatus for Determining the Quality of A Coating", USA Patent No. 5,373,734, December 20, 1994.

[2] Hong Shih and M. S. Mekhjian, "Apparatus for Measuring A Coating on A Plate", USA Patent No. 5,493,904, February 27, 1996.

[3] Hong Shih, N. Han, S. Mak, and G. Yin, "Boron Carbide Parts and Coatings in A Plasma Reactor", USA Patent No. 6,120,640, September 19, 2000.

[4] Hong Shih, Nianci Han, Jie Yuan, Danny Lu, and Diana Ma, "Ceramic Composition for An Apparatus and Method for Processing A Substrate", USA Patent No. 6,352,611, March 5th, 2002.

[5] N. Han, Hong Shih, J. Yuan, D. Lu, and D. Ma, "Ceramic Composition for An Apparatus and Method for Processing A Substrate", USA Patent No. 6,123,791, September 26, 2000.

[6] Hong Shih, Joe Sommers, and Diana Ma, "Method for Monitoring the Quality of A Protective Coating In A Reactor Chamber", USA Patent No. 6,466,881, October 15, 2002.

[7] N. Han, Hong Shih, J. Y. Sun, Li Xu "Diamond Coated Parts in A Plasma Reactor", USA Patent No. 6,508,911, January 21, 2003.

[8] Hong Shih, Nianci Han, Jie Yuan, Joe Sommers, Diana Ma, Paul Vollmer, "Corrosion-Resistant Protective Coating for An Apparatus And Method for Processing A Substrate", USA Patent No. 6,592,707, July 15, 2003.

[9] Nianci Han, Hong Shih, Jie Yuan, Danny Lu, and Diana Ma, "Substrate Processing Using a Member Comprising An Oxide of A Group IIIB Metal", USA Patent No. 6,641,697, November 4th, 2003.

[10] Hong Shih and Nianci Han, "Coating Boron Carbide on Aluminum", USA Patent No. 6,808,747, October 26, 2004.

[11] N. Han, Hong Shih, and Li Xu, "Process Chamber Having Component with Yttrium-Aluminum Coating", USA Patent No. 6,942,929, September 13, 2005.

[12] Hong Shih, Duane Outka, Shenjian Liu, and John Daugherty, "Extending Lifetime of Yttrium Oxide As A Plasma Chamber Material", Lam Research Patent Application, Publication Number: US20080169588, Publication Date : 07/17/2008: Database : US Patents.

[13] Harmeet Singh, John Daugherty, Vahid Vahedi, and Hong Shih, "Components for Use in A Plasma Chamber Having Reduced Particle Generation and Method of Making", Lam Research Patent Application, Publication Number: US20090261065, Publication Date : 10/22/2009: Database : US Patents.

[14] Hong Shih, "Method for Refurbishing Bipolar Electrostatic Chuck", Assignee: Lam Research Corporation, Publication Number US 2010/0088872 A1, April 15, 2010.

[15] Hong Shih, Saurabh Ullal, Tuochuan Huang, Yan Fang, and Jon McChesney, "System and Method for Testing An Electrostatic Chuck", Assignee: Lam Research Corporation, Publication Number US 20100090711 A1, April 15, 2010.

[16] Hong Shih, Qian Fu, Tuochuan Huang, Raphael Casaes, and Duane Outka, "Extending Storage Time of Used Y2O3 Coated Ceramic Window before Wet Cleaning", US Patent No 7,976,641, July 12, 2011.

[17] Josh Cormier, Fangli Hao, Hong Shih, Allan Ronne, John Daugherty, and Tuochuan Huang, "Automatic Hydrogen Bubble Detection and Recording System for the Study of Acidic Corrosion Resistance for Anodized Aluminum and Surface Coatings", Lam Research Patent Invention Disclosure P2232, Filed for US Patent on January 4, 2011.

[18] John M. Kerns, Yan Fang, Hong Shih, and Allan Ronne, "Silicon Gas Line Coating for Semiconductor Applications", Lam US Patent Application, Docket No: 2287, May 5, 2011.

[19] Hong Shih, Duane Outka, Shenjian Liu, and John Daugherty, "Extending Lifetime of Yttrium Oxide As A Plasma Chamber Material", Lam Research Patent Application, Publication Number : (WO 2008/088670), Publication Date: 07/17/2008

[20] Hong Shih, Saurabh Ullal, Tuochuan Huang, Yan Fang, and Jon McChesney, "System and Method for Testing An Electrostatic Chuck", Assignee: Lam Research Corporation, Publication Number: WO 2010/042908, April 15, 2010.

[21] Hong Shih, Nianci Han, Jennifer Y. Sun, and Li Xu, "Diamond Coated Parts in A Plasma Reactor", Assignee: Applied Materials, Publication Number: WO2001/013404, February 2001.

[22] N. Han, Hong Shih, Danny Lu, and Diana Ma, "A Ceramic Composition of An Apparatus and Method for Processing A Substrate", Applied Materials, Publication Number: WO2000/007216, February, 2000.

[23] Hong Shih, N. Han, S. Mak, and G. Yin, "Boron Carbide Parts and Coatings in A Plasma Reactor", European Patent Number. EP0849767, Publication Date: June 24, 1998.

[24] Hong Shih, N. Han, S. Mak, and G. Yin, "Boron Carbide Parts and Coatings in A Plasma Reactor", European Patent Number. EP0849767, Publication Date: March 21, 2001.

[25] Tuochuan Huang, Daxing Ren, Hong Shih, Catherine Zhou, Chun Yan, Enrico Magni, Bi Ming Yen, Jerome Hubacek, Danny Lim and Dougyong Sung, "Methods for Silicon Electrode Assembly Etch Rate and Etch Uniformity Recovery", European Patent Publication No. 1848597, October 31, 2007.

[26] Hong Shih, Nianci Han, Jie Yuan, Danny Lu, and Diana Ma, "Ceramic Composition for An Apparatus and Method for Processing A Substrate", Applied Materials, July 2001: Korea 1020017001225.

[27] Tuochuan Huang, Daxing Ren, Hong Shih, Catherine Zhou, Chun Yan, Enrico Magni, Bi Ming Yen, Jerome Hubacek, Danny Lim and Dougyong Sung, "Methods for Silicon Electrode Assembly Etch Rate and Etch Uniformity Recovery", Lam Research Corporation, China Patent Application No: CN200580044362, March 2008.

[28] Hong Shih, Duane Outka, Shenjian Liu, and John Daugherty, "Extending Lifetime of Yttrium Oxide As A Plasma Chamber Material", Lam Research Patent Application, China Patent Number CN200880002182, November 2009.

[29] Hong Shih, N. Han, S. Mak, and G. Yin, "Boron Carbide Parts and Coatings in A Plasma Reactor", Assignee: Applied Materials, Singapore Patent Application Number: 9704073-7, November 17, 1997.

[30] Hong Shih, Duane Outka, Shenjian Liu, and John Daugherty, "Extending Lifetime of Yttrium Oxide As A Plasma Chamber Material", Lam Research Patent Application, Taiwan Patent Number TW097100617, December 2008.

[31] Hong Shih, N. Han, S. Mak, and G. Yin, "Boron Carbide Parts and Coatings in A Plasma Reactor", Assignee: Applied Materials, Taiwan Patent Application Number: TW086114934, November, 1999.

[32] Hong Shih, Joe Sommers, and Diana Ma, "Method for Monitoring the Quality of A Protective Coating In A Reactor Chamber", Applied Materials, Taiwan Patent Application No: TW089107053, May 2002..

[33] Hong Shih, Nianci Han, Jie Yuan, Joe Sommers, Diana Ma, Paul Vollmer, "A Corrosion-Resistant Protective Coating for An Apparatus and Method for Processing a Substrate", Applied Materials, Taiwan Patent Application No: TW089106928, November, 2011.

[34] Hong Shih, Nianci Han, Jie Yuan, Danny Lu, and Diana Ma, "Ceramic Composition for An Apparatus and Method for Processing A Substrate", Assignee: Applied Materials, Taiwan Patent Publication Number: TW088112796, August 2005.

[35] Hong Shih, Duane Outka, Shenjian Liu, and John Daugherty, "Extending Lifetime of Yttrium Oxide As A Plasma Chamber Material", Lam Research Patent Application, Taiwan Patent Number TW097100617, December 2008.

[36] Hong Shih, Saurabh Ullal, Tuochuan Huang, Yan Fang, and Jon McChesney, "System and Method for Testing An Electrostatic Chuck", Assignee: Lam Research Corporation, Taiwan Patent Publication Number: TW096115736, June 2008.

[37] Hong Shih, Saurabh Ullal, Tuochuan Huang, Yan Fang, and Jon McChesney, "System and Method for Testing An Electrostatic Chuck", Assignee: Lam Research Corporation, Taiwan Patent Publication Number: TW098142359, March 2010.

6. References

[1] M. A. Lieberman and A. J. Lichtenberg, "Principles of Plasma Discharges and Materials Processing", John Wiley & Sons, Inc., Second Edition, April 14, 2005.

[2] J. W. Coburn, "Plasma Etching and Reactive Ion Etching: The Fundamentals and Applications", A Short Course on Plasma Etching, 2008.

[3] M. A. Lieberman, "A Short Course on the Principle of Plasma Discharges and Materials Processing", A Short Course on Plasma Etching, 2008.

[4] M. A. Lieberman and R. A. Gottscho, in Physics of Thin Films, edited. By M. Francombe and J. Vossen, Academic Press, 1993.

[5] R. A. Gottscho, C. W. Jurgensen, and D. J. Vitkavage, "Microscopic Uniformity in Plasma Etching", J. Vac. Sci. Technol. B10(5), 2133, 1992.

[6] J. W. Coburn, J. Vac. Sci. Technol. B12, 1384, 1994.

[7] C. Lee, D. B. Graves, and M. A. Lieberman, and D. W. Hess, J. Electrochem. Soc. 141, 1994.

[8] C. Lee, D. B. Graves, and M. A. Lieberman, Plasma Chemistry and Plasma Processing, Vol. 16, No. 1, 1996.

[9] B. N. Chapman, "Glow Discharge Processes", John Wiley & Sons, Inc., 1980.

[10] O. A. Popov, "High Density Plasma Sources", Noyes Publications, 2000.

[11] R. J. Shul and S. J. Pearton, editors, "Handbook of Advanced Plasma Processing Techniques", Springer-Verlag, 2000.

[12] H. Singh, J. W. Coburn, and D. B. Graves, J. Vac. Sci. Technol. A17(5), 2447, 1999.

[13] M. Sugawara, "Plasma Etching: Fundamental and Applications (Semiconductor Science and Technology), Oxford Science Publications, 1998.

[14] J. E. Daugherty and D. B. Graves, J. Vac. Sci. Technol. A11, 1126, 1993.

[15] Hong Xiao, "Introduction to Semiconductor Manufacturing Technology", Pearson Education International, Prentice Hall, New Jersey, 2001.

[16] C. Y. Chang, "Semiconductor Manufacturing Equipment", Wunan Publishing Co., Ltd., Taiwan, November 2000.

[17] C. Y. Chang, "Deep Submicron Silicon Processing Technology", Wunan Publishing Co., Ltd., Taiwan, May 2002.

[18] Masanori Kikuchi, "Graphics of Semiconductor", Nippon Jitsagyo Publishing, Co., Ltd., 2000.

[19] "Zukai Handotai Guide", edited by Toshiba Semiconductorsha, Toshiba Corporation, 2001, Seibundo Shinko-sha Publishing Co., Ltd., Toyko, 2001, translated by Zhou Yong Xu and published by Princeton International Publishing Co., Ltd., Taiwan, 2004.

[20] "Arute 21 Handoutai Device", edited by Hiromu Haruki, Published by Ohmsham Ltd., 1999.

[21] H. Shih, "Materials Characterization under High Density Plasma", Keynote presentation on American Ceramic Society, Coconut Beach, Florida, February 26, 2004.

[22] H. Shih, "A Materials Study and Characterization of Semiconductor Wafer Fabrication", Presentation at the Pennsylvania State University for the 2004 McFarland Award, April 24, 2004.

[23] H. Shih, "Technology Development of Materials Characterization in Wafer Fabrication Equipment", Technical Presentation on 2001 IC Equipment Supply Chain Symposium and Tainan Manufacturing Center Opening", Hosted by Applied Materials, Taiwan, May 4th, 2001.

[24] H. Shih, N. Han, S. Mak, and G. Yin, "Development and Characterization of Materials for Sub-Micron Semiconductor Etch Application under High Density Plasma", Presentation on the 13th International Symposium on Plasma Chemistry", June 22, Beijing, China, 1997.

[25] H. Shih and D. Ma, "Revolutionary Chamber Materials for Metal Etch", SEMICON West Oral Presentation, Moscone Convention Center, San Francisco, California, June 1998.

[26] H. Shih, "Defect Density Reduction of 0.18μm and Beyond", Presentation on SEMICON Korea, Seoul, Korea, February 15-17, 2000.

[27] H. Shih, "Defect Density Reduction and Productivity Enhancement for 0.18μm and Beyond", Presentation on AMSEA 7th Annual Technical Seminar at Singapore, May, 2000.

[28] H. Shih and J. Daugherty, "Systematic Study of Yttrium Oxide Coating on Anodized Aluminum Surfaces", Keynote Presentation on International Thermal Spray Coating (ITSC) Conference & Exposition, May 4-7, 2009, Las Vegas, Navada, USA.

[29] H. Shih, "Can Semiconductor IC Equipment Survive without Thermal Spray Coatings", Guest Editorial in Advanced Materials & Processes, Vol 168, No 5, May 2010.

[30] H. Shih, N. Han, S. Mak, E. Polar, G. Yin, "Significant Lifetime Enhancement of Dielectric Materials for Next Generation Etch Chamber Materials", Applied Materials ET Conference Paper No. 455, Oral Presentation at San Jose State University, San Jose, California, 1998.

[31] H. Shih and Patrick Barber, "Materials Laboratory Development for Anodized Aluminum Study and Supplier Qualification - Part one and Part Two", Lam Research Confidential Technical Report, May 31, 2003.

[32] H. Shih, T. C. Huang, S. Wu, and J. Daugherty, "Summary of Corrosion Study During 180 days Immersion in 3.5wt% NaCl Solution", Lam Research Confidential Technical Report, Lam Research Corporation, September 28, 2006.

[33] H. Shih and J. Daugherty, "EIS Data Explanation of Anodized Aluminum 6061-T6 After Thermal Cycling", Lam Research Confidential Technical Report, Lam Research Corporation, January 26, 2009.

[34] Mike Kerns, Yan Fang, and Hong Shih, "Summary of Corrosion Tests on Silicolly Coated SS316L Gas Lines", Lam Research Confidential Technical Report, Lam Research Corporation, July 29, 2010.
[35] H. Shih, "Precision Wet Cleaning at Lam Research Corporation", Technical Presentation at SMIC, Beijing and Shanghai, August, 16 to 19, 2011.
[36] H. Shih, "Precision Wet Cleaning and Wet Cleaning Technical Support for TSMC", Technical Presentation at TSMC, Taiwan, July 5th, 2011, Hsinchu, Taiwan.
[37] H. Shih, "Advanced Chamber Materials for Plasma Etching Applications", Lam Research Confidential Technical Report, February 6, 2011.
[38] Y. L. Huang, H. Shih, T. C. Huang, J. Daugherty, S. Wu, S. Ramanathan, C. Chang, F. Mansfeld, "Evaluation of the Properties of Anodized Aluminum 6061-T6 Using Electrochemical Impedance Spectroscopy (EIS)", Journal of Corrosion Science, Vo. 50, Issue 12, p. 3569-3575, 2008.
[39] Y. L. Huang, H. Shih, J. Daugherty, F. Mansfeld, " Evaluation of the Properties of Anodized Aluminum 6061 Subjected to Thermal Cycling Treatment Using Electrochemical Impedance Spectroscopy", Journal of Corrosion Science, Vo. 51, Issue 10, p. 2493-2501, October, 2009.
[40] D. Outka and H. Shih, "Surface Analysis of Electrostatic Chuck Surface and Wet Cleaning Development", Lam Confidential Technical Report, August 20, 2008.
[41] Y. Takakura, T. Miyauchi, T. Ono, J. Hachiya, S. Kitamura, A. Endo, S. Park, N. Han and H. Shih, "Wafer Defect Reduction with DPS Metal Etch", Applied Materials ET Paper 471, Canada, 2000.
[42] H. Shih, "Intel Anodized Aluminum GDP Failure Analysis through Eddy Current, Surface Resistivity, Admittance, Hardness, HCl Bubble and Electrochemical Impedance Spectroscopy", Applied Materials Confidential Technical Report, January 4, 1994.
[43] H. Shih, J. Sommers, S. Lin, and D. Ma, "Analysis and Improved Specification for Critical Anodization on Metal Etch Process Parts", Applied Materials ET Conference Paper No. 453/484, San Jose State University, 1998.
[44] H. Shih, D. Outka, and J. Daugherty, "Specification for Hard Anodized Aluminum Coatings Using Mixed Acid for Critical Chamber Components", Lam Research specification 202-047671-001, January 17, 2006.
[45] MIL-A-8625, "Military Specification : Anodic Coatings for Aluminum and Aluminum Alloys.
[46] ASTM B117 - Method of Salt Spray (Fog) Testing.
[47] H. Shih, N. Han, S. Mak, and G. Yin, "Overview of A-Coat Development for DPS Metal Etch", Applied Materials ET Conference Oral Presentation at San Jose State University, 1997.
[48] H. Shih, N. Han, J. Yuan, and Q. Li, "DPS Chamber Lifetime Enhancement - A Coating Characterization and Burn-in", Applied Materials ET Conference Paper No. 588, San Jose State University, 1997.
[49] H. Shih, R. Xie, R. Koch, X. K. Wang, H. Chen, G. W. Ding, C. Sun, L. Chen, S. Arias, E. Chiang, K. Kawaguchi, M. Jain, A. Jiang, J. Papanu, R. Hagborg, B. Hatcher, B. Aeaia, B. Ching, R. Hartlage, V. Todorov, P. Leahey, N. Arboiuz, B. Dodson, S. Mak, R. Kerns, C. Lane, J. Holland and M. Barnes, "Summary of Process and Productivity Results of 1,000 Wafer Alpha Release Burn-in of Metal Etch New 300mm System", Applied Materials ET Conference Paper No. 599, San Jose State University, 2001.

[50] D. D. Macdonald, Private communication, 1991.
[51] A. W. Brace, "Anodic Coating Defects - Their Causes and Cure", Published by Technicopy Books, England, 1992.
[52] "Aluminum and Magnesium Alloys", Vol. 02.02, Annual Book of ASTM Standards, 1993.
[53] S. Wernick, R. Pinner, and P. G. Sheasby, "The Surface Treatment and Finishing of Aluminum and It's Alloys", 5th edition, ASM International, 1987.
[54] G. E. Thompson and G. C. Wood, "Treatise on Materials Science and Technology", 23, 205-329, 1983.
[55] Hong Shih and John Daugherty, "Systematic Study of Yttrium Oxide Coating on Anodized Aluminum Surfaces", Keynote Presentation on International Thermal Spray Coating (ITSC) Conference & Exposition, May 4-7, 2009, Las Vegas, Nevada, USA.
[56] Hong Shih and John Daugherty, "Chamber Materials for Current and Future Etchers", Lam Technical Report, December 10, 2011.
[57] Hong Shih, Shun Wu, and David Song, "Coordination of HCl Bubble Test and Electrochemical Impedance Spectroscopy of Lam Worldwide Anodizers", Lam Technical Report, December 20, 2010.
[58] Lin Xu and Hong Shih, "Secondary Phase Distribution of Al6061-T6 Alloys", Lam Technical Report, September 7, 2011.
[59] Hong Shih, "Electrochemical Impedance Spectroscopy Study of AI and Altefco D-Chamber Using Alcan Ravenswood 11" Thick Block", Lam Technical Report, February 14, 2011.
[60] Hong Shih and John Daugherty, "EIS Data Explanation of Anodized Aluminum 6061-T6 After Thermal Cycling", Lam Technical Report, January 26, 2009.
[61] Hong Shih, "Aluminum Alloy Anodic Oxidation - Theory, Common Failure, and Techniques for Anodization Study in Semiconductor Equipment Applications", Lam Technical Report, December 16, 2007.
[62] F. Mansfeld and M. W. Kendig, J. Electrochem. Soc. 135, 828, 1988.
[63] H. Shih, T. C. Huang, and J. Daugherty, "Lam Research Confidential Technical Report, October 10, 2006.
[64] F. Mansfeld, H. Shih, H. Greene, and C. H. Tsai, " Analysis of EIS Data for Common Corrosion Processes", in "Electrochemical Impedance : Analysis and Interpretation", ASTM STP 1188, J. R. Scully, D. D. Silverman, and M. W. Kendig, Eds., ASTM, p23, 1993.
[65] H. Shih and F. Mansfeld, in "New Methods for Corrosion Testing of Aluminum Alloys", ASTM 1134, edited by V. S. Agatwala and G. M. Ugianksy, ASTM, 180-195, 1992.
[66] F. Mansfeld, "Analysis and Interpretation of EIS Data for Metals and Alloys", Schlumberger Technical Report 26, 1993.
[67] F. Mansfeld, H. Shih, and C. H. Tsai, "Software for Simulation and Analysis of Electrochemical Impedance Spectroscopy (EIS) Data", in Common Modeling in Corrosion", ASTM STP 1154, R. S. Munn, ed, ASTM, p.186, 1992.
[68] H. Shih, "Electrochemical Impedance Spectroscopy and Its Application for the Characterization of Anodic Layers of Aluminum Alloys on Semiconductor Manufacturing Industry", Presentation on Corrosion Asia, 94, September 26-30, Marina Mandarin, Singapore, 1994.

[69] H. Shih and H. W. Pickering, "Some Aspects of Potential and Current Distributions During AC Polarization in Electrochemical Systems", Presentation on Corrosion Asia, 94, September 26-30, Marina Mandarin, Singapore, 1994.
[70] Y. L. Huang, H. Shih, and F. Mansfeld, "Concerning the Use of Constant Phase Element (CPE) in the Analysis of Impedance Data", Materials and Corrosion, 2009, 60, No. 9999.
[71] H. Shih, H. J. Chen, and F. Mansfeld, "Data Analysis of Electrochemical Impedance Spectroscopy (EIS) in Corrosion Monitoring and Detection", Presentation on Corrosion Asia, 94, September 26-30, Marina Mandarin, Singapore, 1994.
[72] F. Mansfeld and H. Shih, "ANALEIS A/S 1.0 Module 4: ANODAL - A Software Library for the Simulation and Analysis of Electrochemical Impedance Data", February, 1992.
[73] H. Shih, T. C. Huang, S. Wu, S. Ramanathan, and J. Daugherty, "The Development of Next Generation Anodized Aluminum - Summary of Corrosion Study during Immersion in 3.5wt% NaCl Solution for 180 Days", Lam Research Confidential Technical Report, September 28, 2006.
[74] H. Shih, S. Wu, T. C. Huang, S. Ramanathan, D. Outka, D. Larson, Y. Fang, C. Zhou, and J. Daugherty, "Summary of DOE of ACME Coated Anodized Aluminum", Lam Research Confidential Technical Report, January 21, 2008.
[75] H. Shih, S. Wu, Y. J. Du, Y. Fang, and J. Daugherty, "Breakdown Voltage Measurements on A 8K Thermal Oxide Si Wafer Using Ball Tip and Needle Tip", Lam Research Confidential Technical Report, February 8, 2008.
[76] H. Shih, S. I. Chou, and R. Casaes, "D-Chamber Evaluation after Running 12,000 Wafers", Lam Research Confidential Technical Report, November 11, 2010.
[77] D. Song, H. Shih, S. Wu, and H. Haruff, "Systematic Study of D-Chamber Anodization", Lam Research Confidential Technical Report, February 14, 2011.
[78] H. Shih, "Electrochemical Impedance Spectroscopy (EIS) Technology and Applications", Lecturer for A 8 Hour Short Course Sponsored by the San Francisco Section of the Electrochemical Society, Crown Plaza Hotel, Milpitas, California, June 15, 2006.
[79] H. Shih, "Aluminum Alloy Anodic Oxidation - Theory, Common Failure and Techniques for Anodization Study in Semiconductor Equipment Application", Lam Research Confidential Technical Report, December 16, 2007.
[80] Yan Fang, Duane Outka, and Hong Shih, "TEM Analysis of Al6061-T6 Type III Anodization and the Mixed Acid Anodization", Lam Research Confidential Technical Report, January 8, 2008.
[81] H. Shih, A. Avoyan, T. C. Huang, D. Outka, and J. Daugherty, "Anodized Aluminum Cleaned with Additional Chemical Solutions to Achieve a Better Surface Cleanliness on Both Anodized Aluminum and Bare Aluminum Surfaces", Lam Research Confidential Technical Report, November 18, 2008.
[82] Siwen Li, Duane Outka, and Hong Shih, "Cracks in Anodized Aluminum Film", Lam Research Confidential Technical Reports, August 26, 2009, September 8, 2009, November 11, 2009, February 5, 2010, September 23, 2010, November 12, 2010, November 30 2010.

2

Corrosion Resistance of Directionally Solidified Casting Zinc-Aluminum Matrix

Alicia Esther Ares[1,2,3], Liliana Mabel Gassa[1,4]
and Claudia Marcela Mendez[3]
[1] *Researcher of CIC, CONICET,*
[2]*Materials, Modeling and Metrology Program,*
Faculty of Sciences, National University of Misiones, Posadas,
[3]*Materials Laboratory, Faculty of Sciences, National University of Misiones, Posadas,*
[4]*INIFTA, National University of La Plata, Faculty of Exact Sciences, La Plata,*
Argentina

1. Introduction

Generally solidification leads to two types of grain morphologies: columnar and equiaxed. The origin of each one has been the subject of numerous theoretical and experimental researches in the field of metallurgy for many years. Columnar grains often grow from near the mold surface, where the thermal gradients are high, and the growth is preferentially oriented in a direction close to the heat flux. When the gradients are reduced near the center of the casting, equiaxed grains grow in all space directions leading to a material with more isotropic macroscopic mechanical properties and a more homogeneous composition field than with columnar structure. Depending on the application, one type of grain is preferred and thus favoured, e.g. equiaxed grains in car engines and columnar grains in turbine blades as reported by Reinhart et al, 2005 and McFadden et al., 2009.

Since the grain structure influences the properties of a casting, a great deal of effort has been devoted in the last decades to understand the mechanism behind the development of the macrostructure during solidification. Thus, equiaxed grains can nucleate and grow ahead of the columnar front causing an abrupt columnar to equiaxed transition (CET) whose prediction is of great interest for the evaluation and design of the mechanical properties of solidified products. As a consequence, it is critical for industrial applications to understand the physical mechanisms which control this transition during solidification (Spittle, 2006).

In order to realize the control of the columnar and equiaxed growth, it is necessary to understand the columnar to equiaxed transition (CET) mechanism during solidification, and make clear the CET transition condition. Fundamentally, it is necessary to have knowledge of the competition between nucleation and growth during solidification. Qualitatively, the CET occurs more easily when an alloy has a high solute concentration, low pouring temperature (for casting), low temperature gradient, high nucleation density in the melt and vigorous melt convection.

However, a quantitative understanding of the CET requires a thorough comprehension of all physical mechanisms involved.

In 1984, Hunt first developed an analytical model to describe steady-state columnar and equiaxed growth, and to qualitatively reveal the effects of alloy composition, nucleation density and cooling rate on the CET. On the other hand, he used a very simple empirical relationship to describe the variation of the undercooling with alloy composite and solidification rate. Cockcroft et al. (1994) used a more recent growth theory for the columnar and equiaxed growth but without considering high velocity non-equilibrium effects under rapid solidification. Recently, based on Hunt's CET model, Gäumann et al. (1997, 2001) developed a more comprehensive model by combining KGT model (Kurtz et. al., 1986) for directional solidification with LKT model (Lipton et. al., 1987) for the undercooling melt growth, with high velocity non-equilibrium effects to be taken into account. Gäumann et al. (2001) succeeded in applying their model to epitaxial laser metal forming of single crystal.

In previous research, the authors of this work carried out experiments in which the conditions of columnar to equiaxed transition (CET) in directional solidification of dendritic alloys were determined The alloy systems in this work include Pb-Sn (Ares & Schvezov, 2000), Al-Cu (Ares et. al., 2011), Al-Mg (Ares et. al., 2003), Al-Zn and Zn-Al alloys (Ares & Schvezov, 2007). These experiments permit to determine that the transition occurs gradually in a zone when the gradient in the liquid ahead of the columnar dendrites reaches critical and minimum values, being negative in most of the cases. The temperature gradients in the melt ahead of the columnar dendrites at the transition are in the range of -0.80 to 1.0 °C/cm for Pb-Sn, -11.41 to 2.80 °C/cm for Al-Cu, -4.20 to 0.67 °C/cm for Al-Si, -1.67 to 0.91 °C/cm for Al-Mg, -11.38 to 0.91 °C/cm for Al-Zn. Two interphases are defined; assumed to be macroscopically flat, which are the liquidus and solidus interphases. After the transition, the speed of the liquidus front accelerates much faster than the speed of the solidus front; with values of 0.004 to 0.01, 0.02 to 0.48, 0.12 to 0.89, 0.10 to 0.18 and 0.09 to 0.18 cm/s, respectively. Also, the average supercooling of 0.63 to 2.75 1C for Pb-Sn, 0.59 to 1.15 1C for Al-Cu, 0.67 to 1.25 1C for Al-Si, 0.69 to 1.15 1C for Al-Mg, 0.85 to 1.40 1C for the Al-Zn and was measured, which provides the driving force to surmount the energy barrier required to create a viable solid-liquid interface (Ares et al., 2005). A semi-empirical model to predict the columnar to equiaxed transition is developed based on experimental results obtained from measurements during solidification of lead-tin alloys directly upwards (Ares et al., 2002). The measurements include the solidification velocities of the liquidus and solidus fronts, and the temperature gradients along the sample in the three regions of liquid, mushy and solid. The experimental data was coupled with a numerical model for heat transfer. With the model, the predicted positions of the transition are in agreement with the experimental observations which show that the transition occurs when the temperature gradient reaches values below 1°C/cm and the velocity of the liquidus front increases to values around 0.01cm/s.

In addition, the thermal parameters, type of structure, grain size and dendritic spacing with the corrosion resistance of Zn-4wt%Al, Zn-16wt%Al and Zn-27wt%Al alloys were correlated (Ares et al., 2008). The polarization curves showed that the columnar structure is the most susceptible structure to corrosion, in the case of the alloy with only 4wt%of Al. The rest of the structures presented currents of peaks in the same order which were independent to the concentration of Al composition presenting in the alloy.

The biggest susceptibility to corrosion of the alloys with columnar structure can be observed by analyzing the values of Rct (charge transfer resistance) obtained using the electrochemical impedance spectroscopy (EIS) technique. In Zn-4wt%Al and Zn-27wt%Al, the corrosion susceptibility depends on the structure of the alloy. The alloy with 16wt%Al is less resistant to corrosion and their susceptibility to corrosion is independent of the structure. The alloy with 27wt%Al and the CET structure is the alloy which has the most corrosion resistance. When the critical temperature gradient becomes more negative, the Rct values increase. In the case of the correlation of Rct values and the structural parameters such as the grain sizes and secondary dendritic spacing, Rct values increase when the grain size and secondary dendritic spacing increase. But this does not happen for ZA16 alloy.

Composite materials obtained by solidification of alloys have made remarkable progress in their development and applications in automotive and aerospace industries in recent decades. Among them the most current applications are the zinc and aluminum base composite materials (Long et al., 1991; Rohatgi, 1991). It is well-known that the corrosion behavior of MMCs is based on many factors such as the composition of the alloy used, the type of reinforcement particles used, the reinforcement particle sizes and their distribution in the matrix, the technique used for the manufacture, and the nature of the interface between the matrix and reinforcement. A very slight change in any of these factors can seriously affect the corrosion behavior of the material.

In short, there is little research related to the study of mechanical and electrochemical properties of Zn-Al alloys as well as Zn-Al alloys MMCs containing SiC and Al_2O_3 particulations with different grain structures in the matrix. Also there is lack of fundamental study on the performance of Zn-Al alloys and their MMCs in corrosive environments when both solidification microstructure and type of particle distribution are in consideration. In the present research, Zn-Al-SiC and Zn-Al-Al_2O_3 composites are prepared and solidified by vertical directional solidification method. By means of voltammograms and electrochemical impedance spectroscopy, the corrosion resistances of Zn-Al matrix composite materials with different types of particles are obtained and analyzed and the results are compared.

2. Materials and methods

2.1 Alloys and metal matrix composites preparation

Zinc-Aluminum (ZA) alloys of different compositions were prepared from zinc (99.98 wt pct), aluminum (99.94 wt pct), and composites were prepared by adding SiC and Al_2O_3 particles to the alloys. The compositions of the alloys and composites prepared and directionally solidified are: Zn-27wt%Al, Zn-50wt%Al, Zn-27wt%Al + 8vol%SiC, Zn-27wt%Al + 15vol% SiC, Zn-50wt%Al + 8vol%SiC, Zn-50wt%Al + 15vol%SiC, Zn-27wt%Al + 8vol%Al_2O_3, Zn-27wt%Al + 15vol%Al_2O_3.

The chemical compositions of the commercially pure metals used to prepare the alloys are presented in Table 1. The molds were made from a 23 mm i.d. and 25 mm e.d. PYREX (Corning Glass Works, Corning, NY) tube, with a flat bottom, a cylindrical uniform section and a height of 200 mm. The sample was a cylinder 22 mm in diameter and 100 mm in height.

Chemical composition of Zn	
Element	Weight percent, wt%
Zn	99.98 ± 0.2
Fe	0.010 ± 0.01
Si	0.006 ± 0.0001
Pb	0.004 ± 0.001
Others	< 0.001 ± 0.0001
Chemical composition of Al	
Element	Weight percent, wt%
Al	99.94 ± 0.2
Fe	0.028 ± 0.0001
Si	0.033 ± 0.001
Pb	0.001 ± 0.0001
Others	< 0.001 ± 0.0001

Table 1. Chemical composition of the Zn and Al used to prepare the alloys.

2.2 Directional solidification

The alloy samples were melted and solidified directionally upwards in an experimental set-up described elsewhere (Ares et. al., 2007). It was designed in such a way that the heat was extracted only through the bottom promoting upward directional solidification to obtain the columnar-to-equiaxed transition (CET), see Figure 1 (a).

In order to reveal the macrostructure, after solidification the samples were cut in the axial direction, polished, and etched using concentrated hydrochloric acid for 3 seconds at room temperature for the zinc-aluminum alloys, followed by rinsing and wiping off the resulting black deposit. The microstructures were etched with a mixture containing chromic acid (50 g Cr_2O_3; 4 g Na_2SO_4 in 100 ml of water) for 10 seconds at room temperature (Vander Voort, 2007). Typical longitudinal macrostructure of different areas of the sample are shown in Figure 1 (b) to (d).

The position of the transition was located by visual observation and optical microscopy. The distance from the chill zone of the sample was measured with a ruler. It is noted in Figure 1 that the CET is not sharp, showing an area where some equiaxed grains co-exist with columnar grains. As was reported before, the size of the transition area is in the order of up to 10 mm (Ares et al., 2007, 2010). The grain structure was inspected by visual observation under Arcano® optical microscopy.

2.3 Corrosion tests

For the electrochemical tests, samples of 20 mm in length of each zone and for each concentration were prepared as test electrodes (see Figure 1), polished with sandpaper (from SiC #80 until #1200) and washed with distilled water and dried by natural flow of air.

All the electrochemical tests were conducted in 3wt% NaCl solution at room temperature using an IM6d Zahner®-Elektrik potentiostat coupled to a frequency analyzer system.

A conventional three-compartment glass electrochemical cell with its compartments separated by ceramic diaphragms was used. The test electrodes consisted of sections of the ZA ingots (see Figure 1) were positioned at the glass corrosion cell kit (leaving a rectangular area in contact with the electrolyte). The potential of the test electrode was measured against a saturated calomel reference electrode (0.242 V vs NHE), provided with a Luggin capillary tip. The Pt sheet was used as a counter electrode.

Voltammograms were run between preset cathodic (open circuit potential ≈ -1.500 V) and anodic (Es,a = -0.700 V) switching potentials at potential sweep rates (v), at 0.002 $V.s^{-1}$. Impedance spectra were obtained in the frequency range of 10^{-3} Hz and 10^{5} Hz at open circuit potential.

For comparison purposes, experiments using pure metals and aluminum-based alloys with different structures were conducted under the same experimental conditions. All the corrosion tests experiments were triplicate and the average values and graphical outputs are reported.

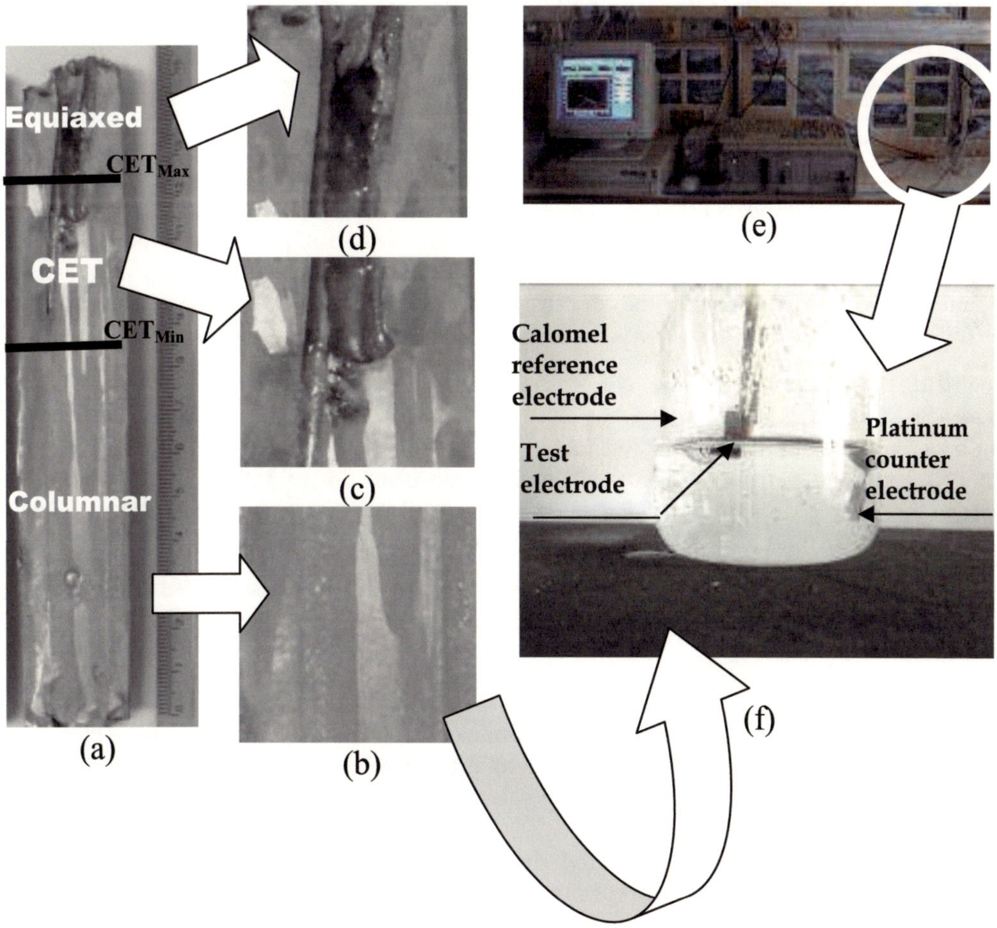

Fig. 1. Experimental device for electrochemical tests. (f) A glass corrosion cell kit with a platinum counter electrode and a sutured calomel reference electrode (SCE).

3. Results and discussion

3.1 Voltammetric data

During the anodic potential scanning, the voltammogram of equiaxed zinc shows that the current is practically zero until it reaches a potential of -1 V, where the current rises sharply, starting the active dissolution of metal (Figure 2 (a)). The negative potential scan shows a hysteresis loop, suggesting that this current increase was due to the start of a process of pitting, and two cathodic current peaks at about -1.2 V and - 1.3 V (called C_1 and C_2). These peaks could be associated with the reduction of $Zn(OH)_2$ and ZnO, respectively (Zhang, 1996). The composition of corrosion products formed on the zinc surface may be not uniformly distributed. The different compositions of the films formed can explain the difference in the outcomes reported in the literature.

The properties of corrosion products are a function of various material and environmental factors and thus vary essentially from situation to situation. For example, only one peak appears in the case of the columnar zinc (Figure 2 (b)). As the concentration of aluminum in the alloy increases, the definition of these reduction peaks is not clear, although the C_2 peak is dominant (Figure 2 (c)).

In the case of the alloys, the values of the anodic currents are similar for the same Al concentration, independently of the structure, and the most important difference is observed in the distribution of the cathodic current peaks, which indicates the different characteristics of the films formed during the anodic scan. These results can be attributed to the aggressive/depassivating action of Cl^- anions (Augustynski, 1978). At present, the mechanism of film formation is still uncertain. For the case of CET structure, profiles are more complex, because the proportion of one or other structure (columnar or equiaxed) can vary from sample to sample (Figure 2 (d)). Also, as the concentration of Al increases, the voltammetric profile of the different structures tends towards the response of pure aluminum (Figure 2 (e)).

Analyzing the response of the composites (Zn-27wt%Al + 8vol%SiC, Zn-27wt%Al + 15vol%SiC (Figure 2(f)), Zn-50wt%Al + 8vol%SiC, Zn-50wt%Al + 15vol%SiC (Figure 4 (g)), Zn-27wt%Al + 8vol%Al_2O_3, Zn-27wt%Al + 15vol%Al_2O_3 (Figure 2 (h))), we observed that when the volume percent of SiC particles increase from 8% to 15% in ZA27 and ZA50 matrix, the rate of dissolution of the alloy increases. In the case of the addition of Al_2O_3 particles to ZA 27 matrix the rate of dissolution is approximately the same.

This different distribution of the peaks in the voltammograms gives rise to surface layers with different corrosion products, as shown in the micrographs of Figure 3, where samples with higher proportion of particles in the matrix show the formation of a thicker layer of corrosion products. Also, it is observed the formation of pitting on the electrode surface.

Three distinctive features in the potentiodynamic curves can be clearly observed (Figure 4): (i) the potential at which the anodic current during the forward anodic bias increases sharply form the passive current level (breakdown or critical pitting potential E_p); (ii) a hysteresis loop (difference between forward and reverse scans) and (iii) the potential at which the hysteresis loop is completed during reverse polarization scan after localized corrosion propagation (repassivation potential E_r). Stable pits form at potentials noble to E_p and will grow at potentials noble to Er (Frankel, 1998). Also, for many years it has been recognized that Ep measurements are applicable to naturally occurring pit initiation on stainless alloys in chemical and marine environments (Wilde, 1972; Bilmes et al., 2005).

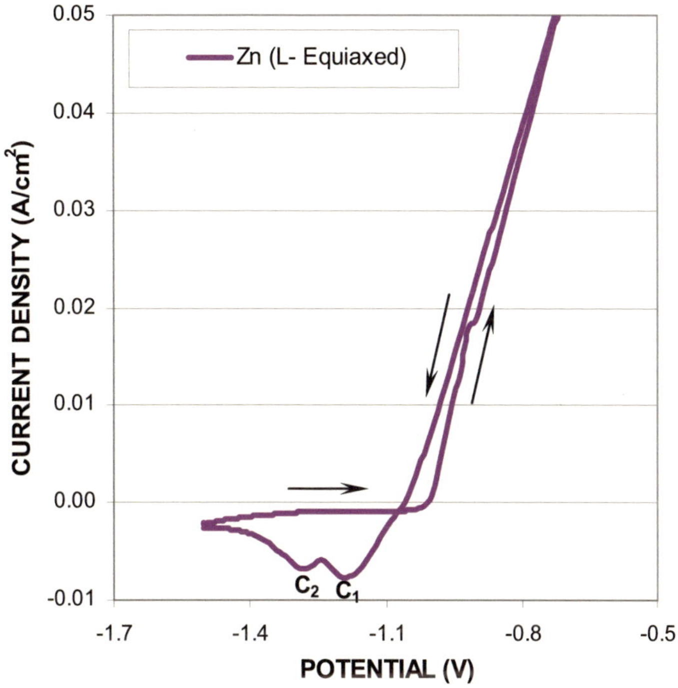
Zn (L- Equiaxed)
CURRENT DENSITY (A/cm^2)
POTENTIAL (V)
0.05
0.04
0.03
0.02
0.01
0.00
-0.01
-1.7
-1.4
-1.1
-0.8
-0.5
C_2
C_1

(a)

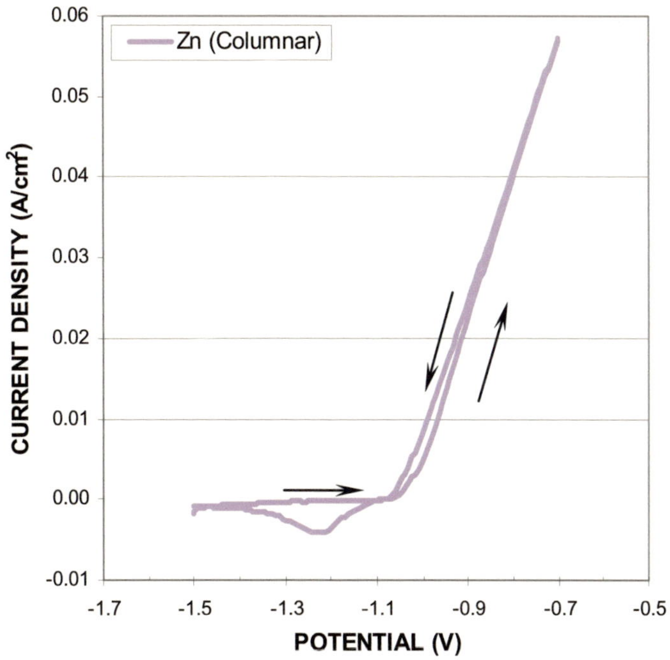
Zn (Columnar)
CURRENT DENSITY (A/cm^2)
POTENTIAL (V)
0.06
0.05
0.04
0.03
0.02
0.01
0.00
-0.01
-1.7
-1.5
-1.3
-1.1
-0.9
-0.7
-0.5

(b)

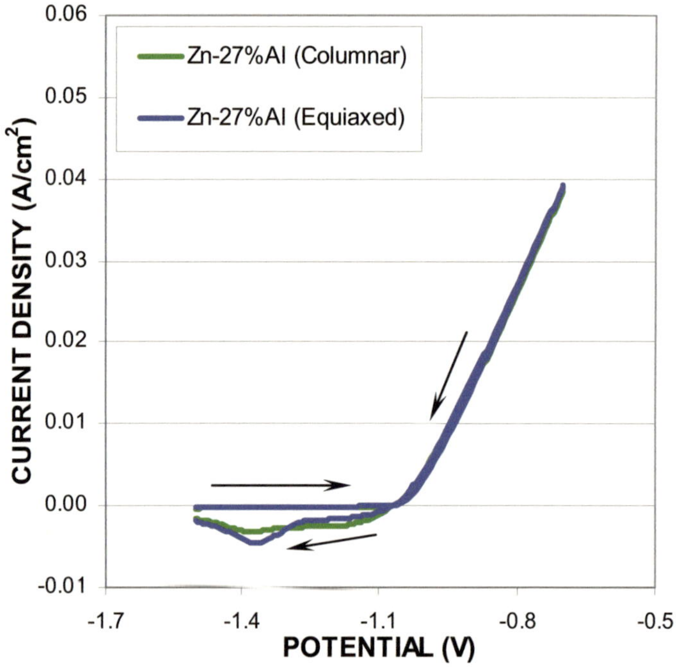

0.06
0.05
0.04
0.03
0.02
0.01
0.00
-0.01
Zn-27%Al (Columnar)
Zn-27%Al (Equiaxed)
CURRENT DENSITY (A/cm^2)
-1.7
-1.4
-1.1
-0.8
-0.5
POTENTIAL (V)

(c)

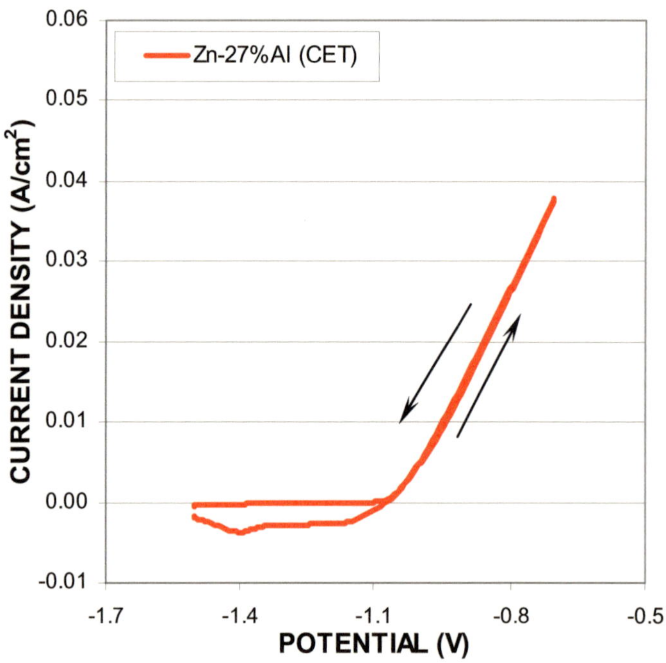

0.06
0.05
0.04
0.03
0.02
0.01
0.00
-0.01
Zn-27%Al (CET)
CURRENT DENSITY (A/cm^2)
-1.7
-1.4
-1.1
-0.8
-0.5
POTENTIAL (V)

(d)

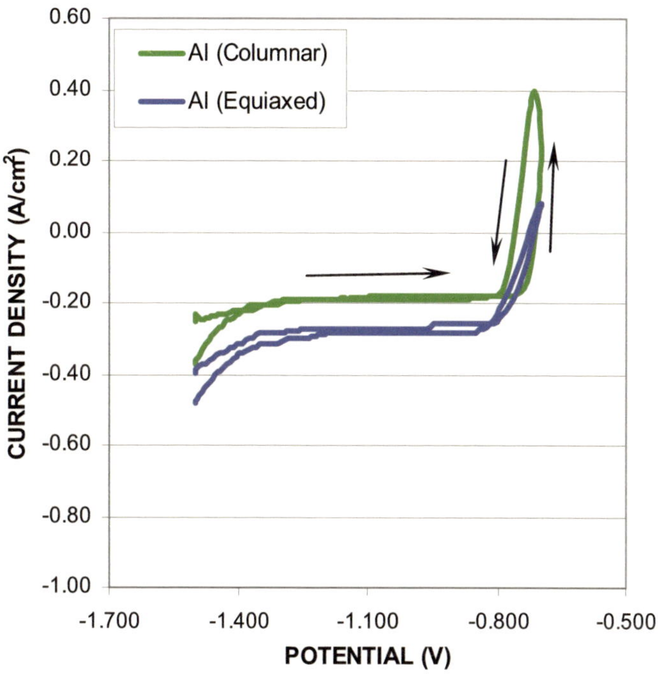
Al (Columnar)
Al (Equiaxed)
CURRENT DENSITY (A/cm²)
0.60
0.40
0.20
0.00
-0.20
-0.40
-0.60
-0.80
-1.00
-1.700
-1.400
-1.100
-0.800
-0.500
POTENTIAL (V)

(e)

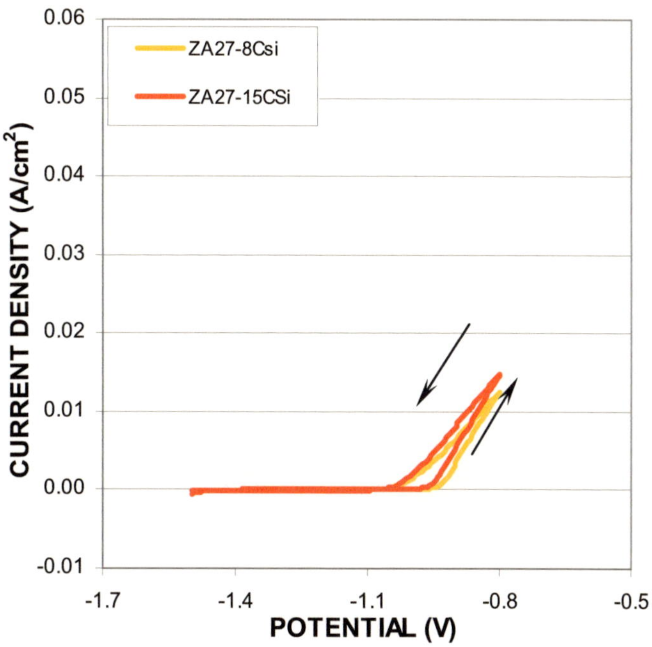
ZA27-8Csi
ZA27-15CSi
CURRENT DENSITY (A/cm²)
0.06
0.05
0.04
0.03
0.02
0.01
0.00
-0.01
-1.7
-1.4
-1.1
-0.8
-0.5
POTENTIAL (V)

(f)

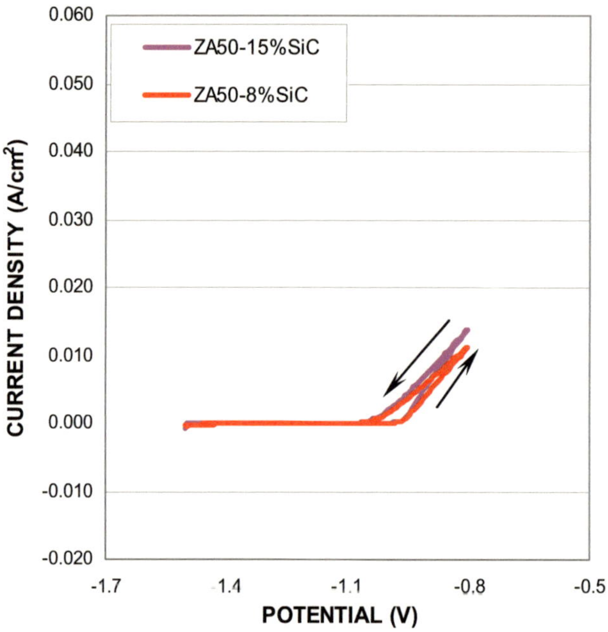
ZA50-15%SiC
ZA50-8%SiC
CURRENT DENSITY (A/cm^2)
POTENTIAL (V)
0.060
0.050
0.040
0.030
0.020
0.010
0.000
-0.010
-0.020
-1.7
-1.4
-1.1
-0.8
-0.5

(g)

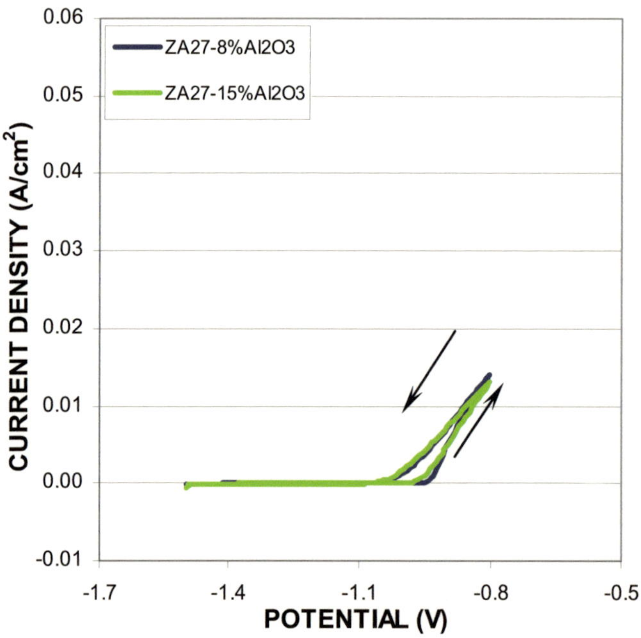
ZA27-8%Al2O3
ZA27-15%Al2O3
CURRENT DENSITY (A/cm^2)
POTENTIAL (V)
0.06
0.05
0.04
0.03
0.02
0.01
0.00
-0.01
-1.7
-1.4
-1.1
-0.8
-0.5

(h)

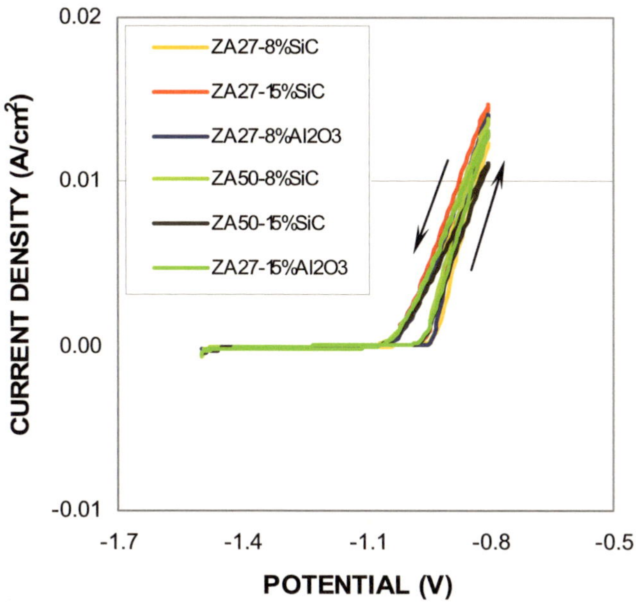

(i)

Fig. 2. Voltammograms of (a) pure Zinc with equiaxed structure, (b) pure Zinc with columnar structure (c) Zn-27wt%Al alloy with columnar and equiaxed structures, (d) Zn-27wt%Al alloy with CET structure, (e) pure Aluminum with columnar and equiaxed structure, (f) Zn-27wt%Al + 8vol%SiC, Zn-27wt%Al + 15vol%SiC, (g) Zn-50wt%Al + 8vol%SiC, Zn-50wt%Al + 15vol%SiC, (h) Zn-27wt%Al + 8vol%Al_2O_3, Zn-27wt%Al + 15vol%Al_2O_3 and (i) All types of composites.

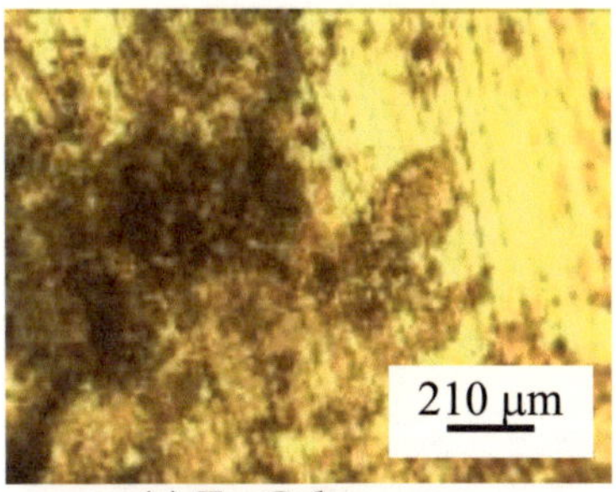

(a) Zn-Columnar

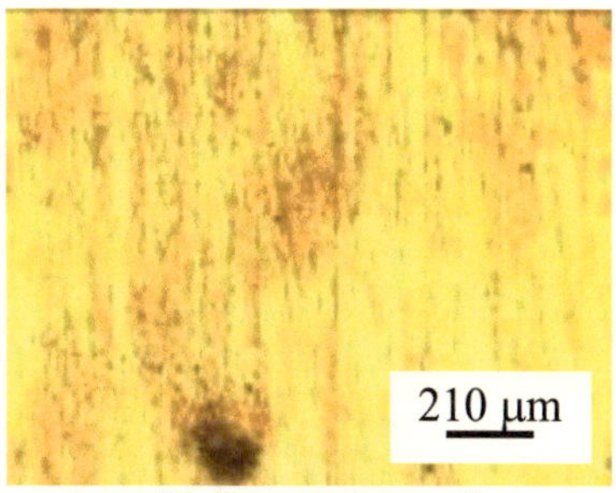

(b) Zn - Equiaxed

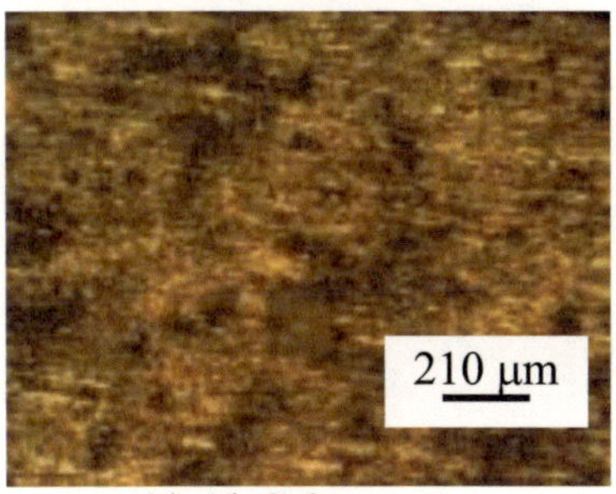

(c) Al-Columnar

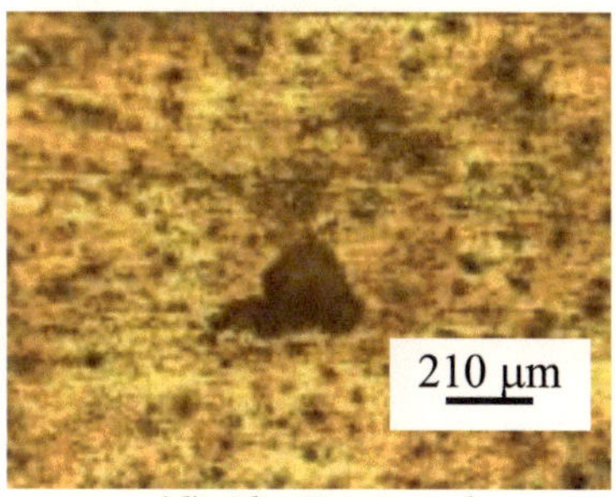

(d) Al - Equiaxed

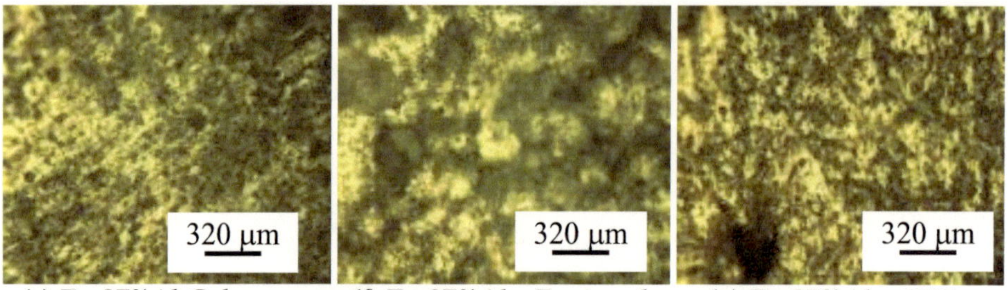

(e) Zn-27%Al-Columnar (f) Zn-27%Al – Equiaxed (g) Zn-27%Al – CET

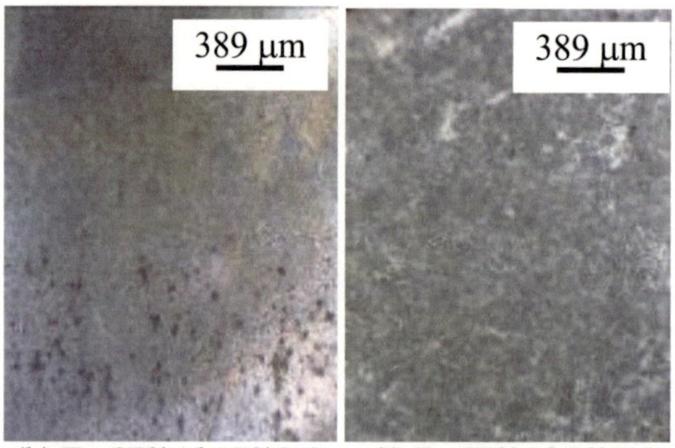

(h) Zn-27%Al-15%SiC (i) Zn-27%Al-8%SiC

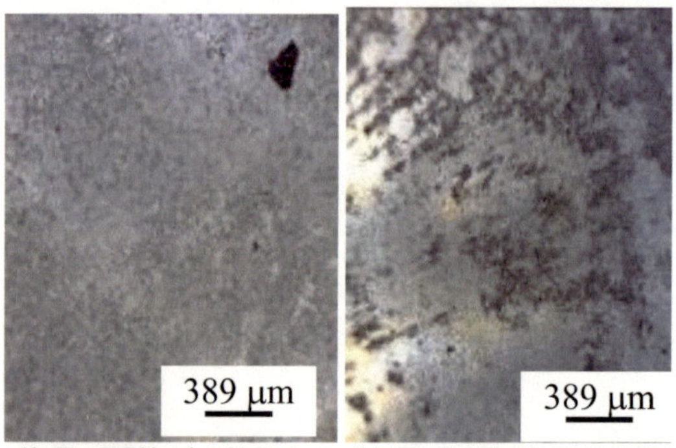

(j) Zn-50%Al-8%SiC (k) Zn-50%Al-15%SiC

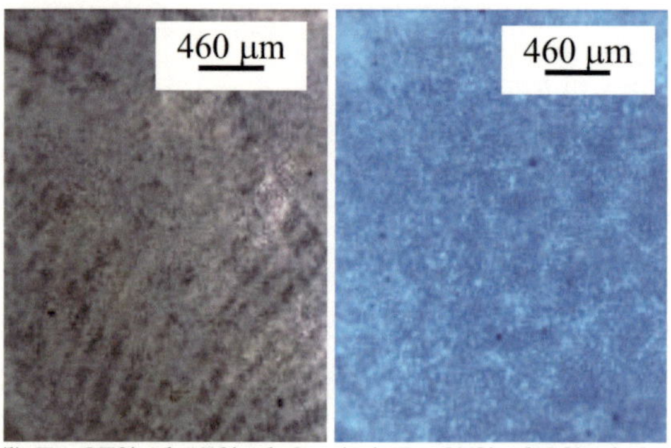

(l) Zn-27%Al-15%Al_2O_3 (m) Zn-27%Al-8%Al_2O_3

Fig. 3. Micrographs of different alloy samples and structures.

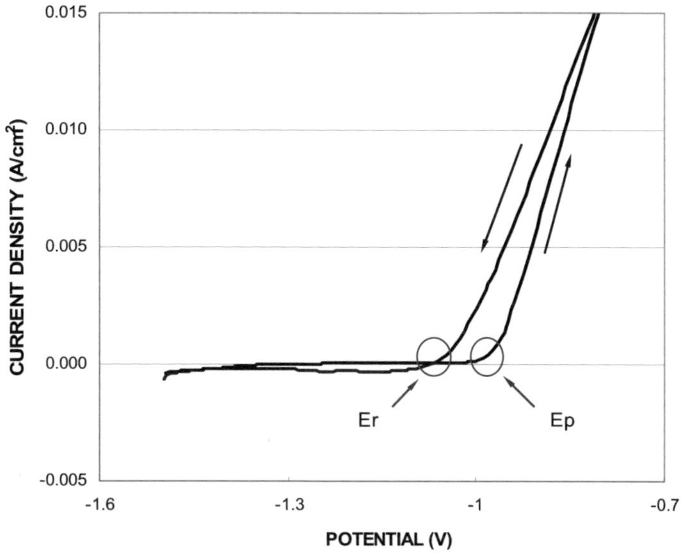

Fig. 4. Representative curve.

In all cases, the E / I response of the alloys shows the typical hysteresis indicates the phenomenon of pitting and found that the more susceptible are the composites than the alloys. The most susceptible are those containing neither SiC nor Al_2O_3 in the matrix, see Table 2 and Figure 5.

Alloy / Composite	Ep (V)	Er (V)	ΔEp-r (mV)	Ecorr (V)	ΔEr-corr (mV)
ZA50-15vol%SiC	-1.002	-1.093	91	-1.271	178
ZA50-8vol%SiC	-0.988	-1.082	94	-1.302	220
ZA27-8vol%Al_2O_3	-0.974	-1.086	112	-1.297	211
ZA27-15vol%Al_2O_3	-0.974	-1.079	105	-1.298	219
ZA27-15vol%SiC	-1.002	-1.086	84	-1.312	226
ZA27-8vol%SiC	-0.981	-1.107	126	-1.245	138
ZA27 CET	-1.02	-1.06	30	-1.102	42
ZA27 Col	-1.028	-1.052	24	-1.071	19
ZA27 Eq	-1.038	-1.068	40	-1.068	0

Table 2. The susceptibility to corrosion, ΔE, was measured as the difference between the potential of pitting, Ep, and the repassivation potential, Er.

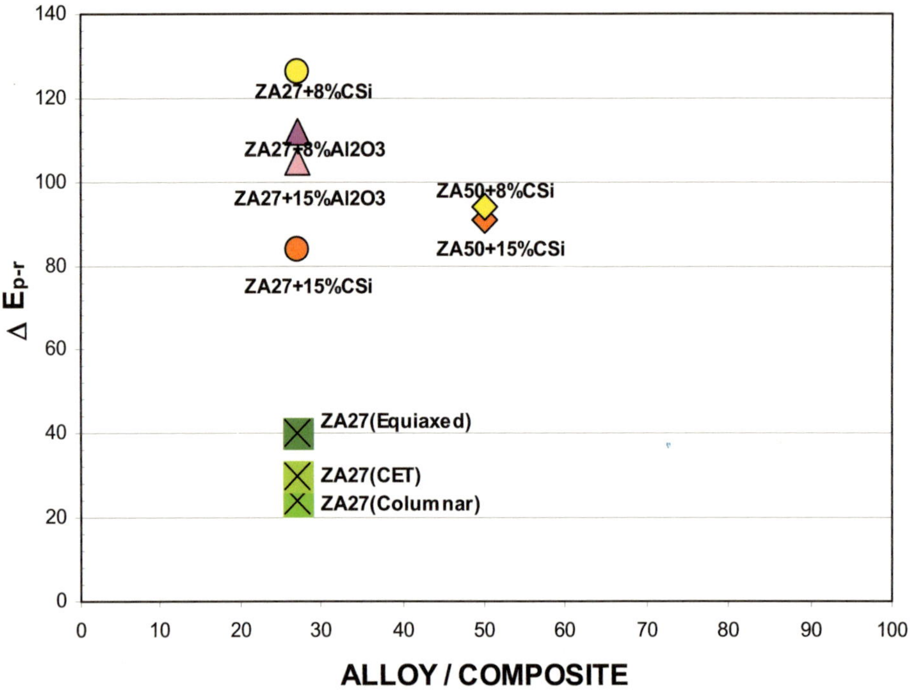

Fig. 5. $\Delta E_{p\text{-}r}$ as a function of concentration and alloy structure.

The susceptibility to corrosion was measured as the difference between the potential of pitting, Ep, and the repassivation potential, Er, as Δ Ep-r and the difference between the repassivation potential and the corrosion potential of each sample through Table 2 it is possible to observe that the values of repassivation potential for materials without particles in the matrix are near the corrosion potential, but not in the case of the the other samples.

3.2 Electrochemical impedance spectroscopy data

Impedance spectra are strongly dependent on the composition and structures of the alloys and composites. Figure 6 shows the experimental Nyquist diagrams for all the alloys and composites used. All the diagrams show one capacitive time constant at high frequencies and a non-well defined time constant at low frequencies, probably associated with diffusion processes also reported in the literature (Deslouis et al., 1984; Trabanelli at al., 1975). It can be seen that as the concentration of aluminum in the alloy increases, the second time constant approximates the response associated with a diffusion process in finite thickness, due to the formation of a more compact oxide.

In some cases, the shape of the Nyquist diagrams for CET structure in alloys resembles that of those with equiaxed structure and in others those with columnar structure, depending on the relative amount of each phase in the CET structure, which in turn depends on the region where the specimen was obtained.

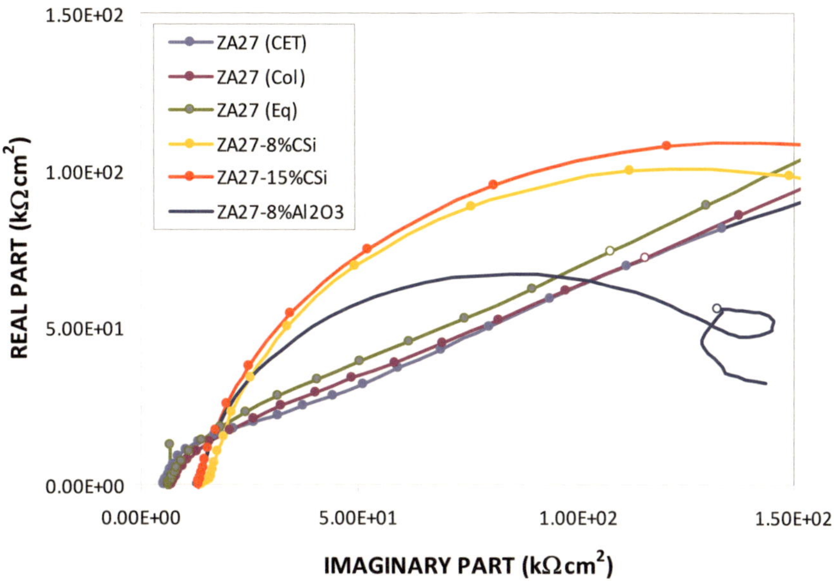

Fig. 6. (a) Nyquist Diagram for different samples.

The whole set of experimental impedance spectra can be discussed according to the following total transfer function.

$$Z_t(jw) = R_\Omega + Z \tag{1}$$

with:

$$\frac{1}{Z} = \frac{1}{R_{ct} + Z_W} + j_W.C \tag{2}$$

where R_Ω is the ohmic solution resistance, ω= 2πf; C_{dl} the capacitance of the electric double layer, R_{ct} the charge transfer resistance and Z_W the diffusion contributions in impedance spectra.

$Z_W = R_{DO}$ (jS) $^{-05}$ for semi-infinite diffusion contribution and $Z_W = R_{DO}$ (jS) $^{-05}$ coth (jS) $^{-05}$ is related to diffusion through a film of thickness d, formed on the electrode, where R_{DO} is the diffusion resistance and the parameter S= d2ω/D, where d and D are the diffusion thickness and diffusion coefficient related to the transport process (Fedrizzi et al., 1992).

The good agreement between experimental and simulated data according to the transfer function given in the analysis of Eqs. 1 and 2 using non-linear least square fit routines is shown in Figure 8 (a) at high frequencies. At low frequencies (less than 10-1 Hz) is not achieved a good fit with this model, since the impedance measurement does not give us enough information to define a new input capacitance, this occurs for samples ZA27-8% SiC, ZA50-8% SiC and ZA27-Al2O3. This process can represent by an equivalent circuit in Figure 8 (a).

For ZA27samples and those containing 15%SiC at high frequencies seems to be defined one second capacitive loop corresponding to corrosion processes controlled by precipitation and dissolution of ions Zn, see Figure 7 (b, c and d). The equivalent circuit corresponds to that showed in Figure 8 (b).

The values of C_{dl}, C_1 and R_{ct} determined from the optimum fit procedure are presented in Table 3.

The analysis of the impedance parameters associated with the time constant at low frequencies is difficult because in some cases the loop it is not complete. However, it was possible to calculate from by fitting an approximate value of diffusion coefficient $D \approx 10^{-10}$ – 10^{-12} cm²/s.

High values of capacity confirm the formation of porous corrosion products, as can be seen in Figure 9. These high values of capacity may also be correlated with an increase in the area.

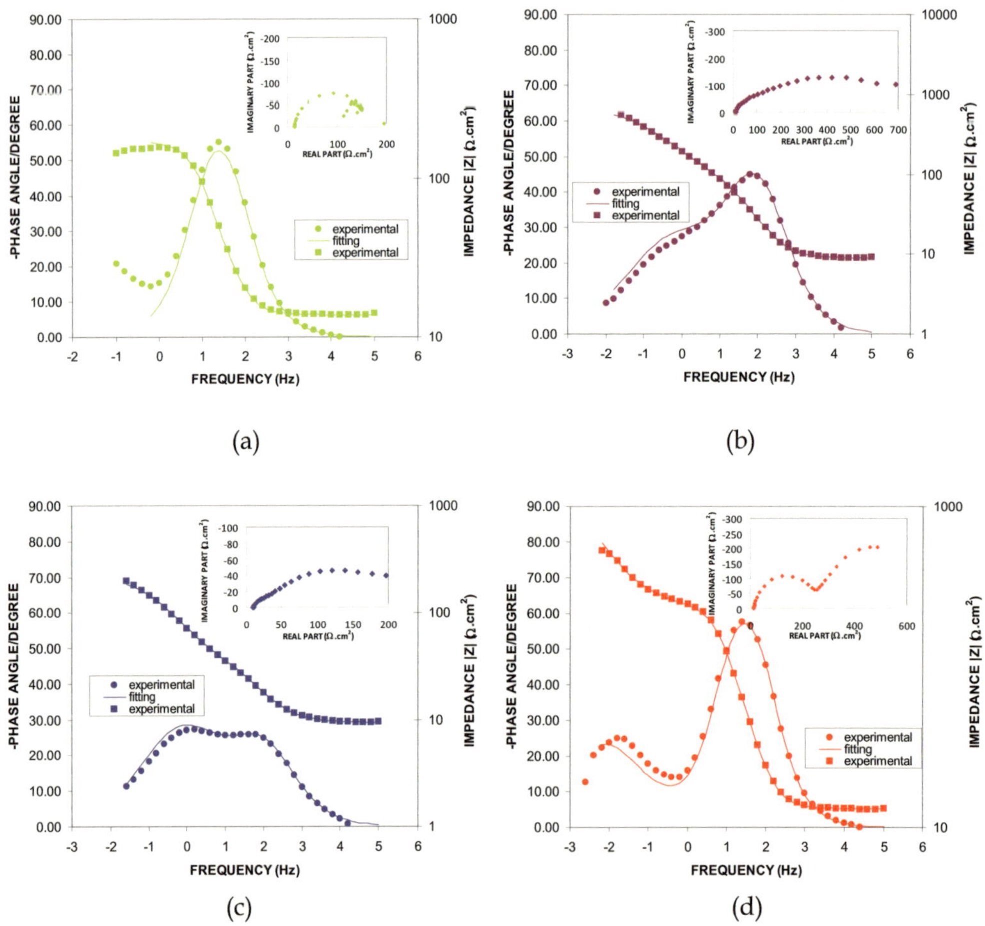

Fig. 7. Bode y Nyquist plot for – (a) ZA50-8%SiC (b) ZA27 L(CET) (c) ZA27 L(Columnar) (d) ZA27-15%SiC

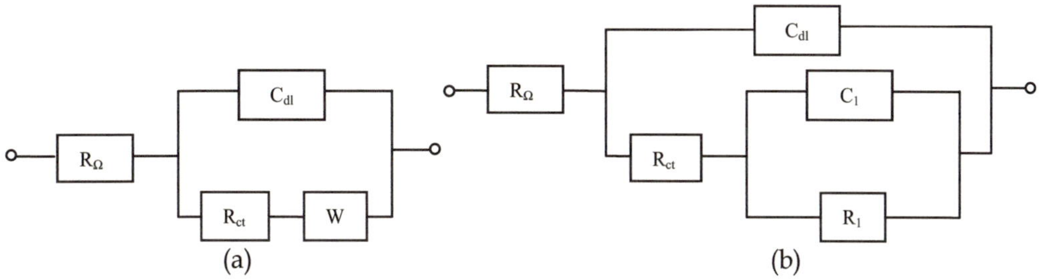

Fig. 8. Equivalent circuit of EIS for different samples.

The corrosion current can be related to the R_{ct} in the case of mixed control (Epelboin et al., 1972), where the polarization resistance technique fails, according to the following expression:

$$R_{ct} = b_a\, b_c / 2.303(b_a + b_c) I_{corr} \quad (3)$$

However, it is important to note that the Rct values are not directly related to the susceptibility to corrosion of the different alloys and composites. They are related to the rate of charge transfer reactions that give rise to the formation of a passive layer on the surface of the samples (the impedance measurements are at open circuit potential only). The protective characteristics of these passive films depend on the preparation conditions of the alloys, the distribution of elements in the alloy and the presence on the surface of active sites for adsorption of chloride ion.

Ions are formed during the anodic dissolution of alloy

$$Zn \rightarrow Zn^{+2} + 2\,e^-$$

which can react with hydroxyl ions

$$H_2O + \frac{1}{2} O_2 + 2\,e^- \rightarrow 2\,OH^-$$

and can be generated $Zn(OH)_2$

The ZA27 alloy with different structures is less resistant to corrosion and its susceptibility to corrosion is dependent of the structure. The ZA27 - 15%CSi composite has the highest corrosion resistance. Also, discriminating by type of composite materials, the MMCs with SiC are more corrosion resistant than those MMCs prepared with alumina particles.

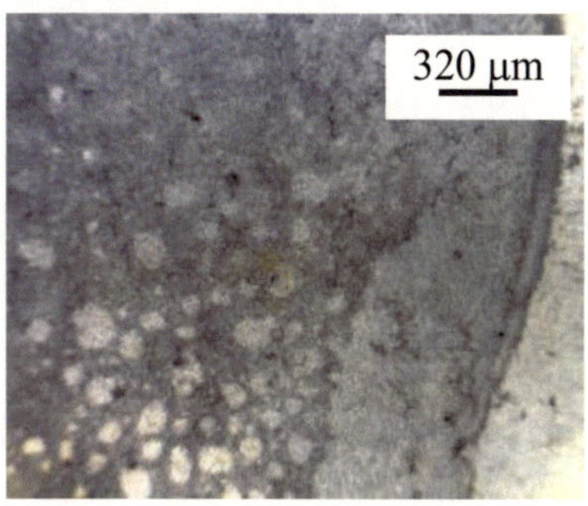

Fig. 9. Micrograph of ZA27 – 8%SiC sample after EIS test.

Alloy / Composite (wt pct)	R_{Ω} (Ω.cm^2)	C_{dl} (F/cm^2)	R_{ct} (Ω.cm^2)	R_1 (Ω.cm^2)	C_1 (F/cm^2)
Zn27-15%SiC	13.03	7.95e^{-5}	1317.23	278.67	1.35e^{-03}
Zn50-15%SiC	16.48	3.65e^{-5}	------	19.33	5.18e^{-05}
Zn-27Al (L-Columnar)	8.85	4.85e^{-5}	94.78	615.29	7.51e^{-05}
Zn-27%Al (L-CET)	9.41	1.04e^{-4}	26.88	193.42	6.51e^{-04}
Zn-27%Al (L-Equiaxed)	8.56	5.75e^{-5}	87.01	304.59	6.05e^{-04}
Zn-27%Al - 8%CSi	14.47	9.22e^{-5}	228.77		
Zn-27%Al - 15%Al_2O_3	12.60	6.94e^{-5}	151.78		
Zn-27%Al - 8%Al_2O_3	11.64	4.22 e^{-5}	139.57		

Table 3. Principal parameters obtained from the EIS analysis.

4. Conclusions

The highest corrosion resistance or susceptibility to corrosion is a complex function of the alloys and composite composition, structure and the exposed surface, all of which determine the protective characteristics of the film that formed on the alloys.

Alloys with a higher aluminum content have a higher corrosion resistance, mainly due to the formation of a protective film.

Even at a higher Al concentration, the corrosion resistance depends on the structure of the alloys.

The results also indicate that the corrosion resistance of Zn-Al-SiC and Zn-Al-Al_2O_3 MMCs composites has demonstrated the improvement in comparison to ZA alloys in 3wt% NaCl solutions.

5. Acknowledgment

The authors would like to thank Consejo Nacional de Investigaciones Científicas y Técnicas (CONICET) for the financial support.

6. References

Ares, A.E.; Schvezov, C.E. (2000) Solidification Parameters During the Columnar-to-Equiaxed Transition in Lead-Tin Alloys. *Metallurgical and Materials Transactions A*, Vol. 31, 1611-1625, ISSN 1073-5623/83

Ares, A.E.; Gueijman, S.F.; Schvezov, C.E. (2010) Experimental Study of the Columnar-to-Equiaxed Transition During Directional Solidification of Zinc-Aluminum Alloys and Composites, *J. Crystal Growth*, Vol. 312, pp. 2154-2170, ISSN 0022-0248

Ares, A.E.; Caram, R.; Schvezov, C.E. (2003) Columnar-to-Equiaxed Transition Studies in Aluminum-Magnesium and Aluminum-Zinc Alloys, *Proceedings of Light Metals 2003*, ISBN, San Diego, California, United States, March of 2003

Ares, A.E.; Gueijman, S.F.; Caram, R.; C.E. Schvezov (2005) Analysis of Solidification Parameters During Solidification of Lead and Aluminum Base Alloys. *J. Crystal Growth*, Vol. 275, pp. 235-240, ISSN 0022-0248

Ares, A.E. ; Schvezov, C.E. (2007) Influence of Solidification Thermal Parameters on the Columnar to Equiaxed Transition of Al-Zn and Zn-Al Alloys. *Metallurgical and Materials Transactions A*, Vol. 38, pp. 1485-1499, ISSN 1073-5623/83

Ares, A.E.; Gueijman, S.F.; Schvezov, C.E. (2002) Semi-Empirical Modeling for Columnar and Equiaxed Growth of Alloys. *J. Crystal Growth*, Vol. 241, pp. 235-240, ISSN 0022-0248

Ares, A.E.; Gassa, L.M.; Gueijman, S.F.; Schvezov, C.E. (2008) Correlation Between Thermal Parameters, Structures, Dendritic Spacing and Corrosion Behavior of Zn-Al Alloys With Columnar to Equiaxed Transition. *J. Crystal Growth*, Vol. 310, pp. 1355-1361, ISSN 0022-0248

Ares, A.E.; Caram, R.; Schvezov, C.E. (2006) Relation between As-Cast Mechanical Properties, Microstructure and Solidification Conditions for Zn-Al Alloys, Proceedings of MCWASP International Conference Modeling of Casting, Welding and Advance Solidification Processes - XI, Opio, France, June of 2006

Ares, A.E.; Gatti, I. P.; Gueijman, S.F.; Schvezov, C.E. (2009) Mechanical Properties of Zinc-Aluminum Alloys versus Structural and Thermal Parameters, Proceedings of MCWASP International Conference Modeling of Casting, Welding and Advance Solidification Processes - XII, Vancouver, Canadá, June of 2009

Augustynski, J. (1978) Etude De La Rupture De Passivite De Certains Metaux Electrochimiquement Actifs, *Corrosion Sci.*, Vol. 13, pp. 955-965, ISSN 0010-938X

Bilmes, P.D., Llorente, C.L., Saire Huamán, L., Gassa, L.M., Gervasi, C.A. (2005) Microstructure and pitting corrosion of 13CrNiMo weld metals. *Corrosion Science*, Vol. 48, p.p. 3261–3270, ISSN 0010-938X

Cockcroft, S. L.; Rappaz, M.; Mitchell, A. (1994) An Examination of Some of the Manufacturing Problems of Large Single-Crystal Turbine Blades for Use in Land-Based Gas Turbines, In: *Materials for Advanced Power Engineering*, Coutsouradis, J. et al., pp. 1161-1175, ISBN 0-7923-3075-7, New York: Kluwer Inc.

Deslouis, C.; Duprat, M.; Tulet-Tournillon, C. (1984) The Cathodic Mass Transport Process During Zinc Corrosion in Neutral Aerated Sodium Sulphate Solutions. *J. Electroanal Chem.*, Vol. 181, pp. 119-136, ISSN 0022-0728

Epelboin, I.; Keddam, M.; Takenouti, H. (1972) Use of Impedance Measurements for the Determination of the Instant Rate of Metal Corrosion. *Journal of Applied Electrochemistry*, Vol. 2, pp. 71-79, ISSN 0021-891X

Frankel, G.S. (1998) Pitting Corrosion of Metals. *Journal of the Electrochemical Society*, Vol. 145, pp. 2186–2198, ISSN 1945-7111

Fedrizzi, L.; Ciaghi, L.; Bonora, P.L.; Fratesi, R.; Roventi, G. (1992) Corrosion Behaviour of Electrogalvanized Steel in Sodium Chloride and Ammonium Sulphate Solutions, a Study by E.I.S. *J. Appl. Electrochem*, Vol. 22, pp. 247-254, ISSN 0021-891X

Gäumann, M.; Trivedi, R.; Kurz, W. (1997) Nucleation Ahead of the Advancing Interface in Directional Solidification, *Mater. Sci. Eng.* A, Vol. 226-228, pp. 763-769, ISSN 0921-5093, ISSN 0921-5093

Gäumann, M.; Bezençon, C.; Canalis, P. et al. (2001) Single-Crystal Laser Deposition of Superalloys: Processing-Microstructure Maps. *Acta Mater.*, Vol. 49, pp.1051-1062, ISSN 1359-6454

Hunt, J. D. (1984) Steady State Columnar and Equiaxed Growth of Dendrites and Eutectic. *Mater. Sci. Eng.*, Vol. 65, pp. 75-83, ISSN 0921-5093

Kurz, W.; Giovanola, B.; Trivedi, R. (1986) Theory of Microstructural Development During Rapid Solidification. *Acta Metall. Mater.*, Vol. 34, pp. 823-830, ISSN 0956-7151

Lipton, J.; Kurz,W.; Trivedi, R. (1987) Rapid Dendrite Growth in Undercooled Alloys. *Acta Metall.*, Vol. 35, pp.957-964, ISSN 0956-7151

Long, T.T.; Nishimura, T.; Aisaka, T.; Morita, M. (1991) Wear Resistance of Al-Si Alloys and Aluminum Matrix Composites. *Materials Transactions JIM*, Vol. 32, N° 2, pp. 181-188, ISSN 0916-1821

McFadden S.; Browne D.J.; Gandin C.A. (2009) A Comparison of Columnar-to-Equiaxed Transition Prediction Methods Using Simulation of the Growing Columnar Front. *Metall Mater Trans A*, Vol. 40, pp. 662-672, 1073-5623

Reinhart G.; Mangelinck-Noël N.; Nguyen-Thi H.; Schenk T.; Gastaldi J.; Billia B.; Pino P.; Härtwig J.;Baruchel J. (2005) Investigation of Columnar–Equiaxed Transition and Equiaxed Growth of Aluminium Based Alloys by X-ray Radiography. *Mater Sci Eng A*, Vol. 413–414, pp. 384–388, ISSN 0921-5093

Rohatgi, P. (1991) Cast Aluminum-Matrix Composites for Automotive Applications. JOM, Vol. 43, pp. 10–15, ISSN 1047-4838

Spittle J.A. (2006) Columnar to Equiaxed Grain Transition in as Solidified Alloys. *International Materials Reviews*, Vol.51, No.4, pp. 247–269, ISSN 0950-6608

Trabanelli, G.; Zucchi, F.; Brunoro, G.; Gilli, G. (1975) Characterization of the Corrosion or Anodic Oxidation Products on Zinc. *Electrodeposition Surf. Treat.*, Vol. 3, pp. 129-138, ISSN 0300-9416

Vander Voort, G.F. (June 2007) *Metallography Principles and Practice* (Fourth Edition) , ASM International, ISBN-10: 0-87170-672-5, New York, United States

Wilde, B.E. (1972) Critical Appraisal of some Popular Laboratory Electrochemical Tests for Predicting the Localized Corrosion Resistance of Stainless Alloys in Sea Water. *Corrosion*, Vol. 28, pp. 283–291, ISSN 00109312

Zhang, X. G. (1996) *Corrosion and Electrochemistry of Zinc* (First Edition), Plenum Press, ISBN 0-306-45334-7, New York and London

3

Corrosion Resistance of High Nitrogen Steels

Roman Ritzenhoff and André Hahn
Energietechnik-Essen GmbH
Germany

1. Introduction

1.1 Some basics about High Nitrogen Steels (HNS)

1.1.1 Nitrogen in steel

Nitrogen as an alloying element has been known and used in technical applications since the 1940s, initially under the premise for nickel substitution in stainless grades.

Nitrogen in low alloy steels is undesirable due to the formation of brittle nitrides. However, the use of nitrogen in high alloy steels has an array of advantages that makes it appear interesting as an alloying element. In references one find this sufficiently researched, so that in this situation only the most important points need to be summarized [Dailly & Hendry, 1998], [Energietechnik-Essen [ETE], 2011], [Allianz Industrie Forschung [AIF], 2003]:

- significant increase of strength without restricting ductility
- Improvement of corrosion resistance
- Increasing the high temperature tensile strength
- Extended / stabilized austenite form
- no formation of tension induced martensite with high cold working rates
- Inhibits the discharge of inter-metallic phases

These as HNS-Alloy (**H**igh **N**itrogen **S**teels) specific material group are characterised through an interesting material profile, i.e. a combination of strength and corrosion resistance.

A state-of-the-art production routine is P-ESR melting (pressurised electro slag remelting) which will be covered within the following section. Austenitic steels as well as martensitic steels can be manufactured and are suitable for forging and hot rolling. Some basic knowledge about the material characteristics is mandatory to avoid any potential issues at plastic deformation, heat treatment and to maintain the excellent corrosion resistance.

1.2 About this book chapter

This chapter will not cover the fundamentals of corrosion – we assume that the reader will have a basic knowledge about the principles of corrosion. This paper will provide an overview about the role of nitrogen related to corrosion of stainless steels. Since a lot of results have been published to date, common knowledge will summarized and topped up with own data and experimental results.

Finally, some typical HNS grades will be discussed with regards to their industrial application.

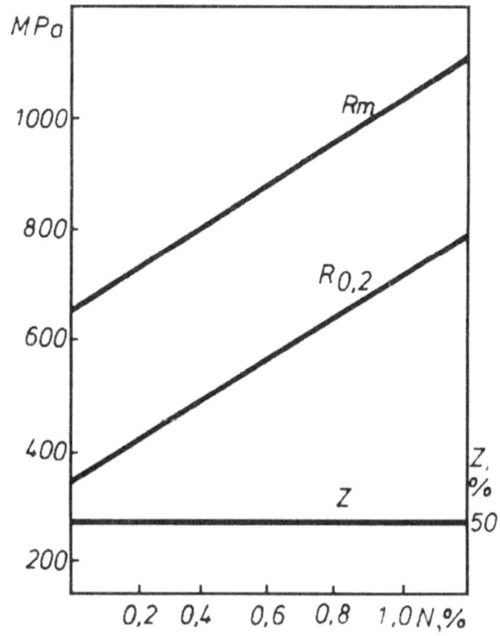

Fig. 1. Mechanical properties in dependency of different nitrogen concentrations after quenching at 1150°C, 2h in water. [Rashev et al.,2003]

2. Melting of HNS alloys

Melting of HNS alloys requires special process techniques as the nitrogen content is above the solubility limit at atmospheric pressure. The equipment development has started in the 1960´s with pressurized induction furnaces (lab scale) and has finally led to the first PESR unit in 1980. Today, PESR is state of the art due to its high process capability, good productivity, large ingot sizes and a save H&S environment [Holzgruber, 1988].

2.1 The PESR-process

Today's biggest PESR unit is located at Energietechnik Essen GmbH, Germany. Its operating pressure is max. 40 bars and can achieve ingot weights up to 20 tons and 1030 mm diameter. The functional principle is shown schematically in fig.2.

Basically, the PESR process is a conventional remelting facility that works in a pressure tank. The process is designed to meet both, an ESR refining and nitrogen pick up. The metallurgical approach is similar to a standard ESR-process, i.e. refining, low segregation, no porosity or shrinkage, defined microstructure and solidification.

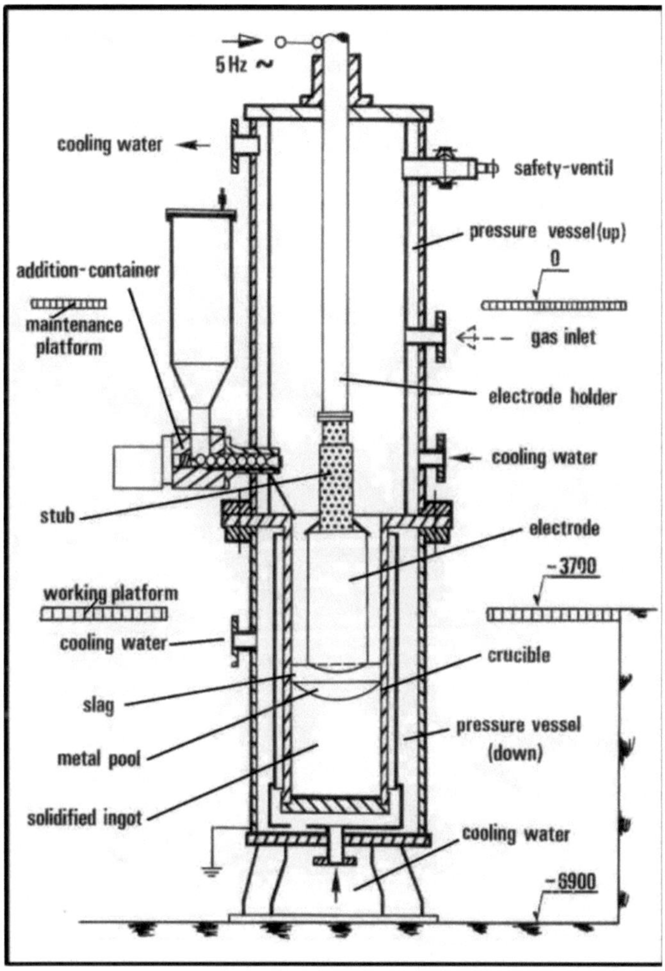

Fig. 2. Schematic design of a pressure electro slag remelting furnace (PESR).

The physical fundamentals of nitrogen pick up are specified over the Sievert square root law, accordingly that the nitrogen solubility is a function of pressure and temperature:

$$[\%N] = k \cdot \sqrt{P_{N2}} \tag{1}$$

With p_{N2}: Nitrogen partial pressure over melt in bar, k: Material constant (temperature and alloy dependent)

In real systems, the actual solubility is additionally codetermined through the alloy composition. Thermodynamic activities are used to describe the effect of the individual elements.

$$[\%N]_{Fe-X} = \frac{[\%N]_{Fe}}{f_N^X} \cdot \sqrt{p_{N2}} \tag{2}$$

With [%N] $_{Fe-X}$: Nitrogen solubility in multi-component systems, $[\%N]_{Fe}$= 0,044% (equilibrium constant in pure Fe at 1600 °C and 1 bar)

The activity coefficient f is thereby defined as

$$\log f_N^X = e_N^X[\%X] \tag{3}$$

With e_N^X : interaction coefficient, [%X]: Concentration of the elements X in %

Fig. 3. View of the industrialized PESR –process at Energietechnik Essen GmbH for ingot weight up to 20 t und ∅1030 mm.

It is obvious as per table 1 that specific elements will increase the nitrogen solubility (e.g. manganese), while others will reduce the solubility (e.g. silicon). This has not only an impact on the nitrogen pick up at remelting but also on the precipitation of inter-metallic phases in the solid state.

Element	Coefficient e_N	
C	+ 0.125	
Si	+ 0.065	Reduction of N-Solubitly
Ni	+ 0.01	
W	- 0.0015	
Mo	- 0.01	
Mn	- 0.02	
Cr	- 0.045	Increase of N-Solubility
V	- 0.11	
Nb	- 0.06	
Ti	- 0.053	

Table 1. Activity coefficients of several elements with effect on the nitrogen solubility in steel at 1 bar.

Due to the general alloy composition, the nitrogen solubility is accordingly larger in austenitic as in ferrite or martensitic steels.

The nitrogen pick up can occur through the gas phase and also as well from a solid nitrogen carrier. The choice of a solid body nitrogen pick up medium is down to the following boundary conditions:

- Nitrogen partial pressure: high enough to allow a dissociation at ~ 40 bar
- Characteristics of the slag or flux are not allowed to change (e.g. electrical conductivity, metallurgical properties, etc)

In practice the standard Si_3N_4 is used, in exceptions CrN as well. A transfer of silicium respectively chromium in this case must be taken into consideration. The following table 2 provides a comparison about advantages and disadvantages of Si_3N_4 and gaseous nitrogen.

	Advantage	Disadvantage
Si_3N_4	• Nontoxic • Ease of operation and storage • Ease of dissociation	• Very abrasive (joints and gaskets, valves) • Kinetic of dissociation of N^{3-} -ion must be regarded. • Non-continiuous allowance on slag • Silicium transfer in melt
N_2 - Gas	• Continuous allowance possible • Simple regulation over the pressure • High equal distribution in ingot • Appropriate for Si-critical steel grades	• Slag composition very important • Sievert´sches law at high pressure not ideally achieved. • Diffusions conditions in system slag-metall must be known.

Table 2. Advantages and disadvantages different nidriding mediums.

The selection of the slag takes place after metallurgical consideration and depends on the alloy. Above all, the slag composition has importance for the nitrogen pick up of the steel.

3. Microstructrual characteristics of HNS

Nitrogen stabilizes the γ-area in a very clear way. It is undesired in low alloyed steels due to the formation of brittle phases; however it is very beneficial in terms of strengthening and corrosion resistance for high alloyed steels. It is considered to be the most efficient solid solution strengthening element. [Pickering, 1988] has reported that nitrogen is approximately twice as efficient as carbon.

[Bernauer & Speidel, 2003] et al. has published results, whereas nitrogen can improve the γ-stability by adjusting a carbon/nitrogen ratio. The high interstitial steels, i.e. containing carbon and nitrogen show a higher thermodynamic stability compared to common Cr-Mn-N-steels.

[Pickering, 1988] has investigated the influence of nitrogen within various steelbased alloys with regards to their microstructure. Therefore, nitrogen has a higher solubility in the lattice than carbon. Its presence is related with the formation of nitrides (or carbon nitrides). These nitrides tend to precipitate as small particles and - this is particular of interest for any hot forming applications - grow significantly slower than carbides. It is obvious that this will have an impact on recovery, grain growth and heat resistance. The nitrides are thermodynamically more stable than the corresponding carbides, i.e. have a lower solubility.

At first sight, the considered HNS- alloys are not significantly distinguishable than the nitrogen free varieties. The following fig. 4 & 5 show exemplary micrographs of a Mn-austenite with approx 0.65% nitrogen as well martensite with approx. 0.4% nitrogen.

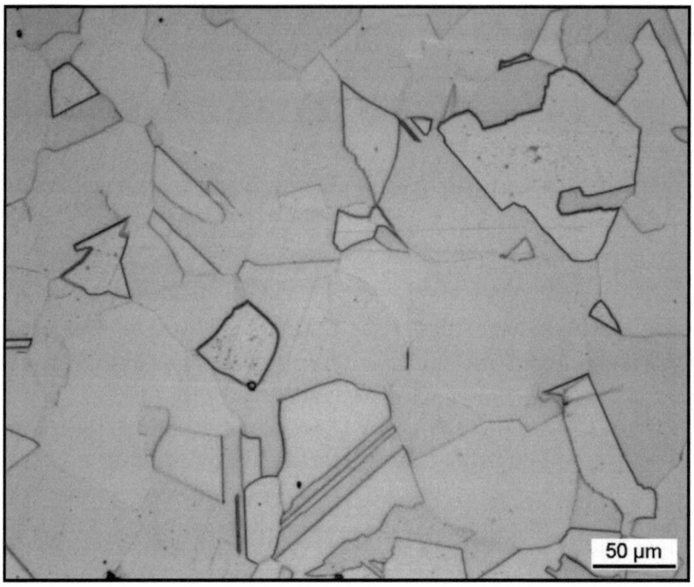

Fig. 4. Microstructure of a unformed and solutions annealed austenite with ~ 0,6 % N.

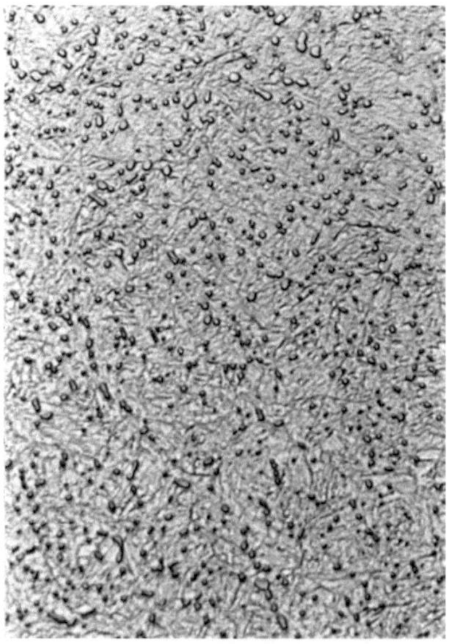

Fig. 5. Tempered microstructure of a nitrogen alloy martensite (1.4108). (M 1000:1)

However - and this is a difference to the conventional nitrogen free alloy variations - one should consider that HNS-alloys have a specific precipitation behavior. This must be kept in mind so that potential difficulties at hot forming at heat treatment can be avoided. Additionally, any precipitation will affect the corrosion resistance so a good understanding of the alloy is mandatory to maintain the alloy characteristic.

3.1 Atomic structure of nitrogen alloyed steels

Much effort has been put into place to understand the beneficial effect of nitrogen in stainless steels over the past years. A major step was the calculation of the atomic structure within the d-band of Fe-C and Fe-N carried out by [Rawers, 2003], [Gravriljuk & Berns, 1999] and [Mudali & Raj, 2004]. Therefore, nitrogen increases the state density on the Fermi surface whereas carbon leads to a decrease of state density. Consequently, a higher concentration of free electrons can be found in austenitic nitrogen alloyed steels – this result in a metallic character of interatomic bonds. This also explains the high ductility in HNS, even at high strengthening. Contrary, interatomic bonds in carbon austenites show a covalent characteristic. This is due to the localization of electrons at the atomic sites [Rawers, 2003]. The preference for different atoms to be nearest neighbors is defined as short range order and is mainly driven by the degree of metallic character of an intermetallic bond. A metallic interatomic bond supports a homogenous distribution as single interstitials, whereas a covalent bond results in clustering of atoms. These clusters can then potentially precipitate secondary phases such as carbides, nitrides etc. A cluster is to be realized as local accumulation of approx. 100 atoms [Berns, 2000]. The high thermodynamic stability of nitrogen stabilized austenites can also be led back on the hindered clustering of atoms [Rawers, 2003]. In summary, the electron configuration is therefore the main driver for an increased corrosion resistance. Due to nitrogen, the allocation of Cr-atoms within the lattice is homogenous so that Cr- clustering and formation of $M_{23}C_6$-carbides is reduced. Since nitrogen delays the precipitation of carbides as seen in fig. 6 & 7 , the likeliness of a local Cr depletion is limited.

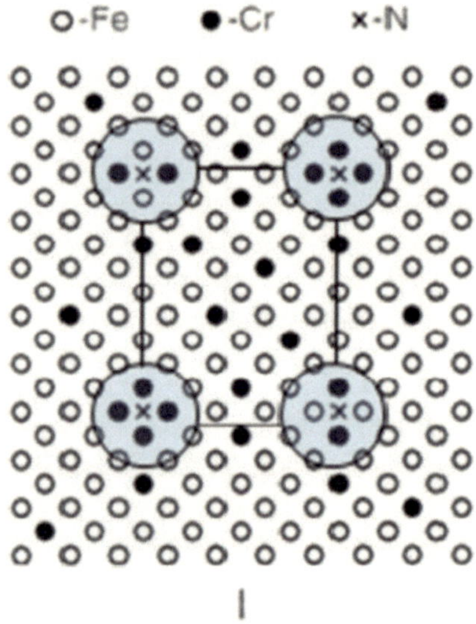

Fig. 6. Schematic of a short range order. Nitrogen increases the concentration on free electrons. Thereby forming a non-directional bonding and an equal distribution of the atom in crystal lattice. [Berns, 2000]

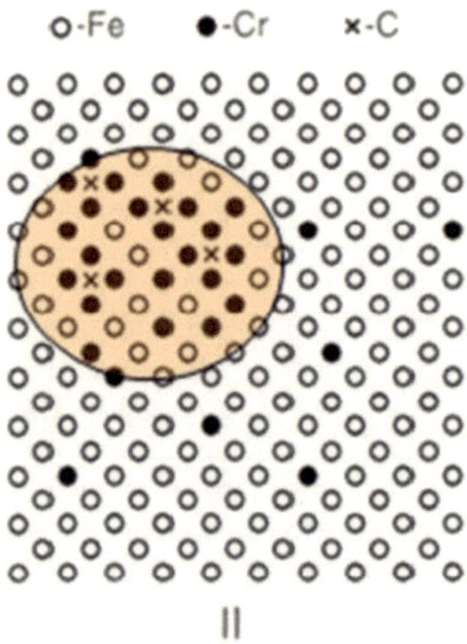

Fig. 7. Schematic of a cluster formation. Carbon decreases the concentration of free electrons. Thereby forming a directional bonding of non equal distribution of the atom in crystal lattice. [Berns, 2000]

3.2 Carbides

In dependency to the carbon content and the tempering time, austentic steels tend to precipitate $M_{23}C_6$-Carbides at the grain boundaries. Through this the ductility and corrosion resistance of the material significantly declines. However, the strength properties have no mentionable change. The susceptibility for intercrystalline corrosion clearly increases.

The precipitation behavior of this carbide can only be prevented through a quick quench in the critical temperature range. Fig.8 shows the location of the precipitation with relation to the alloy composition. Fine carbides are beneficial with regards to the corrosion resistance as the local chrome depletion is less in comparison to coarser carbides. The Cr depletion can be balanced out through an extended homogenisation (i.e. holding time) within the precipitation area.

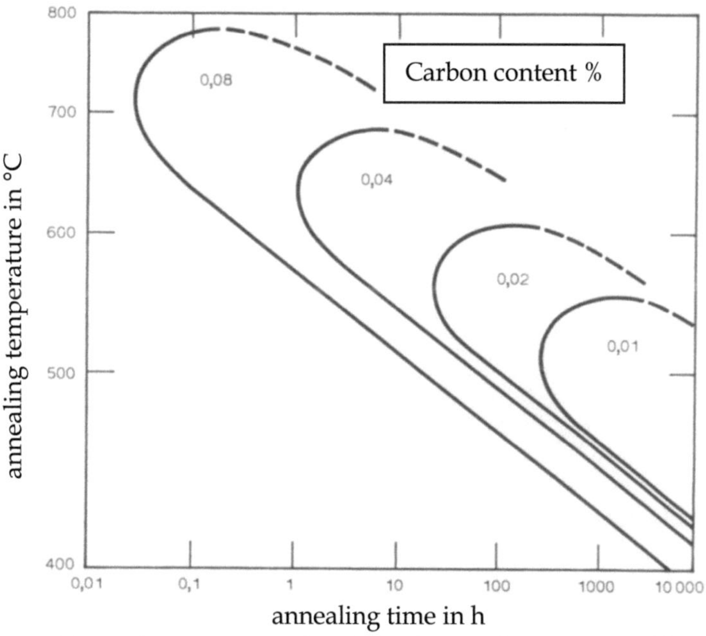

Fig. 8. Influence of the carbon content in location of grain decay in unstable austentic steels with circa 18 % Cr und 8 % Ni. Examination in Strauß-Test. [Thyssen, 1989]

For 12% Cr-steels, containing nitrogen it has been observed by [Pickering, 1988] that nitrogen lowers the martensite start temperature M_S; 1% nitrogen lowers M_S by 450 °C.

[Pickering, 1988] has investigated the influence of nitrogen on the carbide morphologies. The main type is as previously described the $M_{23}C_6$ type. In Nb-containing alloys, M_4X_3 have been observed where nitrogen can occupy interstitial dislocations. It also can be solutioned within M_6C.

HNS martensitic steels are also characterised with good high temperature strength and show an according hot forming behavior. Under circumstances these steels are found in thermo mechanical forging and rolling applications. The last forming step will effectively increase the dislocation density so that adequate nucleation for a desired precipitation exists. For example, the precipitation of carbides, nitrides as well as carbon nitride could be finely distributed. This could be of interest to the high temperature strength. The previously mentioned effects of fine carbides concerning the dissolution and corrosion resistance are also valid here.

3.3 Nitride and nitrogen perlite

In the case of the austenitic steels it should be considered, that in the temperature range of approx. 500- 900 °C and in connection with the alloy composition a precipitation of nitrides (Type Cr_2N) occurs. This nitrogen perlite identified microstructure raises significantly the susceptibility to cracking of the steel but can also support intergranular cracking. Depending on the alloy composition, the precipitation window for nitrogen perlite or other nitrides are adjusted to higher or lower temperatures. The figures 9 & 10 show exemplary an austenitic structure with beginning and advance nitrogen perlite precipitation. Clear to recognize at what speed that the precipitation occurs.

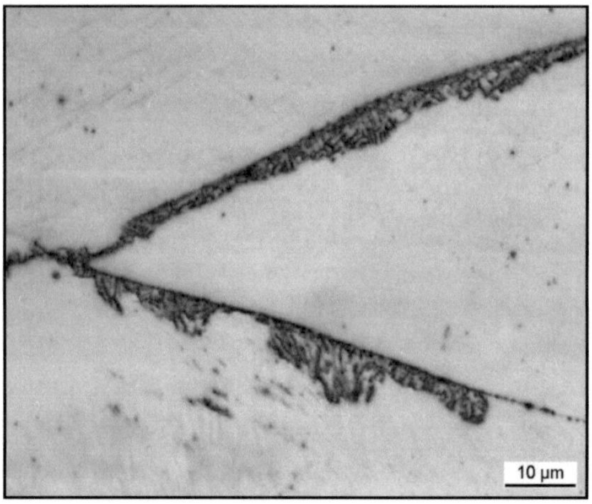

Fig. 9. Beginning of precipitation of nitrogen perlit cold worked austenitic structure 1.3816. 800 °C/15 min.

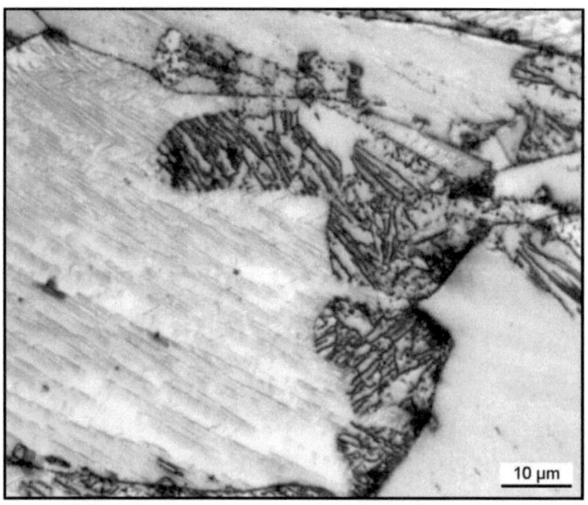

Fig. 10. Advanced precipitation of nitrogen perlit cold worked austenitic structure 1.3816. 800 °C/30 min.

For prevention of such brittle phases the precipitation area of the hot forming must be followed through fairly quick. The nitrogen level has obviously an impact on the precipitation kinetics of Cr_2N, see figure 11 for details. Best corrosion resistance can be achieved if all nitrogen is in solid solution, i.e. no nitrides are precipitated.

As seen in fig. 11 the precipitation depends on both, the alloy composition and holding time. The Cr_2N-formation has been reported by [Pickering, 1988] to be a major issue to high Cr and/or high Ni-alloys

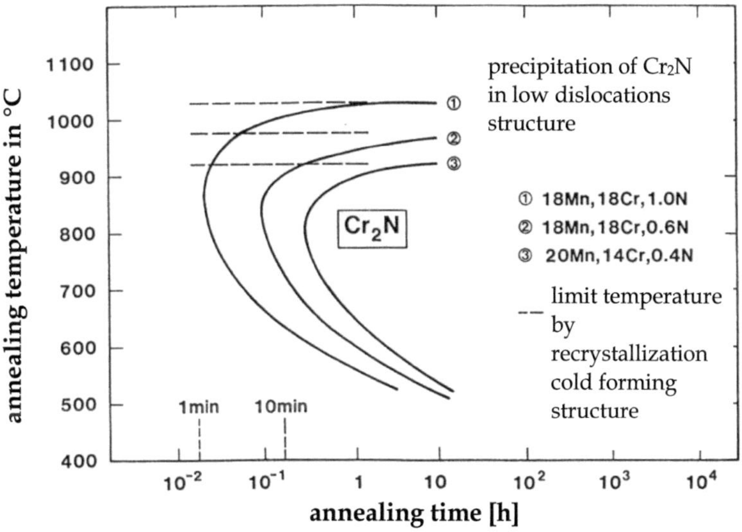

Fig. 11. Precipitation of Cr_2N in 18Mn18Cr at various nitrogen levels. [Uggowitzer, 1991]

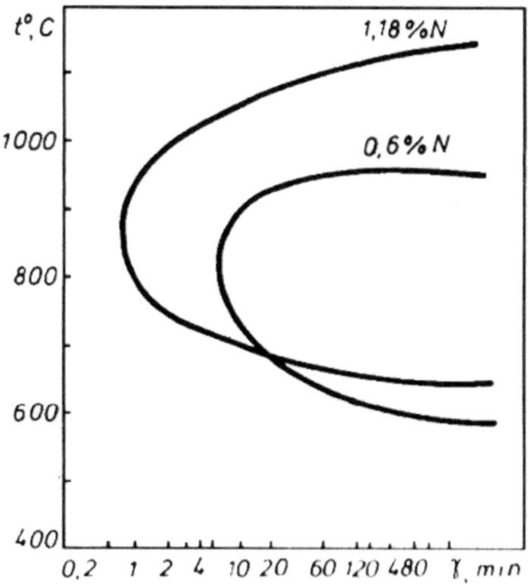

Fig. 12. TTT-diagram for the beginning of Cr_2N-precipitation at different nitrogen contents. Curve is based on X5CrMnN18-12. [Rashev et al., 2003]

Other intermetallic phases, such as Laves-phase, Z-phase and χ-phase have been investigated with regards to nitrogen alloying within literature. Generally, nitrogen seems to delay the formation of intermetallic phases. The underlying mechanism has been discussed controversially; the enhanced solubility of Chromium and Molybdenum due to nitrogen or its influence on Gibbs free energy for phase formation [Mudali & Rai, 2004]. The following table will provide an overview about the influence of nitrogen on some intermetallic phases [Mudali & Rai, 2004], [Heino et al., 1998].

Description	**Influence of nitrogen...**	**Remark**
σ	Suppresses formation of σ	
χ	Shifted to longer times Narrows temperature range for precipitation	M6C appears instead of χ
$M_{23}C_6$	Suppresses formation of $M_{23}C_6$	Can be replaced by M_6C with increasing nitrogen contents
Laves	Precipitation is shifted to higher temperature but accelerated	
R		See Laves

Table 3. Overview about the role of nitrogen alloying on the precipitation behavior of some intermetallic phases. [Mudali & Raj, 2004], [Heino et al., 1998]

4. Corrosion resistance of HNS

The role of nitrogen in stainless steel with regards to the corrosion resistance has been previously reported within literature [Pleva, 1991], [Truman, 1988], [Pedrazzoli &Speidel, 1991], [Dong et al., 2003], [Mudali & Rai, 2004]. This chapter will review today´s knowledge and present own data; corrosion fatigue and high temperature corrosion will not be covered.

4.1 General

It is well known that nitrogen in high alloyed steels improves the corrosion resistance; this is especially true for pitting and crevice corrosion. Additionally, nitrogen helps to prevent the alloy from stress corrosion cracking, though in an oblique way. Generally, the beneficial effect of nitrogen can be led back to an enrichment of nitrogen at the oxide/metal-interface and its influence on passivation. It has been reported that the enrichment increases with increasing potentials. However, these mechanisms have been discussed controversially [Mudali & Rai, 2004]:

- a formation of nitrides or mixed nitride layer, e.g. Ni_2Mo_3N
- enrichment of negatively charged N-ions, i.e. $N^{\delta-}$. These ions will lower the potential gradient of the passivation film and reject Cl^--ions
- formation of Cr_2N. A high local Cr-concentration should improve the corrosion resistance. However, this approach is unlikely as nitrogen does not change the matrix composition underneath the passivation film. No Cr depletion whatsoever has been reported.

Nitrogen does not show any influence on the thickness of the oxide layer, investigations of various alloys with different nitrogen contents have confirmed an average thickness of 12-22 Å [Mudali & Rai, 2004] within all varieties.

A synergism of nitrogen and molybdenum is suggested by many authors [Mudali & Rai, 2004], [Pickering, 1988], [Pedrazzoli & Speidel, 1991]. Molybdenum shifts the metal dissolution to higher potentials which will consequently lead to an increased enrichment of nitrogen at the metal/oxide-interface. In this case, nitrogen can lower the current density below the critical value for pitting corrosion [Mudali & Rai, 2004].

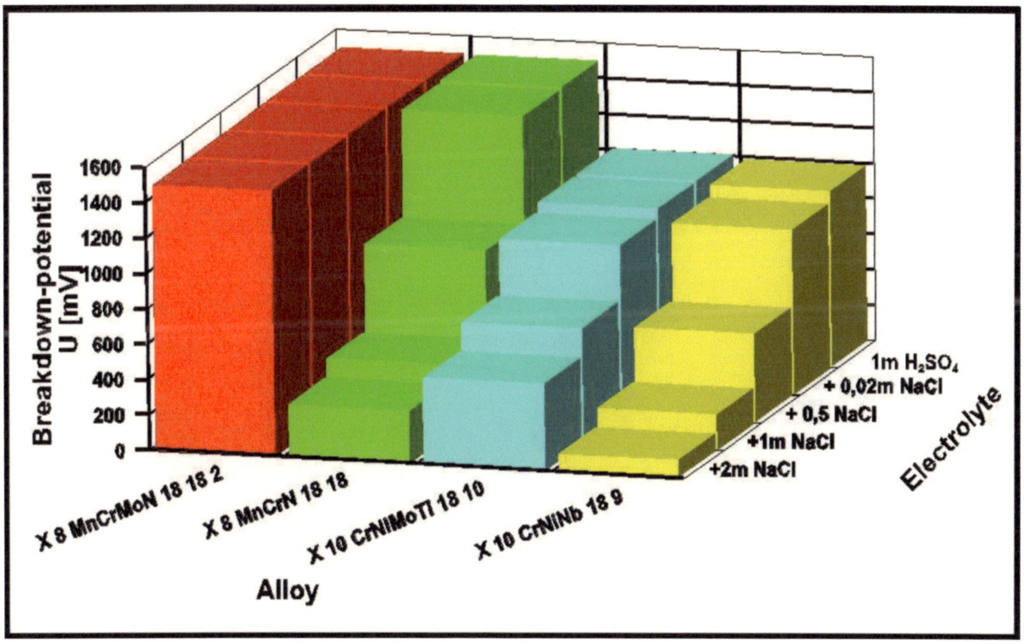

Fig. 13. Breakdown potential of HNS and commercial stainless steels in various electrolytes. [ETE-11]

It has been suggested that Molybdenum and nitrogen support the formation of highly mobile ions that interact with the passivation film. Addtionally, nitrogen seems to have a buffer effect by reacting as follows in oxidizing corrosive media:

$$[N] + 4H^+ + 3e^- \rightarrow NH_4^+ \quad (4)$$

The formation of NH_4^+ - ions helps to increase the pH value which results in an improved repassivation and reconditioning of the base material [Mudali & Rai, 2004]. It also has impact on the depronotation, which might explain the good performance in acids and halide containing liquids such as Cl^-, Br^- and I^- [Truman, 1988].

4.2 The role of nitrogen on pitting and crevice corrosion

Pitting corrosion is a very serious and harmful type of corrosion and is classified as local corrosion, characterized by small holes or pits. Usually, a repassivation cannot be achieved so that these pits can initiate cracks. This is the main reason why pitting often comes along with stress corrosion cracking. Pitting can be determined either by current-density-curves or a critical pitting temperature. It has been reported that nitrogen lowers the passivity in the current-density diagram. In austenitic steels, 1 % nitrogen improves the pitting potential by 600 mV [Pedrazzoli & Speidel, 1991]. Crevice corrosion follows generally the same principles; however the conditions are significantly tougher due to the geometric impact (electrolyte concentration in crevice). This will be covered at a later stage.

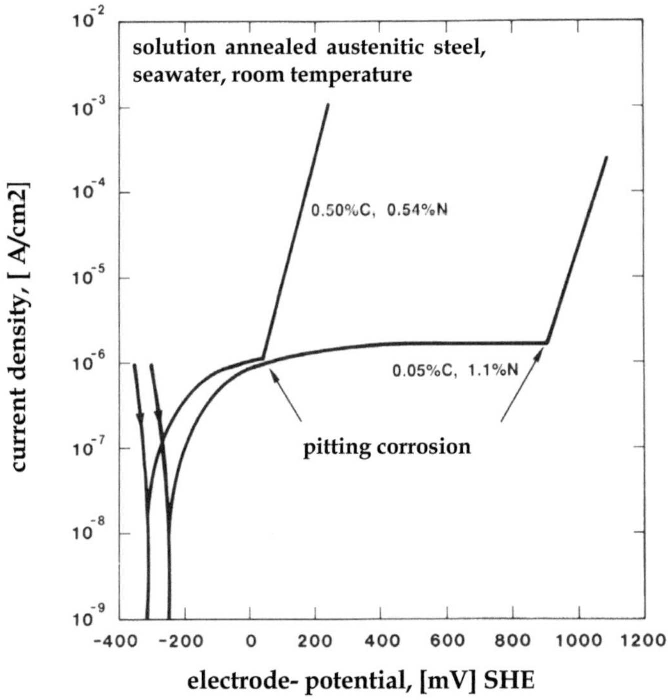

Fig. 14. Influence of nitrogen on Current-density depending of a given alloy. [Pedrazzoli & Speidel, 1991]

Pitting corrosion is a well-known corrosion problem for stainless steels. It can come along with sensitivation, i.e. a local Cr depletion can support pitting corrosion. Therefore, any segregation, welding joint, heat treatment etc. can have an impact on pitting corrosion.

The critical pitting temperature (CPT) is defined at what temperature pitting occurs. A common range for stainless steels is 10-100 °C and obviously depends on the alloy composition. [Pedrazzoli & Speidel, 1991] has reported that the critical temperature for crevice corrosion (CCT) is approx. 20 ° lower compared to pitting, see fig.16 & 17 for details.

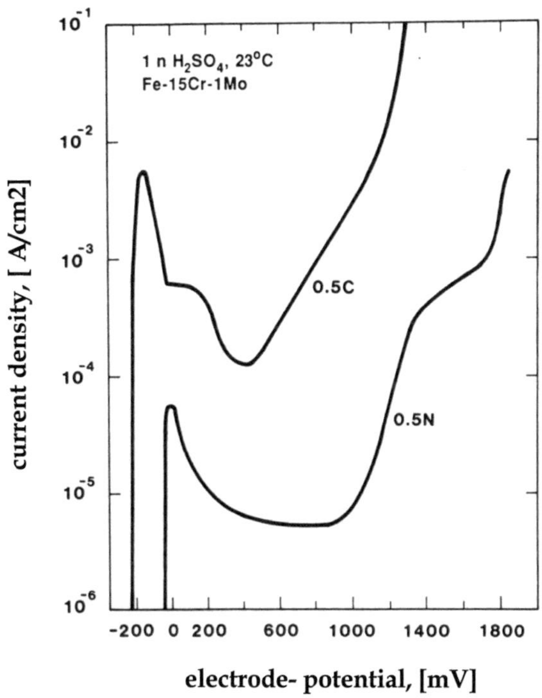

Fig. 15. Comparison of nitrogen and carbon on Current-density depending of a given alloy. [Pedrazzoli & Speidel, 1991]

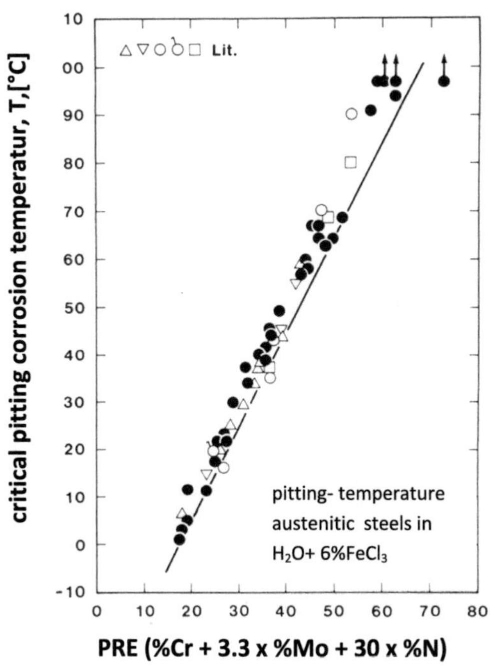

Fig. 16. Critical Pitting Temperature (CPT) as a function of PRE. [Pedrazzoli & Speidel, 1991]

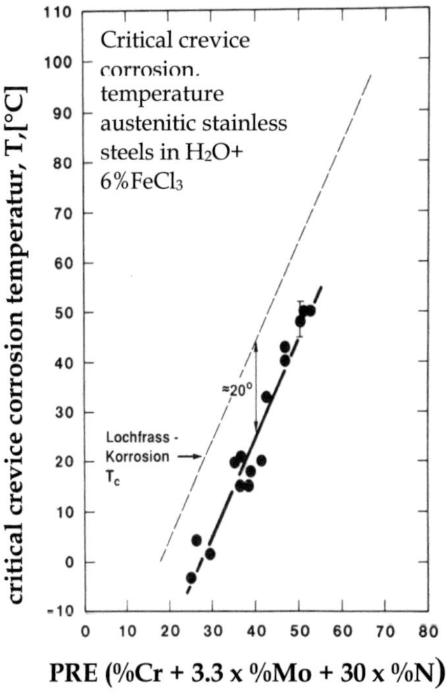

Fig. 17. Critical Crevice Temperature (CCT) as a function of PRE. [Pedrazzoli & Speidel, 1991]

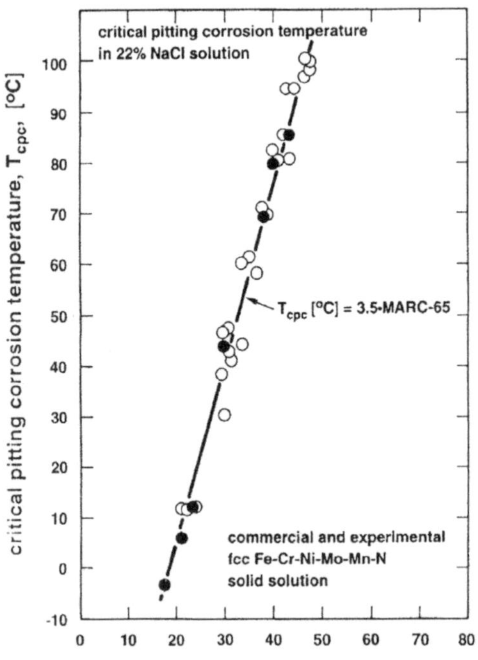

Fig. 18. MARC equation to rank various alloying element in regards to the alloy pitting resistance. [Speidel & Theng-Cui, 2003]

A commonly accepted ranking of alloys in terms of their pitting addiction is the Pitting resistance equivalent (PRE). It is defined as

$$PRE = Cr\,(\%) + 3{,}3\,Mo\,(\%) + x\,N\,(\%) \text{ whereas } x = 13\ldots30 \quad (5)$$

It has been suggested by [Pleva, 1991] to use x = 16 for steels containing Mo < 4,5 % and x = 30 for steels containing Mo 4,5 – 7,0 %.

The PRE does not take any other elements but Cr, Mo and N into account. [Speidel & Theng-Cui, 2003] has suggested a new figure to include also C, Mn and Ni into the equation and has defined MARC (**M**easure of **al**loying for **r**esistance to **c**orrosion):

$$MARC = Cr\,(\%) + 3{,}3\,Mo\,(\%) + 20\,C\,(\%) + 20\,N\,(\%) - 0{,}5\,Mn\,(\%) - 0{,}25\,Ni\,(\%) \quad (6)$$

The MARC-equation is the first formula that considers carbon to be beneficial against pitting. [Bernauer & Speidel, 2003] has suggested a high carbon + nitrogen alloyed steel with improved pitting resistance. This is due to the higher thermodynamic stability of Cr-Mn-N-C systems compared to carbon-free Cr-Mn-N steels. However, both carbon and nitrogen must not form any precipitations but stay into solid solution.

A very global description of the influence of alloying elements on pitting potential was published by [Pedrazzoli & Speidel, 1991]. As seen in fig. 19, nitrogen and molybdenum have a significant impact on the potential shift.

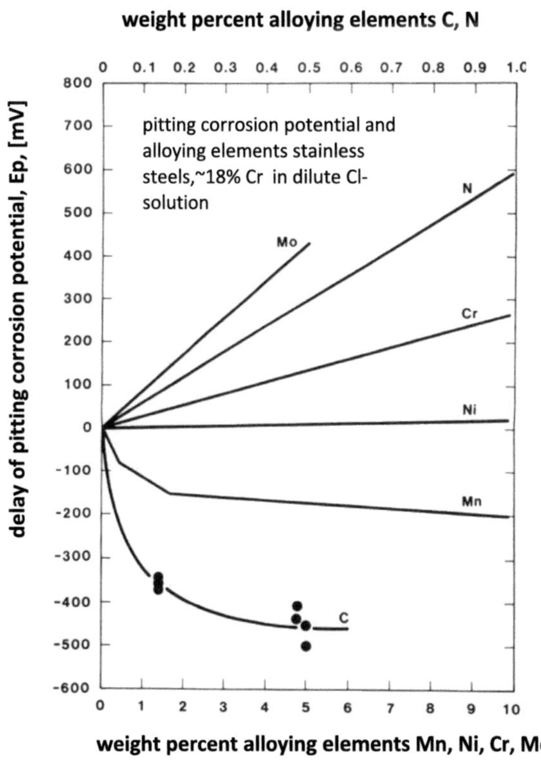

Fig. 19. Influence of various alloying elements on the pitting potential. [Pedrazzoli & Speidel, 1991]

Cold forming is supposed to have an impact on corrosion resistance; however the role of nitrogen in this case is not fully clear. Within Literature, the following has been reported [Mudali & Rai, 2004]:

- cold forming in stainless steels: no significant influence on pitting potential
- cold forming in nitrogen alloyed steels: a cold forming degree up to 20 % improves the critical pitting potential (CPP). A drop in CPP at higher deformation rates has been reported.

The improved pitting potential at low deformation rates i.e. below 20 % is due to the decreased tendency for twin formation. At higher deformation rates, deformation bands will appear which will be influenced by nitrogen (width and dislocation configuration). [Pleva, 1991] reports that the degree of cold working does not show any influence on the pitting corrosion. This has been also confirmed by own data on X8CrMnN 18-18 material, see fig.20 & 21 [ETE-11].

Various investigations have tried to explain the mechanism of nitrogen in terms of pitting. It has been agreed, that nitrogen stabilizes the γ-range in a very clear way. This is important to prevent δ-ferrite, esp. in Mo containing alloys. It also supports a homogenous, single-phase microstructure and avoids carbides to precipitate.

4.3 Ammonium theory

Nitrogen and Molybdenum obviously show a synergism in regards to pitting. Molybdenum seems to support the Cr_2O_3-formation by acting as an electron acceptor. This also leads of a depronotation of hydroxides. In addition, nitrogen reacts as follows:

$$[N] + 4H^{+} + 3e^{-} \rightarrow NH_4^{+} \tag{7}$$

The NH_4^+ - formation will increase the pH-value which support the repassivation. [Mudali & Rai, 2004] reports that NH_4^+ - ions have been confirmed by XPS within the passivation layer.

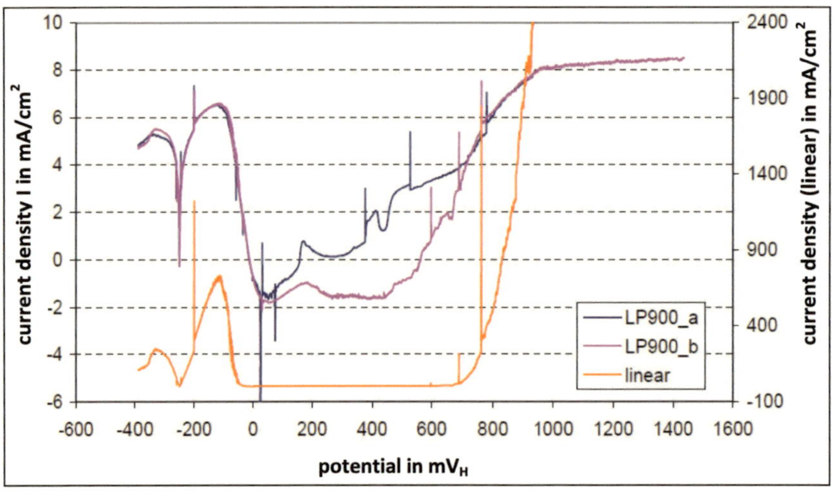

Fig. 20. Influence of cold working on current-density of X8CrMnN 18-18 (0 % cold work). UR: -250 / UL: 722 mV / ΔU: -972 mV / 1m H_2SO_4+ 0,5m NaCl [ETE-11]

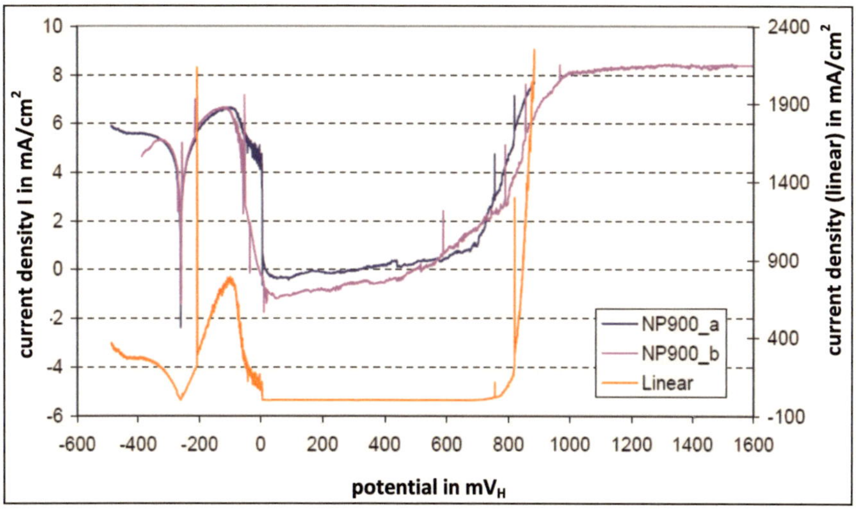

Fig. 21. Influence of cold working on current-density of X8CrMnN 18-18 (37 % cold work). UR: -250 / UL: 780 mV / ΔU: -1030 mV / 1m H_2SO_4+ 0,5m NaCl [ETE-11]

4.4 Surface enrichment theory

The surface enrichment theory is based on the general idea, that nitrogen is build into the lattice underneath the passive layer in solid solution. This nitrogen rich layer shall avoid dissolution of the substrate. Tentatively, there are chemical reactions with Cr and Mo who might change the local potentials as well. The formation of various N-rich phases has been reported, such as Cr_2N or Ni_2Mo_3N [Mudali & Rai, 2004], [Pickering, 1988]. Negatively charged N-ions, i.e. $N^{\delta-}$ are supposed to enrich at the metal/oxide interface. These ions will lower the potential gradient of the passivation film and reject Cl- -ions.

4.5 Inhibitive nitrate formation theory

This theory covers the formation of pit growth inhibiting species. It is basically linked with the ammonia formation theory. The NH4+-formation in the pit tip appears to happen quicker than the OH—formation due to oxide reduction at the pit entrance.

$$[N]+4H^{+}+3e^{-}\rightarrow NH_4^{+} \tag{8}$$

The repassivation by NH4+ can be described as follows:

$$NH_4^{+}+H_2O\rightarrow NH_4OH+H^{+} \tag{9}$$

$$NH_4OH+H_2O\rightarrow NO_2^{-}+7H^{+}+6e^{-} \tag{10}$$

$$NO_2^{-}+H_2O\rightarrow NO_3^{-}+2H^{+}+2e^{-} \tag{11}$$

$$NH_4^{+}+2H_2O\rightarrow NO_2^{-}+8H^{+}+6e^{-} \tag{12}$$

$$NH_4^{+}+3H_2O\rightarrow NO_3^{-}+10H^{+}+8e^{-} \tag{13}$$

Crevice corrosion underlies basically the same principles than pitting; due to the geometry of the crevice the corrosion conditions are believed to be more challenging. [Pedrazzoli & Speidel, 1991] has reported that the critical temperature for crevice corrosion is approx. 20 ° lower compared to pitting.

4.6 The role of nitrogen on Intergranular corrosion (IGC)

IGC is mainly driven by the depletion of Cr at the grain boundaries and/or the precipitations of carbides, usually $M_{23}C_6$. Therefore, Carbon is supposed to be the main driver for IGC but also grain size, cold working and heat treatment have a significant influence on IGC.

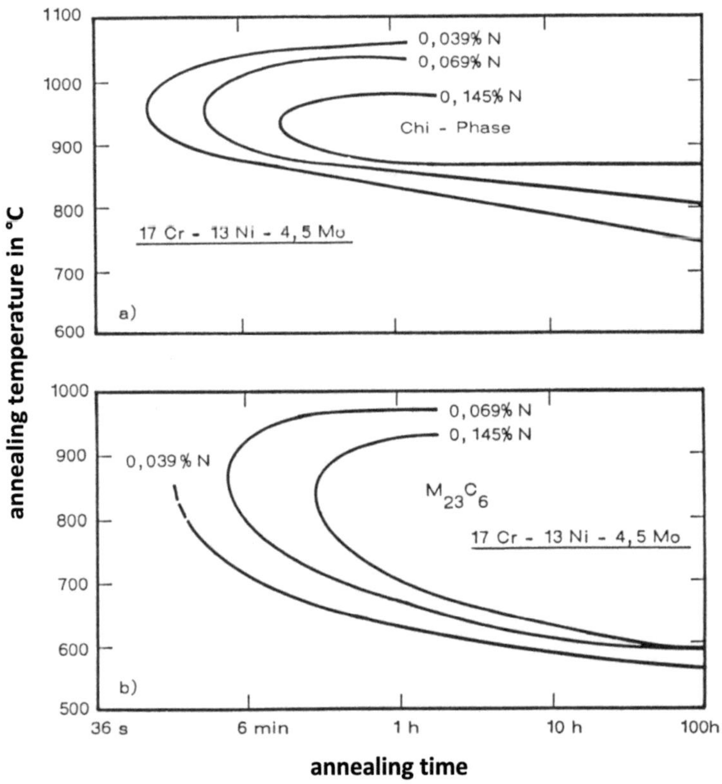

Fig. 22. Precipitation of M23C6 and χ-phase of a given alloy (17Cr-13Ni-4,5Mo) depending on annealing times and temperatures. [Gillessen et al., 1991]

As discussed in the previous chapter (microstructure), nitrogen tends to delay the $M_{23}C_6$ formation as it changes the Cr activity within the carbide. It also increases the passivity (i.e. lowering the current-density) and avoids the formation of α΄- martensite at grain boundaries. However, this is only valid as long as nitrogen is in solid solution. It has been reported [Truman, 1988], [Dong et al., 2003] that an excess of nitrogen can lead to Cr_2N precipitations on the grain boundaries which can significantly decrease the intergranular corrosion resistance.

4.7 The role of nitrogen on Stress corrosion cracking (SCC)

It is commonly known that high alloyed steels are generally sensitive to SCC. The role of nitrogen against SCC has been discussed over the years but it appears that the positive effect of nitrogen is to be seen in a more oblique way. As previously discussed, nitrogen delays the carbide precipitation and avoids a local Cr depletion. Additionally, the crack growth velocity does not only depend on the actual Cr content, but also on the C content. The crack growth is much higher with increased C levels. It has been reported that nitrogen doesn´t have any influence on crack growth velocity at C > 0,5 % . The impact of carbon is therefore higher compared to that of nitrogen. [Pedrazzoli & Speidel, 1991]

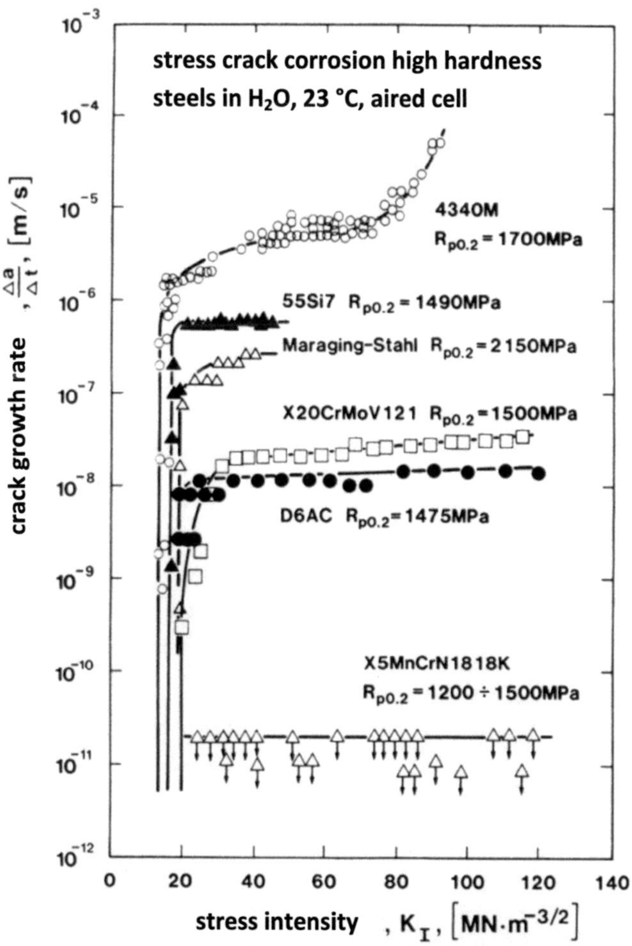

Fig. 23. Crack growth velocity for various alloys as a function of stress intensity. SCC for stainless steels, Water, 23 °C, ventilated. [Pedrazzoli & Speidel, 1991]

[Pickering, 1988] reports that the role of nitrogen is somehow inconsistent. In principle, nitrogen reduces the stacking fault energy which would be diametric to corrosion resistance. Low stacking fault energy in fcc-lattice is undesired in terms of SCC. Elevated nitrogen levels above 0,3 % are reported to be beneficial as the support of passivity is obviously higher than the influence on stacking fault energy. [Mudali & Raj, 2004] confirms that increased stacking fault energy does improve the SCC resistance. As nitrogen decreases the stacking fault energy it should detrimental to SCC. Carbon, due to the formation of wavy slip bands, should be theoretically beneficial. However, it appears that the role of nitrogen on SCC is fairly complex and depends on the alloy design and corrosion media.

The benefit of nitrogen alloying appears to be more oblique - the delay in $M_{23}C_6$ precipitation and improved pitting corrosion resistance has been recognized to be beneficial against SCC since pits are likely to initiate SCC.

5. Applications of HNS

High nitrogen steels are characterized through an interesting material profile that has led to a variety of demanding applications. The following will provide an overview about today´s use of HNS alloy, please note it reflects only pressurized alloys, i.e. PESR alloys. Any air melted nitrogen alloy steels will not be part of this chapter.

5.1 Martensitic steels

Since its development in the early 1990s, Cronidur 30 (X30CrMo 15-1 plus 0,4 % nitrogen) has been approved as high performace alloy for aerospace applications such as spindles, shafts and bearings. It is used as material for helicopter bearings, flap traps, fuel pumps etc. It excellent corrosion resistance combined with a high hardenability of 60 HRC makes it unique. The alloy has become an important material to industries as cutleries and knifes, general engineering, medicals (since it is free of nickel) and powder metallurgy.

Fig. 24. High precision bearing of Cronidur 30 [ETE-11].

Fig. 25. Extruder screw, Cronidur 30 [ETE-11].

A heat resistant alloy is based on X15CrMoV12-1 and contains 0,2 % nitrogen (trade name HNS 15). It precipitates fine V(C,N) and provides a good creep resistance and heat resistance up to 650 °C. Above this temperature the appearance of Laves-phase restricts its usage.

A Molybdenium-free version is known as HNS 28 and consists of X28Cr13 with 0,5 % nitrogen. Its purpose was the closed-die casting industry where a good polish and corrosion resistance is required.

5.2 Austenitic steels

A main driver for the development of HNS austenitic steel was the power generation in the 1980´s. A material for retaining rings was required that could resist the mechanical loads but also stress corrosion cracking. This has finally led to the introduction of X8CrMnN 18-18 also known as P900 or 18-18. This alloy combines superb mechanical properties, e.g. high ductility at elevated strength level with good corrosion resistance.

Further developments have also come up with Mn-stabilized austenites, e.g. X13CrMnMoN 18-14-3 (P2000). This alloy can achieve strength level (YS) beyond 2000 MPa, still with good ductility and corrosion resistance. One should consider heat treatment conditions and corresponding part dimensions with regards to the precipitation of Cr_2N.

Another market is powder metallurgy, i.e. thermal spraying and metal injection moulding (M.I.M.). The powders are gas atomized and very homogenous in terms of nitrogen and chemical composition. Main consumers are jewelry and general engineering.

Fig. 26. Generator shaft with retaining rings, X8CrMnN 18-18 [ETE-11].

Fig. 27. Retaining ring. X8CrMnN 18-18 [ETE-11].

6. References

AIF (2003) AIF-Abschlußbericht „ Untersuchungen zur wirtschaftlichen Warmumformung neuer hoch stickstofflegierter nichtrostender Stähle in Abhängigkeit vom Stickstoffgehalt, des Oberflächenzustandes und der Ofenatmosphäre", AiF-Vorhaben Nr. 13888N/II, 01.09.2003 bis 31.08.2007

Berns, H (2000). Stickstoffmartensit, Grundlage und Anwendung, HTM Härtereitechnische Mitteilungen, Ausgabe 1/2000, Bd. 55, Hansa Verlag, p.10

Bernauer, J., Speidel, M.O (2003). Effects of carbon in high-nitrogen corrosion resistant austenitic steels, High Nitrogen Steels 2003 Conference proceedings, Vdf Hochschulverlag AG ETH Zürich, Switzerland, 2003, pp. 159-168,

Dailly, R.; Hendry, A (1998). The Effect of Nitrogen on the mechanical behavior of cold-worked austenitic stainless steel rod, High Nitrogen Steels 1998 Conference proceedings, TransTechPublications Ltd, Switzerland, 1998, pp.427-435

Dong, H., Lin, Q., Rong, F., Su, J., Xin, C., Lang, Y., Kang, X (1998). Development and applications of nitrogen alloyed stainless steels in China, High Nitrogen Steels 2003 Conference proceedings, Vdf Hochschulverlag AG ETH Zürich, Switzerland, 2003, pp. 53-61

ETE (2011). Energietechnik Essen GmbH, internes Firmenarchiv, 2010

Gillessen, C., Heimann, W., Ladwein, T (1991). Entwicklung, Eigenschaften und Anwendung von konventionell erzeugten hochstickstoffhaltigen austenitischen Stählen, Ergebnisse der Werkstoffforschung, Verlag Thubal-Kain, Schweiz, Zürich, 1991, pp.167-187

Gavriljuk, V., Berns, H (1999). High nitrogen steels, Springer Berlin, 1. Auflage, 1999

Heino, S, Knutson-Wedel, M., Karlsson, B (1998). Precipitation in a high nitrogen superaustenitic stainless steel, High Nitrogen Steels 1998 Conference proceedings, TransTechPublications Ltd, Switzerland, 1998, pp. 143-148

Holzgruber, W (1988). Process technology for high nitrogen steels, High Nitrogen Steels 1988 Conference proceedings, The Institute of Metals, London, Brookfield, 1989, pp. 39-48

Mudali, U.K., Raj, B (2004). High nitrogen steels and stainless steels - manufacturing, properties and applications, New Delhi, Chennai, Mumbai, Kolkata, Narosa Publishing House, ASM International, 2004

Pedrazzoli, R., Speidel, M.O (1991). Korrosion und Spannungsrisskorrosion von stickstoffhaltigen Stählen, Ergebnisse der Werkstoffforschung, Verlag Thubal-Kain, Schweiz, Zürich, 1991, pp.103-121

Pickering, F.B. (1998). Some beneficial effects of nitrogen in steel, High Nitrogen Steels 1988 Conference proceedings, The Institute of Metals, London, Brookfield, 1989, pp.10-31

Pleva, J (1991). Korrosionsfeste stickstofflegierte Stähle - Eigenschaften und Erfahrungen, Ergebnisse der Werkstoffforschung, Verlag Thubal-Kain, Schweiz, Zürich, 1991, pp. 153-165

Rashev, T., Andreev, C., Manchev, M., Nenova, L. (2003). Creation and development of new high nitrogen steels in the Institute of Metal Science at Bulgarian Academy of Science, High Nitrogen Steels 2003 Conference proceedings, Vdf Hochschulverlag AG ETH Zürich, Switzerland, 2003, pp.241-257

Rawers, J (2003). Preliminary study into the stability of interstitial nitrogen and carbon in steels, High Nitrogen Steels 2003 Conference proceedings, Vdf Hochschulverlag AG ETH Zürich, Switzerland, 2003, pp. 273-280

Speidel, M.O., Theng-Cui, M (2003). High-Nitrogen austenitic stainless steels, High Nitrogen Steels 2003 Conference proceedings, Vdf Hochschulverlag AG ETH Zürich, Switzerland, 2003, pp.63-73

Thyssen Technische Berichte (1989). N.N.; Thyssen Technische Berichte, 15. Band 1989, Heft 1

Truman, J.E (1988). Effects of nitrogen alloying on corrosion behaviour of high alloy steels, High Nitrogen Steels 1988 Conference proceedings, The Institute of Metals, London, Brookfield, 1989, pp. 225-239

Uggowitzer, P. (1991). Uggowitzer, P.; Ultrahochfeste austenitische Stähle, Ergebnisse der Werkstoffforschung, Verlag Thubal-Kain, Schweiz, Zürich, 1991, pp. 87-101

4

Tribocorrosion: Material Behavior Under Combined Conditions of Corrosion and Mechanical Loading

Pierre Ponthiaux[1], François Wenger[1] and Jean-Pierre Celis[2]
[1]*Ecole Centrale Paris, Dept. LGPM, Châtenay-Malabry,*
[2]*Katholieke Universiteit Leuven, Dept. MTM, Leuven,*
[1]*France*
[2]*Belgium*

1. Introduction

1.1 Definition of tribocorrosion

Tribocorrosion can be defined as the study of the influence of environmental factors (chemical and/or electrochemical) on the tribological behavior of surfaces. In other words, the process leading to the degradation of a metallic and/or non-metallic material resulting from a mechanical contact (sliding, friction, impact, ...) combined to a corrosive action of the surrounding environment.

The origin of tribocorrosion is closely related to the presence of a passive film on material surfaces subject to wear and the modifications of these surfaces by friction or any other form of mechanical loading. In very general terms, the passive film (mainly oxide) is considered to be snatched in the contact area.

Oxide particles, referred to as 'debris", are released from the contacting materials. Then, the debris can be removed from the contact zone or on the contrary trapped in it. In the case of removal, the debris dissolve chemically or are dragged out by a hydraulic flow along the material surface. In this case, the tribocorrosion mechanism is based on a repeated tearing off of the oxide after each contact and eventually a removal of some of the underlying material depending on the intensity of mechanical stress acting on the contacting materials. The major concern is then to quantify and eventually to model the kinetics of repassivation as accurately as possible. This type of tribocorrosion process can be classified as an oxidative wear mechanism as, for example, the 'mild oxidative wear model' (Quinn, 1992). In the case of debris trapping, one has to consider that under appropriate hydrodynamical, chemical, and thermal contact conditions and relative speed of the two contacting bodies, the debris will remain temporarily in the contact zone mainly as colloids with a diameter usually in the range of a few hundred nanometers. Two cases may then be distinguished: (a) the debris accelerates the wear in comparison to the case of debris-free contacts is accelerated by an abrasive effect, or (b) the debris slows down the wear compared to the case where the contact zone is free of any debris, resulting in a protective effect.

Tribocorrosion may take place in practice in a large number of very different tribological systems consisting of mechanical devices containing metallic parts that are in contact with counterparts and exhibiting a relative movement placed in an environment revealing itself to be corrosive to at least one of the contacting materials. A non-limitative list of examples might contain machinery pumps, bearings, gears, ropes, electrical connectors, hinges, microelectromechanical systems (MEMs), and orthopedic implants like hip and knee implants.

1.2 Synergism between mechanical and chemical loading

To understand the importance and complexity of the phenomena taking place under tribocorrosion, one has to consider that the corrosiveness of a medium (liquid or gas) towards a material is highly dependent on the mechanical stresses that act onto a material, particularly at its surface exposed to that environment.

In tribocorrosion, five mechanisms may explain the synergism noticed between mechanical and chemical factors acting on contacting materials, namely:

1. the debris can speed up or reduce wear compared to what happens in the same environment where debris does not exist like e.g. in sliding contacts polarized at a large cathodic potential,
2. a galvanic coupling is established between the worn and unworn areas. It accelerates the anodic dissolution in the area where the metal is depassivated,
3. a galvanic coupling may be established between the two contacting counterparts,
4. an accumulation of dissolved species may take place in the liquid surrounding the contact. This may render the medium chemically or electrochemically more aggressive,
5. the mechanical loading in the contact area and its nearby zone may causes a work hardening of the materials. This work hardening can alter the kinetics of corrosion and/or repassivation processes.

A synergistic effect occurs in tribocorrosion when the mechanical process affects the corrosion process acting in a tribological system or vice versa. In these cases, the wear, W, found on a given component in a tribological system subjected to a mechanical loading in a given corrosive environment, will be very different and often much greater than the sum of the mechanical wear, W_{mo}, measured as a material loss under a given mechanical load in the absence of a corrosive environment, and the material loss induced by corrosion, W_{co}, in the absence of any mechanical contact (see Equation 1):

$$W \neq W_{mo} + W_{co} \tag{1}$$

This result is partly explained by the fact that the corrosion resistance in the case of a metal depends on the presence at its surface of reaction layers, sometimes only a few atom layers thick, resulting from an interaction between the material and the surrounding environment. Such layers can be classified as oxides, solid precipitates, adsorbed layers, or passive surface films. Some of them like dense oxide layers, precipitates, or passive films play a protective role by isolating the underlying metal from a direct contact with a surrounding corrosive environment. This is particularly true in the case of stainless steels and other alloys containing chromium. Their passive surface film formed in ambient air or in contact with an aqueous solution is a few nanometers thick but gives them a high resistance to corrosion. The sliding of a hard counterbody material on such a surface is likely to damage that passive film what is known as a "depassivation" process by which the bare material is exposed to the corrosive environment. Various but essentially electrochemical processes can then compete on these bare surfaces, namely:

- the dissolution of the metal in the corrosive medium,
- the formation of a new compound that may contribute to the breakdown process, and
- the restoration of the protective film known as "repassivation" process.

1.3 Complexity of the tribocorrosion process

The following examples taken from literature illustrate quite well the numerous parameters and interactions that govern the tribocorrosion process. Lemaire & Le Calvar, 2000, described the wear of a cobalt-based alloy coating generally referred to as "stellite 6" applied on the gripper latch arms of the control rods command mechanisms in pressurized water reactors (PWR). The downwards movement of the control rods is controlled by gripper latch arms of which the protruding teeth are coated with Stellite 6. The teeth block the movement once they come in contact with the control bar at the circular grooves lining their surface. At each blocking step, there is a contact between teeth and inner part of the grooves at a moderate pressure estimated at 150 MPa. Subsequently a sliding takes place over a distance of approximately 0.1 mm before the control rods come to rest. In the middle of the primary cooling circuit stellite 6 does not undergo any significant corrosion in the absence of any mechanical stress, thanks to the protective action of the passive film on stellite 6 consisting of chromium oxides. However under field operating conditions where impact and sliding of the teeth on the control bar take place, corrosion is evident. The wear observed on the teeth was found not to depend only on the number of blocking steps as would be the case in absence of corrosion. But the wear was found to depend also on the time interval between two successive blocking steps. The wear rate for a given number of blocking steps appeared to increase with the latter.

A plausible hypothesis to interpret this behavior is to consider that between two successive blocking steps corrosion takes place on parts of the surface where the passive film was mechanically damaged in the preceding step. The wear progress is correlated with the time interval between successive blocking steps by the following simple empirical equation (Bom Soon Lee et al., 1999):

$$I(t) = I_0 \left(\frac{t}{t_0} \right)^{-n} + I_P \tag{2}$$

in which I(t) is the evolution of the dissolution current of a metal with time starting at the time the metal becomes depassivated due to a mechanical action and extending during the film restoration where dissolution and repassivation are competitive surface processes. The parameters I_0 and t_0 are constants, while I_p is the passivation current under steady state, and n has a value between 0.3 and 1.

However, the effect of sliding on the electrochemical reactivity of the surface of a metal is not always confined to the partial destruction of surface layers. Other phenomena resulting directly or indirectly from contacts between parts can influence the corrosion behavior of their surfaces. For example, under reciprocating sliding conditions at small displacement amplitude, known as fretting (Carton et al., 1995; Godet et al, 1991) cracks may appear at the rim of the contact zone even after only a small number of contact events. Another parameter to be considered in tribocorrosion is the stirring of the corrosive environment along the surfaces of contacting parts caused by their relative movement. It affects tribocorrosion since such a stirring modifies the transport kinetics of chemical species that are generated in the vicinity of the surfaces due to the corrosion-related reactions.

The effect of the environment (mechanical or physico-chemical) on the crack propagation is evident. This returns us, somehow, to the notion of stress corrosion with the nuance that the cracks are induced at a mechanical loading which is not constant in the fretting test under consideration.

In addition one has to consider the possible role of strain hardening and/or structural transformations induced by the sliding action on the electrochemical reactivity of the surface, speeding up or slowing down some reactions. The resulting material transformation may end up in the most stable phase of the material considered being a supersaturated solid solution obtained by the gradual dissolution of pre-existing precipitates. It may also become a nano-crystalline network of a few tens of nanometers in average size that contains a high density of dislocations with no preferred orientation. This structure is very hard but also very fragile. Its intrinsic reactivity with the environment differs necessarily from the one of the original surface. Moreover, starting with the formation of a network of micro-cracks on it, wear particles (known as 'debris') are generated.

In the case of tribocorrosion, it is important to consider the galvanic coupling that might result from the heterogeneity of the electrochemical state of non-rubbed surfaces, and rubbed surfaces undergoing a strain hardening and on which the surface layers are altered. This galvanic coupling causes the polarization of non-rubbed and rubbed areas, and modifies the kinetics of reactions in these areas.

One has also to consider the existence of a third body that consists of wear particles as visible in Figure 1. At first, this third body may interact with the environment to form oxides or hydroxides. If they are ejected out of the contact zone, they become strictly speaking 'debris' and contribute to a material loss. If they remain in the contact zone, they can modify the mechanical response of the system by favoring a sliding action between counterparts, by acting as an abrasive agent promoting the so-called "abrasive wear", or by affecting the reactivity of material surfaces.

In return, a corrosion process can modify the surface states of materials and in consequence the contact conditions. In that way corrosion can affect the sliding conditions (coefficient of friction, wear regime, ...). The interaction between friction and corrosion therefore induces a complex phenomenon of synergy.

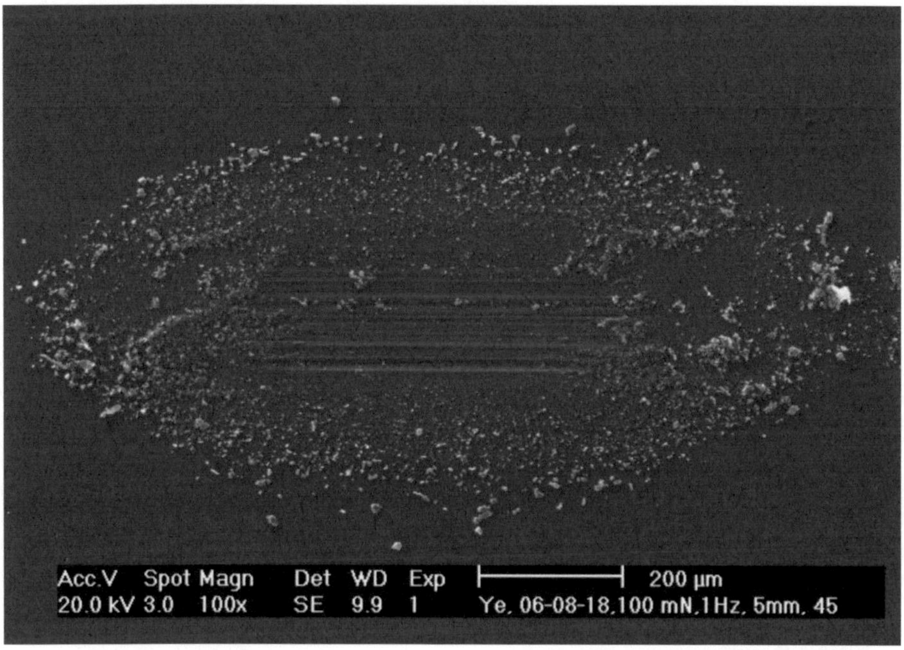

Fig. 1. Example of a third body resulting from a tribological process during reciprocating sliding. Some abrasive grooves are also visible in the central area of the sliding track.

2. The *in situ* study of tribocorrosion processes by electrochemical techniques

When performing a tribocorrosion test, one has to implement not only the following traditional concepts of tribological testing, namely:

- the representativeness of the contact (e.g. sphere-on-flat, sphere, cylinder-on-flat, flat-on-flat,...),
- the choice of the type of relative movement (continuous or reciprocating),
- the applied forces and the relative velocities, accelerations or displacements,
- the influence of the characteristics of the tribometer used on the wear induced, and
- the similarity of the wear mechanisms active in the laboratory test and in the field,

but he has also to take into account and to control simultaneously a large number of testing parameters, like:

- mechanical ones (e.g. contact frequency, noise, vibrations, surface roughness and residual stress ...),
- metallurgical ones (e.g. composition and structural state of contacting materials, microstructure, surface film composition and structure ...), and
- environmental ones (e.g. composition of the corrosive environment, pH, aggressiveness, viscosity, ionic conductivity, temperature, solid particles in suspension, stagnant or stirred,..).

These parameters determine the electrochemical reactivity of the surfaces and in consequence influence the contact conditions (wear regime, existence of a third body, friction ...).

Such tests therefore need to be instrumented to control and/or to record the contact conditions like the normal or tangential force, the relative displacement, velocity, acceleration and contact frequency ...). They need also to be instrumented with electrochemical techniques enabling the control and/or the recording of electrochemical parameters like the polarization of the contacting materials.

The choice of electrochemical techniques that can be implemented in a tribocorrosion test and the development of relevant models for the interpretation of the tribocorrosion mechanism are determined by the mechanical contact conditions being continuous or reciprocating. Electrochemical measurements can be performed with both types of tribometers. However, to be implemented under conditions that allow the interpretation of results, some methods require stationary electrochemical conditions, at least prior to starting up the measurements. In the case of continuous sliding, a quasi-stationary electrochemical surface state can often be reached, and all the electrochemical techniques available for corrosion studies (polarization curves, impedance spectroscopy, electrochemical noise,...), can be used. On the contrary, when reciprocating contact conditions prevail, the interpretations of experimental results are more complex due to the non-stationary electrochemical conditions. Measuring techniques suitable for the recording of current or potential transients will be used preferentially (Mischler et al., 1997; Rosset, 1999).

2.1 Open circuit potential measurements

Under sliding conditions, the measured open circuit potential is a mean value which depends on the electromotive force induced by the surface heterogeneity resulting from the coexistence of non-rubbed and rubbed areas which are in different electrochemical states, and on the areas of these zones and their spatial distribution that determines the non-uniform distribution of potential over the whole surface (Oltra et al, 1986). When applying for example a continuous sliding, this open circuit potential responds to the mechanical loading imposed, as shown in Table 1 and Figure 2.

Fn (N)	0	2.2	2.2	4.2
V (cm s-1)	0	0.5	0.8	0.8
Eoc (V vs SCE)	-0.15	-0.29	-0.32	-0.42

Table 1. Variations of the open-circuit potential E_{OC} of 316L stainless steel in artificial sea water as a function of varying tribological contact conditions. S.C.E.: saturated calomel electrode. F_n: normal force. V: sliding speed.

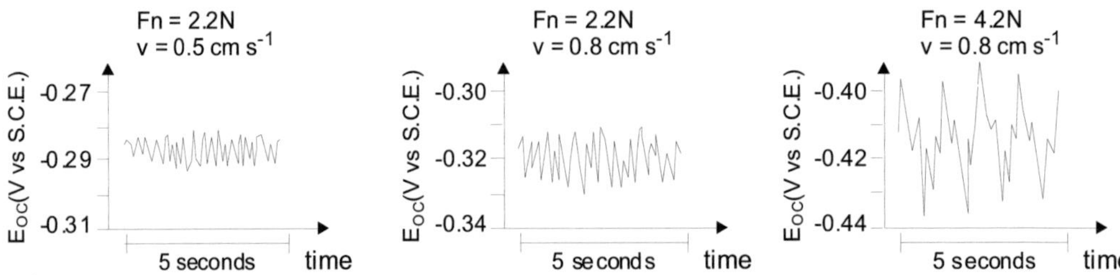

Fig. 2. Electrochemical noise at the open circuit potential E_{OC} of 316L stainless steel in artificial sea water for different tribological contact conditions. S.C.E.: saturated calomel electrode. F_n: normal force. V: sliding speed.

Table 1 gives the evolution of the mean value of the open circuit potential of a 316 L stainless steel in contact with alumina immersed in artificial seawater for different normal forces and velocities (Ponthiaux et al., 1997). Alumina is taken since it has a high electrical resistance and is chemically inert in the liquid used. At a zero normal force and speed, the measured open circuit potential corresponds to the passive state of the entire stainless steel surface. When the normal force or the speed increases, the open circuit potential shifts towards lower potential values. This shift can be explained by considering the following phenomena:

- on the rubbed area, the surface layer on materials can be partially destroyed (also called "depassivation") and that rubbed area acquires a potential that corresponds to the one of an "active material" undergoing dissolution or on which reduction reactions can take place, and
- an increase of the normal force or velocity tends to increase the area of active material.

In Figure 2, the rapid fluctuations of the open circuit potential arising during sliding are represented schematically. They constitute an 'electrochemical noise' representing the electrochemical response to the rapid and stochastic fluctuations of the new bare metal surface generated in the real contact area by sliding friction. This noise can provide information on the mechanism of friction as well as on the mechanism of electrochemical reactions involved in the depassivation - repassivation process (Déforge et al.). A more detailed analysis of open circuit potential measurements under sliding requires a more precise knowledge of the local surface state of contacting materials. Experimentally, microelectrodes can be used to determine potential values of rubbed and non-rubbed areas. Such techniques have already been used to study localized corrosion, and models have been proposed (Lillard et al., 1995). Note that the interpretation of local potential measurements or the development of theoretical models describing the potential distribution, can only be obtained with realistic assumptions on the reaction kinetics of reactions occurring at rubbed and non-rubbed surface areas. Such Information can only be obtained by using complementary methods.

2.2 Potentiodynamic polarization measurements

Potentiodynamic polarization curves obtained at increasing and decreasing potential scan in absence of any sliding is schematically shown in Figure 3. In the case the current measured originates from the whole surface of the tested sample that might be considered as being uniform.

Under sliding conditions, the currents measured during potentiodynamic polarization are in a first approach the sum of two components, namely the current originating from the rubbed area, and the one linked to the non-rubbed area. Under such conditions, the maximum dissolution current, I_M, varies with the mean contact pressure and sliding speed. However, these two test parameters do not necessarily affect in the same way the electrochemical behavior of the alloy:

- at constant speed an increase of the mean contact pressure may cause a decrease of the maximum dissolution current, I_M. This reflects then a slowing down of the dissolution of the alloy which can be due to the strain hardening of the material in the rubbed area. Moreover, the potential, E_M, at which this maximum current is noticed, remains almost unaffected. This indicates that the dissolution process is the same on rubbed and non-rubbed materials,

- at a fixed mean pressure, an increase of the sliding speed may cause, in turn, an increase of the maximum dissolution current. This increase of I_M is then a consequence of the increased bare surface area generated per unit of time. It indicates in that case that the dissolution of the bare material is faster than the non-rubbed material.

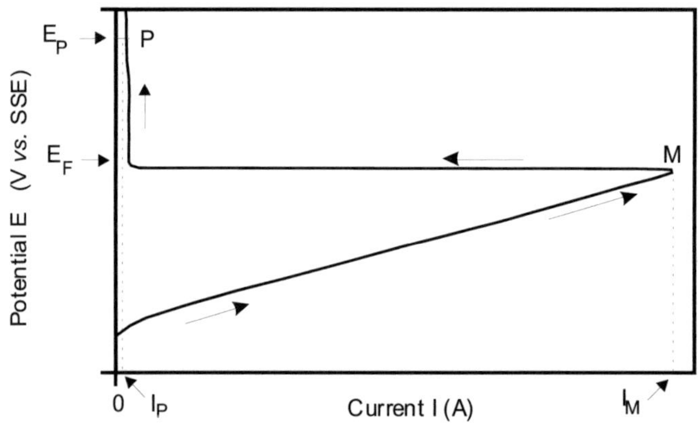

Fig. 3. Schematic potentiodynamic polarization curve recorded on a passivating metallic material (Fe-30%Ni alloy) in absence of any sliding: E_F = Flade potential, E_M = potential at maximum dissolution current I_M, I_P = passivation current.

The possible evolution of the passivation current, Ip, under sliding friction with the mean contact pressure, P_M, and the sliding velocity is schematically shown in Figure 4. That passivation current is increasing with these two testing parameters. That increase can be related to an increase in bare surface area due to:

- an increasing contact area with increasing normal force, and
- an increasing depassivated surface area generated per unit time with increasing sliding speed.

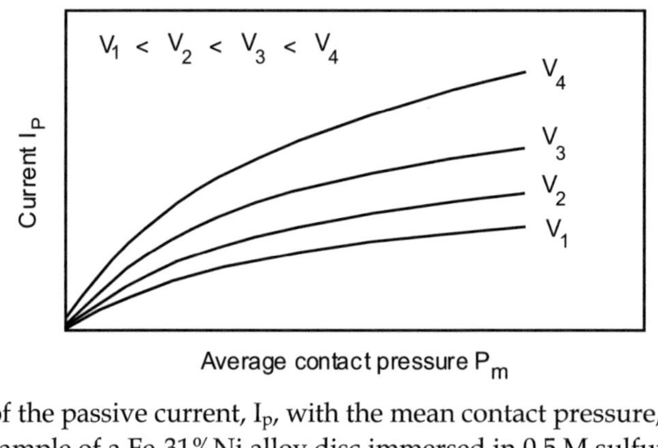

Fig. 4. Evolution of the passive current, I_p, with the mean contact pressure, P_m, at increasing sliding speeds. Example of a Fe-31%Ni alloy disc immersed in 0.5 M sulfuric acid and polarized at a fixed passive potential, under continuous sliding against alumina.

The extent of the passivation range is in general affected by the sliding. In some cases the Flade potential, E_F, shifts towards more anodic values at increasing contact pressure. This means that the stability of the passive film decreases. When the potential E_F is reached, the entire surface becomes active and dissolves. However, under these experimental conditions, the disruption of the passive film is only a local disturbance in the sliding track in a relatively small area that represents generally not more than 10% of the total sample area. Therefore sliding does not only affect the electrochemical state of the rubbed surface, but also the state of the whole surrounding sample surface.

Finally it should be stressed that the coefficient of friction varies with the applied potential during a potentiodynamic polarization. Changes in the value of the coefficient of friction reveal a possible change in the surface state of the materials in the sliding track (Ponthiaux et al., 1999).

The in-depth interpretation of the polarization curves frequently faces difficulties related to the non-uniform distributions of current and potential on the sample surface. This non-uniformity originates from the intrinsic effect of the sliding that causes an heterogeneity of the electrochemical surface reactivity, combined with the ohmic drop in the electrolyte. A full exploitation of the polarization curves in terms of local behavior is possible only if one can model the current and potential distributions under sliding conditions. This brings back to the same approach as in the case of the interpretation of open circuit potential measurements. Note that the effect of non-uniform distributions on the interpretation of polarization curves was already investigated in the absence of any sliding (Law & Newman, 1979; Ponthiaux et al., 1995; Tiedemann et al., 1973).

2.3 Repassivation current transients

Methods based on analysis of potential or current transients (Ponthiaux et al., 1995) are particularly well suited to study reciprocating sliding tests (Mischler et al., 1997). These methods are used to study between successive contact events, the rebuild of damaged surface layers (oxide, passive film ...). Under imposed polarization e.g. in the passivation range, at each stop-start event, a transient variation of current is noticed with time (see Figure 5). The charge corresponding to this transition can be attributed to the re-growth of a uniform film in the damaged area.

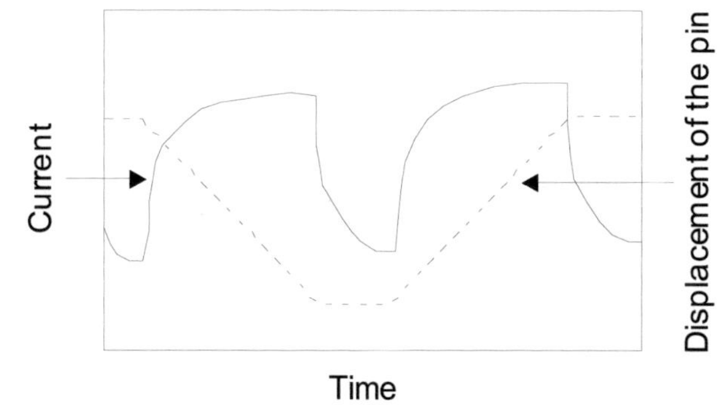

Fig. 5. Schematic evolution of the current transients obtained under reciprocating sliding.

In order to study in more details the mechanism and kinetics of a re-growth of the passive film, the "potential jump" method can be used. However, such a method cannot be applied on metals such as aluminum, for which the passive film can not be reduced by a cathodic polarization, or for some steels for which the contamination of the surface by reduction products can affect the initial current increase at the potential jump.

2.4 Electrochemical impedance measurements

This method requires quasi-stationary electrochemical conditions of currents and potentials. The impedance measurements allow the study of the influence of sliding on the elementary processes involved in the corrosion mechanism. By performing a systematic analysis of the changes in the impedance diagrams with the sliding parameters like normal force and sliding speed or contact frequency, a model can be developed incorporating the sliding effects in the mechanism.

Impedance measurements recorded on a non-rubbed and a rubbed Fe-31% nickel alloy are shown in Figure 6. The measurements were done in 0.5 M sulfuric acid at a potential of -675 mV / SSE located in the active region (Boutard et al., 1985). The impedance diagrams were recorded in a limited range of measurement frequencies. In particular, under sliding, the impedance was not measured at frequencies below 0.01 Hz, to limit the duration of the measurements and to avoid in this way the influence of the long-term evolution of the electrochemical state of the surface due to wear - induced changes in the "rubbed area/unrubbed area" ratio. By taking these precautions, we can consider that the condition of average electrochemical steady-state required for impedance measurements is fulfilled.

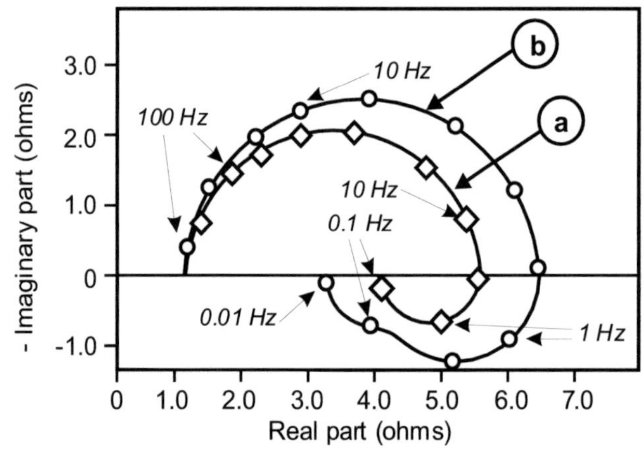

Fig. 6. Electrochemical impedance plots recorded on Fe-31% Ni immersed in 0.5 M sulfuric acid polarized in the active region (-0,675 V/SSE): (a) in absence of any sliding, and (b) under continuous sliding against alumina at a mean contact pressure of 2,6 MPa and a sliding speed of 3,4 cm s^{-1} (Ponthiaux et al., 1999).
SSE: 'mercury/mercurous sulfate/saturated potassium sulfate' reference electrode (E_{SSE} = +0.65 V/NHE).

In absence of friction, the plot consists of two successive semi-circles, a capacitive one reflecting the dielectric properties of the electrochemical double layer, and an inductive one reflecting the contribution of an elementary adsorption step in the reaction mechanism. Under sliding, the shape and size of the plot are modified as follows:

- at high frequencies, the capacitive arc increases reflecting a lower dissolution rate. This might be correlated to a strain hardening effect mentioned above, and highlighted in the polarization curves, and
- at low frequencies, an additional inductive arc appears which highlights the presence of a second adsorbate in the dissolution mechanism.

This can be interpreted by considering that in a non-rubbed condition the semi-circle linked to a possible second adsorbate does not appear because its concentration at the surface of the sample does not vary substantially with the potential. Under sliding conditions, the kinetics of some dissolution steps evolves, in particular the kinetics of steps in which the adsorbate interacts. The variation of the concentration at the surface becomes then detectable.

The in-depth interpretation of these impedance diagrams encounters however some difficulties as those already reported for open circuit potential and polarization measurements which are related to the heterogeneous state of the tested surfaces.

A first difficulty results from the fact that the impedance data reflect the overall state of the tested surface integrating the contributions of non-rubbed and rubbed surfaces. Such data must thus be de-convoluted in order to obtain the specific impedances of these two types of surface states. A first approach to this problem might be to use models similar to those describing the impedance of a sample undergoing a localized corrosion (Oltra & Keddam, 1990). In that specific case, the overall impedance can be considered as the result of two impedances in parallel, namely the impedance of the non-rubbed surface and the one of the rubbed surface. A strict interpretation requires further an evaluation of the areas of these surfaces, e.g. by using profilometry and surface observations by light optical or scanning electron microscopy.

The second difficulty results from the non-uniform state of the rubbed surface. Behind the slider the sample surface can be laid bare for some time before some new surface layers are rebuilt. The restored surface increases gradually with the distance behind the slider along the sliding track. Even if the first difficulty was already solved and the overall impedance of the rubbed surface is obtained, it can not be used as such to characterize the non-uniform distribution of the electrochemical states behind the slider. However, it is expected that impedance measurement procedures already developed for analyzing non-uniform distributions of surface states and the electrochemical models developed for the interpretation of such measurements (Zhang et al., 1987) could be transferred to tribocorrrosion test conditions. Such a study could allow a localized characterization of dissolution and passivation kinetics.

If theoretically this approach seems promising, experimental data analyses show that impedance measurements under sliding are often disturbed at low frequencies due to the random fluctuations of potential or current. This "electrochemical noise" limits unfortunately to some extent the application of impedance measurements. Limitations frequently originate from the sliding action itself and more specifically from the localized damages induced in the contact area by the mechanical interaction. Notwithstanding that, the in-depth analysis of the electrochemical noise will surely in the future be fruitful since it will contain useful information on the progress of the process at both spatial and time scales.

2.5 Electrochemical noise analyses

In the case of tribocorrosion as in cases involving stochastic processes of local surface damages like pitting (Uruchurtu & Dawson, 1987), stress corrosion (Cottis & Loto, 1990), and corrosion-abrasion (Oltra et al., 1986), sliding destroys locally to some extent surface layers. The bare areas created by the slider generate fluctuations of potential or current which consist of the superposition of elementary transients. The sudden increase in current occurs at the time that a bare surface is brought in contact with the electrolyte. The subsequent current decrease reveals the restoration of a protective surface film. The analysis of the transient characteristics (shape, amplitude, duration) provides information on the mechanisms and kinetics of the reactions involved (e.g. dissolution, passivation). The frequency of the transients depends in turn on the number of contacting points in the sliding track at a given time and on the sliding rate of the slider. It is therefore a quantity that provides useful information on the nature of the contact.

Sliding conditions affect the amplitude of the electrochemical noise, namely fluctuations of the open circuit potential as shown in Figure 2 (Ponthiaux et al., 1997). It is usually impossible to isolate the elementary transients. However, the spectral analysis of such a noise allows characteristic quantities to be derived such as the mean amplitude or duration and the average frequency at which transients occur. These characteristics are essential for getting a better understanding of the nature of the contact and the dissolution and passivation kinetics on the bare surface.

3. Modeling approaches

Given the complexity of the tribocorrosion phenomena there is currently no universal predictive model of the wear-corrosion process available. Such a problem solving is largely empirical, and designers rely on expert systems fed by experimental feedback to select the material couples for a given tribological system. In parallel to this technical approach, scientists are developing the modeling elements needed to unravel the phenomena. These models are discussed hereafter.

3.1 Formalism of the wear-corrosion synergy

It was shown here above that electrochemical methods allow under certain conditions to measure in real time the current I related to the corrosion reaction in a sliding track. At the end of a tribocorrosion experiment (e.g. sliding or erosion tests), the electrochemical mass loss, C, can be calculated using Faraday's relationship from the total charge consumed by the corrosion process:

$$C = \frac{M}{nF} Q \tag{3}$$

in which M is the atomic mass of the metal, n is the valency of the oxidized metal in the environment studied, F is the Faraday's constant and Q is the total charge related to the corrosion process, namely

$$Q = \int_0^T I(t)\, dt$$

with T the total duration of the test. However, the electrochemical mass loss due to corrosion under sliding, C, is generally not equal to the mass loss of the metal, C_0, obtained under similar test conditions but in absence of any sliding. The electrochemical mass loss under sliding C has thus to be expressed as:

$$C = C_0 + C_M \quad (4)$$

with C_M the amount of corrosion induced by wear.

The total mass loss, *W*, can be determined at the end of a sliding test by some *ex situ* technique like surface profilometry or gravimetry. That total mass loss can be compared to the electrochemical mass loss under sliding, *C*, from the equation:

$$W = C + M \quad (5)$$

in which M is the mechanical mass loss. This mechanical mass loss M can also be compared to the mass loss, M_0, in a non-corrosive environment as:

$$M = M_0 + M_C \quad (6)$$

in which M_C represents the excess mechanical mass loss due to corrosion. This formalism now allows a general definition of the synergy, S, between corrosion and mechanical processes in case of tribocorrosion as:

$$W = C_0 + M_0 + S \quad (7)$$

In which:

$$S = C_M + M_C \quad (8)$$

The term *S* defined as such reflects the fact that the material mass loss in a corrosive environment cannot be predicted simply by the sum of the mass loss due to corrosion in the environment in absence of any mechanical interaction, and the material loss due to wear recorded under similar testing conditions but in an non-corrosive environment. There is a synergy between these two processes. The formalism of tribocorrosion originally proposed (Watson et al., 1995) has surely an educational value since it allows a diagnostic on the "origin of evil" under the given set of experimental conditions. The wear M_0 is by some authors measured in a test under cathodic polarization in either dry air, de-ionized water or in the presence of corrosion inhibitors (Smith, 1985; Stemp et al, 2003; Takadoum, 1996). There is still some controversy about the validity of such procedures. In practice, the results often depend on the method used which limits the overall benefit of such a decoupling. Moreover, the concepts used do not have a physical meaning, and the synergies between wear and corrosion cannot really be simplified to a summing up. Other approaches that were developed over the past decades are briefly reviewed hereafter.

3.2 Models of oxidative wear and application in aqueous environment

3.2.1 Quinn's model of mild oxidative wear

A two-step model of mild oxidational wear for steel in air was developed (Quinn, 1992, 1994). The author observed that at sliding speeds below 5 ms^{-1} the wear debris consists only of iron oxides and that the particle size is quite constant in the range of several micrometers. In a first step oxides grow on surface asperities as a result of local heating at contact points. When the oxide reaches a critical thickness, x_c, the mechanical stresses generated in the material become too large and a detachment of the oxide layer which reaches the critical thickness takes place on the passage of the slider. The worn volume by unit of sliding distance, w, can be written as:

$$w = \frac{x_c}{Vt_c} \cdot A_r \tag{9}$$

with A_r the real contact area, V the sliding speed, and t_c the time necessary to reach the critical oxide thickness. Quinn's model is thus a law that corresponds to the Archard's wear law with:

$$K_{Archard} = \frac{x_c}{V\,t_c} \tag{10}$$

It is possible to connect x_c and t_c through a thermally activated oxidation kinetics in air which can be considered in first instance as a parabolic function with time:

$$x(t) = \alpha \sqrt{k_0 \exp\left(-\frac{Q_{act}}{R\,T_f}\right) t} \tag{11}$$

where:

- x(t) is the oxide thickness at time t,
- Q_{act} is the activation energy of the oxidation process considered,
- k_0 is the Arrhenius constant associated with the process considered,
- T_f is the flash contact temperature at the interface between two asperities,
- R is the gas constant,
- t is time, and
- α is a constant depending on the density of the oxide and its oxygen volume fraction.

The critical thickness of the oxide, x_c, can only be obtained experimentally by characterizing the debris. The model is thus not a predictive one. Moreover as pointed out (Smith, 1985) the oxide growth laws based on mass can hardly be used under the conditions of contact characterized by a low air supply, and a poor knowledge of the real contact temperature. In practice, the use of oxidation constants leads to wear rates that are several orders of magnitude different from the experimentally verified ones.

3.2.2 Model of Lemaire and Le Calvar

In conclusion, the Quinn's model can not be easily adapted to analyse the depassivation - repassivation process determining the wear laws observed in tribocorrosion conditions. However, the main idea of this model, namely that wear proceeds by a succession of growth and delamination of an oxide layer, can be retained. It was at the origin of the tribocorrosion model presented to explain the the wear of a cobalt-based alloy coating applied on the gripper latch arms of the control rods command mechanisms in PWR (Lemaire & Le Calvar, 2000). In this model, the wear law is given by the following expression:

$$W = W_0\ N\left(\frac{t}{t_0}\right)^{(1-n)} \qquad (12)$$

Where W is the total worn volume, W_0 a constant, and N the number of sliding steps applied to the alloy surface inducing removal of the passive film. t is the mean time interval between two successive sliding steps, and t_0 is a characteristic repassivation time constant. n is a positive exponent whose value was found experimentally close to 0.65 for the cobalt-based alloy coating. The authors explain the wear law expressed by equation (12) by the evolution of the repassivation current Ip given by expression (2). This model implies that the growth of the oxide film between two sliding steps is proportional to $t^{(1-n)}$. The Quinn's law appears as a particular case of such a model for n = 0.5. In triborrosion studies, different values of n (between 0.6 and 0.9) where found depending on the metal, the environment and the electrochemical conditions.

3.2.3 Application to the synergy formalism

Studies in corrosive aqueous solutions (Garcia et al., 2001; Jemmely et al., 2000) are suitable to follow *in situ* the growth of passive films by electrochemical methods, and allow thus the development of more sophisticated models. In these studies performed under continuous or reciprocating sliding conditions, a modeling of currents measured at an applied potential is done. It is then assumed that a unit area of depassivated material repassivates according to a simple repassivation transient, J_a (t), which is not affected by the electrochemical conditions on the areas surrounding the rubbed area. The measured total current is then the sum of the contributions of the different surface areas. A freshly depassivated area produces a large current while a area depassivated some time before produces lower currents. In the case of a reciprocating tribometer operated at a sliding frequency, f, the steady state current I can be expressed as follows assuming that each contact event depassivates the surface:

$$I = A_a \cdot f \int_0^{1/f} j_a(t)dt + (A_t - A_a)f \int_0^{1/f} j_p(t)dt \qquad (13)$$

with A_t the total area in contact with the solution, A_a the depassivated area on the sample during one cycle, f the frequency at which the surface is depassivated, j_p the passive current density at the applied potential, and j_a (t) the transient repassivation current density of a unit area at the applied potential.

Taking into consideration the synergy formalism developed above, the components of mass loss per cycle, C_M and C_0 in Equation (4) can now be written as follows:

$$C_M = \frac{M}{nF} \cdot A_a\ (\int_0^{1/f} j_a(t)dt - \int_0^{1/f} j_p(t)dt) \qquad (14)$$

$$C_0 = \frac{M}{nF} A_t \int_0^{1/f} j_p(t)dt \qquad (15)$$

When the component j_p related to passive zones can be neglected, C_0 becomes zero and equation (4) expressing the mass loss by corrosion under sliding becomes then:

$$C = C_M = \frac{M}{nF} A_a \int_0^{1/f} j_a(t)dt \qquad (16)$$

The depassivated area during one cycle, A_a, can hardly be assessed. Some authors assume in first instance that it is equal to the apparent area of the sliding track. However, it is well known that the contact takes place only on a fraction of that area. An evaluation of the depassivated area from currents resulting from an electrochemical depassivation achieved by a potential jump was proposed (Garcia et al., 2001). This method also allowed them to evaluate the oxide thickness formed in between two successive depassivation events. They obtained oxide layer thicknesses in the range of a few nanometers.

Another approach was developed (Jemmely et al., 2000). The authors proposed to express the depassivated area in terms of a depassivation ratio per unit of time, R_{dep} :

$$R_{dep} = f.A_a \qquad (17)$$

with f the contact frequency. The currents can then be expressed as:

$$I = R_{dep}.Q_{rep} \qquad (18)$$

with Q_{rep} the charge density for repassivation. The rate at which an active area is generated per unit of time depends on the morphology and hardness of the surfaces in contact. A derivation of R_{dep} from the scratching of a ductile material by a hard abrasive one was proposed (Adler & Walters, 1996). That approach was taken over (Jemmely et al., 2000) and extended in more general terms (Mischler et al., 1998) in the following expression:

$$R_{dep} = K.V\frac{F_N^{\beta}}{H} \qquad (19)$$

with K an empirical constant, V the sliding velocity, F_N the applied load, H the hardness of the tested material, $\beta = 0.5$ in the case of a contact between two counterparts with a similar roughness, $\beta = 0.5$ in the case of a rough and hard body against a smooth and ductile counterpart, $\beta = 1$ in the case of a hard and smooth body against a rough and ductile counterpart, and β between 0.5 and 1 in the general case.

One empirical constant K remains in this model which approximates Archard's constant and which is related to the probability that a given contact becomes depassivated. The mass loss by corrosion under sliding, C, can thus finally be written as:

$$C = \frac{M}{nF} KV \frac{F_N^{\beta}}{H} \cdot Q_{rep}(f) \qquad (20)$$

Current measurements performed at different loads and sliding speeds for materials with different hardness, allow the validation of the general form of this law.

4. Tribocorrosion testing

4.1 Specificity of laboratory and industrial tribocorrosion tests

Similarly as in classical mechanical testing, tribocorrosion tests can be classified into two categories based on their different but complementary purposes, namely fundamental and technological tests.

Fundamental tests are implemented in research laboratories and their objective is to clearly identify and to understand under well defined testing conditions, the basic mechanisms and their synergy that govern the phenomena of tribocorrosion. These tests require the development of experimental methodologies for both the test themselves and the techniques to be used for analyzing and measuring data and other experimental outcomes. Concerning friction in particular, two types of tests can be considered:

- tests at low displacement amplitude referred to as "**fretting tests**". These tests provide information on the response of materials with respect to the solicitation (displacement amplitude, load, frequency, and environment). One can differentiate stick-slip, partial slip or gross slip contact conditions. The information collected is on the nature of the degradation, its location in the contact, the kinetics of crack initiation and crack growth, and the size and shape of the degradation products that may appear in the contact during testing, and
- tests at large displacement amplitude referred to as either "**reciprocating sliding tests**" or "**continuous sliding tests**", provide information on the nature and kinetics of the wear process in connection with the synergies resulting from the mechanical, chemical or electrochemical coupling taking place on contacting surfaces in relative motion.

These tests allow the following analyses based on *in situ* and *ex situ* measurements, like the determination of the mean and local coefficient of friction, the identification of and study of the interactions between surfaces and environment, the nature of the mechanical-chemical coupling, the electrochemical or galvanic coupling due to a heterogeneous structure, the shape and location of rubbed and non-rubbed areas, or the establishment of local wear laws and their spatial distribution on the surface in view of a modelling of wear aiming at a future predictive approach.

Technological tests are designed to reproduce at lab scale mechanical loading and/or environmental conditions corresponding to actual operating conditions, or to mimic particular conditions intending to accelerate material degradation processes. These tests are widely used to predict precisely the behaviour of mechanical devices in actual conditions of service and to improve their reliability and durability. In that respect, they are very useful tools.

The full investigation of the tribocorrosion tests requires generally the use of *in situ* tools like open circuit measurements, polarization measurements, current transients, impedance spectroscopy, and noise measurements, and *ex situ* tools like elemental surface analysis techniques, optical or electron microscopy, micro-topography, micro and nanohardness measurements texture and internal stress analyses.

4.2 Testing protocol: A multiscale analysis of tribocorrosion phenomena

A promising approach of synergy in tribocorrosion has been proposed (Diomidis et al., 2009) based on the fact that the surface state of a wear track evolves with time in a cyclic manner. That evolution is due to the repeated removal and subsequent re-growth of a passive surface film when a mechanical loading is applied. During the latency time, t_{lat}, defined as the time between two successive contacts at a given point in the sliding track, the passivation reaction tends to restore the passive film. The fraction of the sliding track surface covered by this re-grown passive film increases with t_{lat}. By controlling the frequency of such depassivation-repassivation events with respect to the time necessary for film growth, it is possible to measure the properties of the surface at different stages of activity and repassivation. For performing tests at different latency times, t_{lat}, two approaches are possible, each approach having own advantages and limitations as detailed hereafter:

- first approach in the case of **continuous sliding** tests: the latency time t_{lat} can be modified by changing the rotation period t_{rot}. Another way to obtain the same t_{lat} is to keep the sliding speed constant and increase the radius of the track. However, for practical reasons, it is not realistic to consider to multiply the radius of the track by a large factor.
- second approach in the case that the latency time is changed by performing **intermittent sliding** tests: during such tests the counterbody slides for one cycle, and then stays immobile for a certain period of time to allow a part of the passive film to re-grow. Thus, an off-time, t_{off}, is imposed at the end of each cycle. As a result, the latency time between two subsequent contact events, t_{lat}, differs from the rotation period, t_{rot}:

$$t_{lat} = t_{rot} + t_{off} \tag{21}$$

It is clear that in continuous sliding tests, Equation (21) is still valid but t_{off} is zero. Such an approach can thus result in a protocol that provides information on the evolution of the surface with testing time, and the identification of the resulting mechanisms of material loss and surface degradation (Pourbaix, 1974).

4.2.1 Selection of test conditions

In order to characterize the sensitivity of the one or more material systems to tribocorrosion, a careful selection of the test conditions has to be done prior to any testing, so as to obtain discriminating results. The following steps are of large importance in that approach:

- Selection of environmental conditions: the electrolyte should be selected in view of its known oxidative or reducing power. The test temperature can be ambient temperature or any temperature relevant for the field application it should reproduce. A decisive parameter in the selection of the electrolyte is its pH. The selection of the pH can be based on pH-potential diagrams (Pourbaix, 1974). In the case of a metallic alloy, a pH range should be selected by preference where at least one of the constituents passivates. An electrolyte that may cause localized corrosion, like pitting corrosion in particular, should be avoided,
- Parameters linked to the sliding tests need to be selected by considering the following recommendations. The normal force, F_N: the normal force should be selected so as to

avoid plastic deformation of the tested materials. The maximum Hertzian contact pressure on the test material before starting sliding should be smaller than the yield strength. Concerning the track radius, R_{tr}, it should be selected in such a way that edge effects are avoided. E.g. in the case of a disc, the track radius should be by preference about half the test sample radius. Finally the number of cycles, n, depends on the type of material tested and the test conditions. It should be selected so that the wear volume is large enough to be measured accurately, while avoiding too long test durations for practical reasons. A preliminary sliding test might be necessary to determine n. In some particular cases, the selection of the number of cycles can also be done so as to reflect the behaviour of the test material in a real life application.

4.2.2 First step in the testing protocol: Electrochemical tests on passive material without any sliding

After selecting the set of appropriate test conditions, measurements are done to collect information on the electrochemical behaviour of a material fully covered by a passive film. This is done by electrochemical tests in absence of any sliding. After immersion in the electrolyte, the open circuit potential, E_{oc}, is measured versus a reference electrode. In general, a stable value of E_{oc} is obtained after some time of immersion. From an electrochemical point of view, a stable E_{oc} is obtained when the long-term fluctuations of E_{oc} are below 1 mV min^{-1} during a minimum of 1 hour. The time necessary to reach such a stationary open circuit potential in the test electrolyte is an important characteristic of a passivating process, and is called in this protocol as the reaction time characteristic, t_{reac}. The evolution of E_{oc} from immersion time on provides useful information on the electrochemical reactivity of the tested material in the test electrolyte (see Figure 7).

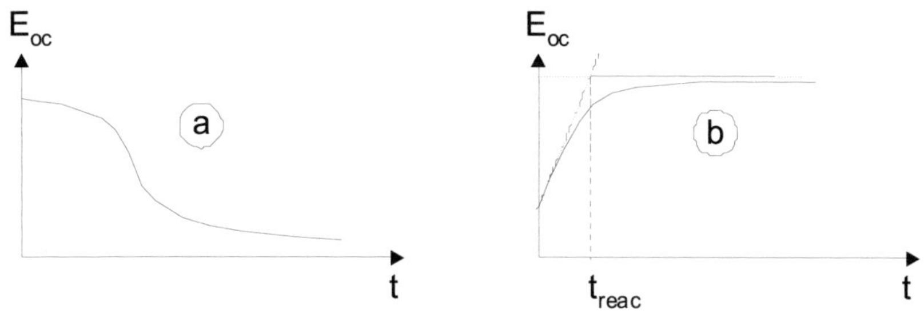

Fig. 7. Schematic representation of the evolution of E_{oc} with immersion time in the case of (a) corrosion, and (b) passivation. The graphical determination of t_{reac} in the case of passivation is shown in (b).

From this figure one can derive that when E_{oc} decreases with time, general corrosion may be suspected, and that when E_{oc} increases with time, passivation or adsorption is probably taking place. In this latter case, t_{reac} can be estimated from the evolution of E_{oc} with immersion time by drawing the tangent to the curve at the point where the slope is maximum, together with a straight line tangential to the data in the part of the curve where E_{oc} is stable. After achieving a long-term stable open circuit potential indicative of passivation, the polarization resistance of the passive material, R_p, is measured by electrochemical impedance spectroscopy. Based on R_p, the specific polarization resistance of passive material, r_{pass}, can be calculated for a test sample with a surface area, A_o, as:

$$r_{pass} = R_p . A_o \tag{22}$$

Specific polarization resistance values for metallic materials of 10^3 Ωcm^2 or lower indicate the presence of an active sample surface, while values around 100×10^3 Ωcm^2 or higher indicate a passive sample surface. The corrosion current density of the material covered by a passive surface film, i_{pass}, is then calculated as follows:

$$i_{pass} = \frac{B}{r_{pass}} \tag{23}$$

with B a constant. For metallic materials, B normally varies between 13 and 35 mV, depending on the nature of the material and the environment. In the protocol worked out hereafter as an example, a value of 24 mV is assumed. This passive current density, i_{pass}, is considered to correspond to the dissolution current of the material through the passive film at stationary state.

4.2.3 Second step in the testing protocol: Electrochemical tests on a fully active sliding track during sliding

The next step is the determination of the corrosion rate of the depassivated material. In order to keep continuously a part of the immersed sample surface in an active state, the passive film has to be removed by mechanical contacts. It is thus necessary to select a rotation period, t_{rot}, which is small compared to t_{reac} so that the passive film has no time to regrow in between two successive contact events. It was proposed (Diomidis et al., 2009, 2010) to take the rotation period t_{rot} equal to:

$$t_{rot} = \frac{t_{reac}}{10000} \tag{24}$$

A first value of t_{lat1} is then taken equal to t_{rot}. It is thus assumed that during such sliding tests the whole wear track area is in an active state, so that:

$$A_{tr} = A_{act} \tag{25}$$

Despite the fact that the width of the sliding track increases progressively due to wear during sliding tests performed against a curved counter-body, a mean sliding track area, A_{tr}, is taken for simplicity, and defined as:

$$A_{tr} = \frac{1}{2}\left(A_{tr\max} + A_{tr\min}\right) \tag{26}$$

with $A_{tr\,max}$ the maximum value measured at the end of the test, and $A_{tr\,min}$ the minimum value at the end of the first cycle. $A_{tr\,min}$ is calculated by multiplying the length of the wear track, L, by the diameter of the Hertzian static contact area, e:

$$e = 2\left(\frac{3\,F_N\,R}{4\,E}\right)^{1/3} \tag{27}$$

with F_N the applied normal load, R the radius of the tip of the curved counter-body, and E the equivalent elastic modulus given by:

$$\frac{1}{E} = \frac{1-\nu_1^2}{E_1} + \frac{1-\nu_2^2}{E_2} \quad (28)$$

with ν_1 and ν_2 the Poisson's ratios, and E_1 and E_2 the elastic moduli of the test sample and counterbody respectively.

Sliding is initiated at the time a stable E_{oc} is achieved. That E_{oc} recorded during sliding is a mixed potential resulting from the galvanic coupling of material inside (A_{tr}) and outside (A_o - A_{tr}) the sliding track. It is assumed that the kinetics of the redox reactions taking place on each of these areas, do not vary with the real potential of the sliding track. In other words, the ohmic drop effect is considered to be negligible in the galvanic coupling between the sliding track and the surrounding area. Electrochemical impedance spectroscopy measurements are performed during sliding to obtain the polarization resistance, R_{ps}, of the sample surface. Similarly to E_{oc}, R_{ps} may be considered as the combination of two polarization resistances, namely R_{act} related to the active area A_{act} which is equal in this case to the wear track, and R_{pass} which corresponds to the surrounding unworn area, (A_o - A_{tr}):

$$\frac{1}{R_{ps}} = \frac{1}{R_{act}} + \frac{1}{R_{pass}} \quad (29)$$

where

$$R_{act} = \frac{r_{act}}{A_{tr}} \quad (30)$$

and

$$R_{pass} = \frac{r_{pass}}{A_0 - A_{tr}} \quad (31)$$

It is then possible to calculate the specific polarization resistance of the active surface, r_{act}, as:

$$r_{act} = \frac{A_{tr}\, R_{ps}\, r_{pass}}{r_{pass} - R_{ps}\left(A_0 - A_{tr}\right)} \quad (32)$$

The corrosion current density of the active material, i_{act}, can now be obtained by substituting r_{pass} with r_{act} in Equation (23):

$$i_{act} = \frac{B}{r_{act}} \quad (33)$$

It can be noted that, the specific resistance of the active bare material in the sliding track, r_{act}, is generally several orders of magnitude lower than the specific resistance of the material covered with a passive film, r_{pass} , outside of the sliding track. Therefore, if A_{tr} is not too small a fraction of the total area A_0 of the sample, the resistance R_{act} related to the sliding track is small compared to the resistance R_{pass} of the area of the sample remained in passive state, outside of the sliding track: if $R_{act} << R_{pass}$, according to equation (29), the measured resistance R_{ps} gives then straight an approximate value of R_{act}.

4.2.4 Third step in the testing protocol: Electrochemical tests on a partially active sliding track during sliding

In the preceding steps, two extreme cases were characterized, namely on the passive material and on the active one repectively. Under tribocorrosion conditions at high latency time, the surface of the material undergoes sequential events of depassivation and repassivation in-between successive contacts. This means that a part of the surface at any given time repassivates progressively. The latency time is then selected so that the regrowth of a surface film between two successive contact events is not anymore negligible as it was the case under sliding at low latency time. To achieve partially active sliding tracks, the latency times can be selected as $t_{lat2} = t_{reac}/1000$ and $t_{lat3} = t_{reac}/100$ (Diomidis et al., 2009). As a result of the increase of the latency time in this step, the wear track can now be assumed to consist of two distinct zones (Diomidis et al., 2010), namely:

- a fraction of the sliding track from which the initial passive film has been removed during sliding. In this area, the test material may be either bare, or covered by a reaction layer different from the initial passive film. This area is named the active area, A_{act}., and
- the remaining sliding track area covered by a surface film that is in the same state as the surface before sliding. This film is either not removed by the counterbody during sliding, or it had the time to restore in its initial state. This area is referred to as the repassivated area, A_{repass} with:

$$A_{tr} = A_{act} + A_{repass} \tag{34}$$

It must be stressed that under continuous sliding, these active and repassivated areas remain constant because of stationary electrochemical state conditions. Under intermittent sliding, these active and repassivated areas on the sliding track evolve with time between two successive contact events since a gradual increase of the coverage of the repassivated area takes place within the "off period". By hypothesis, in both cases, the fraction of the sliding track surface covered by the passive film, A_{repass}/A_{tr}, is assumed to be constant and given by the ratio t_{lat}/t_{reac}:

$$\frac{A_{repass}}{A_{tr}} = \frac{t_{lat}}{t_{reac}} \tag{35}$$

and:

$$\frac{A_{act}}{A_{tr}} = 1 - \frac{t_{lat}}{t_{reac}} \tag{36}$$

At the latency times $t_{lat2} = 0.001\ t_{reac}$ and $t_{lat3} = 0.01\ t_{reac}$, the relationship between repassivated and total wear track area are respectively $A_{repass\,2} = 0.001 A_{tr}$ and $A_{repass\,3} = 0.01\ A_{tr}$, and thus the active wear track areas are $A_{act\,2} = 0.999\ A_{tr}$ and $A_{act\,3} = 0.99\ A_{tr}$.

4.3 Analysis and interpretation of sliding test results

The total wear W_{tr} in a wear track of area A_{tr} can be expressed as a sum of components related to both types of areas present on the track, the area A_{act} in active state, and the area A_{repass} in re-passivated state:

$$W_{tr} = W^{c}_{act} + W^{m}_{act} + W^{c}_{repass} + W^{m}_{repass} \tag{37}$$

with :

- W_{tr} material loss in wear track,
- W^{c}_{act} material loss due to corrosion of active material in wear track,
- W^{m}_{act} material loss due to mechanical wear of active material in wear track,
- W^{c}_{repass} material loss by corrosion of repassivated material in wear track, and
- W^{m}_{repass} material loss due to mechanical wear of repassivated material in the wear track.

In order to assess the values of these different components and to compare them to determine the characteristics of the wear mechanism, the following analyses have to be performed:

- experimental determination of the mass losses due to corrosion and mechanical wear of active material during sliding tests performed at low latency times,
- experimental determination of the mass losses due to corrosion and mechanical wear of active material during sliding tests performed at large latency times. In this case one must differentiate between continuous and intermittent sliding tests,
- experimental determination of the mass losses due to corrosion and mechanical wear of repassivated material during sliding tests performed at large latency times since they allow the study of the periodic removal and re-growth of a passive surface film,

Detailed information on the analysis and related interpretation of sliding tests can be found in a Handbook on Tribocorrosion (Celis & Ponthiaux, 2011).

5. Conclusions

Tribocorrosion is the degradation of material surfaces by the combined action of corrosion, electrochemical passivation, and external mechanical interactions. It is essentially a surface related process, but some events like hydrogen evolution and absorption by the material, can modify the mechanical properties of the sub-surfaces on materials. Under conditions where tribocorrosion is active, the material loss depends in a complex way on many parameters like the tribological conditions in the contact, the composition of the environment, the temperature, the flow rate, the pH and eventually the applied potential. By analogy it is possible to extend this concept when, in the process previously described, a chemical or physical adsorption of inhibiting species strengthens or replaces the electrochemical passivation process. The successive repetition of some of these processes can lead to a possible synergism between mechanical stress and the effect of the environment what results in a damage of surfaces and systems through an accelerated loss of functionality.

Electrochemical methods used for corrosion studies are of interest in tribocorrosion since they allow the *in situ* monitoring and analysis of the interactions between surfaces and their environment as well as changes induced by a mechanical action like sliding or impact. In combination with conventional tribological measurements, they allow to understand the evolution of the surface state in a time space. The choice of methods to be used depends on the type of mechanical action. The data interpretations must be adapted to the heterogeneous state of surfaces undergoing a mechanical interaction, and also to the contact conditions.

Taking into account the surface state heterogeneity that results from successive mechanical and corrosive interactions, the development of an analysis of local phenomena on surfaces is needed to predict the impact of varying environmental conditions, tribological, and even geometrical ones on the operating behavior of a tribological system. Besides the methods already mentioned above, *ex situ* techniques to characterize surfaces, like the determination of residual stresses, micro-and nano-hardness, topography by 3D profilometry, chemical and structural micro-analysis of surfaces, microscopy at different length scales, must be implemented to acquire the spatial response of materials subjected to combined corrosion and mechanical loadings.

6. References

Adler T.A. & Walters R.P. (1996). Corrosion and wear of 304 stainless steel using a scratch test. *Corrosion science,* Vol 33, No. 12, pp. (1855-1876), ISSN 0010-938X.

Bom Soon Lee, Han Sub Chung, Ki-Tae Kim, Ford F.P. & Andersen P.L. (1999). Remaining life prediction methods using operating data and knowledge on mechanisms. *Nuclear Engineering and Design,* Vol. 191, No. 2, pp. (157-165), ISSN 0029-5493.

Boutard D., Wenger F., Ponthiaux P., & Galland J. (1985). Incidence de méthodes expérimentales diverses sur les diagrammes d'impédance électrochimique d'un alliage fer-31% nickel en cours de corrosion. *Proceedings of 8th European Corrosion Congress,* ISBN 2-88074-228-5, Nice (France), November 1985.

Carton J. F., Vannes A. B. & Vincent L. (1995). Basis of a coating choice methodology in fretting. *Wear,* Vol 185, No.5, 1995, pp. (47-57), ISSN 0043-1648.

Celis J.P. & Ponthiaux P. editors. (2011), *Testing tribocorrosion of passivating materials supporting research and industrial innovation: Handbook", EFC Green Book N° 62,* Maney, ISBN 978 1 907975 20 2, Leeds (UK).

Cottis R. A. & Loto C. A. (1990). Electrochemical noise generation during SCC of a high-strength carbon steel. *Corrosion,* Vol. 46, No.1, pp. (12-19), ISSN 0010-9312.

Déforge D., Huet F., Nogueira R.P., Ponthiaux P., and Wenger F. (2006). Electrochemical noise analysis of tribocorrosion processes under steady-state friction regime. *Corrosion,* Vol. 62, No 6, pp. (514-521), ISSN 0010-9312.

Diomidis N., Celis J.-P., Ponthiaux P. & Wenger F. (2009). A methodology for the assessment of the tribocorrosion of passivating metallic materials. *Lubrication Science,* Vol. 21, No 2, pp. (53-67), ISSN 0954-0075.

Diomidis N., Celis J.P., Ponthiaux P. & Wenger F. (2010). Tribocorrosion of stainless steel in sulfuric acid: Identification of corrosion–wear components and effect of contact area. *Wear,* Vol. 269, No 1-2, pp. (93-103), ISSN 0043-1648.

Garcia, I., Drees, D. & Celis, J.-P. (2001). Corrosion-wear of passivating materials in sliding contacts based on a concept of active wear track area. *Wear,* Vol. 249, No 5-6, pp. (452-460), ISSN 0043-1648.

Godet M., Berthier Y., Lancaster J. & Vincent L. (1991). Wear modelling : using fundamental understanding or practical experience ?, *Wear*, Vol. 149, No 1-2, pp. (325-340), ISSN 0043-1648.

Jemmely, P., Mischler, S. & Landolt, D. (2000). Electrochemical modelling of passivation phenomena in tribocorrosion. *Wear*, Vol. 237, No. 1, pp. (63-76), ISSN 0043-1648.

Law C. G., & Newman J. (1979). A model for the anodic dissolution of iron in sulfuric acid. *Journal of the Electrochemical. Society.*, Vol. 126, No.12, pp. (2150-2155), ISSN 0043-1648.

Lemaire E. & Le Calvar M. (2000). Evidence of tribocorrosion wear in pressurized water reactors. *Wear*, Vol. 249, No. 5-6, pp. (1-7), ISSN 0043-1648.

Lillard R. S., Kruger J., Tait W. S. & Moran P. J. (1995). Using local electrochemical impedance spectroscopy to examine coating failure. *Corrosion*, Vol. 51, No. 4, pp. 251-259 ISSN 0010-9312.

Lim, S.C. & Ashby, M.F. (1987). Wear-mechanism maps. *Acta Metallurgica*, Vol. 35, No. 1, pp. (1-24), ISSN 1359-6454.

Mischler S., Ayrault S., Debaud S., Jemmely Ph., Rosset E. & Landolt D. (1997). Aspects physico-chimiques de la tribocorrosion. *Matériaux et Techniques*, Vol. HS, No. July, pp. (5-10), ISSN 0032-6895.

Mischler, S., Debaud, S. & Landolt, D. (1998). Wear-accelerated corrosion of passive metals in tribocorrosion systems. *Journal of the Electrochemical. Society.*, Vol. 145, No. 3, pp. (750-758), ISSN 0043-1648.

Oltra R. & Keddam M. (1990). Application of E.I.S. to localized corrosion. *Electrochimica Acta*, Vol. 35, No. 10, pp. (1619-1629), ISSN 0013-4686.

Oltra R., Gabrielli G., Huet F. & Keddam M. (1986). Electrochemical investigation of locally depassivated iron. A comparison of various techniques. *Electrochimica Acta*, Vol. 31, No. 12, pp. (1501-1511), ISSN 0013-4686.

Ponthiaux P., Wenger F. & Galland J. (1995). Study of the anodic current-voltage curve of an iron-nickel alloy in normal sulfuric acid. *Journal of the Electrochemical. Society.*, Vol. 142, No. 7, pp. (2204-2210), ISSN 0043-1648.

Ponthiaux P., Wenger F., Galland J., Lederer G. & Celati N. (1997). Utilisation du bruit électrochimique pour déterminer la surface dépassivée par frottement. Cas d'un acier Z2 CND 17-13 en milieu chloruré (NaCl 3%), *Matériaux et Techniques*, Vol. HS, No. July, pp. (43-46), ISSN 0032-6895.

Ponthiaux P., Wenger F., Galland J., Kubecka P. & Hyspecka L. (1999). Effets combinés du frottement et de la corrosion dans le cas d'un alliage fer-nickel en milieu sulfurique. *Matériaux et Techniques*, Vol. HS, No. December, pp. (11-15), ISSN 0032-6895.

Pourbaix M. (1974), *Atlas of Electrochemical Equilibria in Aqueous Solutions*, National Association of Corrosion Engineers, ISBN 0915567989, Houston (USA).

Quinn, T.F.J. (1992). Oxidational wear modelling: part I. *Wear*, Vol. 153, No. 1, pp. (179-200), ISSN 0043-1648.

Quinn, T.F.J. (1994). Oxidational wear modelling: part II. The general theory of oxidational wear. *Wear*, Vol. 175, No. 1-2, pp. (199-208), ISSN 0043-1648.

Rosset E. (1999). Tribologie Systémique. *Oberflächen-Polysurface*, Vol. 1, pp. (7-9), ISSN 1422-3511.

Smith A.F. (1985). Sliding wear of AISI 316 stainless steel in air, 20 - 500°C. *Tribology International*, Vol. 18, No. 1, pp. (35–43), ISSN 0301-679X.

Stemp M., Mischler S. & Landolt D. (2003). The effect of mechanical and electrochemical parameters on the tribocorrosion rate of stainless steel in sulfuric acid. *Wear*, Vol. 255, No.1-6, pp. (466–475), ISSN 0043-1648.

Takadoum J. (1996). The influence of potential on the tribocorrosion of nickel and iron in sulfuric acid solution. *Corrosion Science*, Vol. 38, No. 4, pp. (643–654), ISSN 0010-938X.

Tiedemann W. H., Newman J. & Bennion D. N. (1973). The errors in measurements of electrode kinetics caused by nonuniform ohmic-potential drop to a disk electrode. *Journal of the Electrochemical. Society.*, Vol. 120, No. 2, , pp. (256-258), ISSN 0043-1648.

Uruchurtu J. C. & Dawson J. L. (1987). Noise analysis of pure aluminium under different pitting conditions. *Corrosion*, Vol. 43, No. 1, pp. (19-25), ISSN 0010-9312.

Watson, S.W., Friedersdorf, F.J., Madsen, B.W. & Cramer, S.D. (1995). Methods of measuring wear-corrosion synergism. *Wear*, Vol. 181-183, No. 2, pp. 476-484, ISSN 0043-1648.

Zhang J., Wenger F. & Galland J. (1987). Contrôle de l'état local de corrosion de structures métalliques de grandes dimensions par les mesures d'impédance électrochimiques. *Comptes Rendus de l'Académie des Sciences, Paris,* Série 2, Vol. 304, No. 14, pp. (797-800), ISSN 12518069.

Corrosion Resistance of Pb-Free and Novel Nano-Composite Solders in Electronic Packaging

L.C. Tsao
*Department of Materials Engineering,
National Pingtung University of Science & Technology, Neipu, Pingtung,
Taiwan*

1. Introduction

Tin-lead (Sn-Pb) alloys for metal interconnections were first used about 2000 years ago. Recently, the use of alloys has become essential for the interconnection and packaging of virtually all electronic products and circuits. Sn-Pb solder alloys have been widely used in the modern electronics industry because of their low melting points, good wettability, good corrosion resistance, low cost, reasonable electrical conductivity, and satisfactory mechanical properties. However, due to health concerns, recent legislation, and market pressures [1], the electronic industry is moving toward green manufacturing as a global trend. In the area of packaging, mainly driven by European RoHS (Reduction of Hazardous Substances), lead was banned effective July 1, 2006, except in some exempt items. In addition, Pb and Pb-containing compounds, as cited by the Environmental Protection Agency (EPA) of the US, are listed among the top 17 chemicals posing the greatest threat to human life and the environment [2] because of lead's toxicity [3]. In the electronics industry, the lead generated by the disposal of electronic assemblies is considered hazardous to the environment. Therefore, developing viable alternative Pb-free solders for electronic assemblies is of principal importance.

2. Lead-free solder systems

Although several commercial and experimental Pb-free solder alloys are available as replacements for Sn-Pb solders, the following families of solders are of particular interest and are the prevailing choices of industry [4]: eutectic Sn-Ag, eutectic Sn-Cu, eutectic Sn-Zn, eutectic Bi-Sn, and Sn–In, as shown in Table 1. Since the properties of the binary Pb-free solders cannot fully meet the requirements for applications in electronic packaging, additional alloying elements are added to improve the performance of these alloys. Thus, ternary and even quaternary Pb-free solders have been developed [5-7], such as Sn-Ag-Cu, Sn-Ag-Bi, and Sn-Zn-Bi solder. However, the knowledge base on Sn-Pb solders gained by experience is not directly applicable to lead-free solders. In other words, the reliability of Pb-free solder joints in consumer products is attracting more interest and concern from both academia and technologists[8-10].

All system	Eutectic composition (wt.%)	Melting point or range (°C)
Sn-In	Sn52In	118(e)
Sn-Bi	Sn58Bi	138(e)
Sn-Zn	Sn9Zn	198.5(e)
Sn-Ag	Sn3.5Ag	221(e)
Sn-Cu	Sn0.7Cu	227(e)
Sn-Ag-Bi	Sn3.5Ag3Bi	206-213
Sn-Ag-Cu	Sn3.8Ag0.7Cu	217(e)
	Sn3.50.5Cu	218
Sn-Zn-Bi	Sn8Zn3Bi	189-199

Table 1. Data showing the enhancement of the mechanical properties of Pb-free solders[9, 10].

3. Nano-composite solders

As electronic devices continue to become lighter and thinner, they require much smaller solder joints and fine-pitch interconnections for microelectronic packaging. For example, portable electronic devices, such as portable computers and mobile phones, have become thinner and smaller while adding more complicated functions. The miniaturization of these electronic devices demands better solder-joint reliability. Hence, in all chip connection and ball grid array (BGA) technologies, solder interconnection through flip-chip assembly has been proven to offer the highest density of input/output (I/O) connections in a limited space. To meet the insatiable appetite for ever-finer I/O pitches and ever-higher I/O densities, C4 (controlled collapse chip connection) technology was developed by IBM in the mid 1960s, and this technology was applied to future microelectronic packaging. According to the International Technology Roadmap for Semiconductors (ITRS), the pad pitch may fall below 20 μm by the year 2016 [11]. In some flip chip packages, solder balls of 20μm in size are used to connect the pads on the chip and the print circuit board (Fig. 1). Furthermore, Thru-Silicon-Via (TSV) technologies are also lurking on the horizon as the next-generation higher-density chip connection technology, and they also require fine-pitch Pb-free solder interconnections.

The conventional solder technology may not guarantee the required performance at such pitches due to characteristics such as higher diffusivity and softening [12]. In order to solve these problems, efforts have been made to develop new Pb-free solders with a low melting point, good mechanical properties, better microstructure properties, and high creep resistance. Recently, Pb-free solders doped with nano-sized, nonreacting, noncoarsening oxide dispersoids have been identified as potential materials that could provide higher microstructure stability and better mechanical properties than the conventional solders [13-24]. Tsao et al. [14-16] studied the influence of reinforcing TiO_2 and Al_2O_3 nanoparticles on microstructural development and hardness of eutectic Sn-Ag-Cu solders. In their work, microhardness measurements revealed that the addition of TiO_2 and Al_2O_3 nanoparticles is helpful in enhancing the overall strength of the eutectic solder. Shen et al. [17] controlled the formation of bulk Ag_3Sn plate in Sn-Ag-Cu solder by adding ZrO_2 nanoparticles to reduce the amount of undercooling during solidification and thereby suppress the growth of bulk Ag_3Sn plates. Zhong and Gupta [18] successfully prepared a nano-Al_2O_3 reinforced nano-composite solder by mechanically intermixing nano-Al_2O_3 particles into Sn0.7Cu Pb-free solder, and this composite solder shows improved mechanical properties. The best tensile strength realized for the composite, which contains 1.5 wt.% alumina, far exceeds the strength of the eutectic Sn–Pb solder. Many authors have studied the effect of adding single-walled carbon nanotubes [19] or multi-walled carbon nanotubes [20, 21] on the mechanical properties of nano-composite solders. The data on the enhancement of the mechanical properties of nano-composite solders collected from some of the literature are listed in Table 2 [13, 14, 16, 22, 23]. Here, it should be stressed that although the addition of nanoparticles into solder matrices can improve the creep behavior[24], the effects on the corrosion resistance and mechanical properties of the nano-composite solders cannot be ignored.

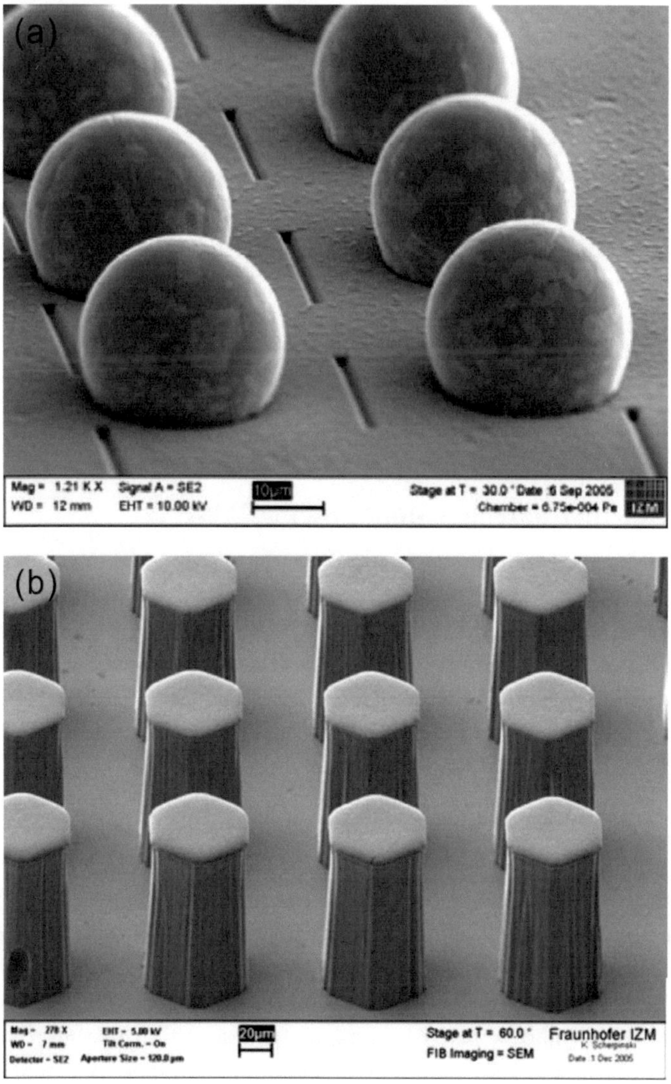

Fig. 1. Micro bump and pillar bump structures for highly reliable chip-to-substrate interconnects: (a) SnAg microbump (20 μm diameter), and (b) Cu pillarbump (height: 80 μm) [11]

Solder matrix	Reinforcement nanoparticles	Mechanical properties			
		0.2%YS (MPa)	UTS (MPa)	Elongation (%)	References
Sn4In4.1Ag0.5Cu	Nil	56±6	60±8	37±7	[22]
	1.0 vol.% Al_2O_3	72±6	75±6	21±3	
	3.0 vol.% Al_2O_3	73±3	77±6	11±3	
	5.0 vol.% Al_2O_3	74±3	76±2	10±0	
Sn3.5Ag0.7Cu	Nil	31±2	35±1	41±8	[23]
	0.01wt.% MWCNTs	36±2	47±1	36±2	
	0.04wt.% MWCNTs	36±4	46±6	37±2	
	0.07wt.% MWCNTs	33±3	43±5	35±4	
Sn3.5Ag0.5Cu	Nil	45.96±1.14	54.34±1.42	49.2±1.3	[16]
	0.25 wt.% Al_2O_3	48.81±1.23	60.20±1.84	47.3±0.8	
	0.5 wt.% Al_2O_3	52.56±1.56	62.44±1.76	44.0±1.2	
	1.0 wt.% Al_2O_3	57.22±1.8	68.05±1.63	43.5±2.1	
	1.5 wt.% Al_2O_3	61.45±2.3	70.05±2.06	32.5±3.2	
Sn3.5Ag0.25Cu	Nil	53.2	55.7	48.6	[14]
	0.25 wt.% TiO_2	59.5	61.5	40.5	
	0.5 wt.% TiO_2	67.6	69.1	32.1	
	1.0 wt.% TiO_2	69.3	70.1	25.2	

Table 2. The data showing the enhancement of the mechanical properties of nano-composite solders[13, 14, 16, 22, 23].

4. The interfacial intermetallic compound (IMC) layers

In connected metals, all the common base materials, coatings, and metallizations, such as Cu, Ni, Ag, and Au, form intermetallic compounds (IMC) with Sn, which is the major element in Sn solders. Cu is the material most frequently used for leads and pads on flip chip substrates and printed wiring boards. It is now known that in the solder/Cu interfacial reaction, Sn reacts rapidly with Cu to form Cu_3Sn (ε-phase) and Cu_6Sn_5 (η-phase) [25]. Other metal substrate/solder interfacial reactions form IMCs, such as Ag_3Sn[26] (Sn solder/Ag), Sn-Ni [27] (Sn solder/Ni), Ag-In[28] (In solder/Ag) and Cu-In IMC[29] (In solder/Cu). These intermetallic compounds are generally more brittle than the base metal, which can have an adverse impact on the solder joint reliability. Excessive thickness may also decrease solder joint ductility and strength [30-34]. Recently, we found that a great number of nano-Ag_3Sn particles form on the Cu_6Sn_5 IMC when the solders contain Ag_3Sn precipitate phase after a Pb-free Sn3.5Ag0.5Cu (SAC) nano-composite solder/Cu substrate interface reaction[30, 31]. These nanoparticles apparently decrease the surface energy and hinder the growth of the Cu_6Sn_5 IMC layer during soldering and aging. All these results indicate that Gibbs absorption theory can be used to explain the formation of these nanoparticles and their effects on the surface energy of the IMC. Many studies have reported that nano-sized, nonreacting, noncoarsening oxide dispersoid particles, such as TiO_2 [30-32], Al_2O_3 [33], Y_2O_3[34], CNTs [35], and ZrO_2[36] can affect the growth rate of interfacial IMC.

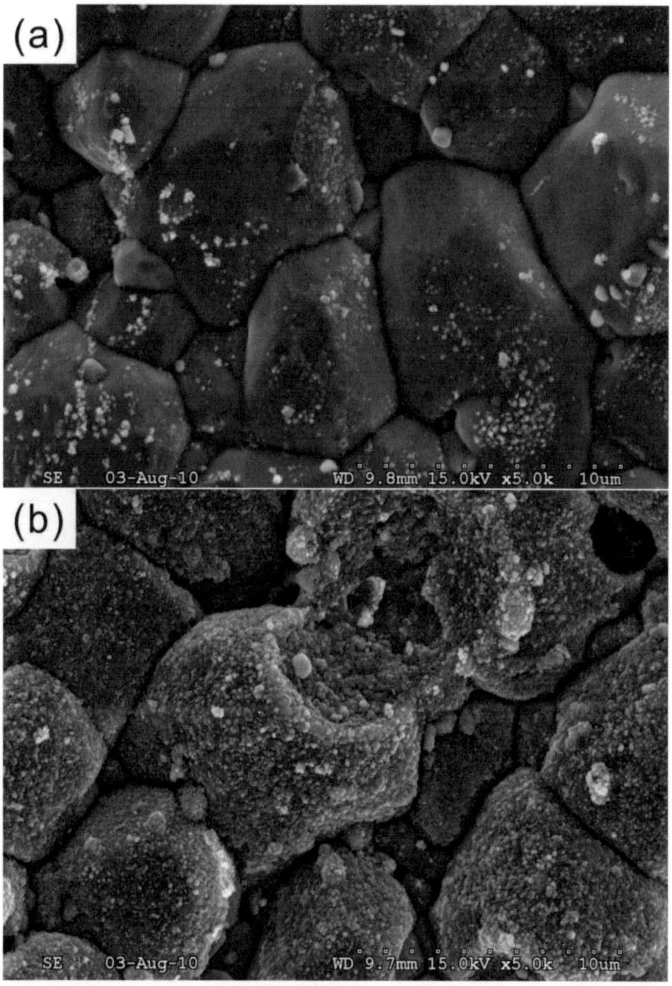

Fig. 2. Top view of the IMC at the interfaces of the nano-composite solder joints on Cu substrate after aging for 7 days at 175°C: (a) SAC and SAC- TiO_2 [31].

5. Corrosion behavior of Pb-free solder joints

The diversity of materials, drive toward miniaturization, and globalization have significantly contributed to the corrosion of microelectronic devices [37]. However, the key point is that solder joints are often exposed to corrosive environments that can accelerate the corrosion process. Although corrosion resistance is an important parameter in choosing solder alloys, the corrosion behavior of Sn-Pb solder joints was rarely of interest because the oxide that forms on the tin-lead alloy is relatively stable. Mori et al. showed that both Pb-rich and Sn-rich phases dissolve when the Sn-Pb solder alloy is immersed in corrosive solution, and the corrosion rate is slower than that of the Sn-Ag solder [38, 39]. Compared to traditional Sn-Pb solders, Sn-Ag-Cu solders are easily corroded in corrosive environments due to their special structures (as shown in Fig. 3). The presence of Ag_3Sn in Sn-Ag-Cu solders accelerates the dissolution of tin from the solder matrix into a corrosive medium

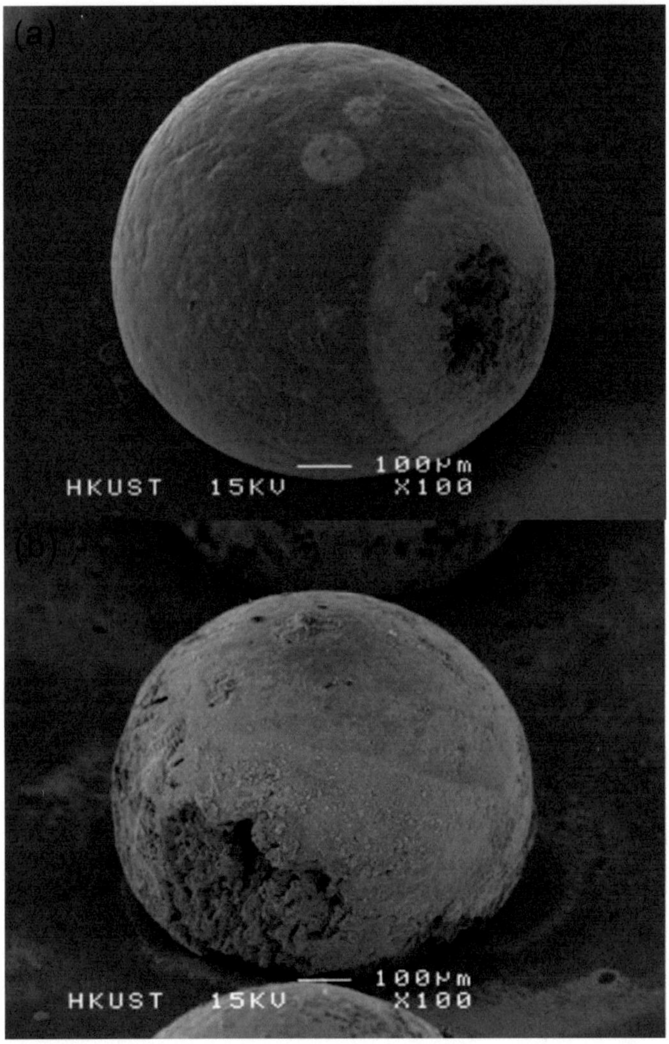

Fig. 3. Surface morphology changes of solder balls after the salt spray test for 96 hrs: (a) Sn-Pb solder, and (b) SAC solder[39].

because of the galvanic corrosion mechanism [39]. When corrosion occurs in the solder joints, it may change the microstructure of corroded regions and provide crack initiation sites, thereby decreasing the mechanical properties of the joints. Lin and Lee have investigated both Sn-Pb and Sn-Ag-Cu solder alloy wafer-level packages, with and without pretreatment by 5% NaCl salt spray, with thermal cycling to failure. The salt spray test did not reduce the characteristic lifetime of the Sn-Pb solder joints, but it did reduce the lifetime of the Sn-Ag-Cu solder joints by over 43% (Fig.4). The characteristic lifetime cycle number was 1384 for the as-assembled and non-salt spray treated components, but it was only 786 for the components which were treated in 5 wt.% NaCl salt spray for 96 h. In addition, the presence of multiple corrosion sites per solder joint poses an additional risk factor to the structural stability of the joint, for corrosion sites are all potential crack initiation sites.

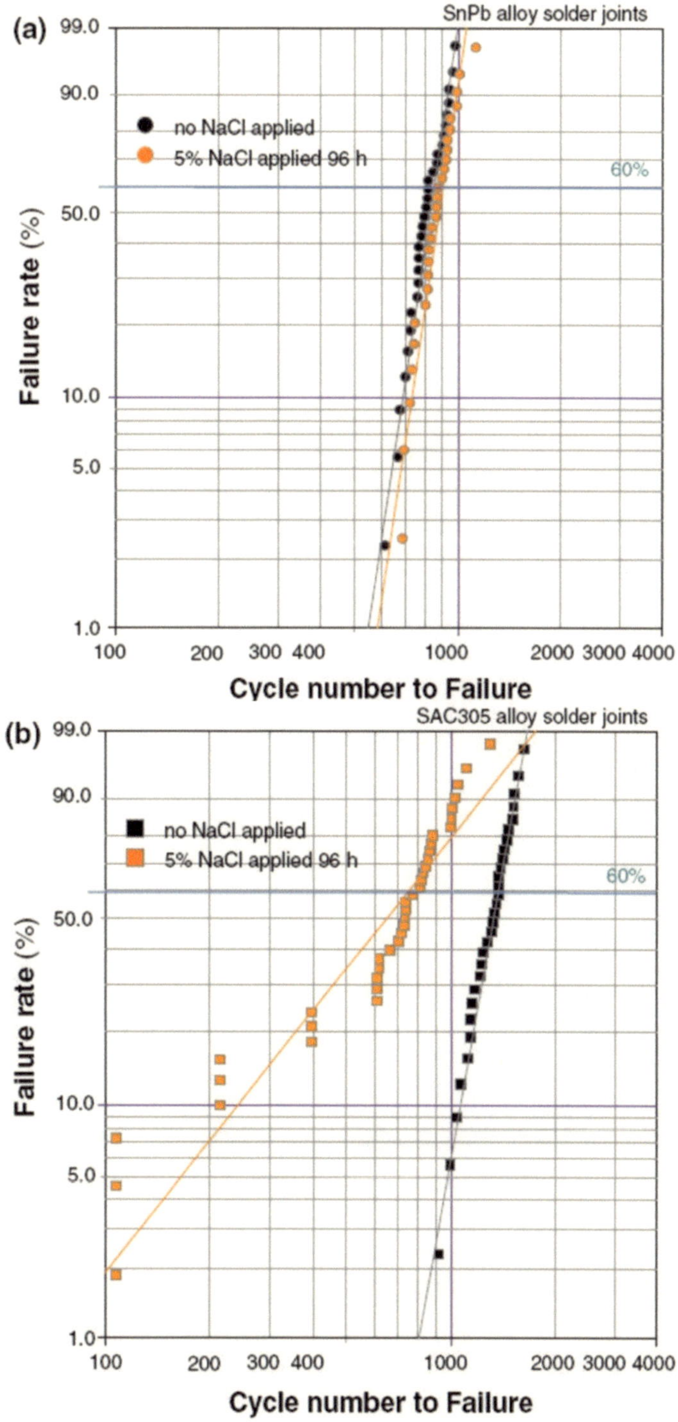

Fig. 4. Weibull plot for the thermal cycling results on 5 wt.% NaCl aqueous solution (salt spray) treated WLCSP: (a) Sn-Pb solder alloy samples and (b) SAC305 solder alloy samples [40].

Unlike Sn-Pb joints, which have a dual phase structure and block the path of corrosion due to the existence of phase boundaries, the SAC305 joint is basically pure Sn with coarse islands of Ag_3Sn and Cu_6Sn_5 intermetallic precipitate (Fig. 5). A corrosion crack can propagate and lead to additional corrosion along the way, without interruption from the Sn phase structure. Although both materials show strong resistance to corrosion, the localized nature of the corroded area at critical locations causes significant degradation in Sn-Ag-Cu solder joints[40].

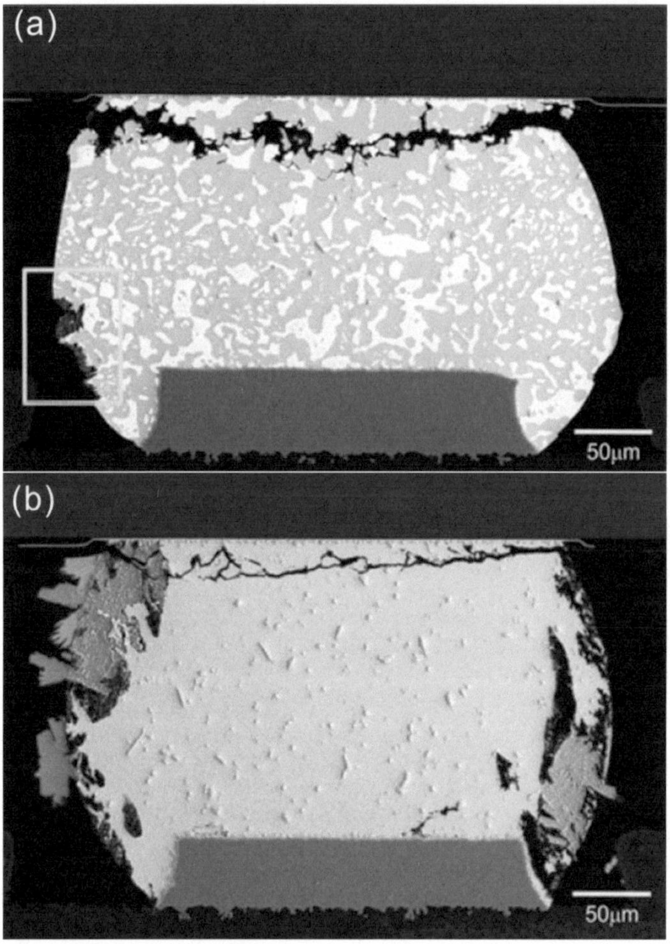

Fig. 5. Cross-section SEM microstructure after salt spray treatment and then thermal cycling: (a) Sn-Pb, and (b) SAC305 solder joint [40].

6. Galvanic corrosion of soldering

Corrosion of solder alloys, in the presence of a suitable electrolyte can occur either due to the potential difference between the major phases in the alloy or galvanic coupling between one or more phases of the alloy and other parts of the microelectronics device. Some metals that are frequently used in microelectronics are Cu, Au, Ag, Ni and Pd. The standard emf for these metals and metals used in solder alloys are listed in Table 3[4]. Especially, advanced packaging technologies make the solder alloy susceptible to corrosion problems

[41]. Thus, in the electronics industry, corrosion has become a significant factor in recent years because of the extremely complex systems that have been developed and the increasing demand on their reliability [42, 43]. For example, using Cu and Sn metals allows fine-pitch interconnections to be fabricated at relatively low cost. These features make Cu-Sn based SLID bonding very appealing for 3D stacked applications (Fig. 6) [44].

Metals used in solder	**Metals used in microelectronics**				
	Au	Ag	Cu	Ni	Pd
Sn	1.636	0.935	0.473	-0.114	1.123
Pb	1.626	0.925	0.463	-0.124	1.113
In	1.842	1.141	0.679	0.092	1.329
Zn	2.263	1.562	1.10	0.513	1.75

Table 3. Δemf values for metals commonly used in microelectronics[4].

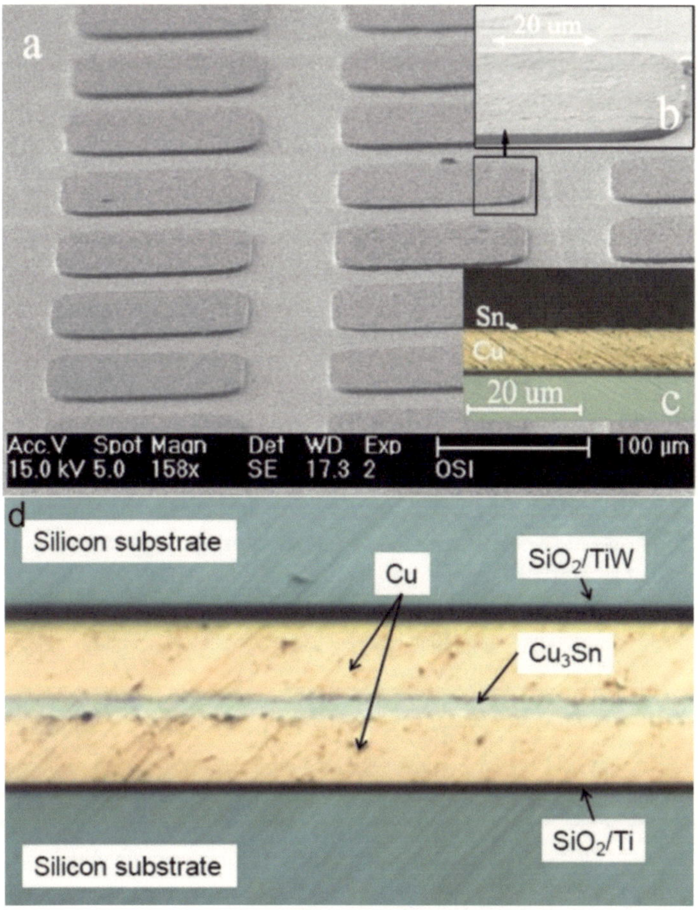

Fig. 6. Electroplated pads of 5 µm Cu and 200 nm Sn: (a) and (b) SEM image with different magnification; (c) Cross-section view under optical microscope; and (d) Cross-section view of a fluxless bonded Cu/Sn Interconnect [44].

The joining of materials with solders generally results in a multi-layer structure in which IMC are formed between substrate and solders. Such a structure in a flip chip package is a galvanic couple. The galvanic corrosion behavior of the solder bump structures have a great effect upon reliability[45]. For instance, the galvanic current densities of the Sn solder with respect to the IMC Cu_6Sn_5 and Cu_3Sn, and base Cu have been investigated (Fig. 7). It appears that Sn solder has a greater galvanic current density and thus is very subject to corrosion, and it is especially so in coupling with the formation of Cu_3Sn layers than with Cu_6Sn_5 layers. The galvanic current densities of the Sn37Pb solders of Cu_3Sn, Cu, and Cu_6Sn_5 are about 38, 16, and 5 (μA/cm²), respectively.

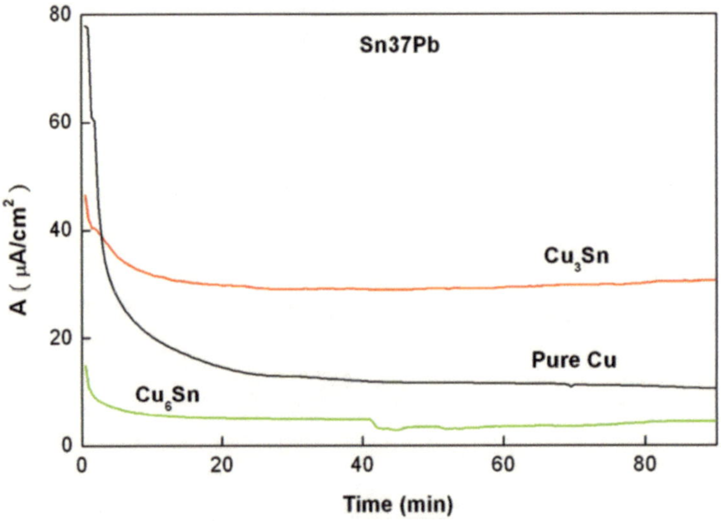

Fig. 7. The galvanic current densities of the solder with respect to intermetallic compounds Cu_6Sn_5 and Cu_3Sn, and Cu substrate, in a 3.5 wt.% solution [45].

Increasing the copper content, which reacts with Sn to form IMC, significantly improves the corrosion resistance of solders and increases the corrosion current density (I_{corr}), as shown in Fig. 8, 9 and Table 4. At above 460 mV_{SCE}, the passivation current densities of all specimens are around 10^{-1}A/cm², with the declining sequence of Sn37Pb ≥ Cu_6Sn_5 > Cu_3Sn > Cu.

Specimens	Φ_{corr} (mV_{SCE})	Φ_b (mV_{SCE})	$\Delta\Phi$ (mV)	I_{corr} (μA/cm²)	I_p (mA/cm²)
Sn37Pb	-584.4	-303.0	281	6.48	67.7
Cu_6Sn_5	-457.7	-45.0	412	2.61	56.9
Cu_3Sn	-309.0	-8.9	300	48.17	18.3
Cu	-192.1	236	428	391.6	6.5

$\Phi_{corr.}$: corrosion potential; $I_{corr.}$: corrosion current density; Φ_b : breakdown potential;
$\Delta\Phi = \Phi_{corr.} - \Phi_b$, Φ_p: passivation range of solder alloy;
I_p: passivation current density at above 460 mV_{SCE}.

Table 4. Corrosion properties in a 3.5 wt.% NaCl solution for the Sn37Pb solder, Cu_6Sn_5 IMC, Cu_3Sn IMC and pure Cu samples [45].

It can be seen that the galvanic corrosion behavior of Cu_3Sn is generally greater than that of Cu_6Sn_5 for the flip chip package in a 3.5 wt. % NaCl solution environment. This indicates that the formation of IMC Cu_3Sn and Cu_6Sn_5 layers causes many problems with corrosion behavior and reliability.

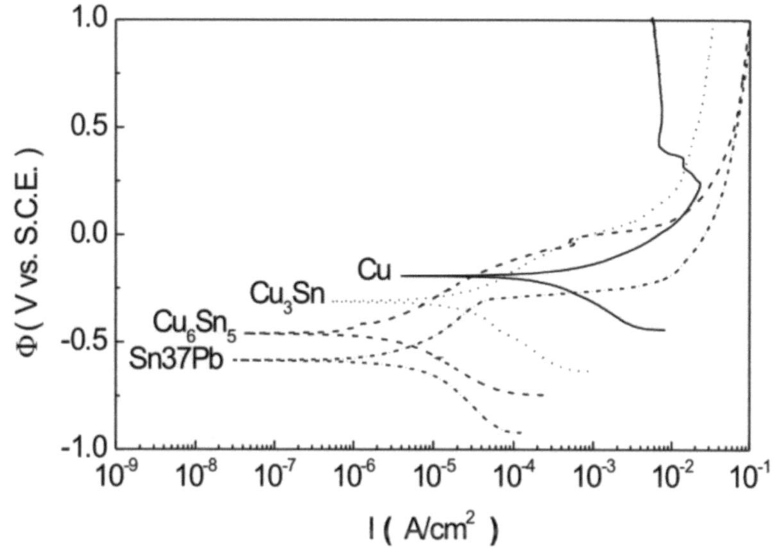

Fig. 8. The potentiodynamic polarization curves of Sn37Pb solder, Cu_6Sn_5 IMC, Cu_3Sn IMC, and pure Cu samples in a 3.5 wt.% NaCl solution [45].

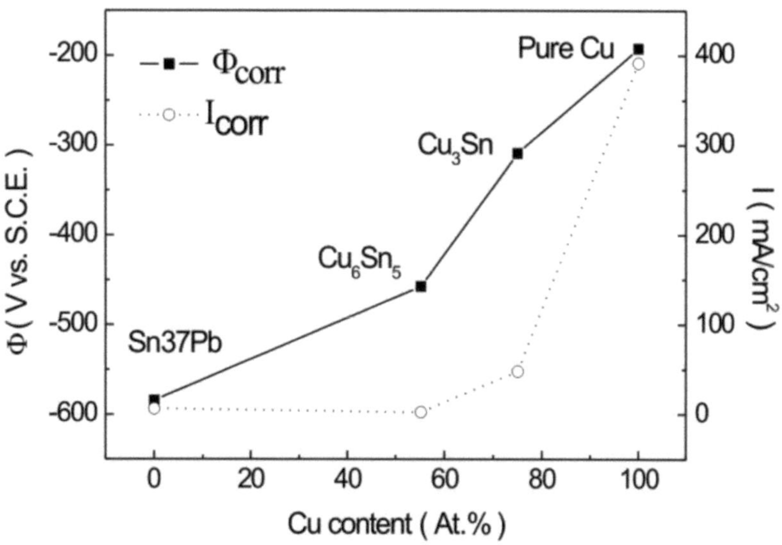

Fig. 9. Effect of Cu content on both Φ_{corr} and I_{corr} during polarization of the Sn37Pb solder, Cu6Sn5, Cu3Sn, and Cu substrate in 3.5 wt.% NaCl solution[45].

7. Corrosion behavior of Pb-free solder

Both the particular design of the electronic system, and the manner in which it is mounted in a substrate or printed wiring board, the solder connection can be exposed to the atmosphere. The solder is thus not only exposed to air, but also moisture and other corrosives such as chlorine and sulfur compounds. The ability of the solder to be able to withstand corrosion property is therefore relevant to the long-term reliability of solder joints [4]. In addition, solder alloys are electrically connected with other metallic components in the electronic device. Some metals that are frequently used in microelectronics are Cu, Au, Ag, Ni and Pd. Therefore, there is also the potential for galvanically induced corrosion of the solder, which could exacerbate any atmospheric corrosion that might be occurring. However, the properties of these lead free alloys in corrosive environments has not been widely reported, though it is of importance in many automotive, aerospace, maritime and defence applications [46]. Some researchers have studied the corrosion behaviour of Sn–Zn–X solders [47, 48] and Sn–Zn–Ag–Al–XGa [49], but few [50, 51] have studied the corrosion properties of Sn-Ag, Sn-Cu and Sn-Ag-Cu solders. Zinc is both metallurgically and chemically active. The presence of Zn in the solder alloy results in poor corrosion resistance, which is an important problem to address before practical application of this material [49]. Hence, the electrochemical corrosion behaviour of Pb-free Sn-Zn binary solder and Sn-Zn-X (X=Bi, Ag and Al) solder alloys have been investigated in NaCl solution by potentiodynamic polarization techniques[52-55]. Lin et al. [47-49] have investigated the corrosion behaviour of Sn–Zn–Al, Sn–Zn–Al–In and Sn–Zn–Ag–Al–XGa solders in 3.5% NaCl solution. They found that Sn–Zn–Al alloy [47] undergoes more active corrosion than Sn–37Pb alloy. Furthermore, they found that 5In–9(5Al–Zn)–YSn and 10In–9(5Al–Zn)–Sn alloys exhibit electrochemical passivation behaviour, and the polarization behaviours of these two alloys are similar to that of 9(5Al–Zn)–Sn alloy. Sn-Ag-M (M=In, Bi) solders exhibit poor corrosion behaviour as compared to that of Sn-Pb eutectic solder (0.1M NaCl solution)[56]. In contrast, increasing the copper content (from 0.8 to 6.7 at.%) enhances the corrosion resistance of Sn-Ag solder alloys, which exhibit improved passivity behaviour as compared to Sn-Pb eutectic solder. EPMA results indicate that the Ag_3Sn IMC is retained after the polarization test. Hence, the Ag_3Sn is more noble than the β-Sn phase. The pit formation on the surface of Sn–Ag–M alloys is due to the dissolution of the tin-rich phase. Wu et al. [51] has studied the corrosion behaviors of five solders in salt and acid solutions by means of polarization and EIS measurements. The Sn3.5Ag0.5Cu solder has the best corrosion resistivity due to the high content of noble or immune elements (Ag and Cu) and theorized stable structure, whereas the Sn9Zn and Sn8Zn3Bi solder have the worst corrosion behavior. Nevertheless, the four Pb-free solders exhibit acceptable corrosion properties, since there is not much difference in key corrosion parameters between them and the Sn37Pb solder. The corrosion data of the solders in 3.5 wt.% NaCl solutions are listed in Tables 5 [46, 51]. Lin and Mohanty et al. [46-49] studied the corrosion properties of Sn–Zn–X and Sn–Zn–Ag–Al–XGa in NaCl solution, and their results showed that the corrosion product on the surface could be SnO, SnO_2, $SnCl_2$ and ZnO, etc., depending on the applied potential. Li et al. [46] confirms that the corrosion product on the Sn–Pb and lead free solders is tin oxide chloride hydroxide ($Sn_3O(OH)_2Cl_2$).

Solder	Scanning rate	*Ecorr* (mV)	*Icorr* (A/cm²)	E_p (mV)	I_p (mA/cm²)	References
Sn37Pb	1 mV/s	-588	1.905×10^{-6}	-201	4.989	[51]
Sn9Zn		-940	2.691×10^{-5}	-326	2.938	
Sn8Zn3Bi		-1291	1.380×10^{-5}	9	8.035	
Sn3.5Ag0.5Cu		-605	5.370×10^{-7}	-236	4.083	
Sn3.5Ag0.5Cu9In		-578	7.413×10^{-6}	-158	1.524	
Sn0.7Cu	30 mV/s	-688	1.78×10^{-7}	-	0.74	[46]
Sn3.5Ag		-705	4.9×10^{-7}	-	0.49	
Sn3.8Ag0.7Cu		-727	0.89×10^{-7}	-	1.07	

Ecorr – corrosion potential, Ip – passivation current density, Icorr – corrosion current density, Ep – passive potential.

Table 5. Experimental data of the testing solders under polarization in 3.5 wt.% NaCl solution.

8. Corrosion behavior of Pb-free nano-composite solder joints

This author has recently worked on the development of nano-composite solders in microelectronic packaging by applying two methods of fabrication: mechanical mixing of inert nano-particles (Fig. 10) and precipitation of nano-IMC in the solder matrix (Fig. 11) [57]. The average size of the nominally spherical nano-Al_2O_3 particles was 100 nm in diameter.

Notably, the addition of nano-particles decreased the size of dendrite β-Sn grains, the needle-like Ag_3Sn grains, and Ag_3Sn phase located between the average spacing. When 1 wt% was added, the superfine spherical nano-Ag_3Sn grains were about 0.16 ± 0.06μm in length and 0.15 ± 0.05 μm in diameter, and the average spacing between them was a significant improvement (0.14 ± 0.05μm), significantly smaller than the sizes found in the SAC composite solder. However, large Ag_3Sn IMCs were not observed in the Pb-free SAC solder. Another, author reported that the effects of nano-TiO_2 particles on the interfacial microstructures and bonding strength of Sn3.5Ag0.5Cu nano-composite solder joints in ball grid array (BGA) packages with immersion Sn surface finishes [58]. It is clearly shown in Fig. 12a, b that the discontinuous Cu_6Sn_5 IMC layer grows with a rough scallop shape (Mark A), and wicker-Cu_6Sn_5 IMC forms on the rough scallop-shaped Cu_6Sn_5 IMC layer (Mark B) and grows into the SAC solder matrix. However, the addition of a small percentage of nano-TiO_2 particles alters the Pb-free Sn3.5Ag0.5Cu composite solder/pad interface morphology after reflowing, as shown in the SEM micrographs in Fig. 12c, d. Only the continuous scallop-shaped Cu_6Sn_5 IMC layer was detected at the interface. However, the wicker-Cu_6Sn_5 IMC disappeared at the interface with the Cu pads. In addition, the number of Ag_3Sn IMC forms increased in the eutectic area when the content of nano- TiO_2 particles was increased to 0.25–1 wt%. It is interesting that the smallest thickness of the IMC layer was achieved with the addition of 1 wt% of nano-TiO_2 particles. The thickness of the Cu_6Sn_5 IMC layer was reduced by 51%. The results indicate that the growth of the Cu_6Sn_5 IMC layer at the solder/pad interfaces of Sn3.5Ag0.5Cu is depressed through the small addition of nano-TiO_2 particles[58]. With the addition of 0.5–1 wt% nano-TiO_2 particles, fracture occurred in all of the solder joints as cracks propagated through the Sn3.5Ag0.5Cu composite solder balls, which ruptured mostly along the submicro Ag_3Sn IMC and solder matrix, as shown in Fig. 13a, b. This phenomenon is similar to that occurring in Pb-free Sn0.7Cu composite solder BGA packages[59].

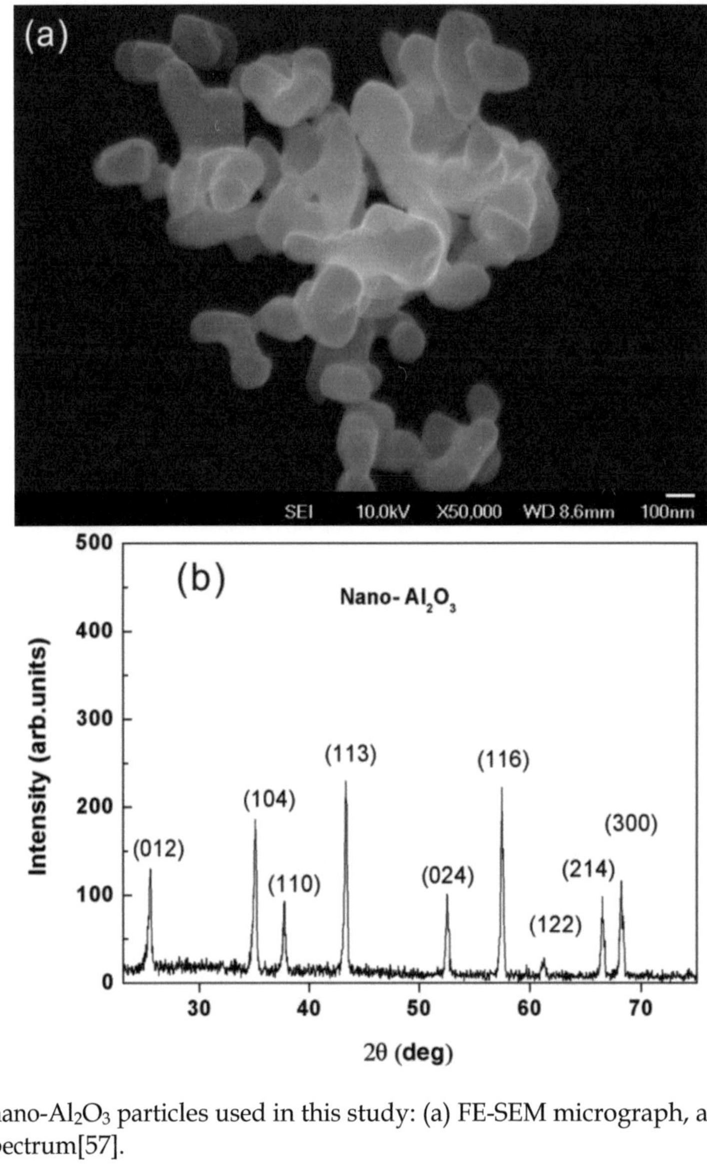

Fig. 10. The nano-Al_2O_3 particles used in this study: (a) FE-SEM micrograph, and (b) X-ray diffraction spectrum[57].

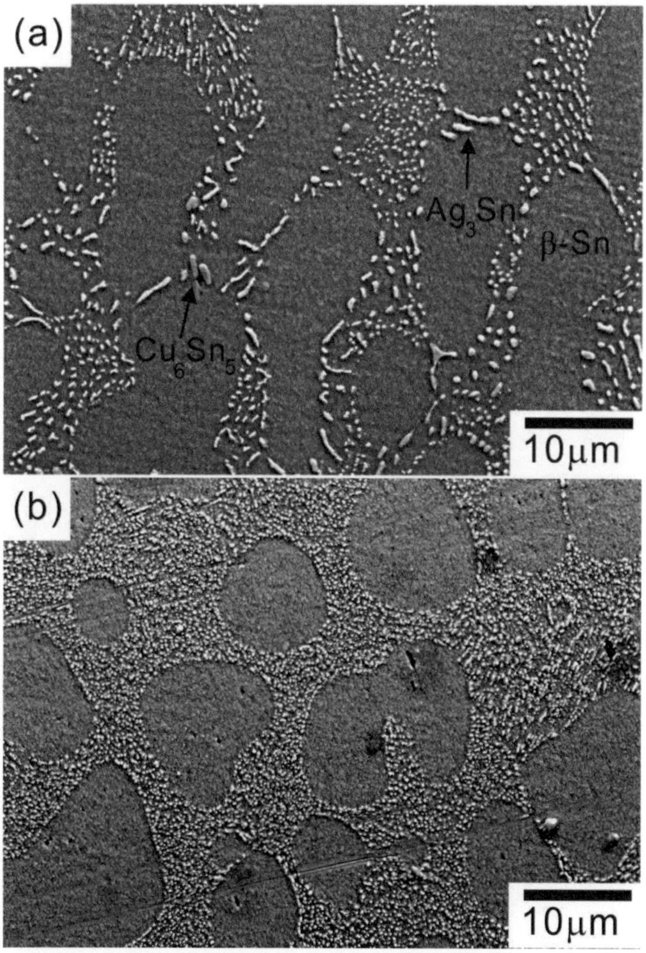

Fig. 11. SEM image of the (a) Sn3.5Ag0.5Cu solder and (b) Sn3.5Ag0.5Cu -1TiO_2 nano-composite solder[57].

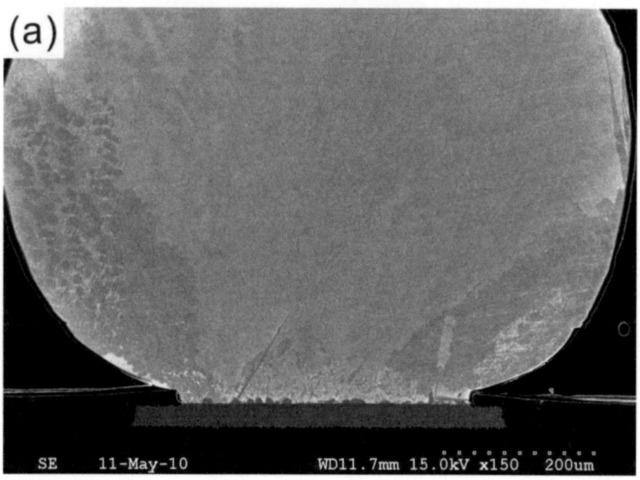

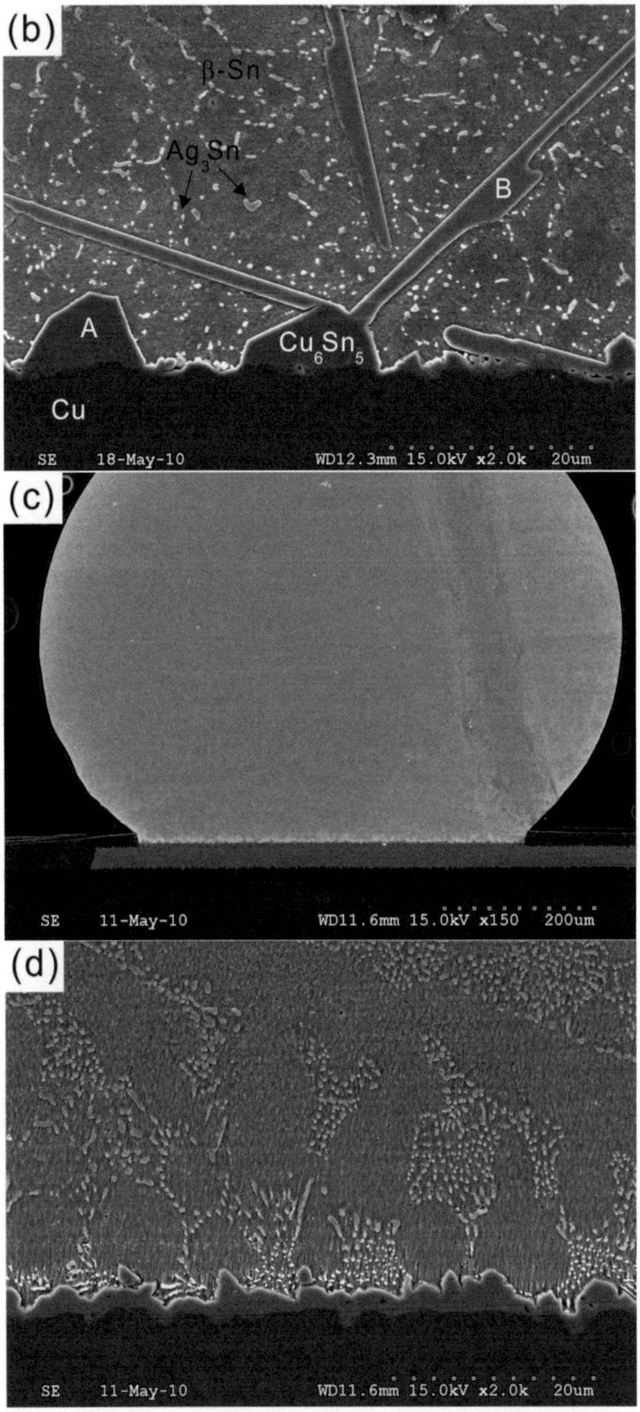

Fig. 12. Morphology of intermetallic compounds formed at the interfaces of the as-reflowed solder joints: (a) Sn3.5Ag0.5Cu, (b) (a) magnifications ; (c) d Sn3.5Ag0.5Cu-0.75TiO_2; (d) (c) magnifications [58].

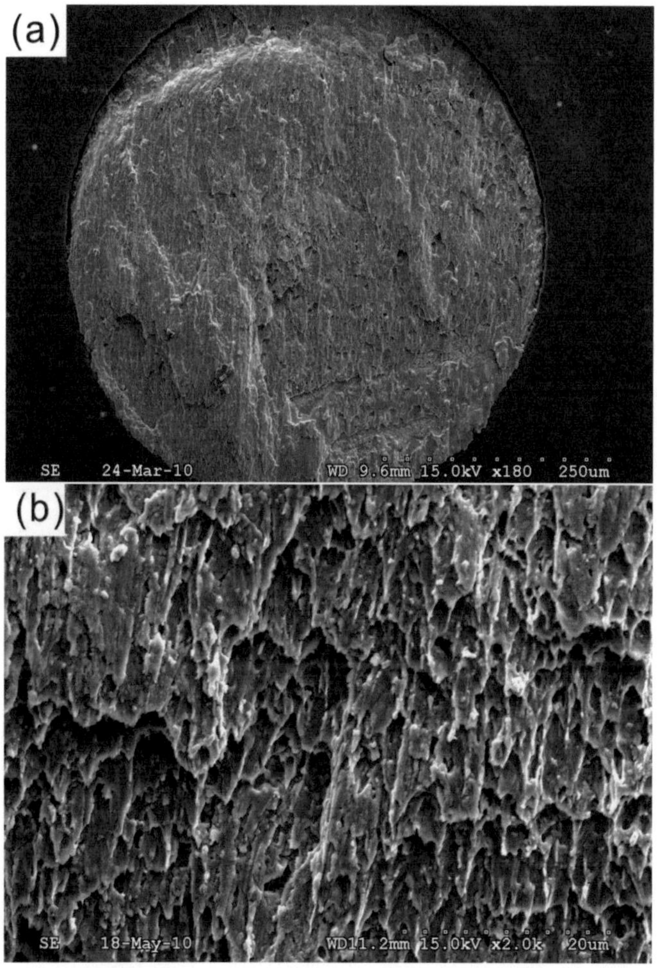

Fig. 13. Fractography of the Sn3.5Ag0.5Cu-1TiO_2 nano-composite solder joints in BGA packages after ball shear tests [58].

To achieve high reliability, solder materials must have high resistance to corrosive conditions such as moisture, air pollutants from industry, and oceanic environments[54]. Although corrosion of solder alloys is not currently a major problem for electronic devices used in normal environments, it may be a problem when they are used in harsh environments, such as oceanic environments. However, there is a lack of information regarding the corrosion resistance of nano-composite solders in corrosive environments.

Figure 14 shows the polarization curves of the Sn3.5Ag0.5Cu solder and the Sn3.5Ag0.5Cu nano-composite solder in 3.5 wt.% NaCl solution[60]. From the polarization curves, the corrosion potential (Φcorr), the breakdown potential (Φ_b), and the dynamic corrosion current density (*I*corr) have been determined (Table 6). The width of the passive region on the anodic polarization curves ($\Delta\Phi = \Phi_b$ - Φcorr) in Table 6 indicates the pitting resistibility or the stability of the passive film on the Sn3.5Ag0.5Cu composite alloy surface. The corrosion potential (*Φ*corr) of the Sn3.5Ag0.5Cu nano-composite solder is slightly more passive than that of the Sn3.5Ag0.5Cu solder. This implies that a finer grain size produces more grain *b*oundaries, which act as corrosion barriers. On the other hand, the breakdown potential (Φ_b) of the Sn3.5Ag0.5Cu nano-composite solders becomes much more passive with the addition of oxide nanoparticles. As Table 6 also indicates, the Sn3.5Ag0.5Cu solders possess a higher pitting tendency (smaller Δ*Φ* value) than the Sn3.5Ag0.5Cu nano-composite solders. Rosalbino et al. reported that the pit formation at the surface of Sn–Ag–M alloys is due to the dissolution of the tin-rich phase [56]. In addition, the corrosion current densities were obtained by using the TAFEL extrapolation method. The corrosion current densities of the Sn3.5Ag0.5Cu solders and Sn3.5Ag0.5Cu nano-composite solders were very similar.

Solder	*Φ*corr (mV_{SCE})	*Φ*b (mV_{SCE})	Δ*Φ* (mV)	*I*corr (μA/cm²)
Sn3.5Ag0.5Cu	-662.1	-284.1	378	0.36
Sn3.5Ag0.5Cu-0.5TiO2	-651.4	-95.1	556	0.27
Sn3.5Ag0.5Cu-0.5Al2O3	-642.1	-146.2	496	0.40

Φcorr : corrosion potential; Icorr: corrosion current density; Φb : breakdown potential;
ΔΦ = Φcorr - Φb.

Table 6. Corrosion properties in a 3.5 wt.% NaCl solution for the nano-composite solder [60].

Many studies have reported that the corrosion behavior of alloys depends on the second phase distribution, shown to be Mg alloy[61, 62] and Al alloy[63]. In the Sn3.5Ag0.Cu nano-composite solder alloys, the microstructure had finer β-Sn grains, a large amount of Ag_3Sn particles, and a small amount of oxidize nanoparticles. This leads to improvement of the corrosion behavior of the Sn3.5Ag0.5Cu nano-composite solder, such as greater corrosion resistance, the lower pitting tendency, and the smaller corrosion current density, respectively.

The corrosion products of Sn3.5Ag0.5Cu and Sn3.5Ag0.5Cu nano-composite solder have similar microstructures (Fig. 15). The corrosion products of Sn3.5Ag0.5Cu solder after polarization have a larger flake-like shape (Mark a) and small mushroom-like shape, and are loosely distributed on the surface, with different orientations (Fig.15a). On the other hand, the corrosion products of Sn3.5Ag0.5Cu nano-composite solder after polarization tests have only a flake-like shape, as shown in Fig. 15b (Mark a). Table 7 shows the surface element concentrations of solder corrosion products from EDS. According to the EDS analysis, the corrosion products of Sn3.5Ag0.5Cu and Sn3.5Ag0.5Cu nano-composite solder contain mainly Sn, O, and Cl (Fig.16). It can be seen that the corrosion products of the Sn3.5Ag0.5Cu solders and Sn3.5Ag0.5Cu nano-composite solders have slightly different compositions.

Solder		Surface element concentration (wt.%)				
		Sn	Ag	Cu	Cl	O
Sn3.5Ag0.5Cu	All area	69.26	4.05	0.37	13.40	12.60
	flake	72.97	0.43	-	15.71	9.58
	mushroom	74.87	1.29	-	17.97	5.76
Sn3.5Ag0.5Cu-0.5TiO$_2$	All area	68.7	3.82	0.68	13.39	13.41
	flake	64.22	0.50		15.03	20.25

Table 7. Surface element concentration of different solders after potentiodynamic polarization tests[60].

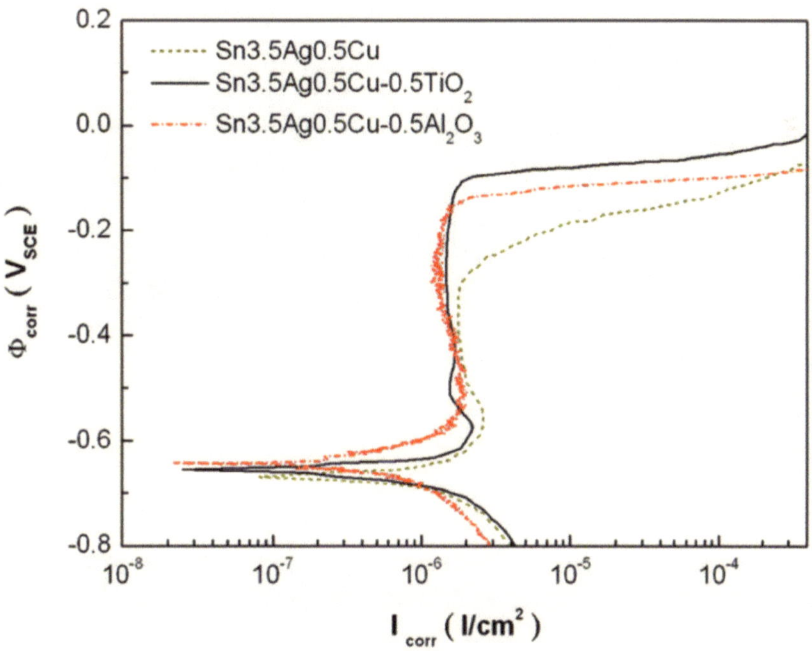

Fig. 14. The potentiodynamic polarization curves of the nano-composite solder in a 3.5wt.% NaCl solution: (a) Sn3.5Ag0.5Cu solder; (b) Sn3.5Ag0.5Cu-0.5TiO$_2$; and (c) Sn3.5Ag0.5Cu-0.5Al$_2$O$_3$ [60].

Fig. 15. Microstructure of the corrosion products on different solders after polarization tests (a) Sn3.5Ag0.5Cu solder, (b) Sn3.5Ag0.5Cu nano-composite solder[60].

During polarization testing in NaCl solution, the only possible cathodic reaction is oxygen reduction [49, 64]:

$$O_2 + 4e^- + 2H_2 \rightarrow 4OH^- \quad (1)$$

When the current density reaches about 10 mA/cm2, many hydrogen bubbles evolve from the cathode due to the hydrogen evolution on the cathode:

$$2H_2O + 2e^- \rightarrow H_2 + 2OH^- \quad (2)$$

The reactions on the anode are quite complicated. Some possible anodic reactions have been reported in the literature [46, 64-66], as displayed below:

$$Sn + 2OH^- - 2e^- = Sn(OH)_2 \quad (3)$$

$$Sn + 2OH^{-} - 2e^{-} = SnO + H_2O \quad (4)$$

$$Sn + 4H_2O = Sn(OH)_4 + 4e^{-} + 4H^{+} \quad (5)$$

The dehydration of $Sn(OH)_2$ and $Sn(OH)_4$ into SnO and SnO_2, respectively, has also been reported [46, 64,65]:

$$Sn(OH)_2 + 2OH^{-} + 2e^{-} = Sn(OH)_4 \quad (6)$$

$$Sn(OH)_2 + 2OH^{-} = SnO + H_2O \quad (7)$$

$$SnO + H_2O + 2OH^{-} + 2e^{-} = Sn(OH)_4 \quad (8)$$

However, Yu et al., after investigating the corrosion properties of Sn9Zn and Sn8Zn3Bi solder in NaCl solution, postulated the formation of a tin oxyhydroxychloride according to the following reaction[66]:

$$3Sn + 4OH^{-} + 2Cl^{-} = Sn_3O(OH)_2Cl_2 + H_2O \quad (9)$$

In addition, Li et al.[46] studied the corrosion properties of Sn-Ag, Sn–Ag–Cu, Sn–Cu, and SnPb solder in 3.5wt.% NaCl solution with different scanning rates, and their results showed that the corrosion product on the surface was tin oxide chloride hydroxide ($Sn_3O(OH)_2Cl_2$). In our case, the presence of such a surface layer, instead of a tin oxychloride layer, cannot be ruled out due to the detection limits of energy-dispersive spectroscopy. In order to understand the reaction during the corrosion products, XRD has been used to analyse the corrosion products on the surface after the polarization tests (Fig. 17). The results show that all the Sn3.5Ag0.5Cu and Sn3.5Ag0.5Cu solder materials have the same corrosion product, $Sn_3O(OH)_2Cl_2$, which is a complex oxide chloride hydroxide of tin[67]. This further confirms that the corrosion product on the Sn3.5Ag0.5Cu composite solders is $Sn_3O(OH)_2Cl_2$.

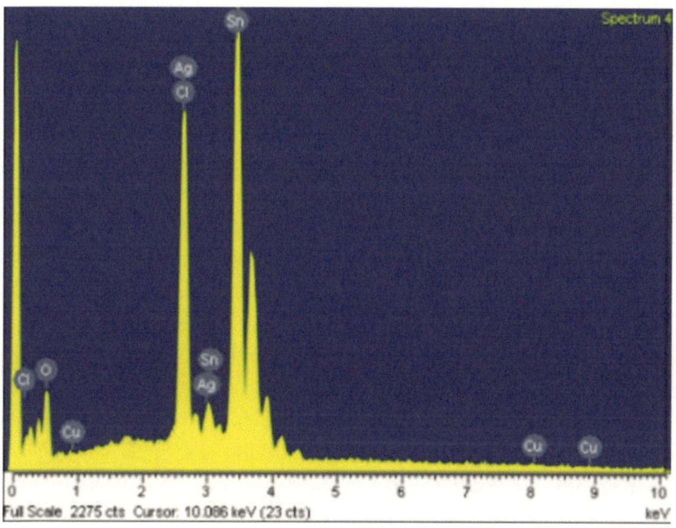

Fig. 16. EDS analysis of corrosion product of the Sn3.5Ag0.5Cu nano-composite solder after polarization tests[60].

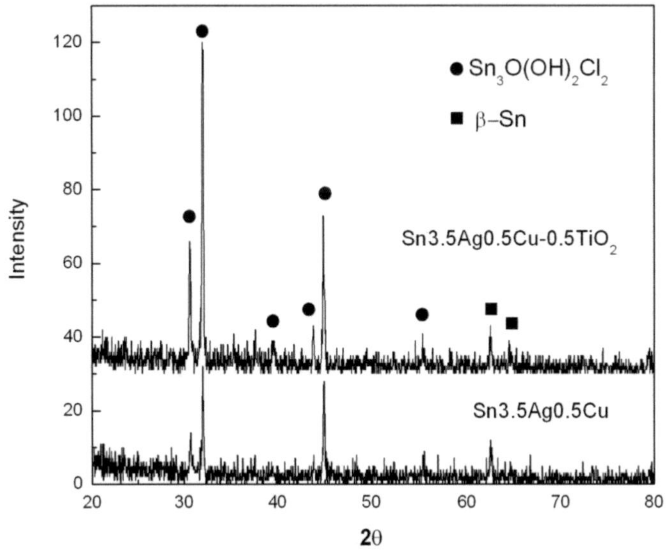

Fig. 17. XRD spectra of different solder materials after polarization tests[60].

9. References

[1] K. N. Tu, K. Zeng, Tin–lead (Sn–Pb) solder reaction in flip chip technology, Mater. Sci. Eng. R 34 (1) (2001) 1–58.

[2] E. P. Wood, K.L. Nimmo, In search of new lead-free electronic solders, J. Electron. Mater. 23 (8) (1994) 709–713.

[3] S. Jin, D. R. Frear, J.W. Morris Jr. Foreword, J. Electron. Mater. 23 (8) (1994) 709-713.

[4] M. Abtew, G. Selvaduray, Lead-free Solders in Microelectronics, Mater. Sci. Eng. R 27 (2000) 95-141.

[5] I. E. Anderson, J. C. Foley, B. A. Cook, J. Harringa, R. L. Terpstra, O. Unal, Alloying effects in near-eutectic Sn–Ag–Cu solder alloys for improved microstructural stability, J. Electron. Mater. 30 (9) (2001) 1050–1059.

[6] M. McCormack, S. Jin, G.W. Kammlott, H. S. Chen, New Pb-free solder alloy with superior mechanical properties, Appl. Phys. Lett. 63 (1) (1993) 15–17.

[7] A. Z. Miric, A. Grusd, Lead-free alloys, Soldering, Surf. Mount Technol. 10 (1) (1998) 19–25.

[8] N. C. Lee, Lead-free Soldering. In Daniel, L. and Wong, C. P. (eds.), Materials for Advanced Packaging, Springer Science Business Media (New York, 2009), pp. 181-218.

[9] K. N. Tu, A.M. Gusak, M. Li, Physics and materials challenges for lead-free solders, J. Appl. Phys. 93 (2003)1335–1353.

[10] H. T. Ma, J. C. Suhling, A review of mechanical properties of lead-free solder for electronic packaging, J. Mater. Sci. 44(2009) 1141–1158.

[11] International Technology Roadmap for Semiconductors—Assembly and Packaging, 2009 ed. (http://www.itrs.net/Links/2009ITRS/2009Chapters_2009Tables/2009_Assembly.pdf).

[12] E. C. C. Yeth, W. J. Choi, K. N. Tu, Current Crowding Induced Electromigration Failure in Flip Chip Technology, Appl. Phys. Lett. 80 (2002) 580-582.

[13] J. Shen, Y. C. Chan, Research advances in nano-composite solders, Microelectronics Reliability 49 (2009) 223–234.

[14] L. C. Tsao, S. Y. Chang, Effects of Nano-TiO_2 additions on thermal analysis, microstructure and tensile properties of Sn3.5Ag0.25Cu solder, Mater. Des. 31(2010) 990–993.

[15] L. C. Tsao, S.Y. Chang, C.I. Lee, W.H. Sun, C.H. Huang, Effects of nano-Al_2O_3 additions on microstructure development and hardness of Sn3.5Ag0.5Cu solder, Mater. Des. 31, (2010) 4831–4835.

[16] T. H. Chuang, M. W. Wu, S. Y. Chang, S. F. Ping, L. C. Tsao, Strengthening mechanism of nano-Al_2O_3 particles reinforced Sn3.5Ag0.5Cu lead-free solder, J. Mater. Sci.: Mater. Electron. 22 (2011)1021–1027.

[17] J. Shen, Y.C. Liu, D. J. Wang, H.X. Gao, Nano ZrO_2 particulate-reinforced lead-free solder composite, J. Mater. Sci. Technol. 22 (2006) 529-532.

[18] X. L. Zhong, M. Gupta, Development of lead-free Sn–0.7Cu/Al_2O_3 nanocomposite solders with superior strength, J. Phys. D: Appl. Phys. 41 (2008) 095403–095409.

[19] K. Mohan Kumar, V. Kripesh, Andrew A.O. Tay, Single-wall carbon nanotube (SWCNT) functionalized Sn–Ag–Cu lead-free composite solders, J. Alloy Compd. 450 (2008) 229–237.

[20] S. M. L. Nai, J. Wei, M. Gupta, Improving the performance of lead-free solder reinforced with multi-walled carbon nanotubes, Mater. Sci. Eng. A 423 (2006) 166–169.

[21] S. M. L. Nai, J. Wei, M. Gupta, Influence of ceramic reinforcements on the wettability and mechanical properties of novel lead-free solder composites, Thin Solid Films 504 (2006) 401–404.

[22] Z. X. Li, M. Gupta, High strength lead-free composite solder materials using nano-Al_2O_3 as reinforcement. Adv. Eng. Mater. 7 (2005)1049–1054.

[23] S. M. L. Nai, J. Wei, M. Gupta, Effect of carbon nanotubes on the shear strength and electrical resistivity of a lead-free solder. J. Electron. Mater. 37(2008) 515–522.

[24] H. Mughrabi, Plastic deformation and fracture of materials. Berlin: Springer-Verlag;. 1993. p. 315–322

[25] L. C. Tsao, 10th International Conference on Electronic Packaging. Technology & High Density Packaging (ICEPT-HDP 2009) (2009)1164 -1166.

[26] R. W. Wu, L. C. Tsao, S. Y. Chang, C. C. Jain and R. S. Chen, Interfacial reactions between liquid Sn3.5Ag0.5Cu solders and Ag substrates, J. Mater. Sci.: Mater. Electron. 22(8) (2011) 1181-1187.

[27] Görlich, D. Baither, G. Schmitz, Reaction kinetics of Ni/Sn soldering reaction, Acta Materialia 58(9) (2010) 3187-3197.

[28] T. H. Chuang, Y. T. Huang, L. C. Tsao, $AgIn_2$/Ag_2In transformations in an In-49Sn/Ag soldered joint under thermal aging, J. Electron. Mate. 30 (2001) 945-950.

[29] D. G. Kim, C. Y. Lee, S. B. Jung, Interfacial reactions and intermetallic compound growth between indium and copper, J. Mater. Sci.: Mater. Electron. 5(2) (2004) 95-98.

[30] L. C. Tsao, Evolution of nano-Ag_3Sn particle formation on Cu-Sn intermetallic compounds of Sn3.5Ag0.5Cu composite solder/Cu during soldering, J. Alloy Compd.509 (2011) 2326–2333.

[31] L. C. Tsao, Suppressing effect of 0.5 wt.% nano-TiO_2 addition into Sn–3.5Ag–0.5Cu solder alloy on the intermetallic growth with Cu substrate during isothermal aging, J. Alloy Compd. 509 (2011) 8441– 8448.
[32] L. C. Tsao, C.P. Chu, S. F. Peng, Study of interfacial reactions between Sn3.5Ag0.5Cu composite alloys and Cu substrate, Microelectron. Eng. 88 (2011) 2964-2969.
[33] S. Y. Chang, L. C. Tsao, M. W. Wu and C. W. Chen, The morphology and kinetic evolution of intermetallic compounds at Sn–Ag–Cu solder/Cu and Sn–Ag–Cu-0.5Al_2O_3 composite solder/Cu interface during soldering reaction, J. Mater. Sci.: Mater. Electron. DOI: 10.1007/s10854-011-0476-9.
[34] X.Y. Liu, M.L. Huang, C. M. L. Wu, Lai Wang, Effect of Y_2O_3 particles on microstructure formation and shear properties of Sn-58Bi solder, J Mater Sci: Mater Electron, J Mater Sci: Mater Electron 21 (2010) 1046–1054.
[35] S. M. L. Nai, J. Wei, M. Gupta, Interfacial Intermetallic Growth and Shear Strength of Lead-Free Composite Solder Joints, J. Alloy Compd. 473 (2009) 100–106.
[36] A. K. Gain, T. Fouzder, Y. C. Chan, Winco K. C. Yung, Microstructure, kinetic analysis and hardness of Sn–Ag–Cu–1 wt% nano-ZrO_2 composite solder on OSP-Cu pads, J. Alloy Compd. 509 (2011) 3319–3325.
[37] R. Ambat, P. Møller, A review of Corrosion and environmental effects on electronics, The Technical University of Denmark, DMS vintermøde proceedings, (2006).
[38] M. Mori, K. Miura, Corrosion of Tin Alloys in ulfuric and Nitric Acids, Corrosion Science, 44 (2002)887-898.
[39] F. Song, S. W. Ricky Lee, Corrosion of Sn-Ag-Cu Lead-free Solders and the Corresponding Effects on Board Level Solder Joint Reliability, 2006 Electronic Components and Technology Conference, (2006) 891-889.
[40] B. Liu, T. K. Lee, K. C. Liu, Impact of 5% NaCl Salt Spray Pretreatment on the Long-Term Reliability of Wafer-Level Packages with Sn-Pb and Sn-Ag-Cu Solder Interconnects , J. Electron. Mater. 40, 2011,doi: 10.1007/s11664-011-1705-y.
[41] Morten S. Jellesen et al., Corrosion in Electronics (Paper presented at 2008 Eurocorr Conference, Edinburgh, Scotland, 7–11 September 2008).
[42] R. Baboian, "Electronics," Corrosion Tests and Standards: Applications and Interpretation, ed. R. Baboian (West Conshohocken, PA: ASTM, 1996), www .corrosionsource. com/events/intercorr/baboian.htm).
[43] V. Chidambaram, J. Hald, R. Ambat, J. Hattel,A Corrosion Investigation of Solder Candidates for High-temperature Applications, JOM , 61 (2009) 59-65.
[44] H. Liu, K. Wang, K. Aasmundtveit, N. Hoivik, Intermetallic Cu_3Sn as Oxidation Barrier for Fluxless Cu-Sn Bonding, 2010 Electronic Components and Technology Conference, 853-857.
[45] L. C. Tsao, Corrosion Characterization of Sn37Pb Solders and With Cu Substrate Soldering Reaction in 3.5wt.% NaCl Solution, 2009 International Conference on Electronic Packaging Technology & High Density Packaging (ICEPT-HDP), (2009) 1164-1166.
[46] D. Li, P. P. C. C. Liu, Corrosion characterization of tin-lead and lead free solders in 3.5 wt.% NaCl solution, Corr. Sci. 50 (2008) 995–1004.
[47] K. L. Lin, T.P. Liu, The electrochemical corrosion behaviour of. Pb-Free Al–Zn–Sn solders in NaCl solution, Mater. Chem. Phys. 56 (1998) 171-176.

[48] K. L. Lin, F.C. Chung, T.P. Liu, The potentiodynamic polarization behavior of Pb-free XIn-9(5Al-Zn)-YSn solders, Mater. Chem. Phys. 53 (1998) 55-59.

[49] U.S. Mohanty, K.L. Lin, The effect of alloying element gallium on the polarisation characteristics of Pb-free Sn–Zn–Ag–Al–XGa solders in NaCl solution, Corros. Sci. 48 (2006) 662–678.

[50] D. Q. Yu, W. Jillek, E. Schmitt, Electrochemical migration of Sn–Pb and lead free solder alloys under distilled water, J. Mater. Sci. :Mater. Electron. 17 (2006) 219–227.

[51] B. Y. Wu, Y.C. Chan, M. O. Alam, Electrochemical corrosion study of Pb-free solders, J. Mater. Res. 21 (2006) 62-70.

[52] A. Ahmido, A. Sabbar, H. Zouihri, K. Dakhsi, F. Guedira, M. Serghini-Idrissi, S. El Hajjaji ,Effect of bismuth and silver on the corrosion behavior of Sn–9Zn alloy in NaCl 3wt.% solution, Mater. Sci. Eng. B 176 (2011) 1032– 1036.

[53] U. S. Mohanty, K. L. Lin, Electrochemical corrosion behaviour of Pb-free Sn-8.5Zn-0.05Al-XGa and Sn-3Ag-0.5Cu alloys in chloride containing aqueous solution, Corrosion Science 50 (2008) 2437–2443.

[54] J. Hu, T. Luo, A. Hu, M. Li, D. Mao, Electrochemical Corrosion Behaviors of Sn-9Zn-3Bi-xCr Solder in 3.5% NaCl Solution, J. Electron. Mater. 40, (2011) 1556-1562.

[55] T. C. Chang, J. W. Wang, M. C. Wang , M. H. Hon, Solderability of Sn-9Zn-0.5Ag-1In lead-free solder on Cu substrate Part 1. Thermal properties, microstructure, corrosion and oxidation resistance, J. Alloy Compd. 422 (2006) 239–243.

[56] F. Rosalbino, E. Angelini, G. Zanicchi, R. Marazza, Corrosion behaviour assessment of lead-free Sn-Ag-M (M = In, Bi, Cu) solder alloys, Mater. Chem. Phys. 109 (2008) 386–391.

[57] T. H. Chuang, M. W. Wu , S. Y. Chang, S. F. Ping, L. C. Tsao, Strengthening mechanism of nano-Al_2O_3 particles reinforced Sn3.5Ag0.5Cu lead-free solder, J Mater Sci: Mater Electron 22 (2011) 1021–1027.

[58] J. C. Leong , L. C. Tsao , C. J. Fang,C. P. Chu, Effect of nano-TiO_2 addition on the microstructure and bonding strengths of Sn3.5Ag0.5Cu composite solder BGA packages with immersion Sn surface finish, J Mater Sci: Mater Electron, (2011)1443-1449.

[59] L. C. Tsao, M. W. Wu, S. Y. Chang, Effect of TiO_2 nanoparticles on the microstructure and bonding strengths of Sn0.7Cu composite solder BGA packages with immersion Sn surface finish, J Mater Sci: Mater Electron DOI 10.1007/s10854-011-0471-1.

[60] L. C. Tsao, T. T. Lo, S. F. Peng, S. Y. Chang, Electrochemical behavior of a new Sn3.5Ag0.5Cu composite solder, 11th International Conference on Electronic Packaging Technology & High Density Packaging, (2010), 1013-1017.

[61] A. Pardo, M. C. Merino, A. E. Coy, R. Arrabal, F. Viejo, E. Matykina, Corrosion Behavior of Magnesium/Aluminum Alloys in 3.5 wt% NaCl, Corr. Sci., 50 (2008)823-834.

[62] G. Ben-Hamu, A. Eliezer, E. M.Gutman. Electrochemical Behavior of Magnesium Alloys strained in Buffer Solutions, Electrochim. Acta, 52 (2006)304-313.

[63] Y. Liu, Y. F. Cheng, Role of Second Phase Particles in Pitting Corrosion of 3003 Al Alloy in NaCl Solution, Mater. Corr., 61 (2010)211-217.

[64] S. D. Kapusta, N. Hackerman, Anodic passivation of tin in slightly alkaline solutions, Electrochim. Acta 25 (1980) 1625–1639.

[65] Q. V. Bui, N. D. Nam, B.I. Noh, A. Kar, J. G. Kim, S.B. Jung, Effect of Ag addition on the corrosion properties of Sn-based solder alloys, Mater. Corr. 61 (2010) 30-33.

[66] D. Q. Yu, C. M. L. Wu, L. Wang, The Electrochemical Corrosion Behavior of Sn-9Zn and Sn-8Zn-3Bi Lead-Free Solder Alloys in. NaCl Solution, 16th International Corrosion Congress, Beijing, P.R. China, (2005)19–24.
[67] F. Rosalbino, E. Angelini, G. Zanicchi, R. Carlini, R. Marazza, Electrochemical Corrosion Study of Sn–3Ag–3Cu Solder Alloy in NaCl solution, Electrochim. Acta, 54 (2009) 7231-7237.

6

Electrochemical Passive Properties of $Al_xCoCrFeNi$ (x = 0, 0.25, 0.50, 1.00) High-Entropy Alloys in Sulfuric Acids

Swe-Kai Chen
Center for Nanotechnology, Materials Science, and Microsystems (CNMM), National Tsing Hua University, Hsinchu, Taiwan

1. Introduction

1.1 Pseudo-unitary lattice with a characteristic parameter as a description of multi-principal alloys – The high-entropy alloys (HEAs)

In the summer of 1995, J.W. Yeh and the author (SKC) started the study of multi-principal-element alloys which was called, then, alloys with high randomness and now the high-entropy alloys (HEAs). SKC checked the first 10 equal-molar alloys, which was designed by Yeh that contained from 6 to 9 elements in the alloys out of one of Al, Cu, and Mo, together with Ti, V, Fe, Ni, Zr, Co, Cr, Pd, and B, with a home-made vacuum-arc remelter, and the author observed that the alloy series containing Mo can be made most easily, while the ones containing 3 at% B are the ones most difficult in melting, and 6 out of 10 can be formed in the water-cooled copper mold of the remelter, i.e., the existence of the HEAs was demonstrated by experiments. The alloys were aimed at that time to design as another kind of bulk glass alloys, and based on the high configurational entropy of R ln(n), n between 5 and 13, similar to the mixing of different gases [1]. No conclusions were drawn with XRD patterns of these alloys that were found two years later to be composed with peaks from a single simple lattice cell like FCC A1 or BCC A2, although some evidence of existence of amorphous phase was observed from TEM diffraction patterns and high resolution images [2,3]. The simple crystalline phases instead of amorphous ones were continuously found in alloys like in AlCoCrCuFeNi during research of HEAs in these 10 to 20 years, and identified with a so-called extended FCC or BCC unit cell that SKC called it a pseudo-unitary lattice in 2010 [4].

As multiple principal element alloys, high-entropy alloys (HEAs) comprise at least five elements whose concentration for each one ranges between 5 at % and 35 at % [5]. Attributes of forming a simple solid solution and nano-particle precipitation, as well as achieving a high hardness and strength, and excellent high-temperature oxidation resistance make HEAs highly promising for application and research and development of these alloys [6-9]. Properties of $Al_xCoCrFeNi$ ($0 \le x \le 1$) HEAs vary significantly with x [10]. For instance, the alloy structure changes from FCC to BCC for increased Al content x. Besides, the coefficient of thermal expansion decreases with x. Both properties are closely related to the bond strength of alloys. Moreover, electrical resistivity of $Al_xCoCrFeNi$ alloys is large, i.e., approximately up to 200 $\mu\Omega$ cm [11].

1.2 Corrosion resistance for HEAs and conventional alloys

Corrosion properties of $AlCoCrCu_{0.5}FeNiSi$ [12,13], $Al_xCrFe_{1.5}MnNi_{0.5}$ [14,15], and $Al_{0.5}CoCrCuFeNiB_x$ [16] HEAs have been extensively studied in recent years. Among these HEAs, $AlCoCrCu_{0.5}FeNiSi$ alloy (HEA 1) displays, at room temperature, a better general corrosion resistance than SS 304 in 1 N H_2SO_4; however, it exhibits a worse pitting corrosion resistance than SS 304 in 1 N H_2SO_4 and in 1 M NaCl, respectively. The general corrosion resistance of each of HEA 1 and SS 304 decreases when exceeding room temperature. The effect of temperature on corrosion resistance of HEA1 is less severe in 1 M NaCl than in 1 N H_2SO_4 [13]. $Al_xCrFe_{1.5}MnNi_{0.5}$ alloys (HEA 2) reveal that in each of the 0.5 M H_2SO_4 and 1 M NaCl solutions, corrosion resistance increases with a decreasing x; in addition, the susceptibility to general and pitting corrosion of HEA 2 increases with an increasing x [14]. $Al_xCrFe_{1.5}MnNi_{0.5}$ alloys (called hereinafter as HEA 2a and 2b for x = 0 and 0.3, respectively) in 0.1 M HCl exhibit different corrosion behaviours for different x values. Although HEA 2a is susceptible to localized corrosion, HEA 2b has a stable passive film on the surface. In 0.1 M HCl, anodized treatment of HEA 2a and 2b alloys in 15 % H_2SO_4 gives higher corrosion resistance than the untreated [15]. In deaerated 1 N H_2SO_4, $Al_{0.5}CoCrCuFeNiB_x$ alloys are more resistant to general corrosion than SS 304, and are not susceptible to localized corrosion. Additionally, the corrosion resistance of $Al_{0.5}CoCrCuFeNiB_{0.6}$ alloy is inferior to $Al_{0.5}CoCrCuFeNi$ alloy [16]. Above HEAs show an extremely close compositional dependence of corrosion behaviour in various solutions.

1.3 Aim of this study

Although many interesting topics have been explored for $Al_xCoCrFeNi$ alloys [10,11], investigation on their corrosion property is still lacking. Therefore, this study elucidates how Al affects their corrosion behaviour. The electrochemical properties of the alloys in sulfuric acids are investigated using the potentiodynamic polarization curve and a weight loss measurement method. Additionally, these alloys are compared with SS 304, especially with respect to the effect of temperature. Moreover, based on use of electrochemical impedance spectroscopy (EIS), the effect of Al on corrosion behaviour is analyzed. Furthermore, the relationship of stability of oxide film with Al content is examined by varying the chloride concentration in a sulfuric solution. Additionally, this study, which extends [17], also attempts to investigate the mechanism of the passive layers influenced by Al content x at various temperatures in detail.

2. Experimental details

2.1 Test materials and conditions for electrochemical and weight loss tests

2.1.1 Test specimens for electrochemical tests and weight loss measurements

As-cast $Al_xCoCrFeNi$ alloys were prepared according to molar ratios of x = 0, 0.25, 0.50, and 1.00 (called C-0, C-0.25, C-0.50, and C-1.00, respectively) in a vacuum arc remelter. Table 1 lists the composition of the alloys. Test specimens were cut in 0.8 cm x 0.8 cm x 0.3 cm and cold-mounted in epoxy with the outside surface from a surface of 0.8 cm x 0.8 cm of specimens. The specimens were subsequently ground and polished with grit #1000 silicon carbide paper, rinsed and dried in preparation for electrochemical tests and weight loss measurements. During determination of the weight loss, six sets of samples were dipped in sulfuric acid for 1, 3, 5, 8, 11, and 15 days, respectively. All tests, except the weight loss test, were performed at least three times to confirm the data reproducibility. Finally, weight loss tests were performed twice and the reproducibility was given in an error bar.

Alloys	Al	Co	Cr	Fe	Ni
C-0	0	27.12	23.74	23.99	25.14
C-0.25	3.05	25.14	22.48	24.15	25.18
C-0.50	5.59	25.25	22.13	22.80	24.22
C-1.00	10.02	23.84	21.11	21.99	23.03
SS 304	0	0	19.40	72.68	7.92

Table 1. Composition (wt %) for alloys C-x and SS 304.

2.1.2 Test solutions and temperatures

The base solution for all tests was 0.5 M of sulfuric acid. Test temperatures were ambient temperature (~25°C). Test solutions bearing chloride ions were with 0.25, 0.50, and 1.00 M sodium chloride in the base solution. To avoid the dissolved oxygen (aeration) affecting the test solutions, deaeration was simultaneously made by a nitrogen gas flow of 120 ml/min in the test solution. The effect of temperature on polarization was examined under thermostatic control at an interval of 15°C in the temperature range of 20°C - 65°C.

2.2 Potentiodynamic polarization curve measurements and electrochemical impedance spectroscopy (EIS)

A three-electrode cell was used for the electrochemical test. The reference electrode was a commercial Ag/AgCl electrode saturated in 3 M KCl electrode (−0.205 V_{SHE} or -0.205 V to standard hydrogen electrode). The auxiliary electrode was made of Pt, and the working electrode was the specimen. Potentiostat was CH Instrument Model-600A. The specimen was cathodically polarized at a potential of −0.4 V_{SHE} for 300 s before the test for the purpose of removing surface oxides. The quasi-steady-state time for an open circuit was 900 s. Scan speed was 1 mV/s for scan potential ranging from −0.6 V_{SHE} to 1.4 V_{SHE}. For EIS, the working potential was that of open circuit at 900 s from the start of immersion with scan amplitude 10 mV and a frequency ranging from 100 kHz to 10 mHz.

2.3 Immersion tests and ICP-AES and XPS analyses

Samples were dipped in sulfuric acid for 15 d to determine the weight-loss rate. Auger electron spectroscopy (AES) and X-ray photoelectron spectroscopy (XPS) analysis were performed with samples after a 0.8 V_{SHE} pretreatment plus a 1-h immersion. Inductively coupled plasma atomic emission spectroscopy (ICP-AES) was performed on the electrolyte after an 8-d immersion of the samples. The effect of temperature on polarization was examined under thermostatic control at an interval of 15°C in the temperature range of 20°C–65°C.

2.4 Scanning electron microscopy (SEM) metallographic examination and energy dispersed X-ray spectroscopy (EDS) analysis

Samples were fine polished, up to 0.05 μm Al_2O_3 powder and, then, examined with SEM (JEOL JSM-840A) equipped with an Oxford EDS for topography and elemental compositions. Finally, samples were examined before and after 3 days immersion of 0.5 M H_2SO_4.

3. Results and discussion

3.1 Potentiodynamic polarization curve and weight loss at 25 °C

Fig. 1 shows the anodic dissolution behaviour of alloys in 0.5 M H_2SO_4, while Table 2 summarizes relevant data. This figure reveals a well-defined passive region of 0 V_{SHE} to 1.2 V_{SHE} in all curves. All curves, except for the one at x = 0.25 (C-0.25), show a secondary passive region at 0.15 V_{SHE}. This passivation is attributed mainly to the further oxidation or hydroxidation of the passive oxide film, thus altering the valence of Cr [18,19]. Fig. 1 also indicates that the secondary passive regions of C-0.50 and C-1.00 are more prominent than those of C-0 and C-0.25. This observation is due to the selective dissolution in the duplex FCC-BCC structure for C-0.50 and in the BCC-ordered BCC structure for C-1.00, as compared with C-0 and C-0.25 which are single FCC phase. In the active-passive transition region, different compositions at different secondary passivation potentials reveal different dissolution rates owing to a selective dissolution. This observation resembles that observed in duplex phase stainless steel [20,21].

Alloys C-x & SS304	E_{corr} (V_{SHE})	I_{corr} ($\mu A/cm^2$)	E_{pp} (V_{SHE})	I_{crit} ($\mu A/cm^2$)	I_{pass} ($\mu A/cm^2$)
C-0	-0.081	15.8	0.002	42.8	4.5
C-0.25	-0.095	16.7	0.008	87.4	7.1
C-0.50	-0.084	13.4	0.017	117.2	6.4
C-1.00	-0.094	13.1	0.010	198.0	13.9
SS 304	-0.185	45.3	-0.071	603.0	19.1

Table 2. Potentiodynamic polarization curve diagram parameters of alloys C-x and SS 304 at 25°C.

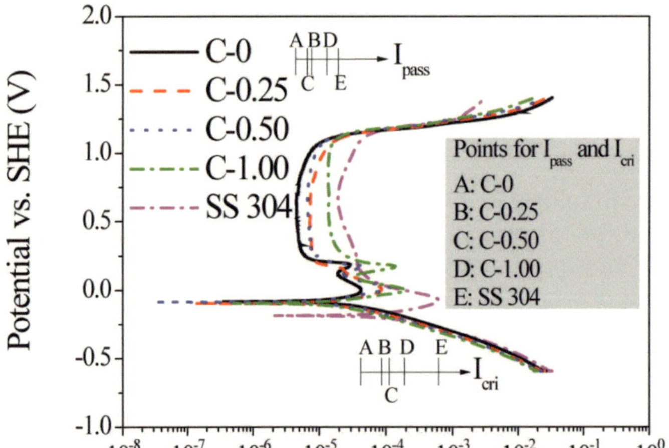

Fig. 1. Potentiodynamic polarization curve diagrams for alloys C-x and SS 304 at 25°C.

Table 2 reveals that the corrosion potential (E_{corr}) and the corrosion current density (I_{corr}) for all of the alloys differ only slightly, and no obvious trends occur for E_{corr} and I_{corr} vs. x variation. The above phenomenon can be attributed to the spontaneous passivation of pure Al in H_2SO_4 [22]. Al metal spontaneously passivates in H_2SO_4, explaining why its corrosion potential is ready in the passive region, i.e., this passivation explains why the polarization curve of Al does not display an apparent active-passive transition region. However, elements such as Cr and Fe exhibit a large critical current density (I_{cri}) for passivation, explaining why Cr and Fe dissolve more than Al before the alloy reaches its passive state. Thus, the variation of Al affects the active region of the polarization curves slightly. Furthermore, in H_2SO_4, all Al, Co, Cr, Fe, and Ni metals show passivity. Among them, Al has a relatively high passive current density (I_{pass}) [22,23] because only Al oxide can easily form a porous film on the metal surface [24]. Therefore, protection by oxide layer on the alloys with higher Al content is inferior to that with lower Al content. Fig. 1 thus reveals that I_{pass} increases with x.

The results of potentiodynamic polarization were compared via performing 15-day-dipping and weight loss experiments. In the 15-day-dipping and weight loss experiments, the corrosion rates for C-0.50 and C-1.00 were markedly higher than those of C-0 and C-0.25 (Fig. 2). This observation differs substantially from the values of I_{corr} obtained from polarization experiment (Fig. 1), in which the two groups only differ slightly, despite the fact that the trend is the same. A previous study found a similar deviation in corrosion current densities obtained from weight loss test and potentiodynamic polarization method [25].

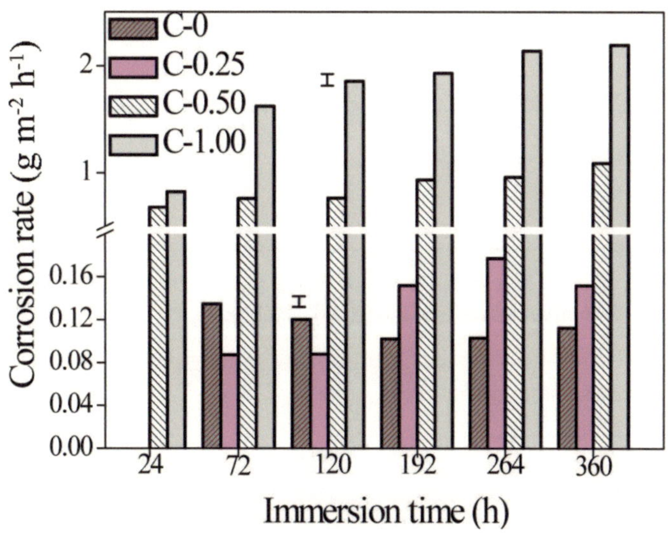

Fig. 2. Diagram showing change in corrosion rate (g m^{-2} h^{-1}) in the 15-day-dipping and weight loss measurement for alloys C-x.

Fig. 1 shows potentiodynamic polarization diagrams for the $Al_xCoCrFeNi$ alloys and SS 304. The alloys have better overall general corrosion behaviour, with a larger E_{corr} and smaller I_{corr}, I_{cri}, and I_{pass} than SS 304.

3.2 Effect of temperature on potentiodynamic polarization

Fig. 3 shows polarization diagrams of $Al_xCoCrFeNi$ alloys at various temperatures. A rising temperature decreased the Tafel slopes of anode (Table 3), increased I_{corr}, and increased E_{corr} and E_t (the transpassive potential) slightly. The corrosion rate is directly related to I_{corr}, according to Arrhenius equation, $I_{corr} = A \exp(-E_a/RT)$ [13,26], where the pre-exponential factor A is generally independent of temperature and is a constant of alloys, where R denotes the gas constant, T denotes temperature, and E_a denotes activation energy for corrosion. In the case of small experimental temperature range, E_a is assumed to be independent of T. Consequently, E_a can be obtained from $\ln(I_{corr})$ vs. 1/T plot. Fig. 4 shows such plots for the alloys and SS 304, indicating that E_a increases with x. This finding suggests that the corrosion rate is more sensitive to temperature for a larger Al content than for a smaller Al content. The $\ln(I_{corr})$ vs. 1/T curves intersect with each other in a range of 23°C – 27°C. Beyond this temperature range, I_{corr} increases with x. The situation is reversed at temperatures lower than 23°C, which is inconsistent with a situation in which E_as for all alloys increase with x from 20°C to 65°C. Hence, E_a, i.e., an intrinsic property of metal, and A, i.e., a surface property of metal, are determinative factors of I_{corr}. While E_a depends only on x, A depends on both x and temperature (Table 4). Therefore, although E_a increases with x, A also increases with x. Combining the effects of E_a and A explains the different corrosion behaviours of the alloys with an increasing x at temperatures exceeding 27°C and lower than 23°C. Thus, the performance of passive films, when Al is added, at higher temperatures becomes inferior to that without addition of Al. In determining I_{corr}, A is more important than E_a at temperatures exceeding 27°C, while E_a is more important than A at temperatures lower than 23°C.

Alloys	20 °C		35 °C		50 °C		65 °C	
C-x	β_a[a]	β_c[b]	β_a	β_c	β_a	β_c	β_a	β_c
C-0	158	218	128	158	134	162	89	158
C-0.25	158	178	103	167	89	168	92	149
C-0.50	94	158	113	178	138	159	89	198
C-1.00	104	148	93	173	98	242	100	220

[a] Anodic Tafel slope β_a in mV/decade, the measured Tafel regions are with 40~50 mV of overvoltage.
[b] Cathodic Tafel slope β_c in mV/decade, the measured Tafel regions are with 150~170 mV of overvoltage.

Table 3. Fit data for Tafel slopes of alloys C-x in 20 °C - 65 °C.

Alloys	A(x, T), A/cm²				E_a, kJ/mol
	20 °C (293 K)	35 °C (303 K)	50 °C (323 K)	65 °C (338 K)	
C-0	1.16×10^{-4}	1.16×10^{-4}	1.16×10^{-4}	1.07×10^{-4}	3.96
C-0.25	1.90×10^{-4}	2.03×10^{-4}	1.77×10^{-4}	1.90×10^{-4}	5.35
C-0.50	7.17	5.64	3.70	8.41	31.24
C-1.00	1.78×10^{9}	7.46×10^{8}	1.89×10^{9}	1.31×10^{9}	78.61
SS 304	1.18×10^{-4}	1.28×10^{-4}	1.70×10^{-4}	2.07×10^{-4}	9.87

Table 4. The fits for A(x, T) and $E_a(x)$ in $I_{corr}(x, T) = A(x, T) \exp(-E_a(x)/RT)$.

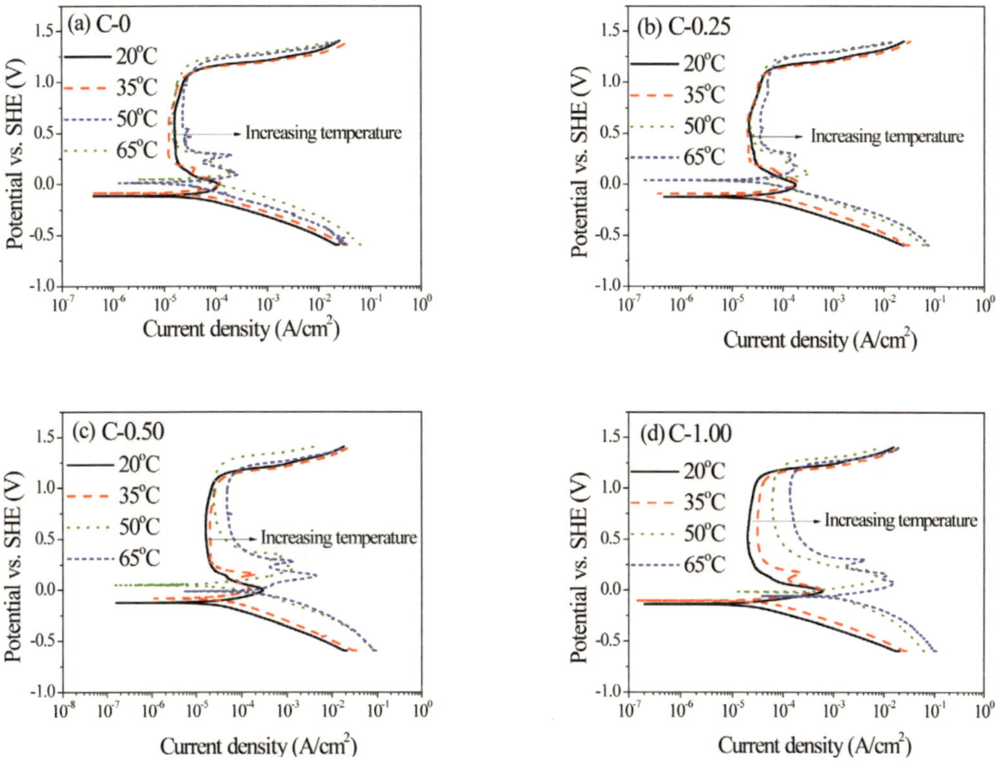

Fig. 3. Potentiodynamic Polarization Curve diagrams for (a) C-0, (b) C-0.25, (c) C-0.50, and (d) C-1.00 at various temperatures.

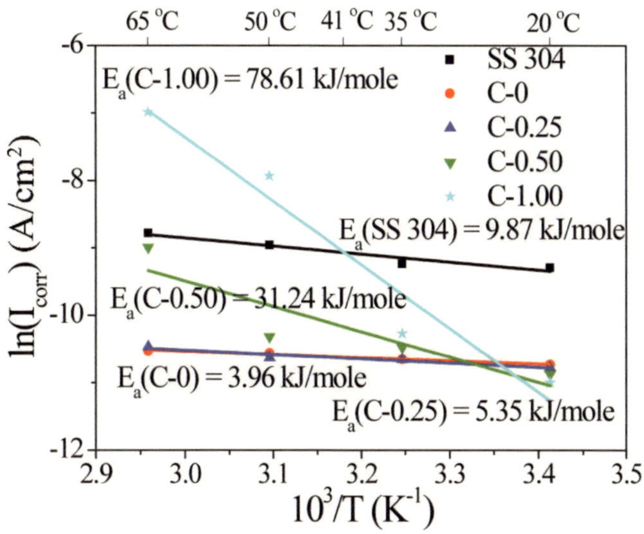

Fig. 4. The Arrhenius plots for alloys C-x and SS 304 at 20°C - 65°C.

3.3 EIS test at 25°C

Figs. 5 and 6 summarize the EIS results of alloys in a sulfuric solution and their schematic equivalent circuit diagrams, respectively. Table 3 lists related parameters of the equivalent circuit diagrams, where R_s, R_f, and R_{ct} denote impedances of the sulfuric solution, oxide layer, and adsorption layer, Q_f and Q_{ad} denote capacitances of constant phase element (CPE) for oxide layer and adsorption layer, respectively. Next, the oxide layer thickness is evaluated by using the Helmholtz model [27] and expressing the layer thickness of the oxide layer, d, as $d = \varepsilon\varepsilon_o S/Q_f$, where ε_o denotes the permittivity of free space (8.85×10^{-14} F/cm), ε denotes the dielectric constant of the medium, and S denotes the surface area of the electrode. Assuming that ε and S for all oxide layers of alloys are the same allows us to compare relative values of d for all samples by $1/Q_f$. Fig. 7 reveals that $1/Q_f$ values are proportional to x, implying that d increases with Al content x. However, according to this figure, the impedance of oxide layer R_f decreases with x and, in Fig. 8, the impedance of the oxide layer is inversely proportional to I_{pass}. Restated, a thinner oxide layer implies a larger value of impedance. To explain this phenomenon, besides the thickness of oxide layer, the density of oxide layer is also considered. As mentioned in Section 3.1, Al oxide easily forms a porous film on the metal surface [24]. Therefore, it is easily understood that in addition to causing a thicker oxide layer, Al element promotes the dispersive oxide layer. Combining these two effects obviously reveals that R_f decreases with x.

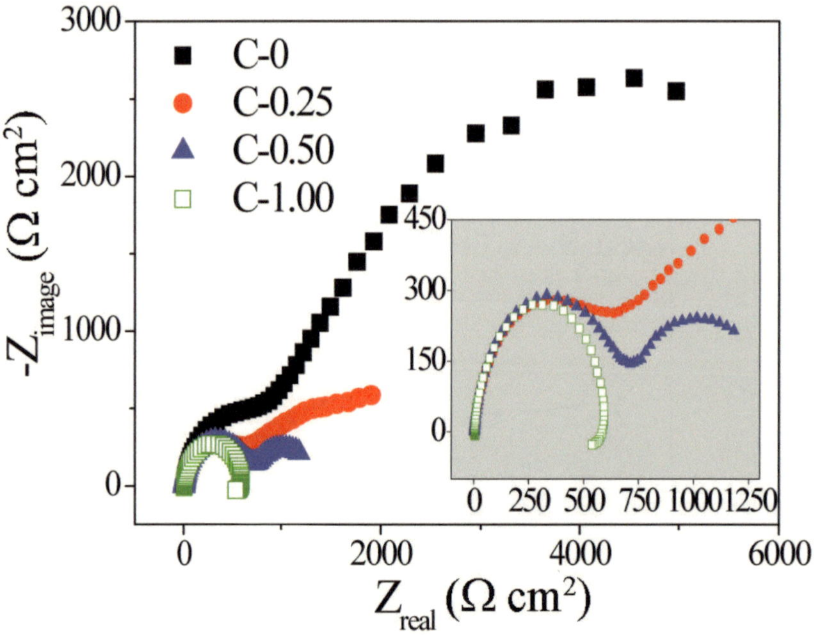

(a)

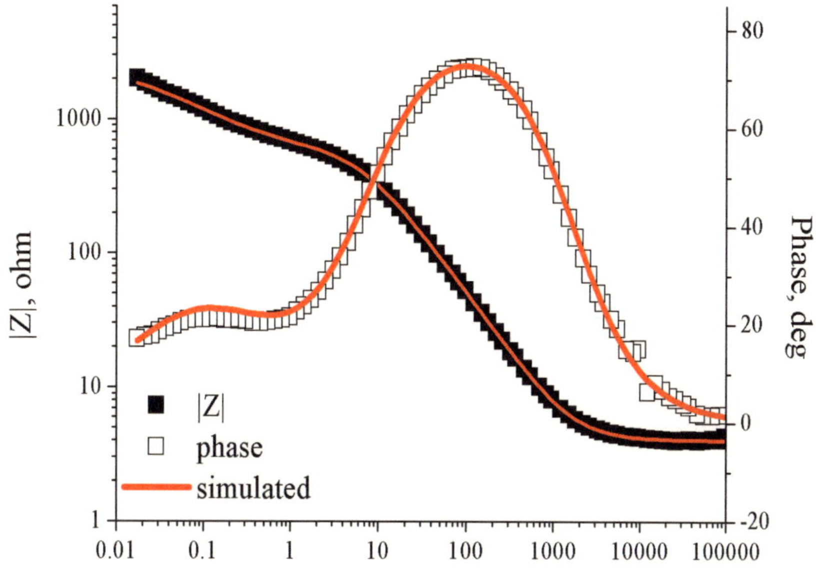

(b)

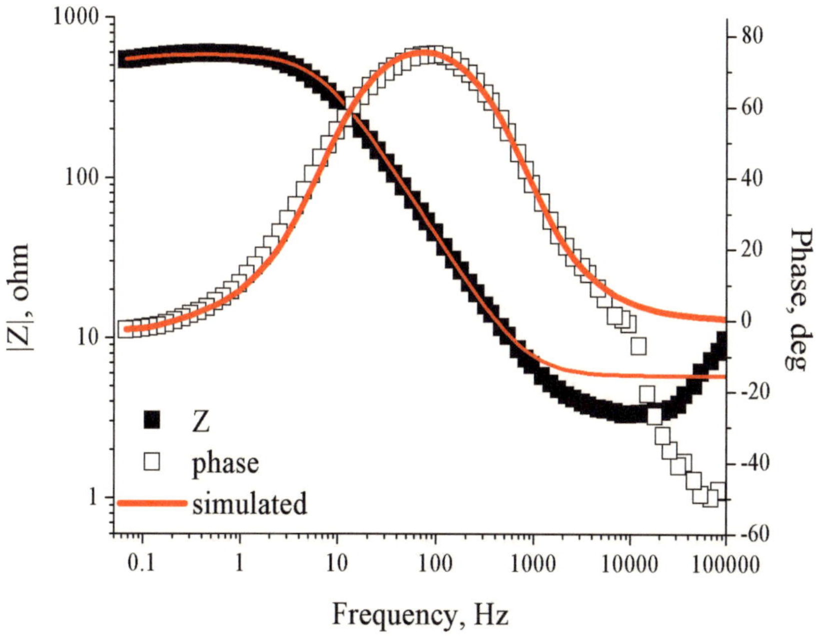

(c)

Fig. 5. (a). The Nyquist plots for alloys C-x at 25°C. (b). The Bode plot for C-0.25. (c). The Bode plot for C-1.00.

According to Fig. 6, only C-1.00 reveals a component of inductance in the equivalent circuit (See Fig. 5(c)). In previous studies [28-31], alloys with Al and Ni readily react with $(OH)^-$ and $(SO_4)^{2-}$ in a sulfuric solution and adsorbed on the surface of the alloys, which increases the amount of the ions in the adsorption layer. Therefore, Q_{ad} increases with x, as listed in Table 5. As x value increases to 1.00, the inductance appears in the equivalent circuit in Fig. 6(b). This effect normally occurs in the case of a severe corrosive condition [32]. Origin of the inductance can generally be influenced by some adsorbed intermediates or can be attributed to a space at the interfaces [33]. In C-1.00, a microstructure with an Al and Ni-rich phase which is seen as a reactive phase from metallograph, not only causes adsorption in these Al and Ni-rich areas in corrosion process, but also decreases the impedance in the low frequency area owing to their continuous dissolution. The fact that R_{ct} decreases with x demonstrates a higher dissolution rate for alloys with a higher Al content.

Alloys C-x	R_s (Ω cm^2)	Q_f (μF/cm^2)	n_f	R_f (Ω cm^2)	Q_{ad} (μF/cm^2)	n_{ad}	R_{ct} (Ω cm^2)
C-0	3.271	54.57	0.9094	992.2	636.7	0.7444	7691
C-0.25	3.758	56.61	0.9081	610.5	1525	0.6347	1932
C-0.50	2.994	46.55	0.9223	642.8	3221	0.6454	819.1
C-1.00	3.462	47.16	0.9614	518.1	L_{ad}	-	66.81

* Lad = 122.4 Henry

Table 5. EIS equivalent circuit parameters for alloys C-x.

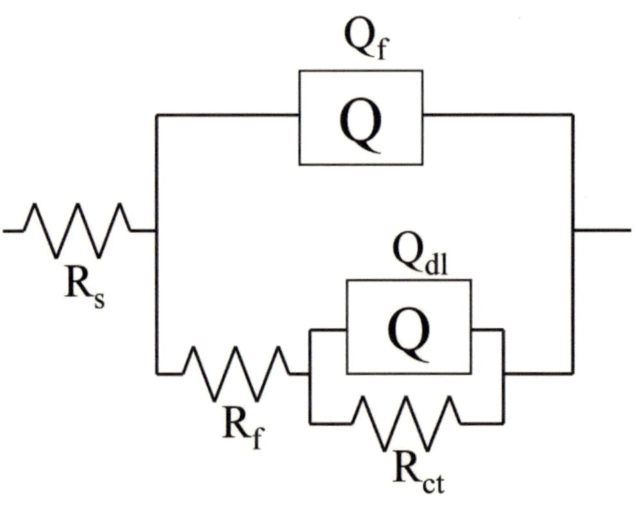

Fig. 6. EIS equivalent circuits for alloys C-0, C-0.25, C-0.50, and C-1.00.

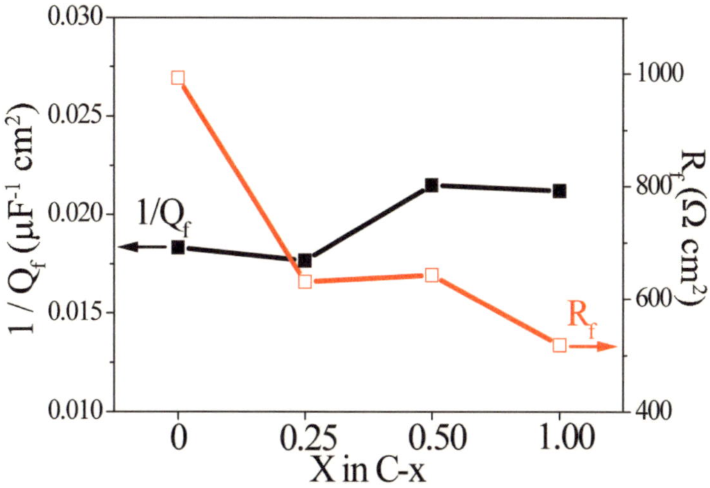

Fig. 7. Impedance and relative thickness ($1/Q_f$) of oxide layer vs. Al content x plots.

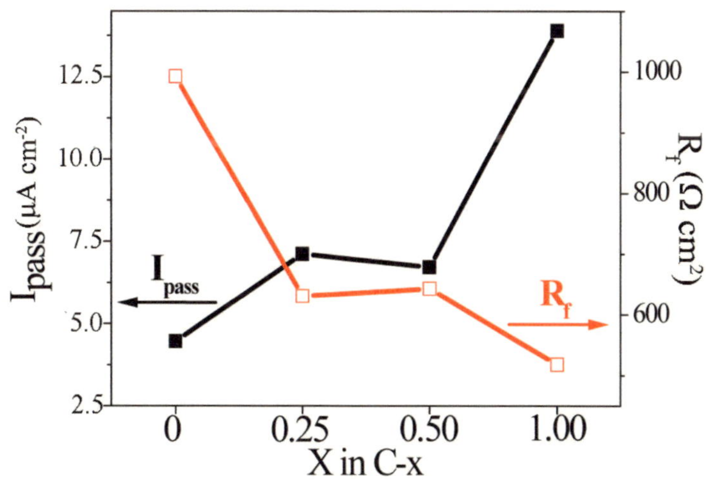

Fig. 8. Impedance and I_{pass} of oxide layer vs. Al content x plots.

3.4 Polarization behaviour for alloys in a chloride-containing H_2SO_4 solution

Fig. 9 shows potentiodynamic polarization curve diagrams for the alloys in 0.5 M H_2SO_4 solution containing various concentrations of chloride ions, as well as in simple 0.5 M H_2SO_4 solution as a comparison. According to Fig. 9(a), oscillation occurs in a passive region for C-0 in 0.5 M H_2SO_4 containing 0.5 M and 1 M of chloride. This phenomenon has been attributed to the cycling process for small pitting and re-passivation with the duration of several seconds for each cycle [34]. Oscillation in the passive region in potentiodynamic polarization curve is a metastable state [35]. This metastable state generally reflects the difficulty of pitting, i.e., alloy C-0 has good anti-pitting ability, while those containing aluminum (C-0.25, 0.50, and 1.00) with no metastable state show an inferior anti-pitting ability (Figs. 9 (b) to (d)).

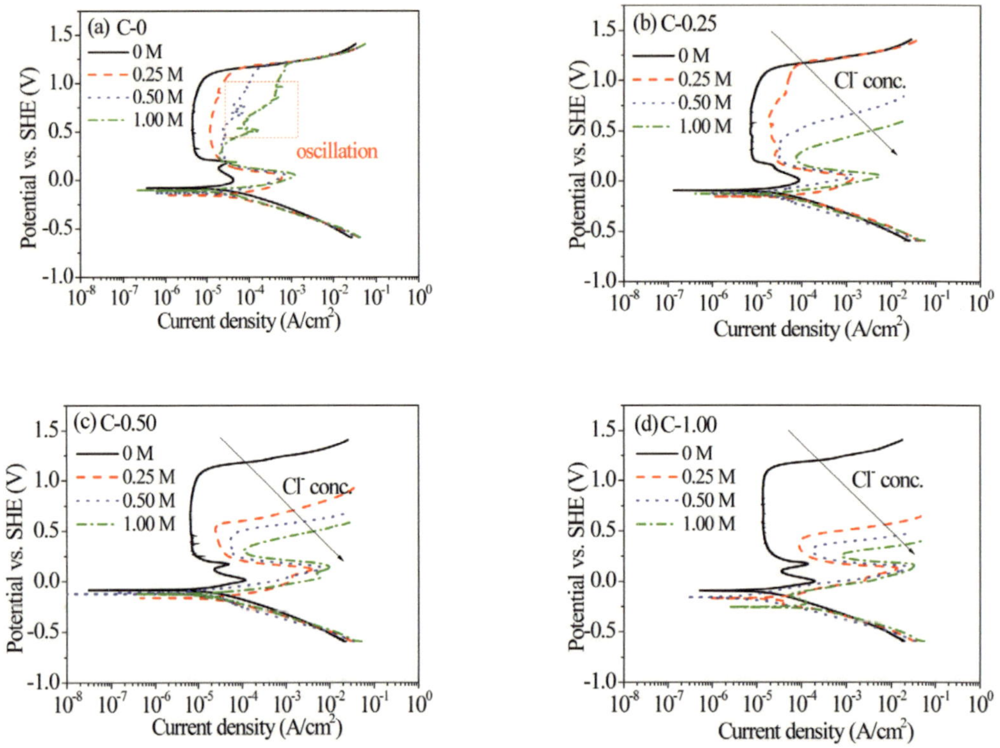

Fig. 9. Potentiodynamic Polarization Curve diagrams for (a) C-0, (b) C-0.25, (c) C-0.50, and (d) C-1.00 at 25 °C in chloride-containing sulfuric acid solution at various Cl^- molarity (M) values.

From an adsorption viewpoint, adsorption competition always prevails on the alloy surface between chloride ions and dissolved oxygen atoms. Notably, no oxide layer forms once chloride ions adsorb on the alloy surface, in which the metal ions readily dissolve. Therefore, the adsorption of chloride ions increases the reacting current density (as indicated by a comparison of Figs. 1 and 9), subsequently increasing the rate of metal dissolution.

Rapid dissolution of alloys in chloride-containing solution is discussed next. When chloride ions are adsorbed on the interface of passive layer and a sulfuric solution, metastable ion complexes gradually form from the anions of a passive layer. These metastable ion complexes enable the anions to dissolve. Once the ion complexes that are on the passive layer/solution interface dissolve into the sulfuric solution, the inner ion complexes of the passive layer move to the passive layer/solution interface in order to correlate with the applied potential. The inability of the anions to form oxide implies the continuous formation of metastable ion complexes and dissolution of ions. Since Al easily forms $[Al(SO_4)]^+$ with $(SO_4)^{2-}$, and $Al(OH)SO_4$ with $(SO_4)^{2-}$ and $(OH)^-$, respectively [36], these metastable ion complexes combine with Cl^- and dissolve afterwards. Therefore, pitting easily occurs on the surface of aluminum alloys. Next, the aluminiferous passive layer and non-aluminiferous passive layer are compared. Fig. 10 shows the pitting potential (E_{pit}) of the alloys and SS 304 in different solutions. The value of E_{pit} for C-0 is almost independent of chloride concentration. The value of E_{pit} for C-0.25 decreases abruptly for a chloride concentration exceeding 0.50 M. This value is close to that of SS 304. The values of E_{pit}, for C-0.25, C-0.50, and C-1.00, decrease to 0.2-0.5 V_{SHE} at a chloride concentration of 0.25 M (Fig. 10). A higher Al concentration in the alloys implies a lower value of E_{pit}. For C-0, deterioration of the passive layer is attributed to the evolution of oxygen. Meanwhile, for C-0.25, C-0.50, and C-1.00, the deterioration of passive layer is attributed to the pitting process. An increasing chloride ion concentration causes the chloride ions to cluster at the defect sites of the passive layer and severely attack the passive layer. Consequently, E_{pit} shifts to a more active region.

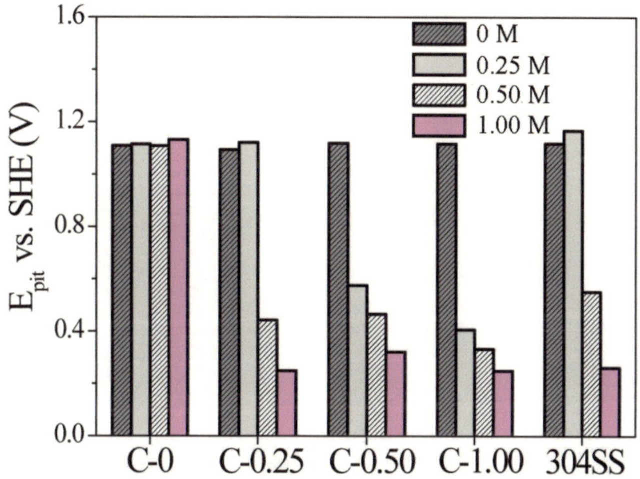

Fig. 10. Histogram of E_{pit} for alloys C-x and SS 304 in solution of different Cl^- ion molarity (M).

3.5 Metallographic examination and EDS analysis

Microstructures for not H_2SO_4-immersed alloys C-0, C-0.25, C-0.50, and C-1.00 are with single FCC, single FCC, duplex FCC-BCC, and BCC-ordered BCC phases, respectively [6]. Table 6 lists the EDS composition for each phase in different alloys.

Alloys	Phases and states	Al	Co	Cr	Fe	Ni
C-0	Overall, not immersed	0	25.93	25.73	24.21	24.13
	Overall, immersed	0	24.45	26.39	24.83	24.32
C-0.25	Overall, not immersed	6.16	23.27	23.58	23.59	23.40
	Overall, immersed	6.18	23.65	24.41	23.04	22.71
C-0.50	Overall, not immersed	11.01	22.77	22.61	21.70	21.92
	FCC matrix, not immersed	8.36	24.74	23.48	22.77	20.65
	BCC, not immersed	**13.94**	21.11	20.48	20.53	**23.94**
	Overall, immersed	8.35	22.75	27.19	23.60	18.11
	FCC matrix, immersed	9.96	22.57	23.16	23.59	20.72
	Wall-shaped BCC, immersed	**3.82**	22.50	**36.33**	25.58	**11.77**
C-1.00	Overall, not immersed	18.88	20.55	20.63	20.01	19.93
	Overall, immersed	12.45	19.80	29.87	23.42	14.45
	BCC, immersed	17.14	20.67	21.96	20.89	19.34
	Ordered BCC, immersed	**3.04**	17.34	**47.53**	27.96	**4.14**

Table 6. EDS analyses (at %) for alloys C-0, C-0.25, C-0.50, and C-1.00.

Figs. 11(a)-(b) show the microstructure of C-0 before and after 3-d immersion in 0.5 M H_2SO_4, respectively. Figs. 11(c)-(d) show the microstructure of C-0.25 before and after 3-d immersion in 0.5 M H_2SO_4, respectively. General corrosion occurs for both C-0 and C-0.25, as revealed by EDS analyses (Table 6).

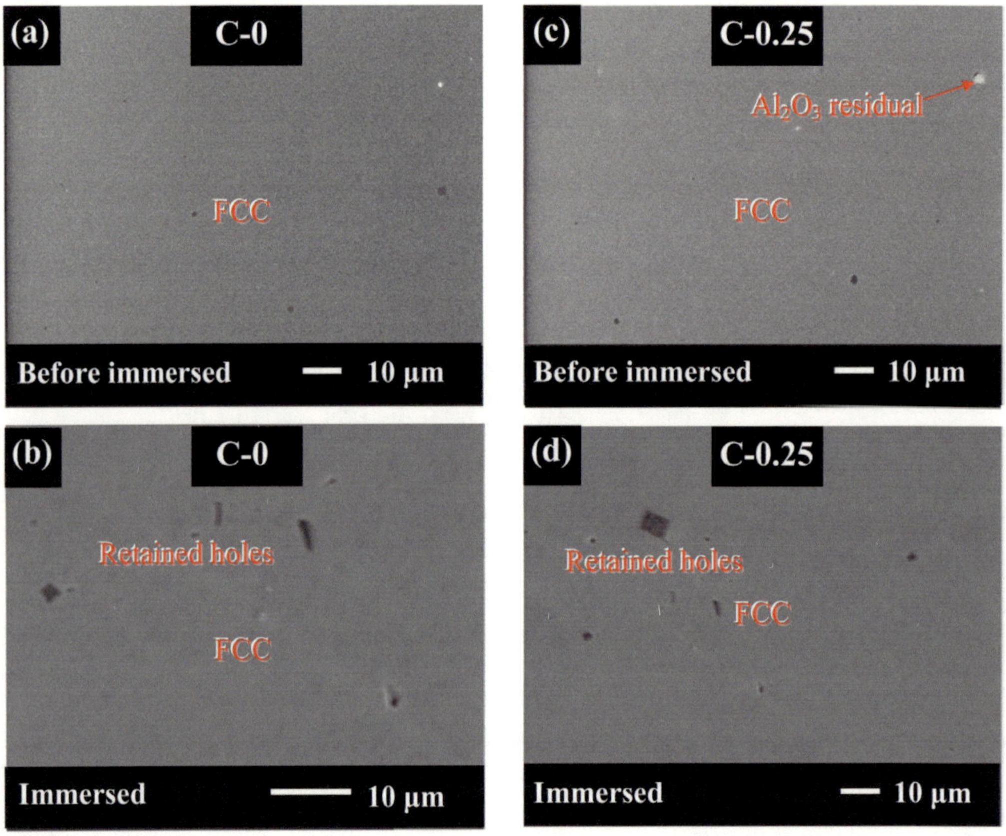

Fig. 11. Metallograph of alloys C-0 ((a) & (b)) and C-0.25 ((c) & (d)). (a) & (c), before; and (b) & (d), after immersion. Retained holes were from cast procedure, and Al_2O_3 residuals were from polishing procedure.

Figs. 12(a)-(b) show the microstructure of C-0.50 before and after 3-d immersion in 0.5 M H_2SO_4, respectively. According to these figures, after immersion the FCC phase remains smooth while the BCC phase shows a rough morphology. Fig. 12(c) shows a line-scanned area across the FCC and BCC phases for an immersed sample. Fig. 12(d) summarizes the line-scanned results, indicating that the BCC phase of C-0.50 before immersion is rich in Al and Ni. However, after immersion, it is poor in Al and Ni and rich in Cr.

Fig. 13 shows the microstructure and line-scan analysis of C-1.00 before and after immersion. Before immersion, BCC and ordered BCC phases cannot be resolved from the microstructure. The composition of BCC phase after alloy immersion is close to the overall alloy composition before immersion, indicating that the BCC phase is a corrosion-resistant phase. Moreover, the change in overall composition after immersion is attributed to the selective dissolution of Al and Ni in the ordered BCC phase of this alloy (Table 6).

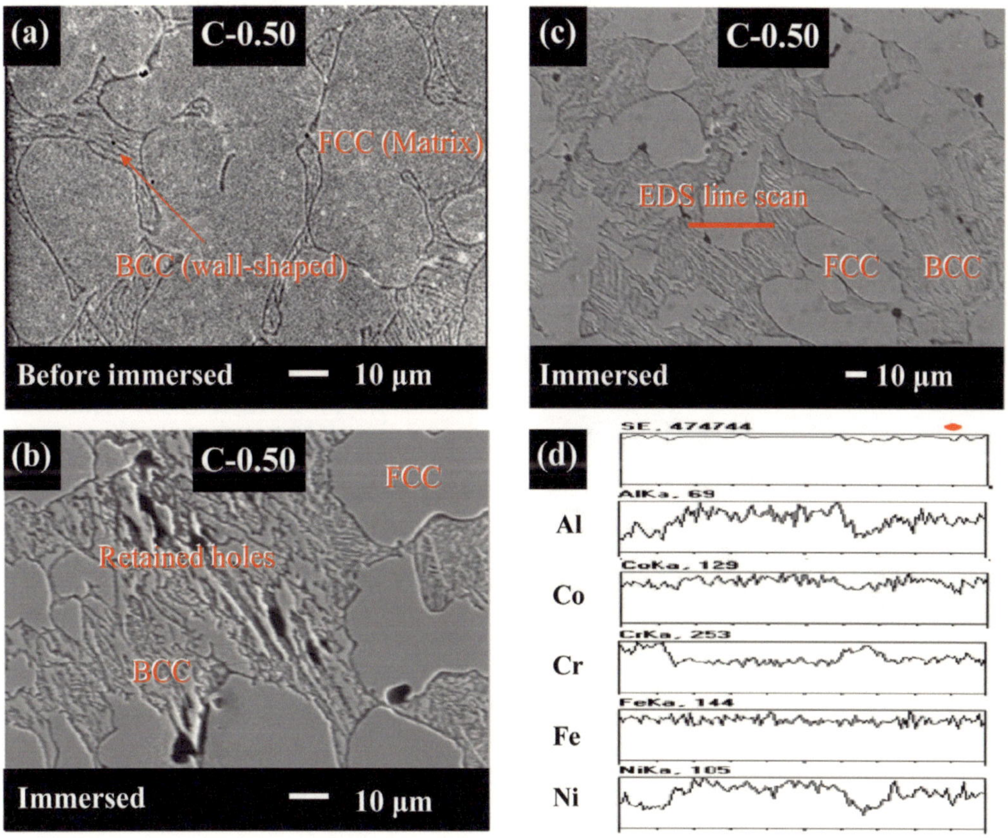

Fig. 12. Metallograph of alloy C-0.50. (a), before; (b) & (c), after immersion; and (d), EDS line-scan results of the location indicated in (c). Retained holes were from cast procedure.

This selective corrosion in Al and Ni-rich phase in C-0.50 and C-1.00, which results in the corrosion attack on Al and Ni, is due to the large bonding in Al and Ni [37]. Alloys containing this bonding readily react with $(OH)^-$ and $(SO_4)^{2-}$ to form Al and Ni complexes and dissolve in a sulfuric solution. Accordingly, after immersion, the remaining compound in the less corrosive-resistant Al and Ni-rich phase is an oxide, rich in Cr, in the residual passive film.

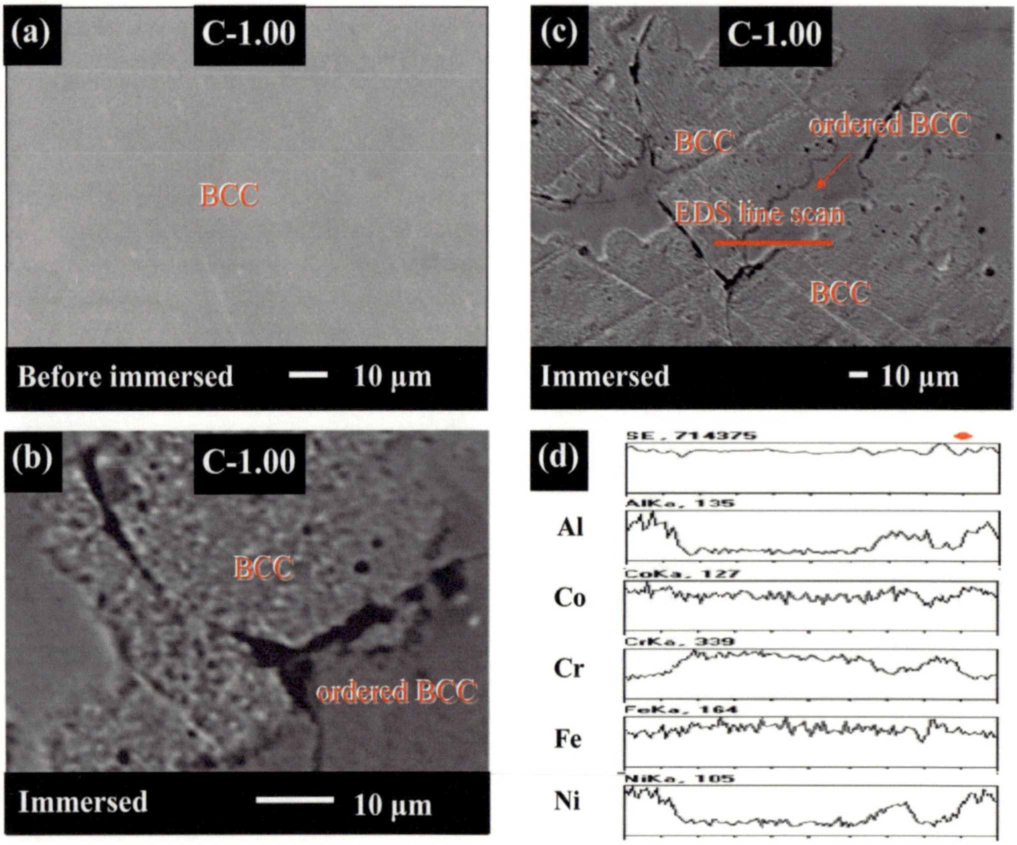

Fig. 13. Metallograph of alloy C-1.00. (a), before; (b) & (c), after immersion; and (d), EDS line-scan results of the location indicated in (c).

3.6 Comparison among potentiodynamic polarization, electrochemical impedance spectroscopy, and weight-loss immersion tests

As discussed above, the corrosion current density (I_{corr}), the critical current density (I_{cri}), and the passive current density (I_{pass}) were obtained from potentiodynamic polarization. The capacitance (Q_f) and the resistance (R_f) of oxide layer were obtained from electrochemical impedance spectroscopy (EIS) equivalent circuits. And the weight-loss rate (W_{loss}) was obtained from weight-loss immersion test. All these data were taken from experiments at ambient temperature (25°C) in 0.5 M H_2SO_4.

Figs. 14(a)-(b), whose data were listed in Table 2, show I_{cri} and I_{pass} vs. Al content x plots, respectively. One can easily see that both I_{cri} and I_{pass} increase with x. This implies that the passive corrosion property of $Al_xCoCrFeNi$ decreases with Al content x. Fig. 14(c) shows W_{loss} vs. x plot. Like I_{cri} and I_{pass}, W_{loss} also increases with x. Notice that, unlike potentiodynamic polarization, immersion weight-loss test is a natural electrochemical reaction, i.e., without applying any voltage on the test sample. On the other hand, I_{cri} and

I_{pass} locate at the passive region of polarization curve. The same tendency for I_{cri}, I_{pass}, and W_{loss} here indicates that the spontaneous passivation occurs for $Al_xCoCrFeNi$, i.e., the open circuit potential (OCP) is readily in the passive region of polarization curve. The above phenomenon can be attributed to the spontaneous passivation of pure Al in H_2SO_4 [24]. EIS equivalent circuits reveal that the passive layers of $Al_xCoCrFeNi$ consist of an oxide layer and an adsorption layer mentioned in Section 3.3. Here, only parameters associated with the oxide layer, i.e., Q_f and R_f, are discussed. The oxide layer thickness is evaluated by using the Helmholtz model mentioned above and denoted by d, as $d = \varepsilon\varepsilon_0 S/Q_f$, where ε_0 denotes the permittivity of free space (8.85×10^{-14} F/cm), ε denotes the dielectric constant of the medium, and S denotes the surface area of the electrode. Assuming that ε and S for oxide layers of alloys are the same allows us to compare relative values of d for all samples by $1/Q_f$. Figs. 14(d)-(e), whose data were listed in Table 5, show the Q_f and R_f vs. x plot, respectively. Both Q_f and R_f decreases with x. This represents that d increases with Al content x, and a thicker oxide layer implies a smaller value of impedance. Therefore, one can explain this phenomenon by considering both the thickness and the density of oxide layer. Related study reported Al oxide easily forms a porous structure [25]. Hence, it is easily understood that in addition to causing a thicker oxide layer, Al promotes the dispersive and porous oxide layer. In summary, Al has a negative effect to the passive parameters, including I_{cri}, I_{pass}, W_{loss}, and R_f, for $Al_xCoCrFeNi$ in H_2SO_4.

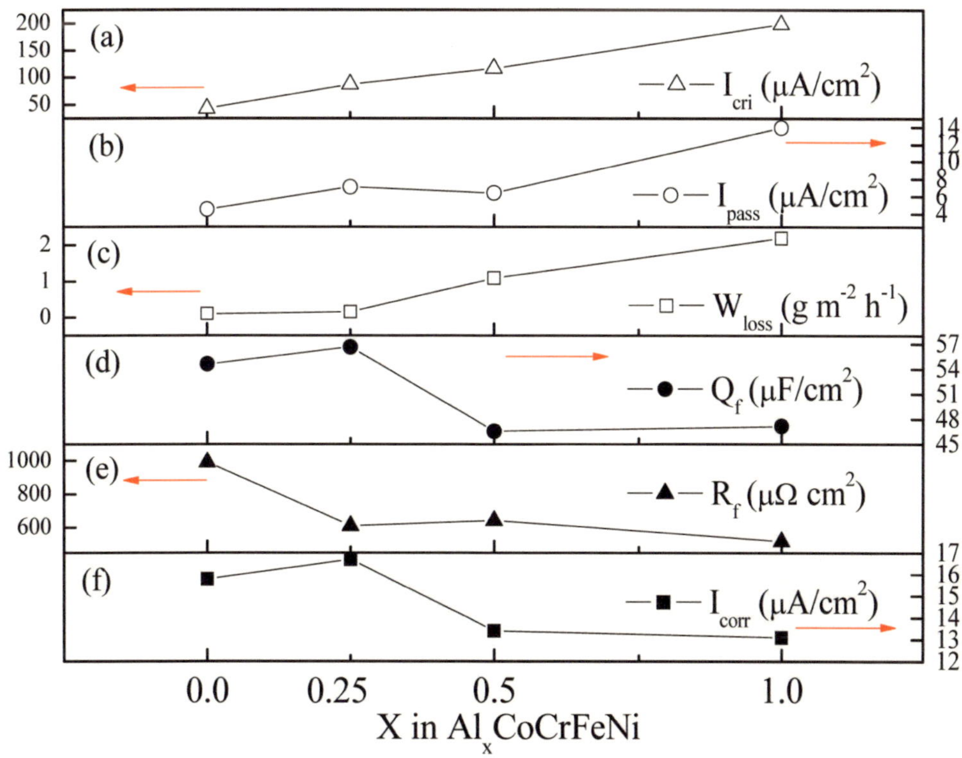

Fig. 14. (a) I_{cri}, (b) I_{pass}, (c) W_{loss}, (d) Q_f, (e) R_f, and (i) I_{corr} values vs. Al content x plots.

Interestingly, Al makes a different effect on general corrosion. Fig. 14(f) shows I_{corr} vs. x plot. One can see that I_{corr} decreases with x. This implies that Al promotes the general corrosion resistance, but degrades the passive one.

3.7 AES, XPS, and ICP-AES analyses of oxide layers

Figs. 15(a)-(d) show the AES results for C-0, C-0.25, C-0.50, and C-1.00, respectively. Owing to the slight difference of atomic number, the signals of Fe, Co, and Ni overlap in AES analysis. Hence, one can see the signals of Co are higher than that of Fe or Ni even for the equal-mole nominal chemical composition of Fe, Co, and Ni. What mentioned above, only the longitudinal composition profiles of O are discussed. A negative and a near-zero slopes are revealed in the relative concentration vs. sputter time profiles in Figs. 15(a)-(d).

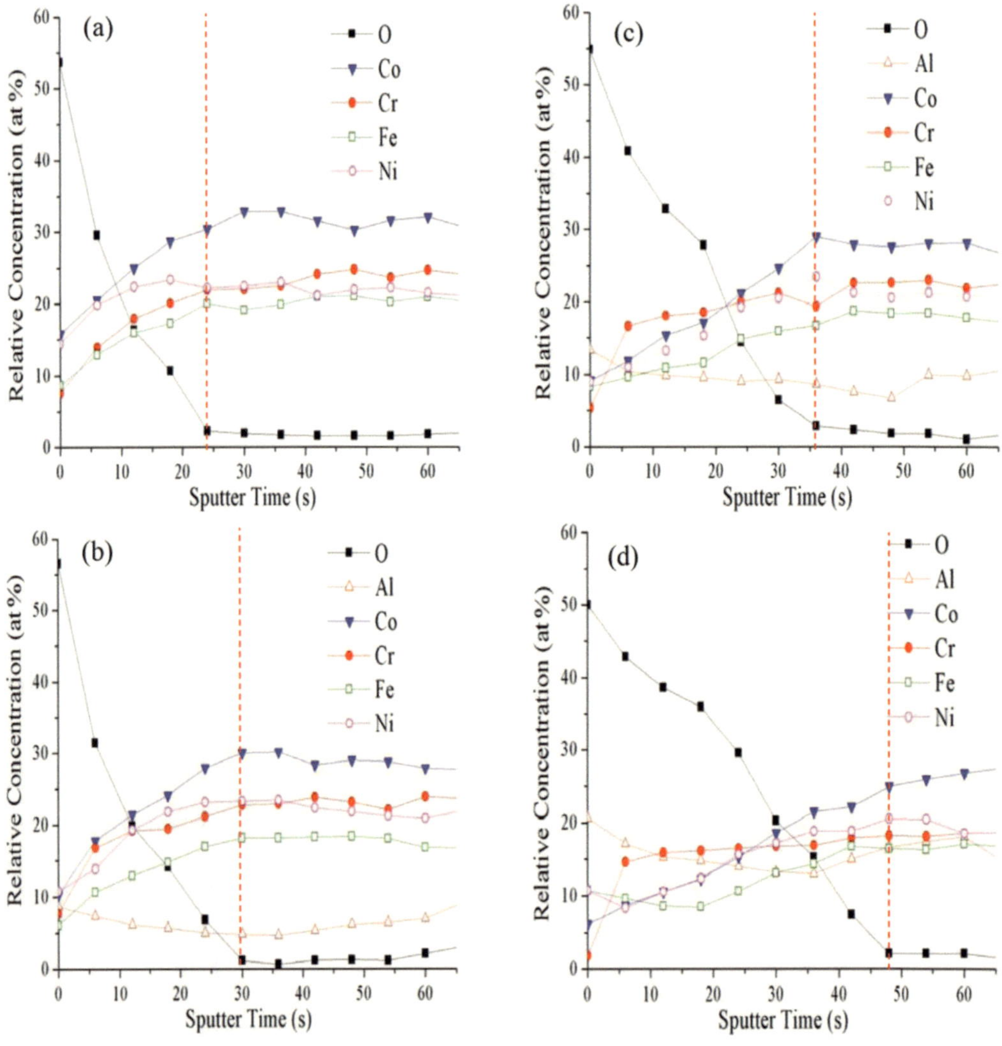

Fig. 15. AES analyses for (a) C-0, (b) C-0.25, (c) C-0.50, and (d) C-1.00.

A value for the identified sputter rate for SiO_2 in this AES device is 7.5 nm per min. Because the immersion for the samples is in the same 1-h period of time, the oxide layers of C-0, C-0.25, C-0.50, and C-1.00 can be distinguished from the terminals of the negative slope shown in each of the profiles. Look at the vertical red-dashed line, i.e., the end-terminal of each profile. It represents the interface between the oxide layer and the intrinsic metal. One can see that the thickness of the oxide layer increases with Al content x that is in accordance with the results of EIS.

XPS analyses attempt to investigate the binding energy profile of 2psub<3/2>sub for Al, Co, Cr, Fe, and Ni. Compared to Co, Cr, Fe, and Ni, Al reveals relatively low atomic sensitivity factor [38]. The signals of Al for alloys C-0 to C-0.50 are too small to identify. Hence, only C-1.00 was used for XPS analysis. Figs. 16(a)-(b) show the Al(2psub<3/2>sub) spectra of C-1.00 after the sputter times of 20 and 35 s, respectively. The raw profile revealing two main peaks represents the exhibition of the selective dissolution. The oxides consists of Al_2O_3, $Al(OH)_3$, and $Al_{25}Ni_{75}O_x$. Al tends to form oxides in H_2SO_4 [24] can explain the formation of Al_2O_3 and $Al(OH)_3$. The existence of $Al_{25}Ni_{75}O_x$ results from the relatively negative enthalpy of Al and Ni. Corresponding to the ICP-AES analysis in the next section, Al-Ni selective dissolution undoubtedly exists for C-1.00. Figs. 16(c)-(d) show the Co(2psub<3/2>sub) spectra of C-1.00 after the sputter times of 20 and 35 s, respectively. The binding types of Co^{2+} and Co^{3+} can be seen. Compare Fig. 16(c) with Fig. 16(d), one can see that the peak intensity of Co_2O_3 is very small in the deep region of the oxide layer. Figs. 16(e)-(f) show the Cr(2psub<3/2>sub) spectra of C-1.00 after the sputter times of 20 and 35 s, respectively. Three kinds of oxide, including Cr_2O_3, $Cr(OH)_3$, and CrO_3, exist for Cr [39]. However, CrO_3 merely forms at high temperatures. Hence, only Cr_2O_3 and $Cr(OH)_3$ are revealed in the profile. One can see that the deep region of oxide layer remains in relatively small amount Cr_2O_3. Figs. 16(g)-(h) show the Fe(2psub<3/2>sub) spectra of C-1.00 after the sputter times of 20 and 35 s, respectively. Similar to references [23,40] Fe_3O_4 and Fe_2O_3 oxides can be found. Figs. 16(i)-(j) show the Ni(2psub<3/2>sub) spectra of C-1.00 after the sputter times of 20 and 35 s, respectively. In resemblance with reference [41], NiO and $Ni(OH)_2$ can be observed. However, a very small amount of $Ni(OH)_2$ appears in our case.

Table 7 lists the results of ICP-AES of immersion solutions for C-x. To trace the ions resulting from the intrinsic metal, one can study the selective dissolution of the alloy elements. Compared with C-0 and C-0.25, C-0.50 and C-1.00 reveal relatively greater Al-Ni selective dissolution. This event is consistent with the results of the XPS analysis.

Alloys		Al	Co	Cr	Fe	Ni	Remarks
C-0	alloy	0	25.93	25.73	24.21	24.13	
	solution	0	24.91	24.89	25.17	25.01	*
C-0.25	alloy	6.16	23.27	23.58	23.59	23.40	
	solution	7.90	23.01	23.07	23.20	22.80	*
C-0.50	alloy	11.01	22.77	22.61	21.70	21.92	
	solution	14.92	21.90	16.51	19.96	26.70	**
C-1.00	alloy	18.88	20.55	20.63	20.01	19.93	
	solution	31.52	20.98	4.86	14.92	27.71	**

*General corrosion, **Selective dissolution in Al and Ni

Table 7. ICP-AES composition (at%) of immersion solution for alloys C-0, C-0.25, C-0.50, and C-1.00.

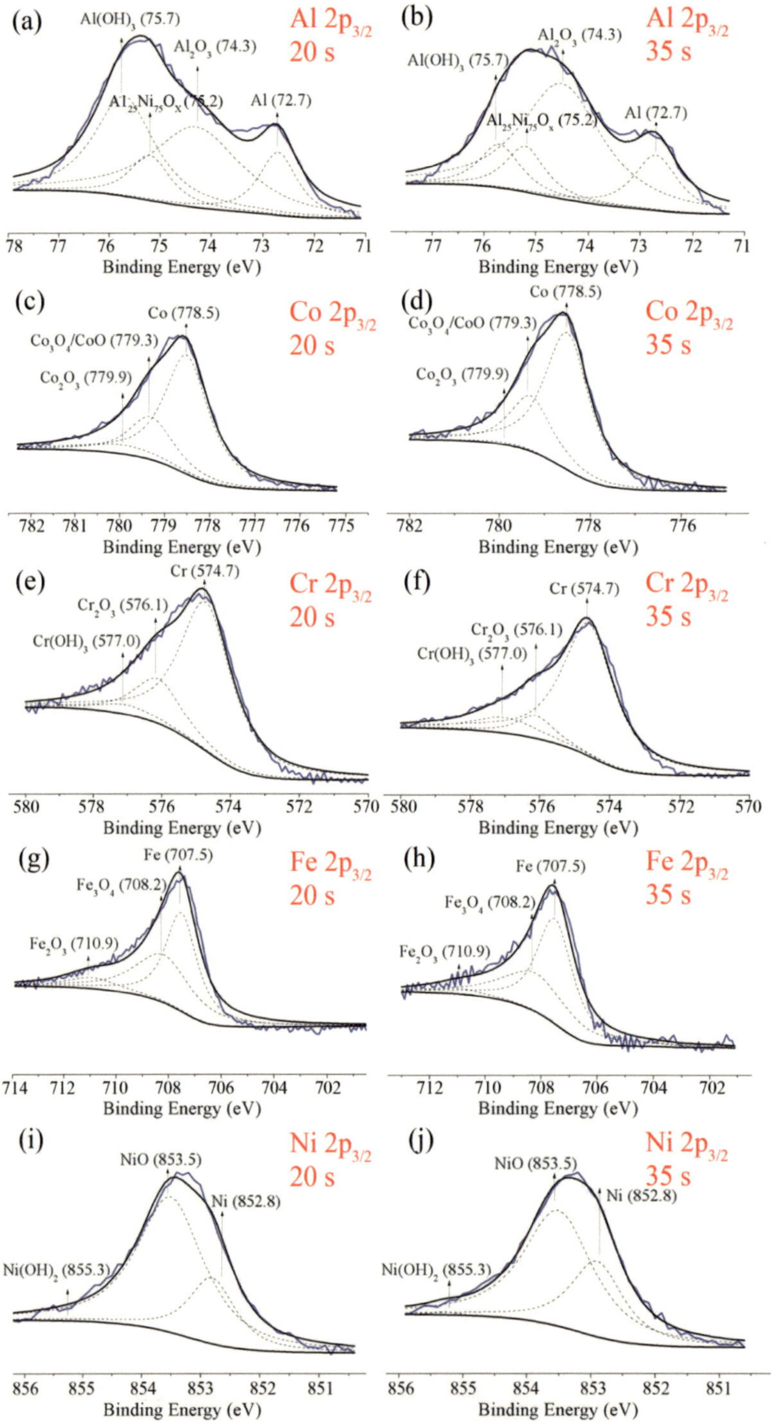

Fig. 16. XPS analyses after pre-sputtering for (a) Al-20 s, (b) Al-35 s, (c) Co-20 s, (d) Co-35 s, (e) Cr-20 s, (f) Cr-35 s, (g) Fe-20 s, (h) Fe-35 s, (i) Ni-20 s, and (j) Ni-35 s.

3.8 Corrosion current density (I_{corr}) at various temperatures

As mentioned in Section 3.1, I_{corr} decreases with Al content at 25°C. However, this differs from temperatures to temperatures. Fig. 17 shows the I_{corr} values of C-x at various temperatures. One can see that I_{corr} decreases with Al content x at low temperatures (< 27°C), and, conversely, at high temperatures (> 27°C). The EIS results (Section 3.1) indicate that more Al content x makes the oxide layers thicker and more dispersive. At low temperatures, the thicker oxide is the dominator for I_{corr}; whereas, at high temperatures, the dispersive oxide dominates. Therefore, this special phenomenon occurs.

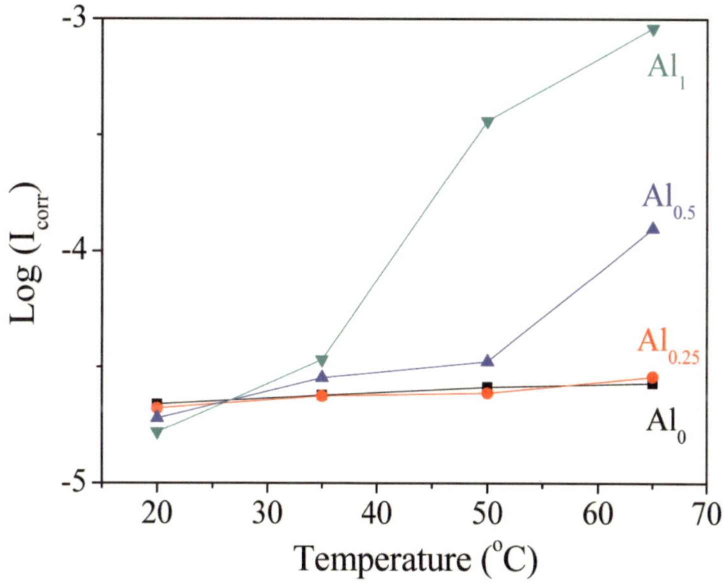

Fig. 17. I_{corr} values for alloys C-x (Al_x) at various temperatures.

4. Conclusions

Owing to the spontaneous passivation of Al element in H_2SO_4, the variation of Al reveals a more apparent effect in a passive region rather than in an active one. Therefore, in contrast with I_{pass}, which increases with x, no obvious trends occur for E_{corr} and I_{corr} vs. x variation. In particular, the weight loss experiment indicates that I_{pass} is a proper index to evaluate the weight loss of samples since $Al_xCoCrFeNi$ alloys are found to have passive behaviour in long-term dipping.

EIS results indicate that the passive films of $Al_xCoCrFeNi$ alloys become increasingly thicker and more dispersive with an increasing x. Therefore, I_{pass} increases with x. As x value increases to 1.00, the inductance effect appears in the equivalent circuit for severe dissolution of Al and Ni-rich phase. As for the effect of chloride on the anti-corrosion property, chloride eases the passive layer to form metastable ion complexes and further dissolve into H_2SO_4. With an increasing chloride concentration and Al content, the metastable ion complexes easily form, allowing E_{pit} to shift to a more active region. Additionally, the microstructure of both C-0 and C-0.25 is single FCC phase, while those of C-0.50 and C-1.00 are duplex FCC-BCC and complex BCC-ordered BCC phase, respectively.

In C-0.50 and C-1.00, the secondary passivation phenomenon in polarization curve results from selective dissolution of the Al and Ni-rich phase.

Moreover, I_{corr} increases with x at higher temperatures (> 27°C), while I_{corr} decreases with x at lower ones (< 23°C). That more closely examining Arrhenius plots of I_{corr} reveals that both pre-exponential factor A and activation energy E_a increase with Al content. However, A affects I_{corr} more significantly than it does so for E_a at higher temperatures (> 27°C) and, conversely, at lower temperatures (< 23°C).

Al is an inferior factor to the passive corrosive resistance but helpful for the general corrosive resistance for $Al_xCoCrFeNi$ in H_2SO_4. The thickness and the density of oxide layers promoted by the addition of Al compete with each other at various temperatures. At ambient temperature, the thick oxide layer dominates I_{corr} value; at temperatures higher than 27°C, the loss oxide layer does. Intuitionally, one may improve the corrosion performance for $Al_xCoCrFeNi$ by adjusting Al content.

5. Acknowledgements

The author would like to thank the financial support of this research from the National Science Council of the Republic of China, Taiwan (NSC96-2221-E007-066-MY3). Mr. Yih-Farn Kao is grateful for his help in compilation of this manuscript. This work is mainly from the 2009 master thesis of the Department of Materials Science Engineering of the National Tsing Hua University by Mr. Tsung-Dar Lee, who was guided by the author.

6. References

[1] See for example, D.R. Gaskell, Introduction to the Thermodynamics of Materials, 3rd ed., Taylor & Francis, Washington D.C., 1995, p. 204.

[2] K.H. Huang, Multicomponent alloy systems containing equal-mole elements, M.S. thesis, Department of Materials Science and Engineering, NTHU, Taiwan, 1996.

[3] K.T. Lai., Microstructure and properties of multicomponent alloy system with equal-mole elements, M.S. thesis, Department of Materials Science and Engineering, NTHU, Taiwan, 1998.

[4] Y.F. Kao, S.K. Chen, J.H. Sheu, J.T. Lin, W.E. Lin, J.W. Yeh, S.J. Lin, T.H. Liou, C.W. Wang, Hydrogen storage properties of multi-principal-component $CoFeMnTi_xV_yZr_z$ alloys, Int. J. Hydrogen Energy 35 (2010) 9046-9059.

[5] J.W. Yeh, S.K. Chen, S.J. Lin, J.Y. Gan, T.S. Chin, T.T. Shun, C.H. Tsau, S.Y. Chang, Nanostructured high-entropy alloys with multiple principal elements: Novel alloy design concepts and outcomes, Adv. Eng. Mater. 6 (2004) 299-303.

[6] P.K. Huang, J.W. Yeh, T.T. Shun, S.K. Chen, Multi-principal-element alloys with improved oxidation and wear resistance for thermal spray coating, Adv. Eng. Mater. 6 (2004) 74-78.

[7] C.Y. Hsu, J.W. Yeh, S.K. Chen, T.T. Shun, Wear resistance and high-temperature compression strength of FCC $CuCoNiCrAl_{0.5}Fe$ alloy with boron addition, Metall. Mater. Trans. A 35A (2004) 1465-1469.

[8] J. Tong, S.K. Chen, J.W. Yeh, T.T. Shun, C.H. Tsau, S.J. Lin, S.Y. Chang, Mechanical performance of the $Al_xCoCrCuFeNi$ high-entropy alloy system with multiprincipal elements, Metall. Mater. Trans. A 36A (2005) 1263-1271.

[9] J.W. Yeh, S.K. Chen, J.Y. Gan, S.J. Lin, T.S. Chin, T.T. Shun, C.H. Tsau, S.Y. Chang, Formation of simple crystal structures in Cu-Co-Ni-Cr-Al-Fe-Ti-V alloys with multiprincipal metallic elements, Metall. Mater. Trans. A 35A (2004) 2533-2536.

[10] Y.F. Kao, T.J. Chen, S.K. Chen, J.W. Yeh, Microstructure and mechanical property of as-cast, -homogenized, -deformed $Al_xCoCrFeNi$ ($0 \le x \le 2$) high-entropy alloys, J. Alloys & Comp. 488 (2009) 57-64.

[11] H.P. Chou, Y.S. Chang, S.K. Chen, J.W. Yeh, Microstructure, thermophysical and electrical properties in $Al_xCoCrFeNi$ ($0 \le x \le 2$) high-entropy alloys, Mater. Sci. Eng. B 163 (2009) 184-189.

[12] Y.Y. Chen, T. Duval, U.D. Hung, J.W. Yeh, H.C. Shih, Microstructure and electrochemical properties of high entropy alloys—a comparison with type-304 stainless steel, Corros. Sci. 47 (2005) 2257-2279.

[13] Y.Y. Chen, U.T. Hong, H.C. Shih, J.W. Yeh, T. Duval, Electrochemical kinetics of the high entropy alloys in aqueous environments—a comparison with type 304 stainless steel, Corros. Sci. 47 (2005) 2679-2699.

[14] C.P. Lee, C.C. Chang, Y.Y. Chen, J.W. Yeh, H.C. Shih, Effect of the aluminium content of $Al_xCrFe_{1.5}MnNi_{0.5}$ high-entropy alloys on the corrosion behaviour in aqueous environments, Corros. Sci. 50 (2008) 2053-2060.

[15] C.P. Lee, Y.Y. Chen, C.Y. Hsu, J.W. Yeh, H.C. Shih, Enhancing pitting corrosion resistance of $Al_xCrFe_{1.5}MnNi_{0.5}$ high-entropy alloys by anodic treatment in sulfuric acid, Thin Solid Films 517 (2008) 1301-1305.

[16] C. P. Lee, Y. Y. Chen, C. Y. Hsu, J. W. Yeh, and H. C. Shih, The Effect of Boron on the Corrosion Resistance of the High-Entropy Alloys $Al_{0.5}CoCrCuFeNiB_x$, J. Electrochem. Soc., 154 (2007) C424-C430.

[17] Y.F. Kao, T.D. Lee, S.K. Chen, Y.S. Chang, Electrochemical passive properties of $Al_xCoCrFeNi$ (x = 0, 0.25, 0.50, 1.00) alloys in sulfuric acids, Corros. Sci. 52 (2010) 1026-1034.

[18] V. Ashworth, P.J. Boden, Potential-pH diagrams at elevated temperatures, Corros. Sci. 10 (1970) 709-718.

[19] Y.Y. Chen, L.B. Chou, L.H. Wang, J.C. Oung, H.C. Shih, Electrochemical polarization and stress corrosion cracking of alloy 690 in 5-M chloride solutions at 25°C, Corros. 61 (2005) 3-11.

[20] M. Femenia, J. Pan, C. Laygraf, In situ local dissolution of duplex stainless steels in 1 M H_2SO_4 + 1 M NaCl by electrochemical scanning tunneling microscopy, J. Electrochem. Soc. 149 (2002) B187-B197.

[21] I.H. Lo, W.T. Tsai, Effect of selective dissolution on fatigue crack initiation in 2205 duplex stainless steel, Corros. Sci. 49 (2007) 1847-1861.

[22] F. D. Bogar, M. H. Peterson, A comparison of actual and estimated long-term corrosion rate of mild steel in seawater, Laboratory Corrosion Test and Standards, ASTM STP 866 (1985) 197-206.

[23] P. Marcus, Corrosion Mechanisms in Theory and Practice, 2nd ed., Marcel Dekker, New York, 2002.

[24] V. Shankar Rao, V.S. Raja, Anodic polarization and surface composition of Fe-16Al-0.14C alloy in 0.25 M sulfuric acid, Corros. 59 (2003) 575-583.
[25] L. Young, Anodic Oxide Films, 1st ed., Academic Press, London, 1961.
[26] G.K. Gomma, Corrosion of low-carbon steel in sulphuric acid solution in presence of pyrazole-halides mixture, Mater. Chem. & Phys. 55 (1998) 241-246.
[27] C.F. Zinola, A.M. Castro Luna, The inhibition of Ni corrosion in H_2SO_4 solutions containing simple non-saturated substances, Corro. Sci. 37 (1995) 1919-1929.
[28] M.R.F. Hurtado, P.T.A. Sumodjo, A.V. Benedetti, Electrochemical studies with a Cu-5 wt.% Ni alloy in 0.5 M H_2SO_4, Electrochimica Acta 48 (2003) 2791-2798.
[29] F.M Reis, H.G. de Melo, I. Costa, EIS investigation on Al 5052 alloy surface preparation for self-assembling monolayer, Electrochimica Acta 51 (2006) 1780-1788.
[30] T.M. Yue, L.J. Yan, C.P. Chan, C.F. Dong, H.C. Man, G.K.H. Pang, Excimer laser surface treatment of aluminum alloy AA7075 to improve corrosion resistance, Surface and Coating Technology 179 (2004) 158-164.
[31] I. Epelboin, C. Gabrielle, M. Keddam, H. Takenouti, Achievements and tasks of electrochemical engineering, Electrochimica Acta 22 (1975) 913-920.
[32] M. Metikoš-Huković, R. Babić, S. Brinić, EIS-in situ characterization of anodic films on antimony and lead-antimony alloys, J. Power Sources 157 (2006) 563-570.
[33] A.R. Trueman, Determining the probability of stable pit initiation on aluminium alloys using potentiostatic electrochemical measurements, Corros. Sci. 47 (2005) 2240-2256.
[34] Y.M. Tang, Y. Zuo, X.H. Zhao, The metastable pitting behaviours of mild steel in bicarbonate and nitrite solutions containing Cl^-, Corros. Sci. 50 (2008) 989-994.
[35] R.T. Foley, T.H. Nguyen, The chemical nature of aluminum corrosion, J. Electrochem. Soc. 129 (1982) 464-467.
[36] S. Van Gils, C.A. Melendres, H. Terryn, E. Stijns, Use of in-situ spectroscopic ellipsometry to study aluminium/oxide surface modifications in chloride and sulfuric solutions, Thin Solid Films 455 (2004) 742-746.
[37] H. Nakazawa, H. Sato, Bacterial leaching of cobalt-rich ferromanganese crusts, International Journal of Mineral Processing 43 (1995) 255-265.
[38] C.D. Wagner, W.M. Riggs, L.E. Davis, J.F. Moulder, G.E. Muilenberg, Handbook of X-Ray Photoelectron Spectroscopy, 1st ed., Perkin-Elmer Corporation, Minnesota, 1979.
[39] A.A. Hermas, M. Nakayama, K. Ogura, Formation of stable passive film on stainless steel by electrochemical deposition of polypyrrole, Electrochimica Acta 50 (2005) 3640-3647.
[40] K. Varga, P. Baradlai, W.O. Barnard, G. Myburg, P. Halmos, J.H. Potgieter, Comparative study of surface properties of austenitic stainless steels in sulfuric and hydrochloric acid solutions, Electrochimica Acta 42 (1997) 25-35.
[41] R. Wang, An AFM and XPS study of corrosion caused by micro-liquid of dilute sulfuric acid on stainless steel, Appl. Surf. Sci. 227 (2004) 399-409.

7

Reinforcement Fibers in Zinc-Rich Nano Lithiun Silicate Anticorrosive Coatings

Carlos Alberto Giudice
UTN (Universidad Tecnológica Nacional),
CIDEPINT (Centro de Investigación y Desarrollo en Tecnología de Pinturas),
La Plata
Argentina

1. Introduction

Well-known the electrochemical nature of most processes of corrosion, the technology of anticorrosive coatings is oriented in the direction of making products that control the development of electrode reactions and that generate the isolating of metal surface by applying films with very low permeability and high adhesion (Sorensen et al., 2011).

The zinc-rich coatings and those modified with extenders and/or metal corrosion inhibitors display higher efficiency than other coatings. A problem that presents this type of primers is the extremely reactive characteristic of metallic zinc; consequently, the manufacturers formulate these coatings in two packages, which imply that the zinc must be incorporated to the vehicle in previous form to coating application (Jianjun et al., 2008 & Lei-lei & De-liang, 2010).

Considering the concept of sacrificial anode (cathodic protection), coatings that consist of high purity zinc dust dispersed in organic and inorganic vehicles have been designed; in these materials, when applied in film form, there are close contacts of the particles among themselves and with the base or metallic substrate to be protected.

The anodic reaction corresponds to the oxidation of zinc particles (loss of electrons) while the cathodic one usually involves oxygen reduction (gain of electrons) on the surface of iron or steel; the "pressure" of electrons released by zinc prevents or controls the oxidation of the metal substrate. Theoretically, the protective mechanism is similar to a continuous layer of zinc applied by galvanizing with some differences because the coating film initially presents in general a considerable porosity (Jegannathan et al., 2006).

In immersion conditions, the time of protection depends on the zinc content in the film and on its dissolution rate. The mechanism is different for films exposed to the atmosphere, because after the cathodic protection in the first stage, the action is restricted substantially to a barrier effect (inhibition resistance) generated by the soluble zinc salts from corrosion by sealing the pores controlling access to water, water vapor and various pollutants. Due to the above, it is necessary to find the appropriate formulation for each type of exposure in service (Hammouda et al., 2011).

With regard to zinc corrosion products, they are basic compounds whose composition varies according to environmental conditions (Wenrong et al., 2009); they are generally soluble in water and can present amorphous or crystalline structure. In atmospheric exposure, zinc-based coatings that provide amorphous corrosion products are more efficient since these seal better the pores and therefore give a higher barrier effect (lower permeability). Fortunately, zinc-rich coatings of satisfactory efficiency in outdoor exposure display in the most cases amorphous corrosion products.

The durability and protective ability depends, in addition to environmental factors, on the relationship between the permeability of the film during the first stage of exposure and the cathodic protection that takes place (Xiyan et al., 2010). The protection of iron and steel continues with available zinc in the film and effective electrical contact; therefore, particularly in outdoor exposure, the time of satisfactory inhibitory action may be more prolonged due to the polarizing effect of the corrosion products of zinc (Thorslund Pedersen et. al., 2009).

A cut or scratch of the film applied on polarized panel allows again the flow of protective electrical current: metallic zinc is oxidized and the film is sealed again. A substantial difference with other types of coatings is that the corrosive phenomenon does not occur under the film adjacent to the cut (undercutting).

With respect to spherical zinc, the transport of current between two adjacent particles is in tangential form and consequently the contact is limited. With the purpose of assuring dense packing and a minimum encapsulation of particles, the pigment volume concentration (PVC) must be as minimum in the order of the critical pigment volume concentration (CPVC).

The problems previously mentioned led to study other shapes and sizes of zinc particles. The physical and chemical properties as well as the behaviour against the corrosion of these primers are remarkably affected by quoted variables and in addition, by the PVC; thus, for example, it is possible to mention the laminar zinc, which was intensely studied by the authors in other manuscripts (Giudice et al., 2009 & Pereyra et al., 2007).

The objective of this paper was study the influence of the content and of the nature of reinforcement fibers as well as the type of inorganic film-forming material, the average diameter of spherical zinc dust and the pigment volume concentration on performance of environmentally friendly, inorganic coatings suitable for the protection of metal substrates. The formulation variables included: (i) two binders, one of them based on a laboratory-prepared nano solution lithium silicate of 7.5/1.0 silica/alkali molar and the other one a pure tetraethyl silicate conformed by 99% w/w monomer with an appropriate hydrolysis degree; (ii) two pigments based on spherical microzinc (D 50/50 4 and 8 μm); (iii) three types of reinforcement fibers used to improve the electric contact between two adjacent spherical zinc particles (graphite and silicon nitride that behave like semiconductor, and quartz that is a non-conductor as reference); (iv) three levels of reinforcement fibers (1.0, 1.5 and 2.0% w/w on coating solids) and finally, (v) six values of pigment volume concentration (from 57.5 to 70.0%).

2. Materials and methods

2.1 Characterization of main components

2.1.1 Film-forming materials

Water-based nano lithium silicate of 7.5/1.0 silica/alkali molar ratio. Previous experiences with these solutions on glass as substrate allowed infer that as silicon dioxide content in the composition increases the film curing velocity also increases and that in addition the dissolution rate decreases.

For this study, a commercial colloidal lithium silicate (3.5/1.0 silica/alkali molar ratio in solution at 25% w/w) was selected; with the aim of increasing the silica/alkali ratio, a 30% w/w colloidal alkaline solution of nanosilica was used (sodium oxide content, 0.32%). The aim was to develop a system consisting of an inorganic matrix (alkaline silicate) and a nanometer component (silica) evenly distributed in that matrix with the objective of determining its behaviour as binder for environment friendly, anticorrosive nano coatings.

Solvent-based, partially hydrolyzed tetraethyl orthosilicate. The tetraethyl orthosilicate is synthesized from silicon tetrachloride and anhydrous ethyl alcohol. This product commercializes as condensed ethyl silicates and usually contains approximately 28% w/w of SiO_2 and at least 90% w/w monomer. The additional purification removes waste products of low boiling point (mainly ethanol) and the dimmers, trimmers, etc.; in some cases, this treatment allows obtaining pure tetraethyl silicate conformed by 99% w/w monomer.

Theoretically, the complete hydrolysis of ethyl silicate generates silica and ethyl alcohol. Nevertheless, the real hydrolysis never produces silica in form of SiO_2 (diverse intermediate species of polysilicates are generated). Through a partial hydrolysis under controlled conditions, it is possible to obtain a stable mixture of polysilicate prepolymers. The stoichiometric equation allows calculating the hydrolysis degree X (Giudice et al., 2007 & Hoshyargar et al., 2009).

The pure or condensed ethyl silicate does not display good properties to form a polymeric material of inorganic nature. In this paper, ethyl silicate was prepared with 80% hydrolysis degree in an acid medium since catalysis carried out in advance in alkaline media led to a fast formation of a gel.

The empirical equation of ethyl silicate hydrolyzed with degree X was used to estimate the weight of the ethyl polysilicate and the hydrolysis degree, through the calculation of the necessary amount of water. The weight was obtained replacing the atomic weights in the mentioned empirical formula; the result indicates that it is equal to 208-148 X.

The percentual concentration of the silicon dioxide in the ethyl polysilicate is equal to the relation molecular weight of SiO_2 x 100 / weight of the ethyl polysilicate; consequently, SiO_2, % = 60 x 100 / (208-148 X). On the other hand, to calculate the water amount for a given weight of tetraethyl orthosilicate and with the purpose of preparing a solution of a predetermined hydrolysis degree, the equation weight of water = 36 (100 X) / 208 was used.

Finally, the amount of isopropyl alcohol necessary to reach the defined percentual level the silica content was calculated. It is possible to mention that after finishing the first hydrolysis stage of tetraethyl orthosilicate that leads to the silicic acid formation, the absence of alcohol would generate the polycondensation of mentioned acid with silica precipitation and the null capacity to conform a polymeric silicic acid (Wang et al., 2009 & Yang et al., 2008).

In a first stage, the pure tetraethyl silicate and the isopropyl alcohol were mixed under agitation. Later, the water and the hydrochloric acid solution selected as catalyst were added (the final pH of the solution was slightly acid, 0.01% w/v, expressed as hydrochloric acid); agitation continued until the end of the dissipation of heat (exothermic reaction).

The conclusions from the experiences indicate that: an excessive amount of water (higher than calculated) generates a rapid gelling in the package, a high pH leads to a fast silica precipitation that reduces the capacity of formation of an inorganic polymer of elevated molecular weight and, in addition, a large quantity of acid retards the condensing reaction due to the repulsion of protonated hydroxyl groups (Giudice et al., 2007).

2.1.2 Pigmentation

In this study, two samples of commercial spherical zinc dust were used; the D 50/50 average particle diameters were 4 μm (fine) and 8 μm (regular), Figure 1. The main features were respectively 98.1 and 98.3% of total zinc and 94.1 and 94.2% of metal zinc, which means 5.0 and 5.1% of zinc oxide; in addition, metal corrosion inhibitors displayed respectively 2282 and 1162 $cm^2.g^{-1}$ values of specific area (BET).

2.1.3 Reinforcement fibers

Nowadays reinforcement fibers are used in many materials to improve their physical and chemical properties (Huang et al., 2009 & Amir et al., 2010). A composite (FRP, fiber-reinforced polymer) is formulated and manufactured with the purpose of obtaining a unique combination of properties; the incorporation of reinforcement fibers to coatings forms a hybrid structure. Fiber is defined as any material that has a minimum ratio of length to average transverse dimension of 10 to 1; in addition, the transverse dimension should not exceed 250 μm.

In anticorrosive coating formulations with hybrid structures, the following reinforcing fibrous materials were used: graphite, silicon nitride and quartz. The levels selected for the experiment were 1.0, 1.5 and 2.0% w/w on coating solids.

Graphite. It is an allotropic form of carbon (hexagonally crystallized). It displays black color with metallic brightness and it is non-magnetic; it has 2.267 $g.cm^{-3}$ density at 25 °C. Usually, it is used as pigment in coatings to give conductive properties to the film. Graphite is formed by flakes or crystalline plates attached to each other, which are easily exfoliated. The electrons that are between layers are those that conduct electricity, and these are what give the quoted brightness (the light is reflected on the electron cloud) (Yoshida et al., 2009 & Abanilla et al., 2005). In perpendicular direction to the layers, it has a low electrical conductivity, which increases with the temperature (it behaves like a semiconductor); on the other hand, throughout the layers, the conductivity is greater and it is increased with the temperature, behaving like a semi-metallic conductor. In this work, graphite in fiber form was used with average values of 1020 μm and 82 μm for length and transverse dimension, respectively (Figure 2.A).

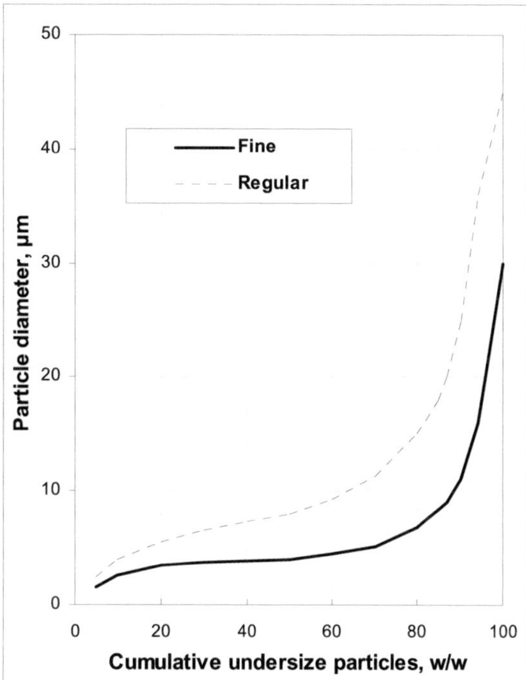

Fig. 1. Commercial spherical zinc dusts. D 50/50 average particle diameters: 4 µm (fine) and 8 µm (regular).

Silicon nitride (Si_3N_4). It displays three different crystal structures (α, β and γ). It is industrially obtained by direct reaction between silicon and nitrogen at temperatures between 1300 and 1400 °C. Silicon nitride is a material frequently used in the manufacture of structural ceramics with high requests of mechanical stress and wear resistance; it displays a moderately high modulus of elasticity and an exceptionally high tensile strength, which makes it attractive for its use in the form of fiber as reinforcement material for coatings films. It behaves like a semiconductor and it has 3.443 $g.cm^{-3}$ density at 25 °C. For this experience, hexagonal β phase in the form of fiber has been selected, with average values of 1205 µm and 102 µm for length and transverse dimension, respectively (Figure 2.B).

Quartz. It is rhombohedral crystalline silica reason why it is not susceptible of exfoliation; chemically it is silicon dioxide (SiO_2). Usually it appears colorless (pure), but it can adopt numerous tonalities if it has impurities; its hardness is such that it can scratch the common steels. It is often used in coatings as extender after being crushed and classified by size (average diameter between 1.5 and 9.0 µm). It is an insulating material from the electrical point of view; it has 2.650 $g.cm^{-3}$ density at 25 °C (Chen et al., 2010 & Lekka et al., 2009). In this experience, quartz fibers were used with average values of 1118 µm and 95 µm for length and transverse dimension, respectively (Figure 2.C).

These reinforcements cannot be classified as nano materials since they no have at least one of the dimensions inferior to 100 nm (Aluru et al., 2003; Radhakrishnan et al., 2009; Behler et al., 2009 & Li & Panigrahi 2006).

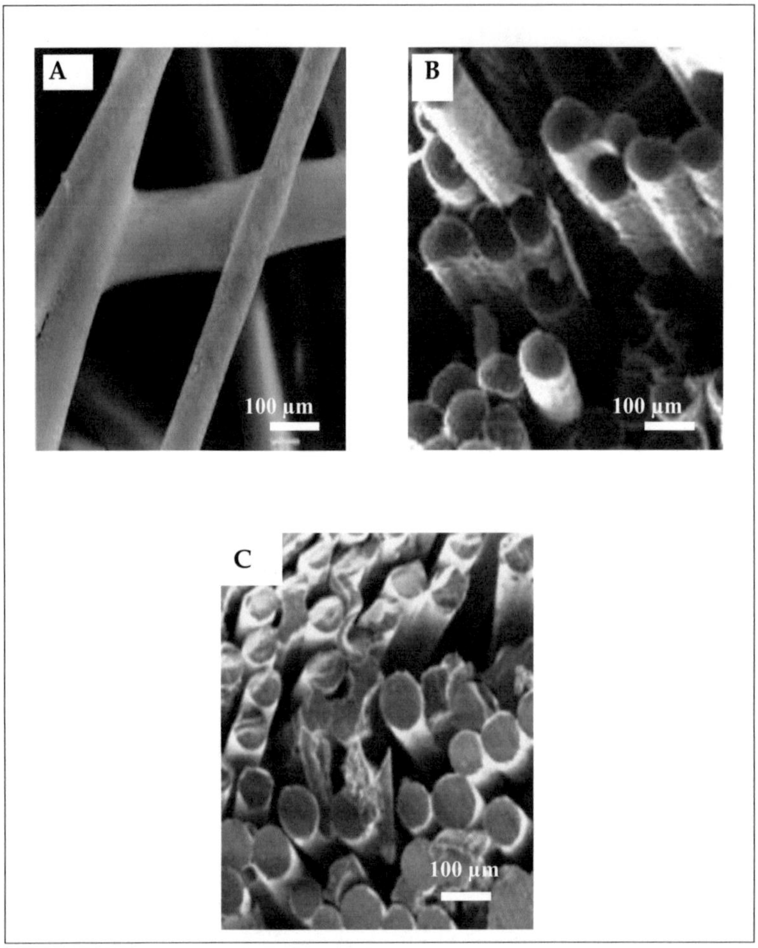

Fig. 2. SEM micrograph of reinforcement fibers: A. Graphite, B. Silicon nitride and C. Quartz.

2.1.4 Rheological agent

Clay modified with amines in gel form was chosen, which was added to the system after finishing the pigment dispersion (1.0% w/w on coating).

2.2 Pigment volume concentration

The physical and mechanical properties of the film and its protective capacity depend on the shape and size of the zinc particles and in addition of the PVC. The critical value (CPVC) is strongly influenced by the ability of binder for wetting the pigment particles.

In this study, PVC values ranged from 57.5 to 70.0%; the variation of two consecutive PVC values was 2.5% in all cases. Preliminary laboratory tests (salt spraying chamber), with values of PVC from 10 to 70% for all formulations, helped to define the range of PVC more convenient to study (Sonawane et al., 2010).

2.3 Manufacture of coatings

In this paper, the manufacture of hybrid coatings was made by ultrasonic dispersion in the vehicle. The efficiency of the fiber dispersion in the polymeric matrix was rheological monitored (viscosity of the system, measured at 10^{-3} sec^{-1}, decreased during the sonification process until reaching a stationary value); the SEM observation corroborated both the efficiency of fiber dispersion and its stability after 3 months in can.

Finally, it should also be mentioned that this type of compositions is provided in two packages with the purpose of avoiding the reaction of metallic zinc with any vestige of moisture in some of the components, which would lead to the formation of gaseous hydrogen. In addition, the system could form a gel because of the reaction between silicic acids and zinc cations. Accordingly, prior to the primer application, the metallic zinc was dispersed for 180 seconds at 1400 rpm in high-speed disperser.

2.4 Preparation of panels

SAE 1010 steel panels were previously degreased with solvent in vapor phase; then they were sandblasted to ASa 2½ grade (SIS Specification 05 59 00/67) obtaining 25 µm maximum roughness Rm. The application was made in a single layer reaching a dry film thickness between 75 and 80 µm. In all cases, and to ensure the film curing before beginning the tests, the specimens were kept in controlled laboratory conditions (25±2 °C and 65±5% relative humidity) for seven days.

The study was statistically treated according to the following factorial design: 2 (type of binder) x 2 (average diameter of spherical microzinc particles) x 3 (type of reinforcement fibers) x 3 (level of reinforcement fibers) x 6 (PVC values), which make 216 combinations. In addition, reference primers (without reinforcement fiber) based on both binders and both spherical microzinc particles for the six mentioned PVC values were also considered, which means in total 24 reference primers. All panels were prepared in duplicate; primer identification is displayed in Table 1.

2.5 Laboratory tests

After curing process, panels of 150 x 80 x 2 mm were immersed in 0.1 M sodium chloride solution for 90 days at 25 °C and pH 7.0. A *visual inspection* was realized throughout the experience; in addition, the *electrode potential* was determined as a function of exposure time; two clear acrylic cylindrical tubes were fixed on each plate (results were averaged).

The tube size was 10 cm long and 5 cm diameter, with the lower edge flattened; the geometric area of the cell was 20 cm^2. A graphite rod axially placed into the tubes and a saturated calomel electrode (SCE) were selected respectively as auxiliary and reference electrodes. The potential was measured with a digital electrometer of high input impedance.

Similar panels were tested in salt spraying (fog) chamber (1500 hours) under the operating conditions specified in ASTM B 117. After finishing the tests, the panels were evaluated according to ASTM D 1654 (Method A, in X-cut and Method B, in the rest of the surface) to establish the *degree of rusting*.

Film-forming material	(A) Water-based nano lithium silicate of 7.5/1.0 silica/alkali molar ratio
	(B) Solvent-based, partially hydrolyzed tetraethyl orthosilicate
Spherical microzinc	(I) Spherical microzinc (fine), D 50/50 4 μm
	(II) Spherical microzinc (regular), D 50/50 8 μm
Reinforcement fibers	Types: (1) Without, (2) Graphite, (3) Silicon nitride (4) Quartz
	Level: (a) 1.0, (b) 1.5 and (c) 2.0% w/w on coating solids
PVC	Values: 57.5, 60.0, 62.5, 65.0, 67.5 and 70.0%

Table 1. Primer identification.

3. Result and discussion

3.1 Visual observation

Immersion test in 0.1 M sodium chloride solution allowed to observe, particularly in those panels with X-cut, that coatings based on nano-structured film-forming material as binder showed greater amount of white products from corrosion of metallic zinc than in those panels protected with primers made with partially hydrolyzed ethyl silicate as binder. This performance would be supported in the less zinc dispersion ability that displays the first binder, which would generate films more porous.

On the other hand, primers that included fine zinc particles (4 μm) also showed a galvanic activity more important than those made from regular zinc particles (8 μm). A similar conclusion was reached with the primers based on microzinc/conducting reinforcement fibers (graphite and silicon nitride) with respect to those based just on spherical microzinc dusts and mixture with insulating reinforcement fiber (quartz). In turn, it was also observed a rise of the galvanic activity of metallic zinc when the amount of conducting fibers was increasing in the film.

For the lower values of PVC studied, the incorporation of conductive reinforcing fibers in increasing levels led to primers with a galvanic activity also increasing (similar amount of white salts than in primers formulated with PVC nearest to CPVC).

3.2 Corrosion potential

Immediately after finishing the immersion of all coated panels in the electrolyte, the potential was inferior to -1.10 V, a value located in the range of protection of the electrode. It is worth mentioning that cathodic protection is considered finished when the corrosion potential of coated panel increased to more positive values (anodic ones) than -0.86 V (referring to SCE) since the characteristic corrosion points of the iron oxides were visually observed.

The electrode potential measurements as a function of immersion time indicates that both types of binders had a significant influence on the electrode potential: in general, more negative values were obtained with nano-structured film-forming materials, which means that the primers based on lithium silicate showed better cathodic protection than those manufactured with ethyl silicate.

On the other hand, slight differences in electrode potential could also be attributed to the average diameter of zinc particles; it was observed greater galvanic activity in samples prepared with 4 µm than with 8 µm (values more negative of electrode potentials for the former than for the latter).

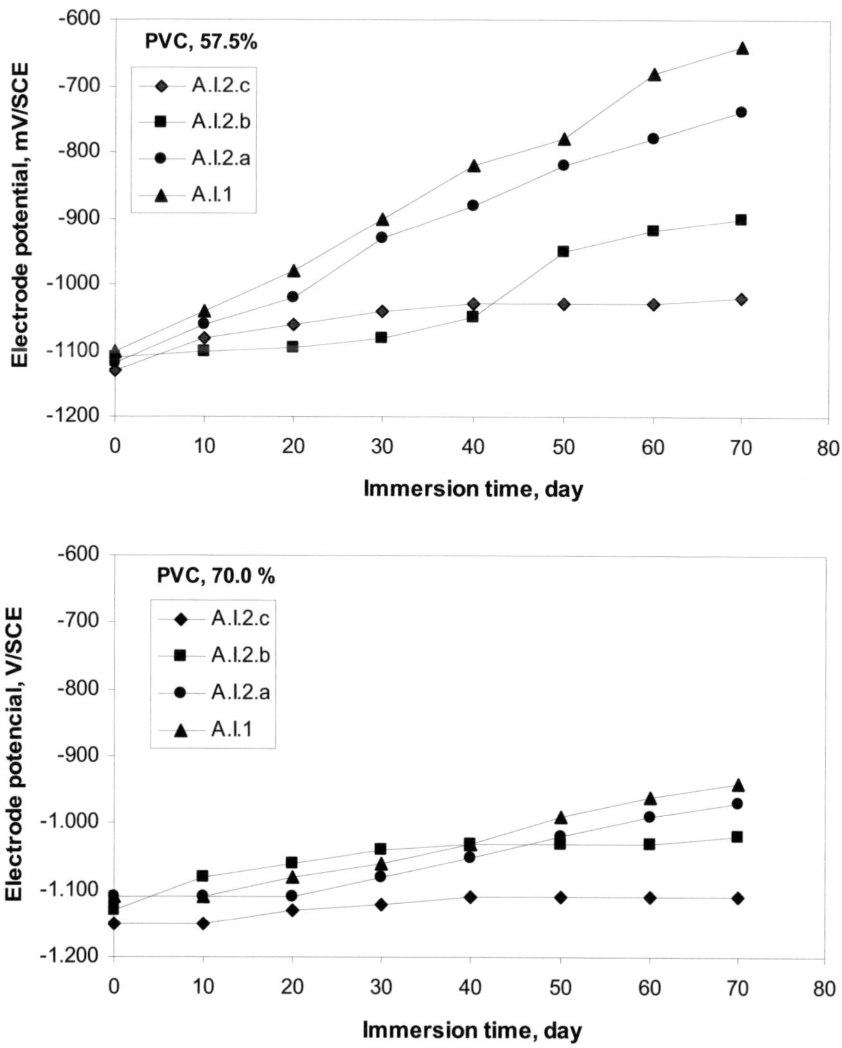

Fig. 3. Electrode potential vs. immersion time in 0.1 M sodium chloride solution at pH 7.0 and 25 °C for primers based on 7.5/1.0 nanosilica/lithium oxide molar ratio, fine microzinc and graphite fibers.

The experimental values indicate a significant shift towards more positive values of potential in those primers with decreasing amounts of conducting reinforcement fibers in their composition (2.0, 1.5 and 1.0% w/w ratio, in that order).

Considering the performance, the worst primers have been formulated both with microzinc dusts alone and mixed with insulation reinforcement fibers.

Finally, there was observed only a slightly decreasing efficiency to the lower values of PVC studied with the incorporation of conductive reinforcing fibers in increasing levels.

Figure 3 includes values of potential versus immersion time in 0.1 M sodium chloride solution at pH 7.0 and 25 °C for primers formulated with 57.5 and 70.0% PVC values and based on 7.5/1.0 nano silica/lithium oxide molar ratio as film-forming material, fine microzinc (D 50/50 4 μm) as pigment inhibiting and graphite as reinforcement fiber in the three levels studied. In addition, this figure displays the corresponding reference primers.

There is a total correlation between conclusions of visual observation and results of the electrode potentials obtained during immersion in 0.1 M sodium chloride solution; therefore, the basis of the quantitative results of electrode potentials are the same that those spelled out in the visual observation.

3.3 Degree of rusting

The performance of panels tested during 1500 hours in salt spraying (fog) chamber (35±1 °C; pH 6.5-7.2; continuous spraying of 5±1% w/w of NaCl solution) are shown in Figure 4 and Figure 5. They include only the average values of the tests performed in the failure in X-cut (Method A) and over the general area of panel (Method B).

The results of Method A were evaluated according to the advance from the cutting area: the value 10 defines a failure of 0 mm while zero corresponds to 16 mm or more. Those results corresponding to Method B were measured taking into account the percentage of area corroded by the environment: the scale ranges from 10 to 0, which means respectively no failure and over 75% of the rusted area.

On the other hand, Figure 6 displays one of the primer with best performance in salt spraying (fog) chamber for 1500 hours: A.I.2.c; applying the Method A, this primer showed a degree of rusting 10 (no failure, which means 0 mm of advance from the cutting area).

To study the variables considered (main effects), a statistical interpretation was carried out. First, the variance was calculated and later the Fisher F test was done.

The results indicated that type of binder, average diameter of microzinc particles, type of reinforcement fibers, level of reinforcement fibers and finally PVC values displayed an important influence on the performance of the protective coatings.

According to results, it was considered desirable for the statistical analysis to take into account all values of PVC studied for allowing a certain margin of safety in the performance since it is possible the generation of heterogeneities in primer composition attributable to poor incorporation of metallic zinc and/or sedimentation in container before applying.

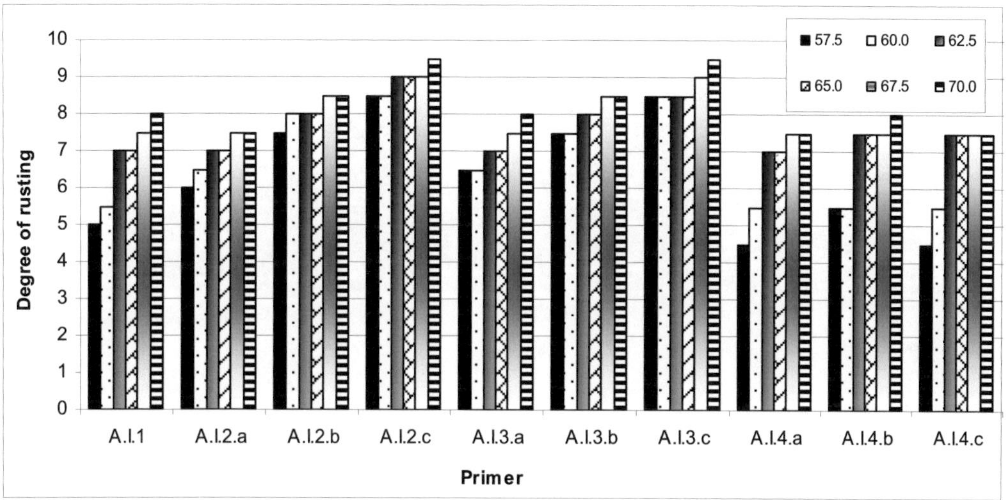

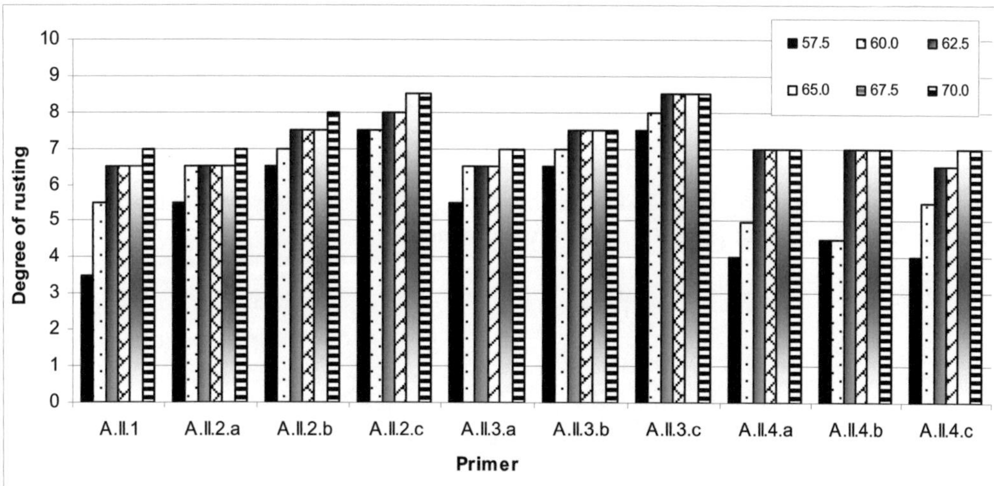

Fig. 4. Coatings based on binder A: Degree of rusting in salt spraying (fog) chamber; average values of failures in X-cut and in general area of panel.

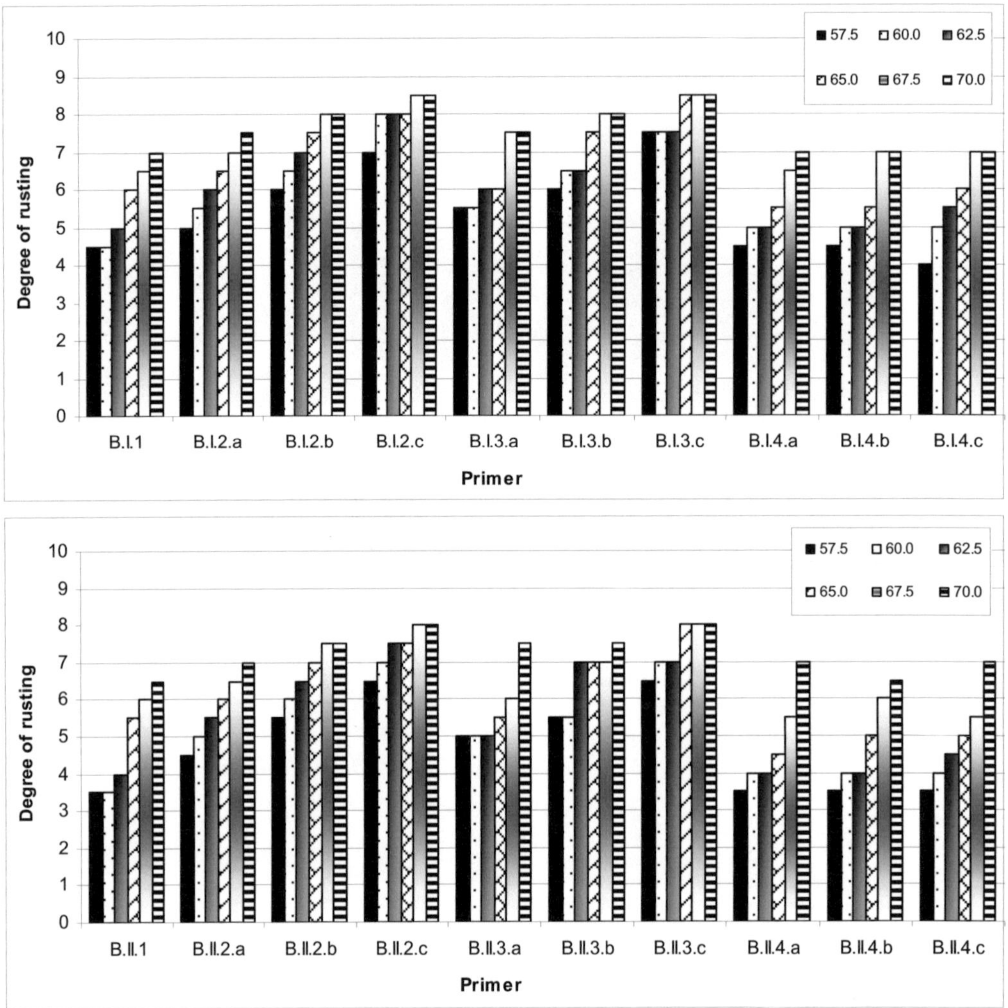

Fig. 5. Coatings based on binder B: Degree of rusting in salt spraying (fog) chamber; average values of failures in X-cut and in general area of panel.

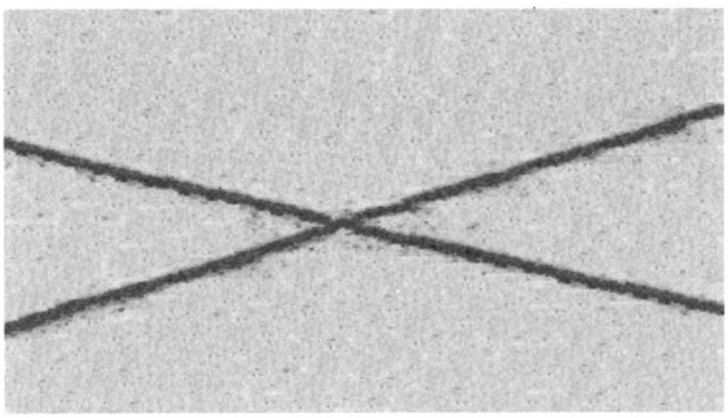

Fig. 6. Primer A.I.2.c, performance in salt spraying (fog) chamber, Method A, degree of rusting: 10.

With the purpose of establishing the efficiency of each protective coating from an anticorrosive point of view, the average value of degree of rusting was calculated for areas with and without cutting. The results of Table 2 confirm the superior performance of water-based nano lithium silicate of 7.5/1.0 silica/alkali molar ratio in relation to solvent-based, partially hydrolyzed tetraethyl orthosilicate, the major efficiency of fine microzinc compared to regular one and the increasing efficiency of primers as the level of conducting reinforcement fibers increased (Ahmed et al., 2010). On this last variable of formulation, it is worth mentioning that primers with 2.0% conducting reinforcement fibers showed the best protective capacity, which would occur due to the improved electric contact between zinc particles and with metallic substrate. On the other hand, the quartz used as reinforcement fiber due to characteristic non-conductive showed a similar performance that the corresponding reference primer (without reinforcement fiber).

Nature of effect	**Type of effect**	**Degree of rusting**
		Average values of failures in X-cut and in general area of panel
Type of binder	A	7.1
	B	6.2
Microzinc D 50/50	I	7.0
	II	6.3
Type of reinforcement fibers	1	5.8
	2	7.2
	3	7.2
	4	5.8
Level of reinforcement fibers	a	6.2
	b	6.8
	c	7.3

Table 2. Average values of the simultaneous statistical treatment of all variables.

On the other hand, Table 3 lists the average values and the standard deviations of statistical processing for each primer; in addition, it also displays the general average values for each type of reinforcement fiber taking into account both binders considered. In this table, the highest value also indicates the best performance in terms of the ability to control the metal corrosion. The analysis of the results obtained by using both types of binder displays that the primers based just on two spherical microzinc (reference primers) and non-conducting reinforcement fibers (quartz) in the three considered levels, formulated with reduced values of PVC, showed a sharp decline in corrosion performance. On the other hand, those that included conducting reinforcement fibers (graphite and silicon nitride), despite having been manufactured with a significantly lower level of pigmentation, maintained their efficiency. Corresponding standard deviation values support this conclusion.

These results would be based on the reduced electrical contact between particles of both types of microzinc and the metal substrate, regardless of the corrosion products could not only increase the electrical resistance of the film but also could decrease the amount of available zinc.

The incorporation of conducting reinforcement fibers seems to have favored the conductivity, which leads to reduction of the efficient PVC, according to the abundant amount of zinc corrosion products visually observed, the results of the electrode potentials and those obtained in the salt spraying (fog) chamber. Figures 7 and 8 display the primer films based on binder A (water-based nano lithium silicate of 7.5/1.0 silica/alkali molar ratio), microzinc I (fine, 4 μm) and fiber 2 (graphite) in level c (2.0% w/w) for 57.5% and 70.0% PVC values respectively, after finishing the accelerated aging test. The analysis of the cited figures reveals that despite having larger distance between the particles of microzinc in the case of 57.5% PVC compared with that formulated with 70.0% PVC, the galvanic activity in the two primers is significant in both cases (as evidenced by the amount of white zinc salts). In addition, results of figures show that the conductive reinforcement fibers linked electrically the microzinc particles each other, even in the primer of less PVC (all particles, despite having no direct contact between them, demonstrated activity like sacrificial anodes).

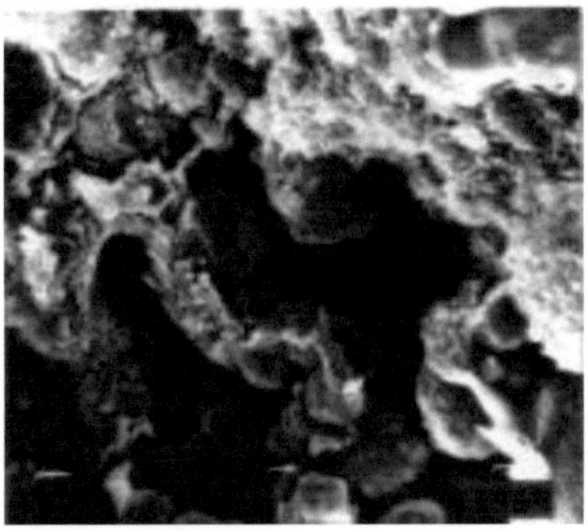

Fig. 7. SEM micrograph of primer A.I.2.c formulated with 57.5% PVC.

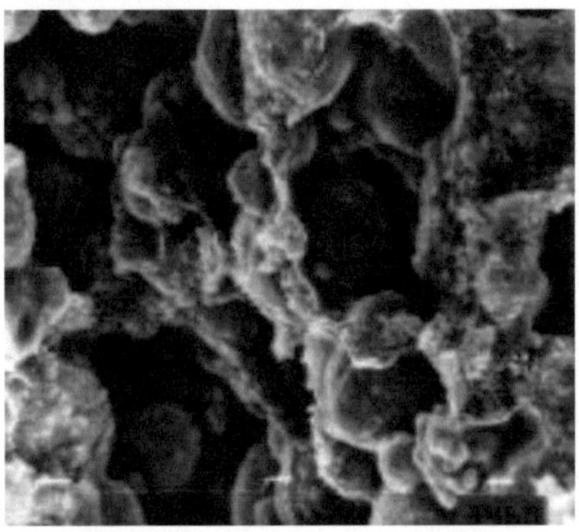

Fig. 8. SEM micrograph of primer A.I.2.c formulated with 70.0% PVC.

Primer	Average values	Standard deviation σ_{n-1}	General average values	Primer	Average value	Standard deviation σ_{n-1}	General average value
A.I.1	6.6	1.16	6.6	**B.I.1**	5.6	1.07	5.6
A.I.2.a	6.9	0.58	8.0	**B.I.2.a**	6.2	0.94	7.1
A.I.2.b	8.2	0.41		**B.I.2.b**	7.2	0.82	
A.I.2.c	8.9	0.38		**B.I.2.c**	8.0	0.55	
A.I.3.a	7.0	0.63	7.9	**B.I.3.a**	6.3	0.93	7.1
A.I.3.b	8.0	0.44		**B.I.3.b**	7.1	0.86	
A.I.3.c	8.8	0.42		**B.I.3.c**	8.0	0.55	
A.I.4.a	6.4	1.20	6.5	**B.I.4.a**	5.6	0.97	5.7
A.I.4.b	6.7	1.13		**B.I.4.b**	5.7	1.08	
A.I.4.c	6.4	1.28		**B.I.4.c**	5.8	1.17	
A.II.1	6.0	1.34	6.0	**B.II.1**	4.8	1.33	4.8
A.II.2.a	6.5	0.55	7.2	**B.II.2.a**	5.8	0.93	6.6
A.II.2.b	7.3	0.52		**B.II.2.b**	6.7	0.82	
A.II.2.c	7.8	0.49		**B.II.2.c**	7.4	0.53	
A.II.3.a	6.5	0.55	7.3	**B.II.3.a**	5.7	0.98	6.6
A.II.3.b	7.2	0.42		**B.II.3.b**	6.6	0.86	
A.II.3.c	8.2	0.41		**B.II.3.c**	7.2	0.66	
A.II.4.a	6.1	1.28	6.1	**B.II.4.a**	4.8	1.29	4.8
A.II.4.b	6.1	1.24		**B.II.4.b**	4.8	1.21	
A.II.4.c	6.1	1.16		**B.II.4.c**	4.9	1.24	

Table 3. Average values and standard deviation.

4. Final considerations

To explain the great tendency of zinc particles to corrode at the film surface of water-based nano lithium silicate primers as comparing with those solvent-based, partially hydrolyzed tetraethyl orthosilicate, it is necessary to consider that the first ones are based on binders, as mentioned, with a higher superficial tension. The last one implies inferior wetting, that means lower adhesion, penetration and spreading during metal zinc incorporation previous to application; consequently, they wet with more difficult the zinc particles while the second ones do it in a better way (more reduced interfacial tension).

The above-mentioned characteristic explains the great porosity of zinc-rich nano lithium silicate films and their high cathodic protective activity as comparing with zinc-rich tetraethyl orthosilicate films.

With regard to average diameter of zinc particle, size diminution increases significantly the surface area for a given weight. Since all surfaces have a given level of free energy, the ratio of surface energy to mass in small particles is so great that the particles adhered strongly themselves. For this reason, a lower particle size in a poor dispersion originates a greater flocculates (a high number of unitary particles are associated), which lead to zinc-rich primer films of high porosity and because of good cathodic protective activity. Moreover, a lower particle size could lead to films with a higher electrical contact (better packaging ability); since the current density is inherent to the chemical nature of zinc dust and the operating conditions of the corrosion cell, the increase in specific area elevates not only the current of protection but also generates a better superficial distribution (more efficient primers). During immersion test in 0.1 M sodium chloride solution, visual inspection of plates protective with zinc-rich primers (both types of binders) showed a more localized steel attack when zinc dust of the higher particle size was used.

Concerning incorporation of reinforcement fibers, the conductive or non-conductive characteristic was a very important variable. The first ones improved notably the primer performance since they increased the electrical contact between particles and with the metallic substrate, particularly in the higher levels in the formulations; the performance is correlated with the higher useful zinc in the film. On the other hand, non-conductive reinforcement fibers did not modify the primer efficiency as compared with reference panels (without reinforcement fibers) and for this reason their incorporation is not justified from technical and economical viewpoints.

Referring to PVC values (zinc content in dry film), previous results of laboratory tests demonstrated that a higher amount of microzinc leads to a longer useful life of primers. Nevertheless, it is important to mention that the choice of zinc content must be made by considering the physical characteristic of the primer film required for each particular case. When pigment volume concentration exceeds largely the CPVC, film properties such as adhesion, flexibility, abrasion resistance, etc. are drastically reduced while when the percentual level is slight under the critical value the efficiency is also considerably diminished.

In the case of primers, which have got incorporated conductive reinforcing fibers, results allow concluding that it is possible to reduce appreciably the PVC without affecting significantly the efficiency in service. In addition, it is important to mention that the quoted diminution of zinc content in the film is direct proportional to decrease the primer cost since it is the most expensive component of the composition.

5. References

Abanilla, M.A.; Li, Y. and Karbhari, V.M. (2005). Durability characterization of wet layup graphite/epoxy composites used in external strengthening, *Composites Part B: Engineering*, Vol. 37, No. 2-3, (August 2005), pp. 200-212, DOI: 10.1016 /j.compositesb.2005.05.016

Ahmed Al-Dulaimi, Ahmed (2010). *Evaluation of polyaniline composites and nanostructures as anti-corrosive pigments for carbon steel*. Masters thesis, Universiti Teknologi Malaysia, Faculty of Chemical and Natural Resource Engineering.

Aluru, N. et al., in: Gooddard, Brenner, Lyshevski, Iafrate Ed., (2003). *Nanostructure Studies of the Si-SiO_2 Interface*, Handbook of Nanoscience, Engineering and Technology, CRC Press., Washington D.C., USA, 2003, Chapter 11.2.

Amir, N.; Ahmad, F. and Megat-Yusoff, P. (2010). *Study on Fiber Reinforced Epoxy-based Intumescent Coating Formulations and Their Characteristics.* International Conference on Plant Equipment and Reliability (ICPER 2010), Kuala Lumpur, Malaysia, June 2010.

Behler, D.K.; Stravato, A.; Mochalin, V.; Korneva, G.; Yushin, G. and Gogotsi, Y. (2009). Nanodiamond-Polymer Composite Fibers and Coatings, *ACS Nano*, 2009, Vol. 3, No. 2, pp 363–369, DOI: 10.1021/nn800445z

Chen, Y. H.; Yu Chu, J. and Zhu, Q.J. (2010). Effects of Coating on Interfacial Fatigue of Fiber-Reinforced Composites, *Advanced Materials Research, Manufacturing Science and Engineering* I, Vol. 97, No. 101, (March 2010), pp. 830-833, DOI: 10.4028/www.scientific.net/AMR.97-101.830

Giudice, C.A. and Pereyra, A.M. (2007). Soluble metallic silicates in the anticorrosive inorganic coating formulation with non-flammable properties. *Pitture e Vernici European Coatings,* Vol. 83, No. 7, pp. 48-57, ISSN 0048-4245

Giudice, C.A. and Pereyra, A.M. (2009). *Tecnología de pinturas y recubrimientos. Componentes, formulación, manufactura y control de calidad,* Ed. edUTecNe, Argentina, 2009, pp. 1-444.

Hammouda, N.; Chadli, H.; Guillemot, G. and Belmokre, K. (2011). The Corrosion Protection Behaviour of Zinc Rich Epoxy Paint in 3% NaCl Solution, *ACES*, Vol.1 No.2, (April 2011), pp.51-60, DOI: 10.4236/aces.2011.12009

Hoshyargar, F.; Ali Sherafati, S. and Hashemi, M. (2009). Short communication: A new study on binder performance and formulation modification of anti-corrosive primer based on ethyl silicate resin. *Progress in Organic Coatings*, Vol. 65, No. 3, (July 2009), pp. 410-413, DOI: 10.1016/j.porgcoat.2009.02.006

Huang, K.M.; Weng, C.J.; Lin, S.Y; Yu, Y.H. and Yeh, J.M. (2009). Preparation and anticorrosive properties of hybrid coatings based on epoxy-silica hybrid materials, *Journal of Applied Polymer Science*, Vol. 112, No. 4, (May 2009), pp. 1933–1942, DOI: 10.1002/app.29302

Jegannathan, S.; Sankara Narayanan, T.; Ravichandran, K.; and Rajeswari, S. (2006). Formation of zinc phosphate coating by anodic electrochemical treatment. *Surface and Coatings Technology*, Vol. 200, No. 20-21, (May 2006), pp. 6014-6021, DOI: 10.1016/j.surfcoat.2005.09.017

Jianjun, Y.; Zhenhua, S.; Xiangsheng, M. and Wenlong, L., Deqiang, J. (2008). Corrosion Protective Performance of Coatings. Study on Zinc Ingredient in Zinc Rich Epoxy Primers. *Paint & Coatings Industry*, Vol. 08-2008, (August 2008), DOI: CNKI:SUN:TLGY.0.2008-08-003.

Lei-lei, M. and De-liang, L. (2010). The Classification and Testing Standards of Zinc Rich Coatings, *Shanghai Coatings*, Vol. 05, (May 2010), DOI: CNKI:SUN:SHTL.0.2010-05-015

Lekka, M.; Koumoulis, D.; Kouloumbi, N. and Bonora N.P. (2009). Mechanical and anticorrosive properties of copper matrix micro- and nano-composite coatings. *Electrochimica Acta*, Vol. 54, No. 9, (March 2009), pp. 2540-2546, DOI: 10.1016/j.electacta.2008.04.060

Li, X.; Tabil, L.G. and Panigrahi, S. (2006). Chemical Treatments of Natural Fiber for Use in Natural Fiber-Reinforced Composites: A Review. *Chemistry and Materials Science, Journal of Polymers and the Environment*, Vol. 15, No. 1, (March 2006) pp. 25-33, DOI: 10.1007/s10924-006-0042-3

Pereyra, A.M. and Giudice, C.A. (2007). Shaped for performance: the combination of lamellar zinc and mica improves the efficiency of zinc-rich primers. *European Coatings Journal*, Vol. 9, pp. 40-45, ISSN 0930-3847

Radhakrishnan, S.; Siju, C.R.; Mahanta, D.; Patil, S. and Madras, S. (2009). Conducting polyaniline–nano-TiO_2 composites for smart corrosion resistant coatings, *Electrochimica Acta,* Vol. 54, No. 4, (January 2009), pp. 1249-1254 DOI: 10.1016/j.electacta.2008.08.069

Sonawane, S.; Teo, B.; Brotchie, A.; Grieser, F. and Ashokkumar, M. (2010). Sonochemical Synthesis of ZnO Encapsulated Functional Nanolatex and its Anticorrosive Performance, *Ind. Eng. Chem. Res.*, (January 2010), Vol. 49, No. 5, pp. 2200–2205. DOI: 10.1021/ie9015039

Sorensen, P.A.; Kiil, S.; Dam-Johansen, K. and Weinell C.E. (2011). Anticorrosive coatings: a review, *Chemistry and Materials Science, Journal of Coatings Technology and Research,* Vol. 6, No. 2, pp. 135-176, DOI: 10.1007/s11998-008-9144-2

Thorslund Pedersen, L.; Weinell, C.; Hempel, A. S; Verbiest, P; Van Den Bosch, J. and Umicore J. (2009). *Advancements in high performance zinc epoxy coatings Zinc,* Chemicals Source Corrosion 2009, (March 2009), Copyright NACE International

Wang, D. and Bierwagen, G. (2009). Sol–gel coatings on metals for corrosion protection. *Progress in Organic Coatings,* Vol. 64, No. 4 (March 2009), pp. 327-338. DOI: 10.1016/j.porgcoat.2008.08.010

Wenrong, J. (2009). The Research Progress on Green Coatings, *Guangdong Chemical Industry,* Vol. 5, (May 2009), DOI: CNKI:SUN:GDHG.0.2009-05-035

Xiyan, L.; Jianming, J.; Sibo, K. and Jing, Z. (2010) Preparation of Flexible Epoxy Anticorrosive Coatings, *China Coatings,* Vol. 7, (July 2010), DOI: CNKI:SUN:ZGTU. 0.2010-07-018

Yang, F.; Zhang, X.; Han, J. and Du, S. (2008). Characterization of hot-pressed short carbon fiber reinforced ZrB_2–SiC ultra-high temperature ceramic composites, *Journal of Alloys and Compounds,* Vol. 472, No. 1-2, (20 March 2009), pp. 395-399, DOI: 10.1016/j.jallcom.2008.04.092

Yoshida, K.; Matsukawa, K.; Imai, M. and Yano, T. (2009). Formation of carbon coating on SiC fiber for two-dimensional SiCf/SiC composites by electrophoretic deposition, *Materials Science and Engineering:* B, Vol. 161, No. 1-3, (April 2009), pp. 188-192, DOI: 10.1016/j.mseb.2008.11.032

8

Comparative Study of Porphyrin Systems Used as Corrosion Inhibitors

Adina-Elena Segneanu, Ionel Balcu, Nandina Vlatanescu, Zoltan Urmosi and Corina Amalia Macarie
National Institute of Research and Development for Electrochemistry and Condensed Matter, INCEMC-Timisoara
Romania

1. Introduction

Fundamental research in durability of materials and structures have shown great potential for enhancing the functionality, serviceability and increased life span of our civil and mechanical infrastructure systems and as a result, could contribute significantly to the improvement of every nation's productivity, environment and quality of life. The intelligent renewal of aging and deteriorating civil and mechanical infrastructure systems includes efficient and innovative use of high performance composite materials for structural and material systems. [Monteiro et al, 2002]

The word corrosion is as old as the earth, but it has been known by different names.

Corrosion is known commonly as rust, an undesirable phenomena which destroys the luster and beauty of objects and shortens their life. A Roman philosopher, Pliny (AD 23-79) wrote about the destruction of iron in his essay 'Ferrum Corrumpitar'. Corrosion since ancient times has affected not only the quality of daily lives of people, but also their technical progress. There is a historical record of observation of corrosion by several writers, philosophers and scientists, but there was little curiosity regarding the causes and mechanism of corrosion until Robert Boyle wrote his 'Mechanical Origin of Corrosiveness.'

Philosophers, writers and scientists observed corrosion and mentioned it in their writings:

- Pliny the elder (AD 23-79) wrote about spoiled iron.
- Herodotus (fifth century BC) suggested the use of tin for protection of iron.
- Austin (1788) noticed that neutral water becomes alkaline when it acts on iron.
- Thenard (1819) suggested that corrosion is an electrochemical phenomenon.
- Hall (1829) established that iron does not rust in the absence of oxygen.
- Davy (1824) proposed a method for sacrificial protection of iron by zinc.
- De la Rive (1830) suggested the existence of microcells on the surface of zinc.

The most important contributions were later made by Faraday (1791-1867) who established a quantitative relationship between chemical action and electric current. Faraday's first and second laws are the basis for calculation of corrosion rates of metals. Ideas on corrosion control started to be generated at the beginning of nineteenth century. Whitney (1903) provided a scientific basis for corrosion control based on electrochemical observation. As early as in eighteenth century it was observed that iron corrodes rapidly in dilute nitric acid but remains intact in concentrated nitric acid. Schonbein in 1836 showed that iron could be made passive. It was left to U.R. Evans to provide a modern understanding of the causes and control of corrosion based on his classical electrochemical theory in 1923. Corrosion laboratories established in M.I.T., USA and University of Cambridge, UK, contributed significantly to the growth and development of corrosion science and technology as a multi disciplinary subject. In recent years, corrosion science and engineering has become an integral part of engineering education globally. [Ahmad, 2006]

The strong damaging effects of corrosion require establishing and taking some control measures. In accordance with the ways in which corrosion manifest, the supporting material and the specific local conditions, corrosion control can take different forms. Surely that here should also be added the use of materials maximum resistant to corrosive environment in order to limit the corrosion effects.

The practice of corrosion prevention by adding substances which can significantly retard corrosion when added in small amounts is called inhibition. Inhibition is used internally with carbon steel pipes and vessels as an economic control alternative to stainless steels and alloys, and to coatings on non-metallic components. One unique advantage is that adding inhibitor can be implemented without disruption of a process. The addition of an inhibitor (any reagent capable of converting an active corrosion process into a passive process) results in significant suppression of corrosion.

Corrosion inhibitors are substances when added in small amounts in a corrosive environment reduces significantly the corrosion rate for metallic material in contact with the environment.

A typical good corrosion inhibitor will give 95% inhibition at concentration of 80ppm, and 90% at 40ppm. Some of the mechanism of its effect are formation of a passivation layer (a thin film on the surface of the material that stops access of the corrosive substance to the metal), inhibiting either the oxidation or reduction part of the redox corrosion system (anodic and cathodic inhibitors), or scavenging the dissolved oxygen.

Some corrosion inhibitors are hexamine, phenylenediamine, dimethythanolamine, sodium nitrite, cinnamaldehyde, condensation products of aldehydes and amines, chromates, nitrites, phosphates, hydrazine, ascorbic acid, and others.

The corrosion inhibitors are added not only in aqueous solutions, but also in oils and fuels, the liquid cooling etc. can also be organic additives and coatings (varnishes, paints) on metallic surfaces.

The presence of a chemical compound in an environment, even in small concentrations, can lead to significant changes in speed and form of corrosion of a metallic material in contact with the environment. The acceleration or inhibition of corrosion processes are specific methods, dependent of metal-corrosive environment characteristics.

Corrosion inhibitors are selected on the basis of solubility or dispersibility in the fluids which are to be inhibited.[Rahimi,2004]

Porphyrins are well-known for their biological, catalytic, and photochemical properties. Considerable effort has been devoted to confining porphyrin molecules in microporous inorganic matrixes because such a hostguest approach can improve the efficiency of photoinduced charge separation by preventing back electron transfer. Moreover, organic-inorganic composite materials sometimes offer unique properties that are not available in any of the individual parts.[Bose et al,2002].

In this paper we intend to test the corrosion resistance of two other types of organic inhibitors and to study in which conditions they behave similarly to an anticorrosive paint. The corrosion resistance was studied by cyclic voltammetry, in 20% Na_2SO_4 electrolyte solution, Tafel tests and in the salt spray chamber, using diverse exposure conditions.

The two types of organic inhibitors used in this study for comparison are:

- 5,10,15,20 tetrakis(1-methyl-4pyridyl)21H,23H-porphine,tetra-p-fosylate salt
- 4,4',4',4'''(porphine-5,10,15,20-tetrayl)-tetrakis (benzeric sulfonic acid)

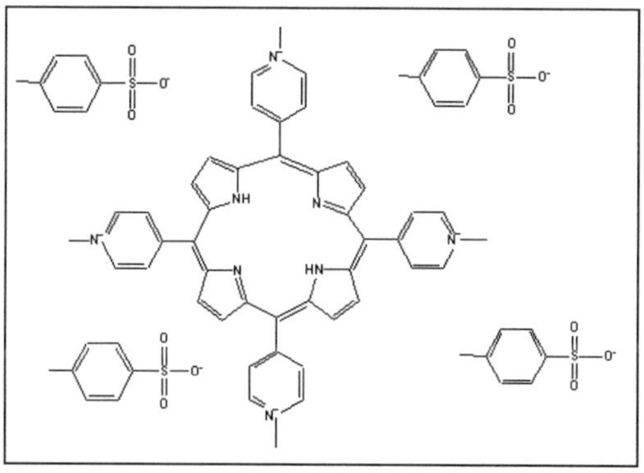

Fig. 1. Structure of 5,10,15,20 tetrakis(1-methyl-4pyridyl)21H,23H-porphine,tetra-p-fosylate salt.

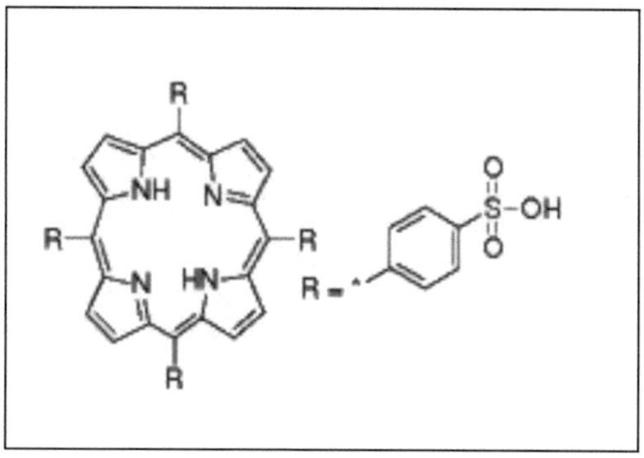

Fig. 2. Structure of 4,4',4',4'''(porphine-5,10,15,20-tetrayl)-tetrakis (benzeric sulfonic acid).

2. Experimental

The initial data consisted of 2 types of porphyrins dissolved KOH, H_2SO_4 and benzonitrile, as presented bellow, as the first set:

a. 0.2 g of Na_4TFP Ac porphyrin ($C_{44}H_{26}N_4Na_4O_{12}S_4 \times H_2O$) dissolved in 40 ml 10% KOH, mentioned as system A.

b. 0.2 g of Na_4TFP Ac porphyrin ($C_{44}H_{26}N_4Na_4O_{12}S_4 \times H_2O$) dissolved in 40 ml 10% H_2SO_4, mentioned as system B.
c. 0.2 g of H_2TPP porphyrin (5,10,15,20 tetrakis 4 phenyl-21H,23H) dissolved in benzonitrile, mentioned as system C.

System C presented the best anticorrosive properties.

The **second set** consists of:

a. 0.2 g of 5,10,15,20 tetrakis(1-methyl-4pyridyl)21H,23H-porphine,tetra-p-fosylate salt dissolved in 40 mL benzonitrile, mentioned from this point forward as **system I**
b. 0.2 g of 4,4',4',4'''(porphine-5,10,15,20-tetrayl)-tetrakis (benzeric sulfonic acid) dissolved in the same solvent namely benzonitrile (40mL) mentioned from this point forward as **system II.**

Various apparatuses were used, like the DCTC 600 salt spray chamber or the Dynamic EIS Voltalab. The results are presented as mm/year corrosion speed, thus evaluating the different coating systems.

The electrochemical studies namely cyclic voltammetry and Tafel curves carried out to test the protective layer were conducted using the PGZ 402 Dynamic EIS Voltalab. For the data acquisition the Voltamaster 4, version 7.08, was used. This specialized software can determine, based on references, from the Tafel test's values, the exact corrosion speed, measured in mm/year.

The voltammetry measurements, the Tafel tests, were conducted between -1000 and 1000 mV potentials at a sweep rate of 100 mV/s. Before each experiment, the working electrodes were polished with a series of wet sandpapers of different grit sizes (320, 400, 600, 800, 1000 and 1200). After polishing, the carbon-steel electrode are washed with ultrapure water and dried at room temperature and then the active part was immersed in porphyrin solution.

The working electrode is the carbon-steel electrode, (prepared as mentioned earlier) with 0,28 cm^2 active surface, (coated or uncoated); platinum counter electrode with 0.8 cm^2 active surface and saturated calomel electrode, (SCE), as reference electrode; all of which connected to the PGZ 402 Dynamic EIS Voltalab, from Radiometer Copenhagen.

20% Na_2SO_4 solution was used as base electrolyte.

The thickness loss and weight loss tests were not conducted, due to the relatively small size of the electrodes.

To test the corrosion resistance of the porphyrin systems eighteen electrodes were used.

For a good repeatability and accuracy the eighteen electrodes are pretreated as follows:

Three electrodes are uncoated/untreated, three electrodes are immersed for 5 minutes in **system I,** three electrodes are immersed for 60 minutes in **system I,** three electrodes are immersed for 5 minutes in **system II,** three electrodes are immersed for 60 minutes in **system II,** three electrodes are coated with anticorrosive paint. The porphyrin systems are dissolved and then applied on the electrodes; the electrodes are immersed in the solution for 5, respectively for 60 minutes, thus simulating a shorter and a longer exposure time.

After testing the porphyrin systems by electrochemical studies, the DCTC 600 dry salt spray chamber, (Figure 2) and the ASTM B 117 method (dry salt spray corrosion test), were used. The method establishes the spraying and drying times respectively the spraying and drying frequencies; in our case, the 5% NaCl solution is sprayed for 5 minutes at 35°C, afterwards is dried at 50°C for 55 minutes. The salt solution was prepared using 1 kg of pure NaCl dissolved in 20 liters of distilled water, resulting the 5% NaCl solution, mentioned earlier.

The exposure time in the salt spray chamber of the probes was 336 hours.

3. Results and discussions

Regarding the electrochemical studies of the corrosion resistance of the protective layers formed from the first set of porphyrin it has been demonstrated that the electrodes which have been treated with the system C (0.2 g of H_2TPP porphyrin (5,10,15,20 tetrakis 4 phenyl-21H,23H) disolved in 40 mL benzonitrile) gave the best results. The immersion time was 5 minutes.

Parameters	Electrodes			
	Uncoated	System A	System B	System C
$i^{\rightarrow}_{peak}$ [mA/cm²]	290	190	90	280
$\varepsilon^{\rightarrow}_{pic}$ [mV]	900	750	600	1300
$i^{\leftarrow}_{peak}$ [mA/cm²]	-	60	100	85
$\varepsilon^{\leftarrow}_{peak}$ [mV]	-	50	100	50
ε_{O2} [mV]	1500	1500	1500	1500
ε_{pas}[mV]	1350	900	850	1600
i_{pas} [mA/cm²]	25	8	0	50

Table 1. Results of cyclic voltammograms.

Parameters	Electrodes			
	Uncoated	System A	System B	System C
i_{cor} [mA/cm²]	0.7666	0.9792	0.6506	0.0718
v_{cor} [mm/year]	8.99	11.48	7.63	0.842
Rp	50.91	129.51	59.57	147.67
C	0.9962	0.9996	0.9997	1.000

Table 2. Results of Tafel test.

The notations from the table: $i^{\rightarrow}_{peak}$ – peak current density for anodic polarization; $\varepsilon^{\rightarrow}_{peak}$ – peak potential for anodic polarization; $i^{\leftarrow}_{peak}$ - peak current density for cathodic polarization; $\varepsilon^{\leftarrow}_{peak}$ - peak potential for cathodic polarization; ε_{O2} - oxygen generation potential; ε_{pas} - passivation potential; i_{pas} - passivation current.

Continuing with the tests, after we have demonstrated that porphyrin (5,10,15,20 tetrakis 4 phenyl-21H,23H) had the best results, we have realized a study to compare this porphyrin with other two types of porphyrins, namely: 5,10,15,20 - tetrakis(1-methyl-4pyridyl)21H,23H-porphine,tetra-p-fosylate salt and 4,4',4'',4'''(porphine-5,10,15,20-tetrayl)-tetrakis(benzeric sulfonic acid); also used as organic inhibitors.

From the obtained voltammograms were determined the anodic i_{peak} and ε_{peak} and from the Tafel curves were determined the corrosion current, polarisation resistance (Rp), corrosion rate and the correlation coefficient.

We continued the studies, cyclic voltammetry and Tafel method for carbon steel electrodes treated in different ways.

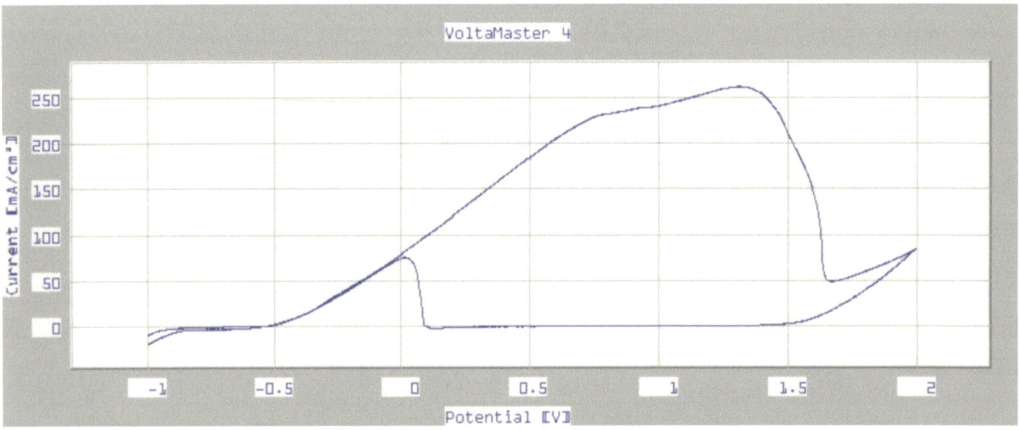

Fig. 3. Cyclic voltammograms of coated electrode with system C.

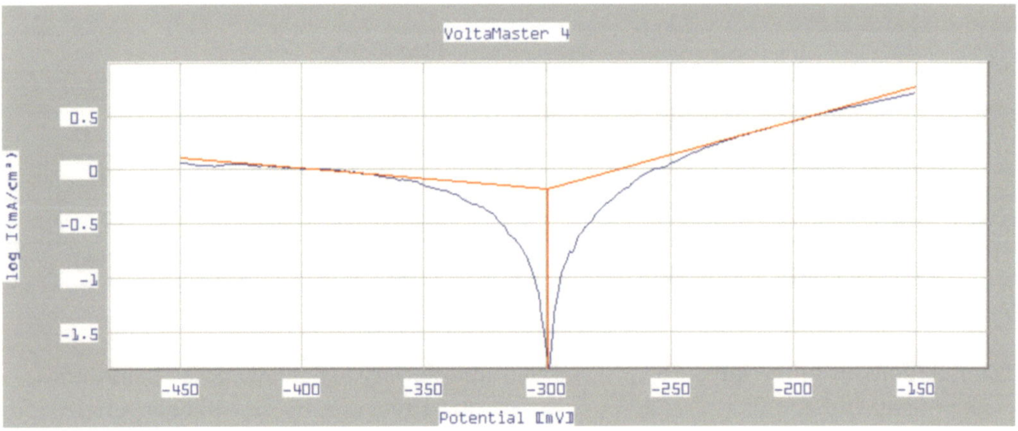

Fig. 4. Tafel tests of coated electrode with system C.

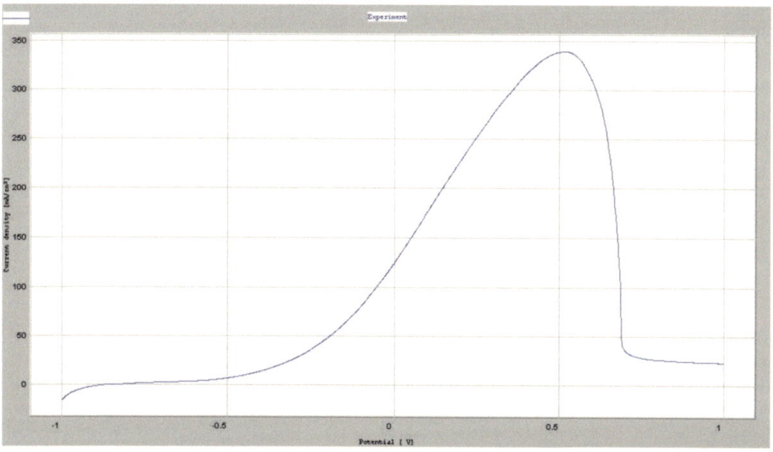

Fig. 5. Cyclic voltammogram of uncoated electrodes.

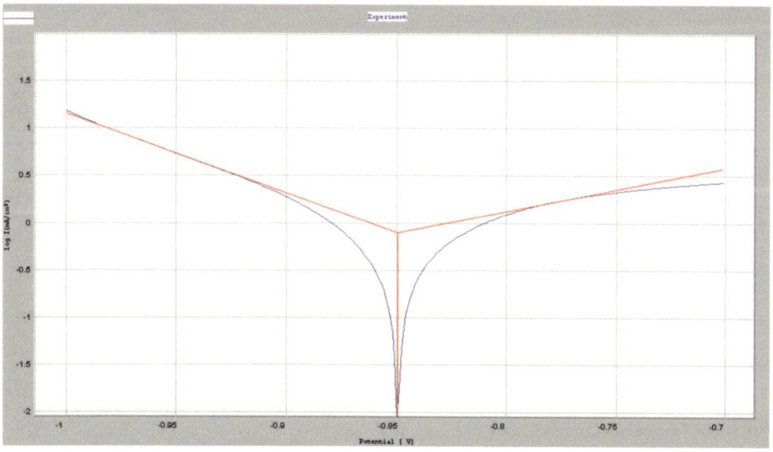

Fig. 6. Tafel tests of uncoated electrodes.

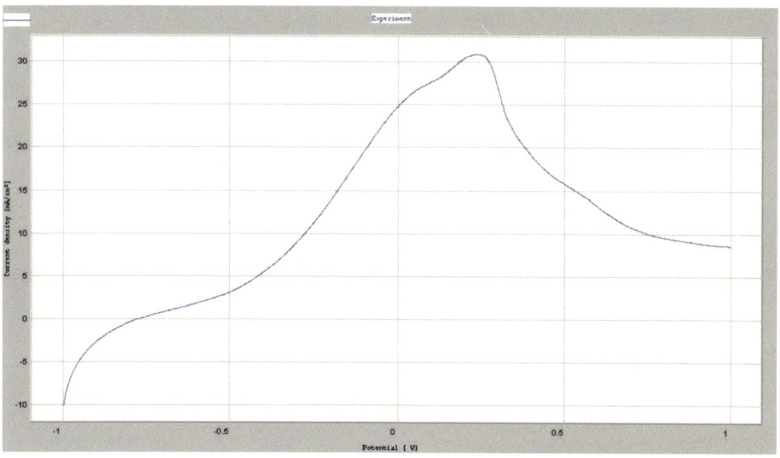

Fig. 7. Cyclic voltammogram of electrodes immersed for 5 minutes in system I.

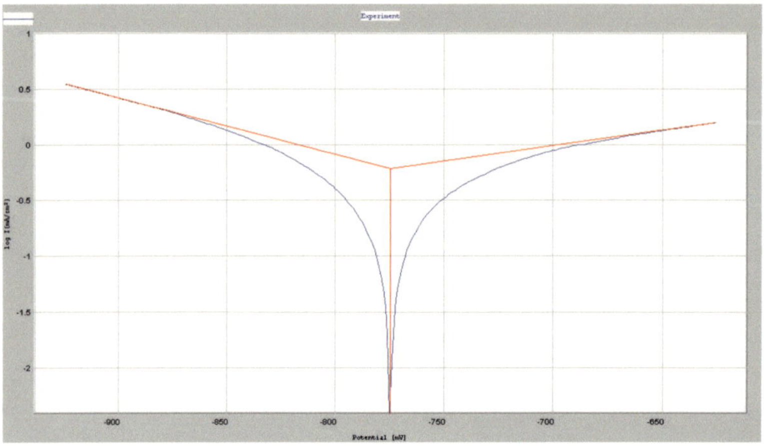

Fig. 8. Tafel tests of electrodes immersed for 5 minutes in system I

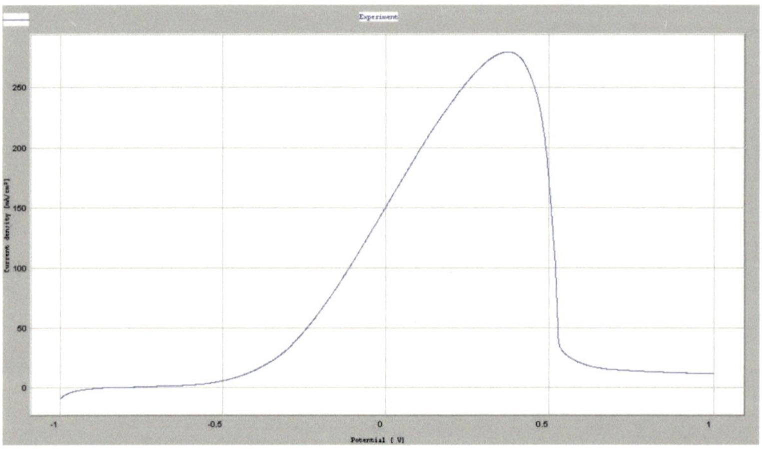

Fig. 9. Cyclic voltammogram of electrodes immersed for 5 minutes in system II

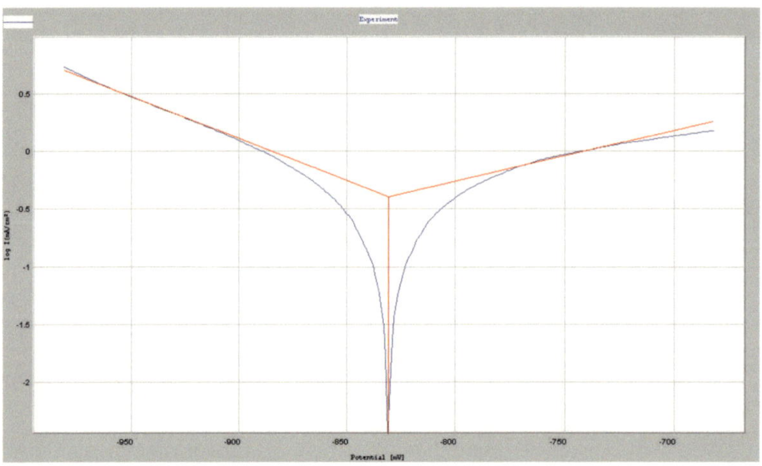

Fig. 10. Tafel tests of electrodes immersed for 5 minutes in system II

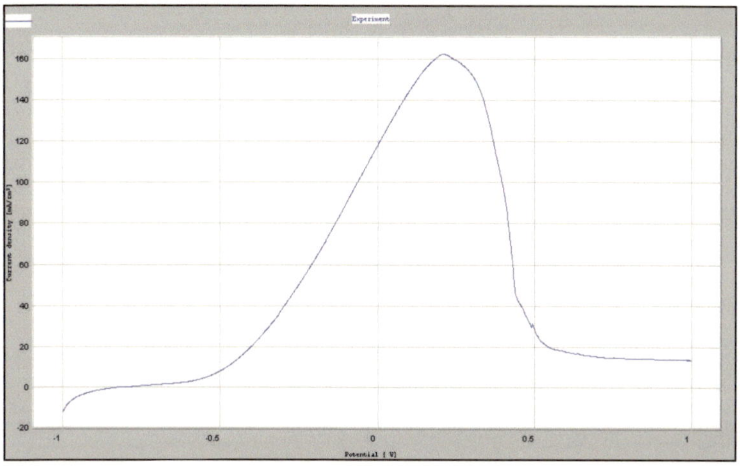

Fig. 11. Cyclic voltammogram of electrodes immersed for 60 minutes in system I.

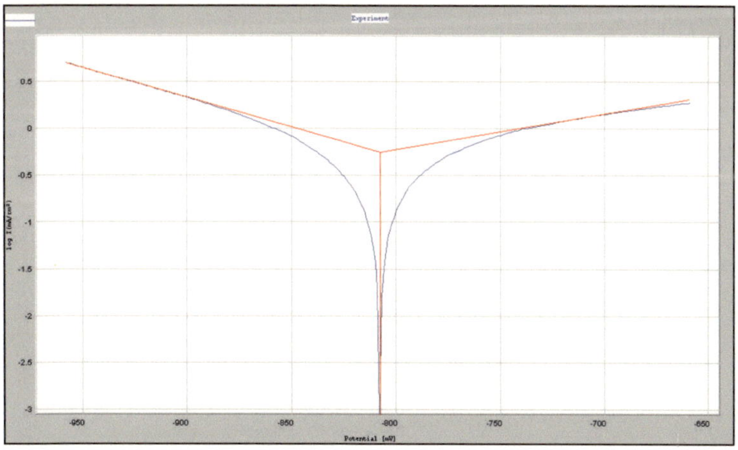

Fig. 12. Tafel tests of electrodes immersed for 60 minutes in system I.

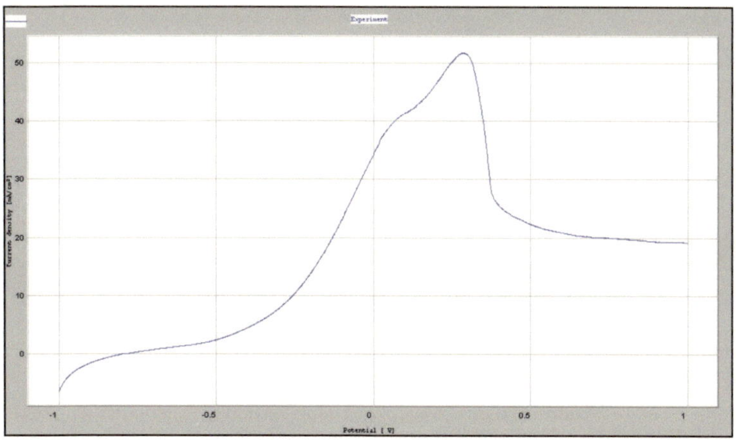

Fig. 13. Cyclic voltammogram of electrodes immersed for 60 minutes in system II.

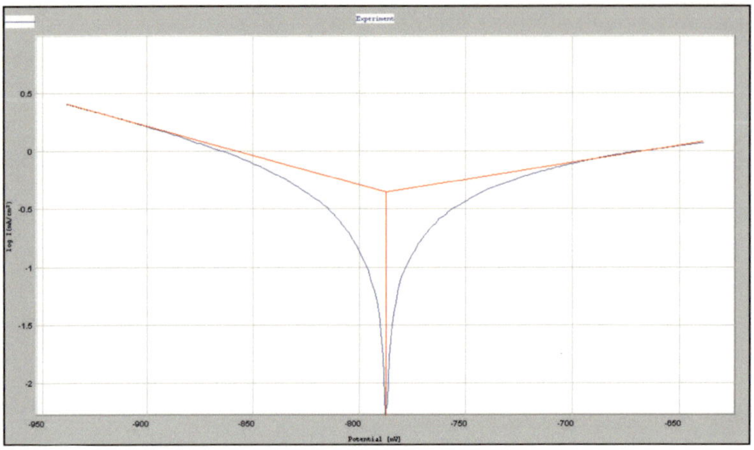

Fig. 14. Tafel tests of electrodes immersed for 60 minutes in system II.

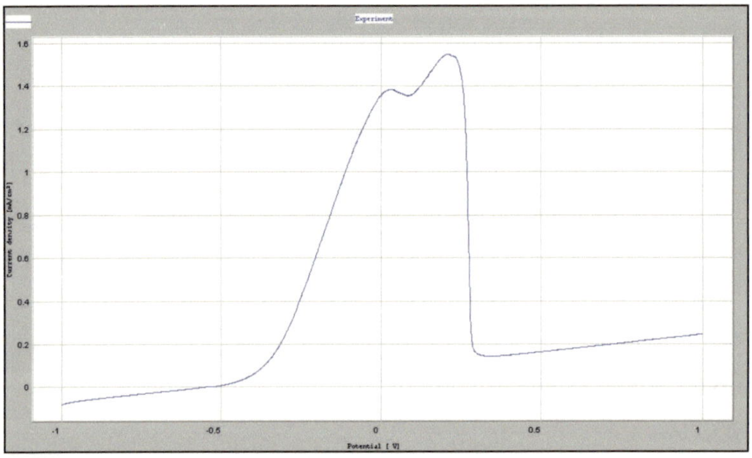

Fig. 15. Cyclic voltammogram of electrodes coated with paint.

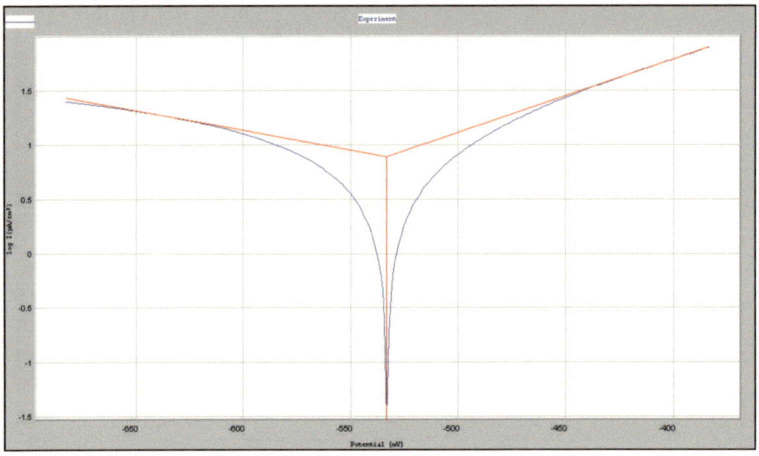

Fig. 16. Tafel tests of electrodes coated with paint.

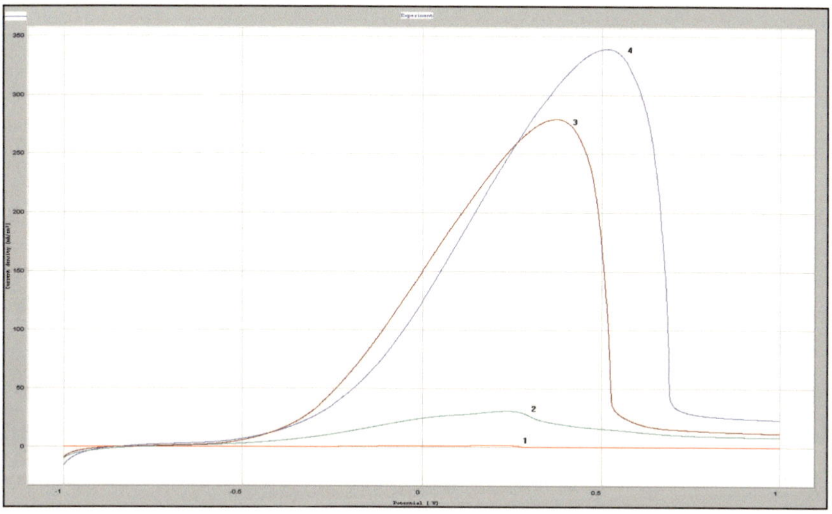

Fig. 17. Cyclic voltammograms of the corrosion process for various electrodes: 1- coated with paint; 2- system I (immersion time 5 minutes); 3- system II (immersion time 5 minute); 4- uncoated.

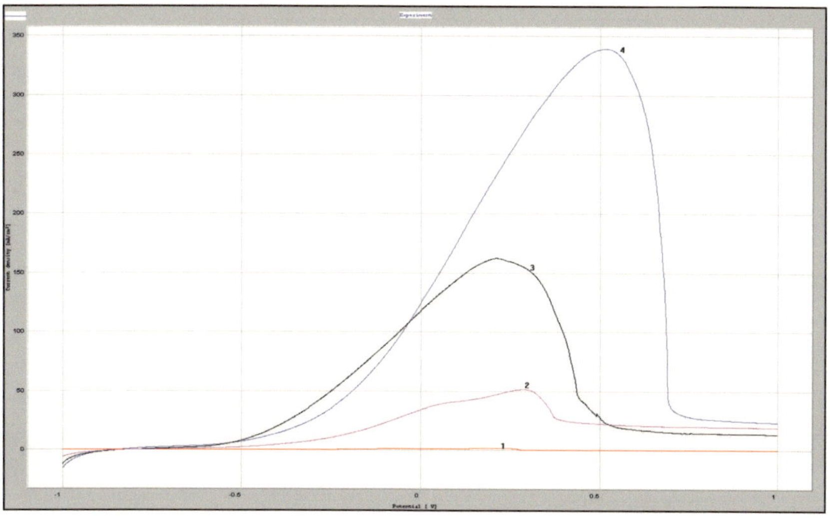

Fig. 18. Cyclic voltammograms of the corrosion process for various electrodes: 1- coated with paint; 2- system I (immersion time 60 minutes); 3- system II (immersion time 60 minute); 4- uncoated.

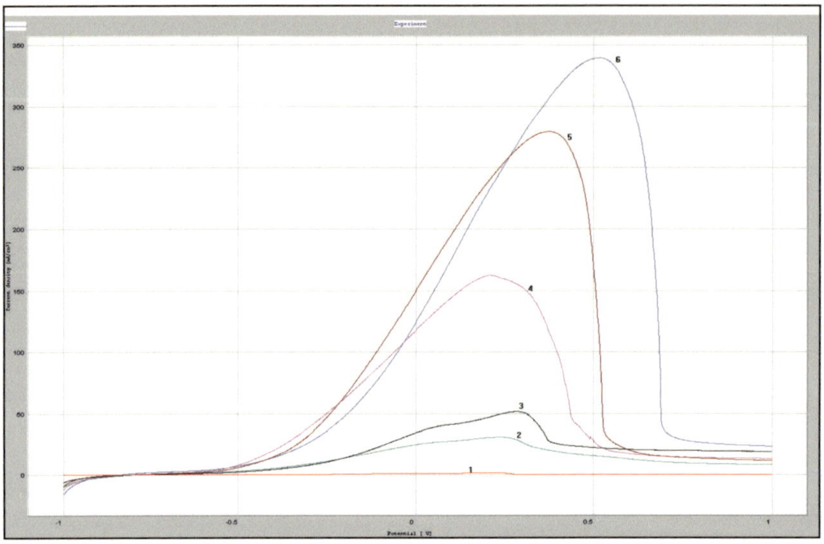

Fig. 19. Cyclic voltammograms of the corrosion process for various electrodes: 1- coated with paint; 2- system I (immersion time 5 minutes); 3- system I (immersion time 60 minute); 4- system II (immersion time 5 minute); 5- system II (immersion time 60 minute); 6- uncoated

Cyclic voltammograms and Tafel tests for 5 minutes and 60 minutes immersion time are presented in Table 3.

Parameters	Electrodes					
	Uncoated	System I		System II		Coated with paint
		Immersion time (5 minutes)	Immersion time (60 minutes)	Immersion time (5 minutes)	Immersion time (60 minutes)	
i_{peak} [mA/cm^2]	337.9	30.68	160.2	279.5	51.67	1.541
ε_{peak} [mV]	0.5178	0.2219	0.2172	0.3747	0.2824	0.21
i_{cor} [mA/cm^2]	0.7791	0.6062	0.5516	0.3983	0.4394	0.7689
v_{cor} [mm/year]	91.37	7.109	6.468	4.670	5.153	9.18
Rp	34.62	68.15	59.81	76.07	93.97	46.42
C	1.0000	1.0000	1.0000	1.0000	1.0000	1.0000

Table 3. Results obtained from cyclic voltammograms and Tafel tests.

The notations from the table:i_{peak} - peak current density; ε_{peak} - peak potential; i_{cor} - corrosion current density; v_{cor} - corrosion rate; C - correlation coefficient; Rp - polarization resistance

Polarization resistance can be related to the rate of general corrosion for metals at or near their corrosion potential, E_{corr}.

These can be obtained from a *Tafel* plot or estimated from the experimental data.

The polarization resistance or R_P is defined by the following equation:

$$R_P = \left(\frac{\Delta E}{\Delta i}\right)_{\Delta E \to 0}$$

where, ΔE variation of the applied potential around the corrosion potential and Δi is the resulting polarization current.

Polarization resistance, R_P, behaves like a resistor and can be calculated by taking the inverse of the slope of the current potential curve at open circuit or corrosion potential. High R_P of a metal implies high corrosion resistance and low R_P implies low corrosion resistance.

From electrochemical studies it was observed that the electrodes treated with system I in a immersion time of 5 minutes gives better results.

For salt spray chamber tests the electrodes were investigated visually; special attention was given to the appearance of the first corrosion signs and a similar or identical evolution of corrosion rates has resulted.

The electrodes were divided into three categories namely, 3 untreated electrodes, 3 electrodes painted with anti-corrosion paint and 12 electrodes treated according with system I and system II.

These 12 fall into four subcategories, namely:

- 3 treated according to system I, with immersion time of 5 minutes, and respectively other three electrodes with an immersion time of 60 minutes.
- 3 treated according to system II, with immersion time of 5 minutes, and respectively other three electrodes with an immersion time of 60 minutes.

Visual observations:

Untreated electrodes:

- After 24 hours there is a brown coloration on the entire surface of the electrode
- The brown coloration, after 48 hours, intensifies, the electrode's surface becomes more rough
- Specific symptoms appear after 120 hours, that is uniform corrosion throughout the surface of the electrode
- Corrosion progresses, symptoms are increasing after 192 hours. Rust formed is still adherent
- After 264 hours, the rust layer becomes more voluminous
- After 336 no major changes occur.

Painted electrodes:

- After 24 and 48 hours respectively, there are no reported signs of corrosion
- Only after 120 hours, there is loss of the initial gloss paint
- After 192 hours localized corrosion can be seen as brown spots on the surface of the electrodes

- The occurrence of pitting corrosion can be observed after 264 hours; pitting spots occur among previously localized brown spots. At one of the three electrodes, the paint swells.
- After 336 hours, in addition to the initial symptoms, few and very small points of pitting are observed

Electrodes treated as system I:

- no changes can be observed after 24 hours or 48
- After 120 hours the surface of electrodes become more matte
- After 192 hours the surface of electrodes become more rough
- After 264 hours signs of localized corrosion occur
- After 336 hours pitting corrosion occurs; the number of pitting spots is very small

Electrodes treated as system II:

- no changes can be observed after 24 hours or 48
- After 120 hours the surface of electrodes become more matte
- After 192 hours pitting corrosion occurs in several points; on the electrodes with immersion time of 60 minutes the number of pitting points is lower
- After 264 hours no major differences can be observed, the pitting points do not multiply
- After 336 hours uniform corrosion is observed on 2 of the 3 electrodes, with an immersion time of 60 minutes; on the electrodes with an immerse time of 5 minutes, the rust is adhering.

4. Conclusion

In conclusion, from the three types of organic inhibitors studied, it can be said that 5,10,15,20 tetrakis(1-methyl-4pyridyl)21H,23H-porphine,tetra-p-fosylate salt, having the immersion time of 5 minutes was almost similar with the paint used and gives an anticorrosive protection much better than the porphyrin previously tested (H2TPP porphyrin (5,10,15,20 tetrakis 4 phenyl-21H,23H). Similar results can be seen, from the visual observations of the salt spray chamber test.

5. References

Ahmad Z., (2006), Principles of Corrosion Engineering and Corrosion Control; *Institution of Chemical Engineers (Great Britain) - Elsevier/BH*, 656 pages ISBN 0750659246;

Bose A., He P., Liu C., Ellman B.D., Twieg R. J.,Huang S.D., (2002), *Journal of the American Chemical Society*, 124,4-5;

Monteiro PJ.M., Chong K.P., Larsen-Basse J., Komvopoulos K., (2001), Long Term Durability of Structural Materials, *Elsevier Science Ltd.*;

Rahimi A., (2004), Inorganic and Organometallic Polymers, *Iranian Polymer Journal* 13 (2), 149-164;

Standard practice for operating salt spray (fog) apparatus, B117-02

9

Properties of Graphite Sinters for Bipolar Plates in Fuel Cells

Renata Wlodarczyk[1], Agata Dudek[2],
Rafal Kobylecki[1] and Zbigniew Bis[1]
[1]*Department of Energy Engineering, Czestochowa University of Technology,*
[2]*Institute of Materials Engineering, Czestochowa University of Technology,*
Poland

1. Introduction

Fuel cell is an electrochemical device which transforms chemical energy stored in fuel directly into electrical energy. The only by-products of this conversion are water and heat. The factors which affect the intensity of electrochemical processes include properties of the materials used for fuel cell components and its working environment. Due to insignificant emissions of pollutants during energy production combined with high efficiency of these generators, and silent operation, fuel cells are an alternative to technologies of energy production from fossil fuels.

Studies on fuel cells today focus on extending their life, limitation of weight and size, and reduction of costs of manufacturing generators. Individual cell is composed of membrane/electrolyte and electrodes at both sides of MEA (membrane electrode assembly) (Fig. 1). The whole component is closed at both sides with bipolar or monopolar plates/interconnectors. Bipolar plates are the key components of generators since they take 80% of weight and 45% of costs of the cell [1].The task of the plates is to evenly distribute the fuel and air, conduct electricity between adjacent cells, transfer heat from the cell and prevent from gas leakage and excessive cooling.

According to DOE (the U.S. Department of Energy), basic requirements for materials for bipolar plates in fuel cells include in particular **corrosion resistance under fuel cell's operating conditions, low contact resistance, suitable mechanical properties, high thermal and electrical conductivity, low costs of manufacturing** [2]. Due to high material and functional requirements, few materials can meet these conditions. Bipolar plates in fuel cells are typically made of non-porous graphite because of its high corrosion resistance [3]. However, low mechanical strength of graphite and high costs connected with processing of graphite elevate the costs of manufacturing of fuel cells. Obtaining graphite-based composites modified with steel will allow for obtaining the material with improved mechanical properties, ensuring suitable corrosion resistance and high thermal and electrical conductivity at the same time. The method of powder metallurgy, which allows for obtaining even complicated shape of components, eliminates the problem of mechanical processing of graphite [4].

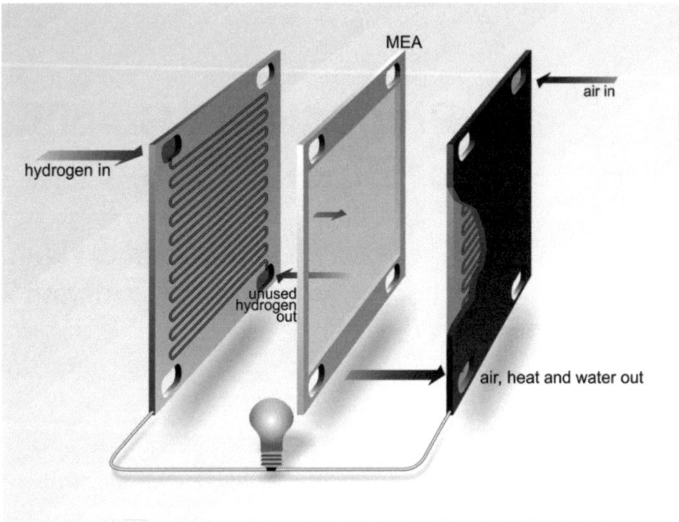

Fig. 1. Elements of fuel cell.

2. Materials for bipolar plates

Metallic materials used for bipolar plates in fuel cells include non-coated stainless steel, aluminum, titanium, nickel and materials coated with conducting, nitrogen- and carbon-based coatings [5-7]. Metals are very good candidate materials for elements of fuel cells because they exhibit very high thermal conductivity, opportunities for repeated processing and are easy to be machined. Alloy steels are the most common materials used for these components (enhanced corrosion resistance, relatively low prices of steel) [8]. The literature data allow for the classification of the materials used for bipolar plates according to the three basic groups:

- nonmetals: non-porous graphite/ electrographite [9-14],
- coated and non-coated metals and nonmetals [15-24],
- composites [25, 26].

The diagram below (Fig. 2) presents the division of materials for bipolar plates according to basic groups. Using the criterion of choice of material for a component, one should decide whether to choose corrosion-resistant materials, which are often more expensive than other available materials or to use cheaper materials. Use of materials which are resistant to corrosion, such as titanium or gold, for bipolar plates substantially improves the cost-efficiency in manufacturing generators, whereas use of generally available stainless steels can decrease the effectiveness of work of the cell because of their properties (passivation of steel under conditions of operation of fuel cells) [27]. Therefore, searching for materials for bipolar plates should involve optimization of all the parameters. Taking into account multifunctional nature of the plates, this is extremely difficult. The table below presents the materials used (graphite [9-14]), suggested (nickel, titanium, stainless steel [15-24, 28]) or being developed (composites [25, 26]).

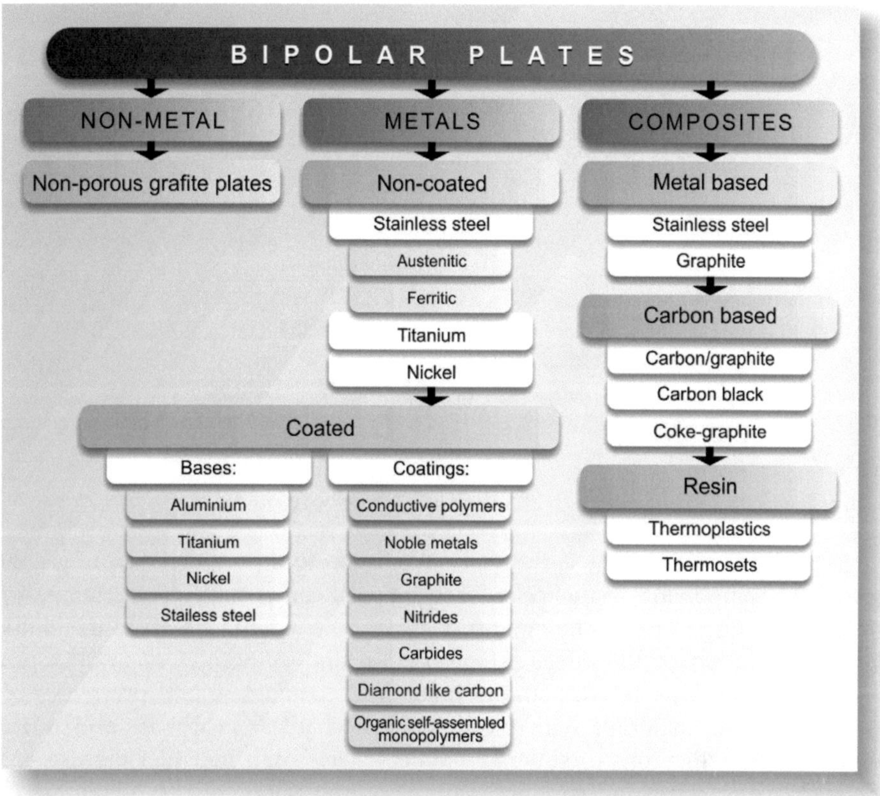

Fig. 2. Materials for bipolar plates in fuel cells.

Bipolar plates/interconnectors in fuel cells typically have channels on their surface to allow for the distribution of media to the electrodes [29]. The shape of channels and direction of flow of media might be different for the plate adjacent to anode compared to the plate near cathode. Media which flow in to both electrodes can be supplied by means of parallel channels, where media flow in one direction or channels where media are supplied to fuel cells with opposite directions. Another possible solution is that the media flow in with the direction transverse to the cell. The choice and optimization of the shape of the channels in bipolar plates affect the operation of the cell, particularly the degree of removal of products and distribution of gases to the surface of electrodes. The figure below presents bipolar plates with channels (Fig. 3). The essential effect on operation of the cell is from the depth of the channels, width of the channels, distance between spirals etc.

Review of the types of channels concerns in particular the geometry which depends on the type of fuel cell and demand for media in a particular cell. The list of opportunities for different channel design is obviously not ended and, apart from finding fundamental geometry, one should also consider the number of channels in the surface and distances between the channels. Proper distance between the channels and the number of channels ensure quick diffusion and effective discharge of water, especially in the cathode. **However, it should be emphasized that among a variety of types of channels used for distribution of media in fuel cells, there are no unequivocal research works which would have provided evidence of which type is the best.**

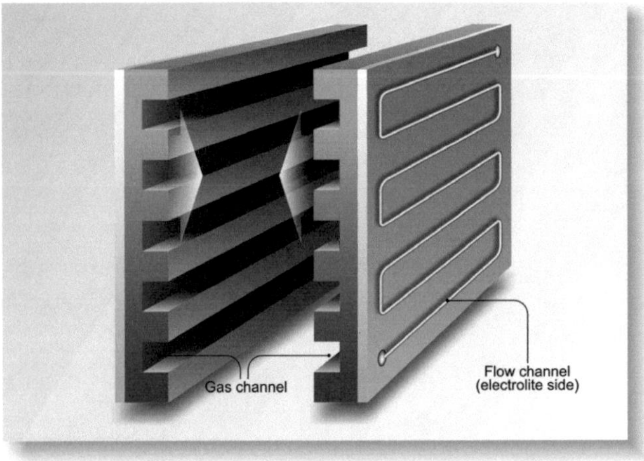

Fig. 3. Bipolar plate in fuel cell with channels which supply media.

The subject of the present study is the analysis of opportunities for the use of graphite-steel composites for components of fuel cells. The proposed composites were obtained by means of powder metallurgy. The technology for obtaining the materials used in the study allows for the determination of the effect of compaction and sintering on product properties. Finding the relationships between the technological parameters and properties of sinters allows for obtaining materials with the desired mechanical properties and resistance to corrosion. The investigations of sintered stainless steel confirmed that the use of suitable parameters of compaction pressure and sintering atmosphere ensures obtaining materials with controllable density, pore and grain size, and that suitable chemical composition of powders allows for obtaining sinters with the desired functional properties [30-35].

3. Research materials

The material composites were obtained from commercial steel powder AISI 316LHD manufactured by Höganäs (Belgium) sprayed with water and graphite powder Graphite FC (*fiber carbon*) manufactured by Schunk Kohlenstofftechnik GmbH (Germany). Chemical composition of the powder 316LHD is presented in Table 1.

Powder	C [%]	S [%]	Mo [%]	Ni [%]	Cr [%]	Si [%]	Mn [%]	O [%]	N [%]	Fe [%]
316LHD	0.025	0.005	2.2	12.3	16.7	0.9	0.1	0.30	0.06	balance

Table 1. Chemical composition of steel powder %.

Bulk densities for the powders used in the study are contained in Table 2.

Powders	Density [g cm^{-3}]
316LHD	2.67
graphite	0.20

Table 2. Bulk densities for the powders used in the study.

Fig. 4 presents the morphology of the powders used for preparation of graphite-steel composites. The values of statistical parameters of the particles of steel and graphite powders are presented in Fig. 5 and Fig. 6 in the form of histograms. Table 3 contains statistical parameters of stereological values of the used powders.

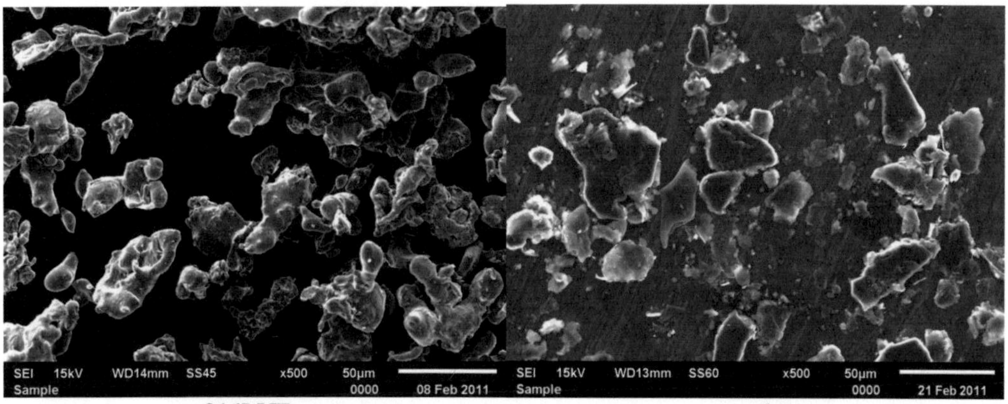

Fig. 4. Powders morphology, magnification x500.

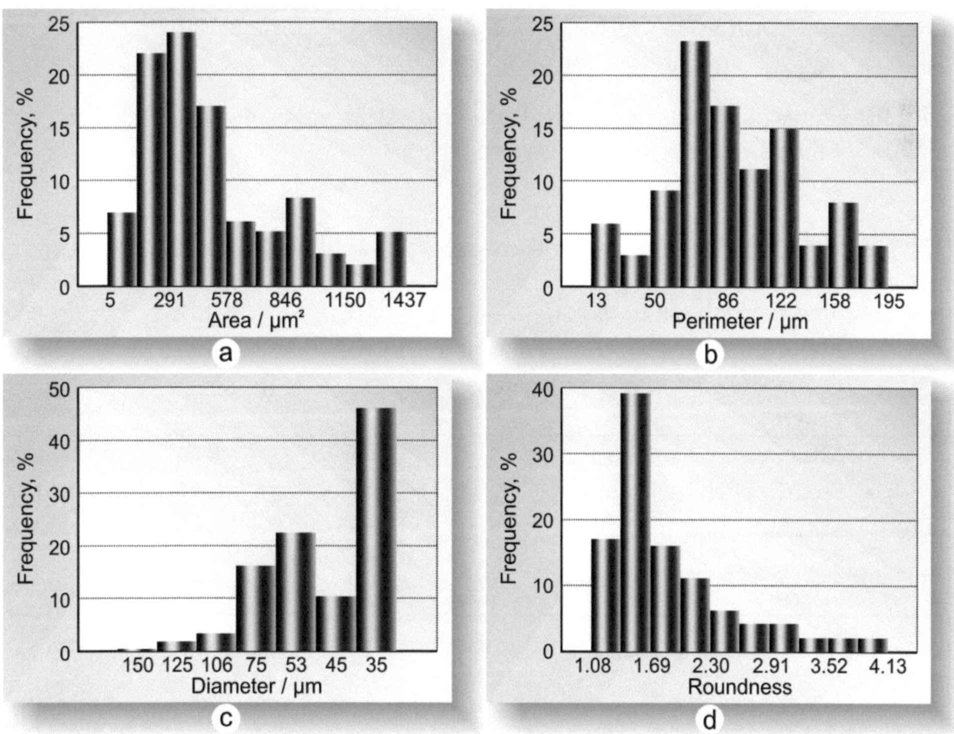

Fig. 5. Histograms of: a) particle surface; b) particle perimeter; c) mean particle diameter; d) roundness of the particle in 316LHD powder.

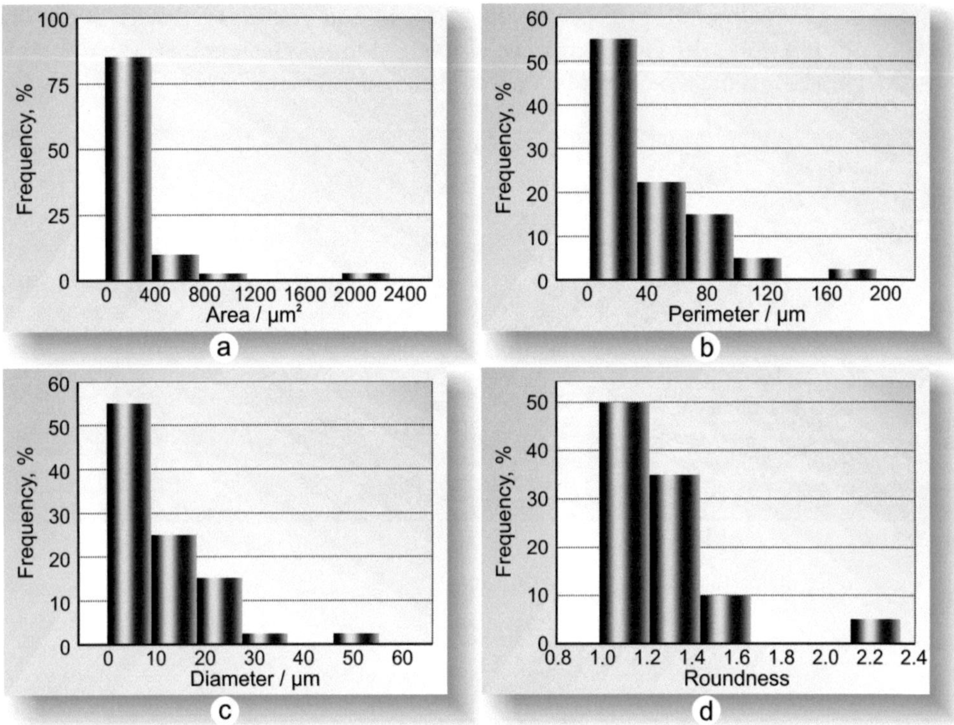

Fig. 6. Histograms of: a) particle surface; b) particle perimeter; c) mean particle diameter; d) roundness of the particle in graphite powder.

Powders	Statistic parameters	Area [μm²]	Perimeter [μm]	Roundness
316LHC	Minimum	5	13	1.081
	Średnia wielkość ziaren	503	101	1.855
	Maksimum	1 437	195	4.131
	Odchylenie standardowe wielkości ziaren	346	40	0.598
Grafit	Minimum	1.27	3.64	1.000
	Średnia wielkość ziaren	186	40.6	1.297
	Maksimum	2 285	91.5	3.456
	Odchylenie standardowe wielkości ziaren	378	36.8	0.420

Table 3. Statistical parameters of stereological values for the powders used in the study.

In order to obtain the sinters, steel and graphite powders were compacted (compaction pressure of 200 MPa), and then sintered in vacuum: sintering parameters: T=1250°C, t=30 min, cooling rate 0.5 °C/min,. Steel and graphite powders were used with the following proportions (expressed in mass percentage):

- 100% 316L;
- 80% 316L + 20% graphite;
- 50% 316L + 50% graphite;
- 20% 316L + 80% graphite;
- 100% graphite.

4. Research methodology

4.1 Phase analysis of graphite-steel composites was carried out with X-ray XRD Seifert 3003 T-T diffractometer. The investigations were carried out using cobalt lamp with the wavelength of radiation of $\lambda_{CoK\alpha}$ = 0.17902 nm. The diffractometer operated with the following parameters:

- power supply: 30 kV,
- current intensity: 40 mA,
- measurement step: 0.2°,
- time of counting impulses: 10s.

4.2 Microstructural analysis of the obtained composites were carried out using Axiovert optical microscope.

4.3 Hardness tests for the graphite-steel composites were carried out by means of Rockwell method in B and F scale.

4.4 Mean grain size for the composites was determined based on comparison of microscopic photographs with the pattern scale (comparative method) according to PN-EN ISO 643 standard [36]. The investigations were also supported by the results obtained based on the research using mercury porosimeter PoroMaster 33.

4.5 Analysis of porosity of graphite-steel composites was carried out using mercury porosimeter PoroMaster 33 equipped in Quantachrome Instruments software for Windows.

4.6 Analysis of wettability of composites was carried out in a following manner: 3μl of water was dropped on the surface of material which had been previously polished with a set of abrasive papers with the finishing paper with grit designation of 2500. Before the examination, the material was degreased and left in the air until dry. The images of the material with a water drop were analyzed by a MicroCapture micro-camera which features software for image analysis. The functionality of angle analysis allowed for the determination of Θ angle.

4.7 Analysis of roughness. In order to determine surface topography and parameters of surface geometry in the composites, the examinations using Hommel T1000 profilometer were carried out. The examinations of sinter geometry were carried out using measurement needle with the ball tip with the radius of 2.5 μm. Using the profilometer allowed for the determination of the parameters which describe height and longitudinal characteristics of the profile [37-38].

4.8 Analysis of contact resistance. Techniques of measurement of interfacial contact resistance have been broadly discussed in the studies [39-40]. Measurements of electrical contact resistance between the surfaces of diffusion layer (GDL, usually carbon composite) and bipolar plates (BP) were carried out according to the methodology used by Wang discussed in the studies [41-42]. For the purposes of the present study, the device for experimental determination of the relationship between contact resistance and unit pressure for a set of pairs of GDL+BP samples of the analyzed materials was designed.

The pressure acting on the sets of samples was generated by pneumatic press with adjustment of pressure force. The pressure force was measured by means of digital force gauge (KMM20 + ADT1U-PC *(Wobit))* with the following metrological parameters:

- measurement range: 200 N cm^{-2};
- non-linearity: 0.5% of full measurement range;
- hysteresis: 0.5% of full measurement range;
- drift error (30 min): 0.2% of full measurement range.

Resistance in the samples was measured by means of 34401A *(Hewlett Packard)* device connected with the samples by means of a measurement system in Kelvin (four-point) configuration. The samples were in the form of the stack composed of two layers of carbon composite (carbon paper) which performs the role of a diffusion layer (GDL) in the cell. A plate made of composite material was placed between the carbon paper in order to ensure even distribution of reactants to the electrodes. The set of studied layers were connected with the resistance meter by means of the electrodes made of polished cuprum. The sample was electrically isolated from the press components by means of the plates made of non-conducting PTFE (polytetrafluoroethylene, Teflon). The diagram which illustrates the method of measurement is presented in Fig. 7.

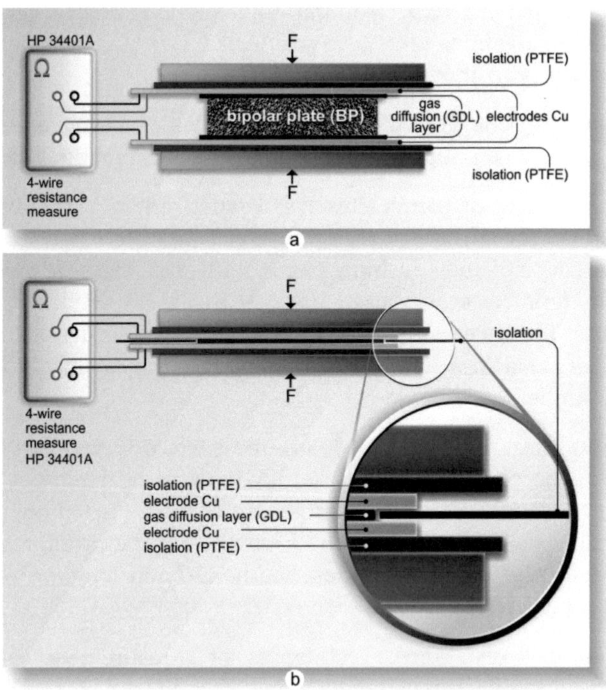

Fig. 7. a) Measurement of contact resistance in the samples which modeled cell components b) measurement of 'inclusion' resistance with cuprum electrodes and diffusion layer.

In order to determine the effect of stress on contact resistance between sintered materials and carbon paper, the analysis of contact resistance was carried out with the following stress values: 20N cm^{-2}, 40N cm^{-2}, 60N cm^{-2}, 100N cm^{-2}, 140N cm^{-2}, 160N cm^{-2}.

4.9 Analysis of corrosion resistance in sinters under operating conditions of fuel cell. Key impact on operation of fuel cell is from the processes which occur simultaneously during reaction on cell electrodes. In the case of the use of metal components for building of individual parts, one should take into consideration the possibility of solubilisation of these components in working environment of the fuel cell.

The process of destruction of metals or metal alloys is intensified by acidification of the environment as a consequence of reactions which occur in the electrodes. Moreover, the process of corrosion is activated through ions from the membrane i.e. F^-, SO_3^-, SO_4^-, because the material which is the most often used for electrolyte in PEM cells is Nafion®. Ions from the membrane intensify the corrosion processes in metal elements, whereas the cations which are created as a result of solubilisation of these components are the cause of 'poisoning' of the membrane. In consideration of the fact that virtually all the cations show higher affinity with sulphonic groups present in the membrane, compared to ion affinity H^+, cations of metals react with polymer and reduce ion conductivity of the membrane. Mechanisms of degradation of the membrane have been broadly discussed in studies by [43-45]. Both phenomena, i.e. 'poisoning' of the membrane and metal corrosion, do not only damage individual parts in the cell but they also impact on reduction in efficiency of the generator.

During operation of the fuel cell, one should additionally consider indirect reactions between the products of corrosion in metallic components with oxidizers and with the fuel. According to the literature data, corrosion products can react with oxygen at the cathode side of the cell, creating oxide layers at the electrode surface [46-48]. This effect results in blocking pores on the electrode, which leads to reduction in the efficiency of the fuel cell. Similarly, on the anode side, hydrogen can reduce metal cations to metallic form. The created metal, which is deposited on the anode, blocks electrochemical processes. Both phenomena (on the cathode and anode) may lead to a reduction in active surface area of catalyst, and, in consequence, to impeding electrochemical processes in fuel cell. Shores and Deluga [49] demonstrated in their study that the environment in the initial phase of operation of H_2/air (*PEMFC*) cell is acid (pH = 1 - 4), whereas after a certain period of time, the environment changes to pH = 6 - 7 [50]. In consideration of the phenomena which occur in fuel cell, the solution of 0.1 mol dm^{-3} H_2SO_4 + 2 ppm $F^-$was proposed in order to evaluate corrosion resistance in materials [51-52]. Since the operating temperature of PEM fuel cell amounts to ca. 80 °C (this cell belongs to low-temperature cells), the corrosion investigations were carried out at the temperature of 80 °C ± 2°C. Thermostat system allowed for maintaining constant temperature of the solution. The proposed corrosion environment allowed for a rough simulation of operating conditions in fuel cell and the evaluation of corrosion parameters in metal components of the cell was possible.

During potentiokinetic measurements, the working electrode was provided by the sintered steel, whereas the reference electrode was saturated calomel electrode, whereas platinum wire was used as auxiliary electrode. The sintered samples had been previously polished with a set of abrasive papers with grit of 60, 80, 100, 180, 400, 800, 1000, with the finishing paper with grit designation of 2500. During electrochemical measurements, corrosion solution was saturated with oxygen or hydrogen. Both gases were obtained by means of an electrolyzer. Before and during measurements, the solution was saturated with a respective gas (ca. 1 hour). Potentiokinetic testing was carried out at a scan rate of 5 mV s^{-1}. This scanning rate prevented too deep etching of the material during a single potentiometric measurement and was sufficient for registration of only Faraday processes in the electrode. Potentiokinetic curves were recorded after 10 seconds from the moment of putting the sample into the solution. The range of potential varied from the cathode values (-0.8 V vs. SCE) to anode values (1.8 V vs. SCE). Polarization curves were recorded by means of electrochemical measurement station CHI 1140 (CH Instruments, USA) connected to the computer. Polarization curves were used for determination or evaluation of the following corrosion parameters:

- corrosion potential (E_{kor}) [V];
- corrosion current density (i_{kor}) [A cm^{-2}];
- current density at anodic potential E= -0.1 V vs. SCE and at cathodic potential E=0.6 V vs. SCE [A cm^{-2}];
- polarization resistance (R_p) [Ω cm^2].

Determination of the polarization resistance R_p allows for the evaluation of the corrosion rate. After the determination of corrosion potential, the sample was subjected to the potential from the range of E_{kor} ± 20 mV. This means the range where the Stern-Hoar relationship is valid: density of external current is linear function of potential. Tangent of slope angle for the relationships of $E = f(i)$ is reversely proportional to the corrosion rate. It should be emphasized that the corrosion rate determined by means of polarization resistance method might differ even by several times from the value of corrosion rate determined through extrapolation of Tafel sections, which, on the other hand, differ from stationary gravimetric measurements. For this reason, in order for the results to be comparable, research station, methodology and conditions of the research was defined in details as above.

Corrosion current density was obtained from extrapolation of tangents to anode potentiokinetic curves with the slope of 0.04 V/decade (it was adopted that the process of anode solubilisation of the sintered steels occurs according to Bockris mechanism [53-54]). The extrapolation method allowed for evaluation of the corrosion rate in composites.

5. Results and discussion

5.1 X-ray examinations

Fig. 8. presents the diffractograms of graphite-steel composites. As results from X-ray examinations, the sinter 316L exhibits austenitic structure (CrFeNi phase). Steel sinters modified with graphite revealed the presence of hexagonal graphite and rhombohedral graphite (unstable thermally), made of deformed hexagonal graphite [55].

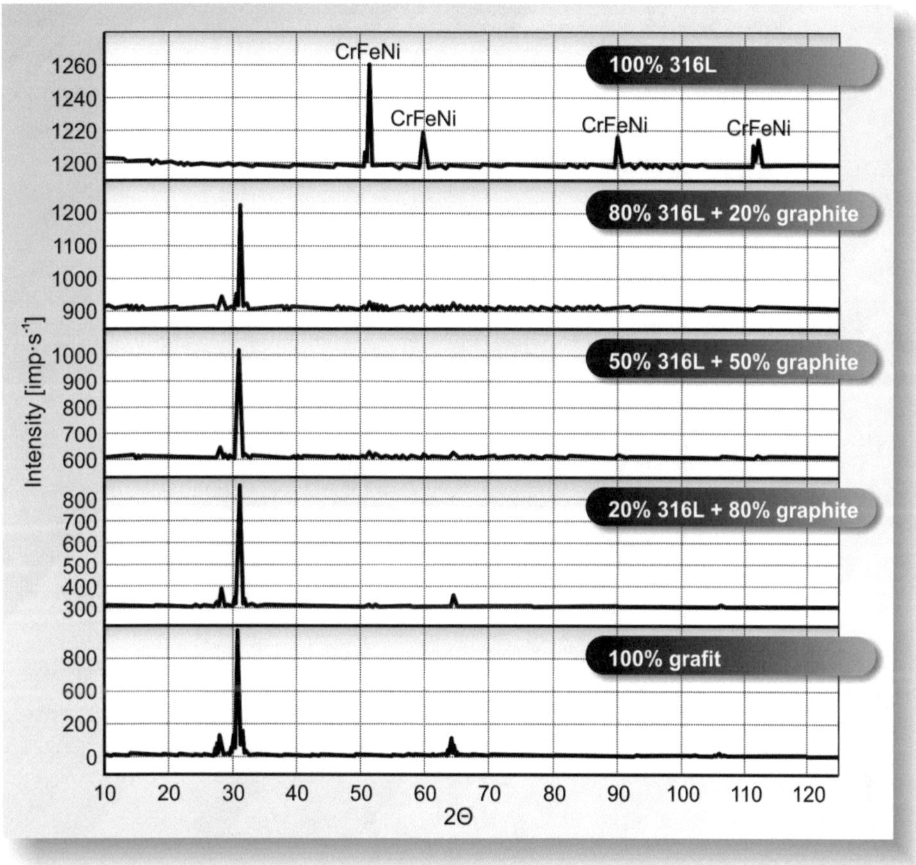

Fig. 8. Diffractograms of graphite-steel composites.

5.2 Microstructural examinations, density and hardness of sinters

Fig. 9 presents microstructures in graphite-steel composites.

50% 316L + 50% graphite **20% 316L + 80% graphite**

100% graphite

Fig. 9. Microstructures of graphite-steel composites, magnitude 100 x.

Table 4 contains the values of density for 316L steel sinter and graphite-steel composites. Addition of graphite to the material reduces density from the level of 7.16 g cm^{-3} for steel to the level of 2.35 g cm^{-3} for the sinter modified with 80% of graphite. Modification of steel sinter 316L with graphite allows for obtaining materials with reduced density. With respect to future applications of these materials for bipolar plates in fuel cells, the use of light materials will allow for achievement of one of the most essential goals of hydrogen technologies, i.e. reduction in generator weight. Addition of graphite to steel sinter impacts also on material hardness. Change in sinter hardness with concentration of graphite in the composite is presented in Table 4.

Composites	Density of sinter [g cm^{-3}]	Hardness
100% 316L	7.16 ± 0.38	79 ± 3.75 HRB
80% 316L + 20% grafit	6.93 ± 0.34	45 ± 4.15 HRB
50% 316L + 50% grafit	3.81 ± 0.19	35 ± 1.75 HRB
20% 316L + 80% grafit	2.35 ± 0.11	86 ± 4.30 HRF
100% grafit	1.97 ± 0.09	97 ± 4.85 HRF

Table 4. Density and hardness of graphite-steel composites.

5.3 Stereology of composite grain size

Table 5 contains mean cross-sectional surface area, mean number of grains per mm^2 of the surface and mean number of grains per mm^3 of graphite-steel composites. Based on the data contained in the table, one should note that no effect of chemical composition of the sinter on mean grain size is observed. Mean grain diameter varies from 48 to 68 mm, whereas the greatest grain diameters are observed for the sinter with 50% proportion of graphite. The data are also presented in Fig. 10.

Composites	Mean grains diameter [mm]	Mean surfach of grains [mm^2]	Mean number of grains per 1mm^2	Mean number of grains per 1mm^3
100% 316L	0.055	0.00346	227	3 633
80% 316L + 20% grafit	0.048	0.00195	512	11 585
50% 316L + 50% grafit	0.068	0.00427	280	4 492
20% 316L + 80% grafit	0.055	0.00346	227	3 633
100% grafit	0.051	0.01275	210	3 369

Table 5. Mean values of grain parameters in graphite-steel composites.

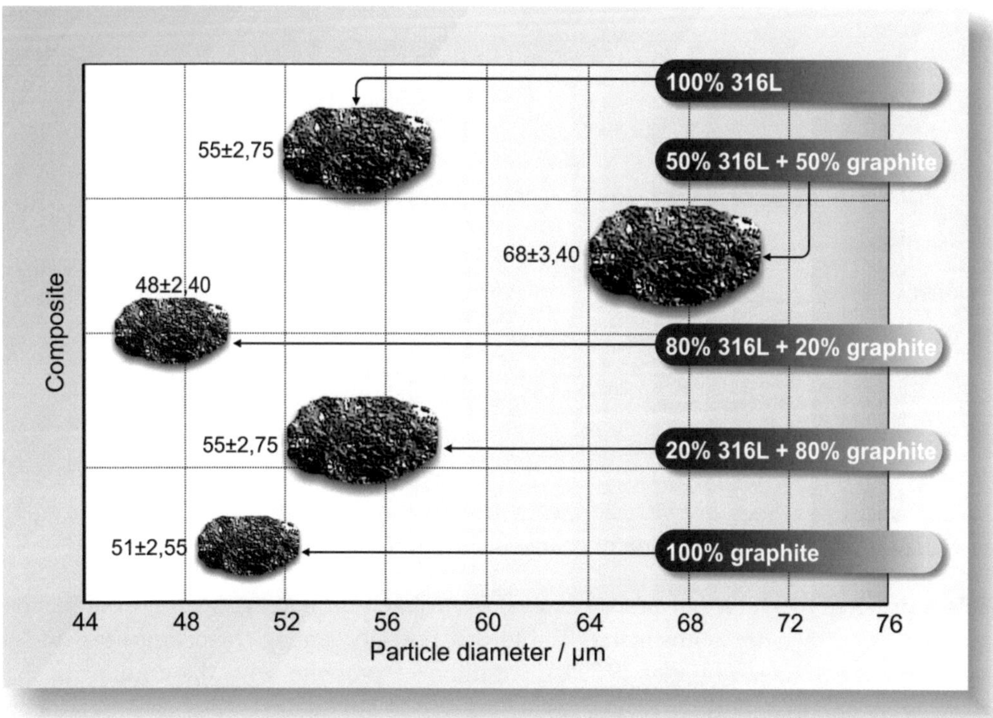

Fig. 10. Relationship between mean grain diameter in graphite-steel composite and proportion of graphite in the composite.

5.4 Composite porosity

A variety of materials and methods of modification of the surface of materials used for bipolar plates points to rising interest in fuel cell technologies, with particular focus on the design of the cell. Main requirements concerning commercial use of materials for manufacturing fuel cells is the relationship between high corrosion resistance and low contact resistance, with low costs of manufacturing. Corrosion rate, contact resistance and wettability of material depend to some degree on material porosity. Therefore, the investigations of functional properties were started from the analysis of pore composition in the sinters included in the study.

Assessment of porosity concerned graphite-steel composites. 316L steel sinters with addition of graphite exhibit varied porosity depending on the proportion of graphite. Fig. 11a presents hysteresis for intrusion and extrusion pressure for mercury in 316L sinter. Narrow pressure hysteresis loop points to the presence of flat pores in the material. Similar profile of hysteresis for mercury intrusion and extrusion pressure was found for other composites included in the study. Fig. 11.b. presents distribution of pore diameters in the sinters included in the study. It should be emphasized that graphite-steel composites show pores with diameters which correspond to mesopores. Only in the sinter with 50% proportion of graphite no pores from the range of diameters corresponding to mesopores were found, whereas macropores with diameters over 0.08 μm were observed.

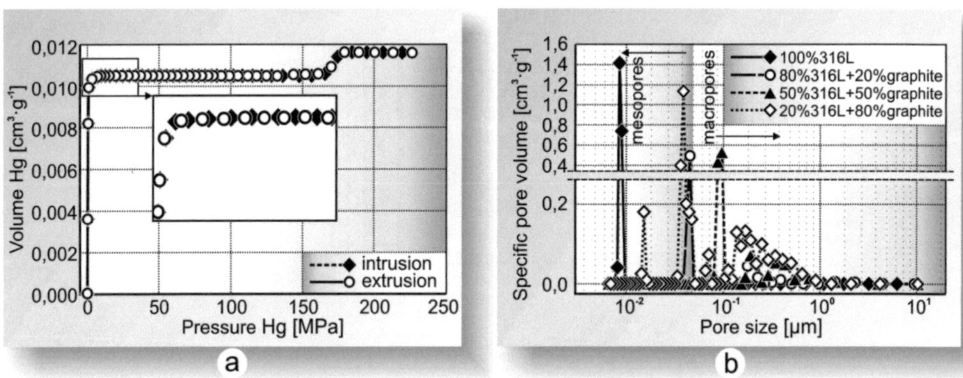

Fig. 11. Hysteresis of mercury intrusion/extrusion in graphite sinters and distribution of pores depending on the proportion of stainless steel in the sinter.

Table 6 presents the values of porosities evaluated based on microstructural examinations and tests using mercury porosimeter. The lowest porosity among the composites studied was found for 316L steel sinter (9.59%). Addition of graphite with the amount of 20% considerably enhances porosity of material compared to steel sinter. Other sinters, enriched with 50% and 80% of graphite, exhibit lower porosity compared to the sinter of 80% of 316L + 20% of graphite, but this is still the value higher than the value of porosity estimated for 316L sinter.

Rodzaj kompozytu	Porosity [%]
100% 316L	9.59
80% 316L + 20% grafit	14.43
50% 316L + 50% grafit	12.17
20% 316L + 80% grafit	11.09
100% grafit	10.73

Table 6. Comparison of porosity in graphite-steel composites.

5.5 Investigations of sinter wettability

In consideration of the degree of wettability, the materials are typically divided into lyophilic materials, which have strong affinity for water (these materials attract water particles) (Fig. 12a) and the materials which repel water particles, termed lyophobic (Fig. 12b). Contact angle Θ provides a measure of wettability, which is an angle between the surface of a solid and tangent going through the point of contact of solid, liquid or gaseous phase determined for the liquid phase. It is conventionally adopted that solid bodies which are characterized by contact angles of $\Theta < 90°$ are wettable; these materials show high surface energy (if the liquid is water, these materials are termed hydrophilic). Materials which exhibit contact angle of $\Theta > 90°$ are regarded to be non-wettable (lyophobic or, alternatively, hydrophobic = low surface energy).

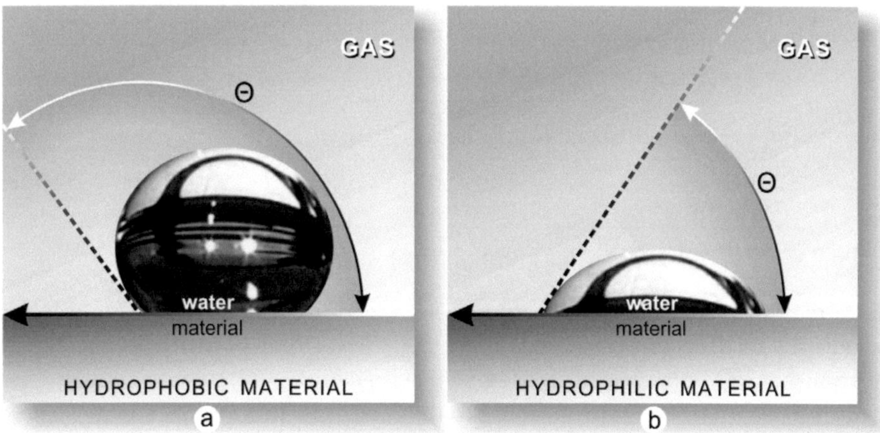

Fig. 12. Diagram of wetting hydrophobic and hydrophilic materials.

In order to determine the effect of chemical composition of a composite on surface wettability, the analysis of wettability was carried out through evaluation of the value of Θ angle. The investigations concerned 316L steel sinter and sinters with addition of graphite. Fig. 13 presents contact angles evaluated for composite materials. A linear relationship between the proportion of graphite and surface wettability: contact angle increases with proportion of graphite in the composite. The highest contact angle was found for the sinter of 100% graphite (102°). Addition of graphite to steel affects surface energy of the material: hence, composites which contain 50% and more of graphite are numbered among a group of materials which are not hydrophobically wettable. The value of Θ angle evaluated for the materials used in the study are contained in the Table 7.

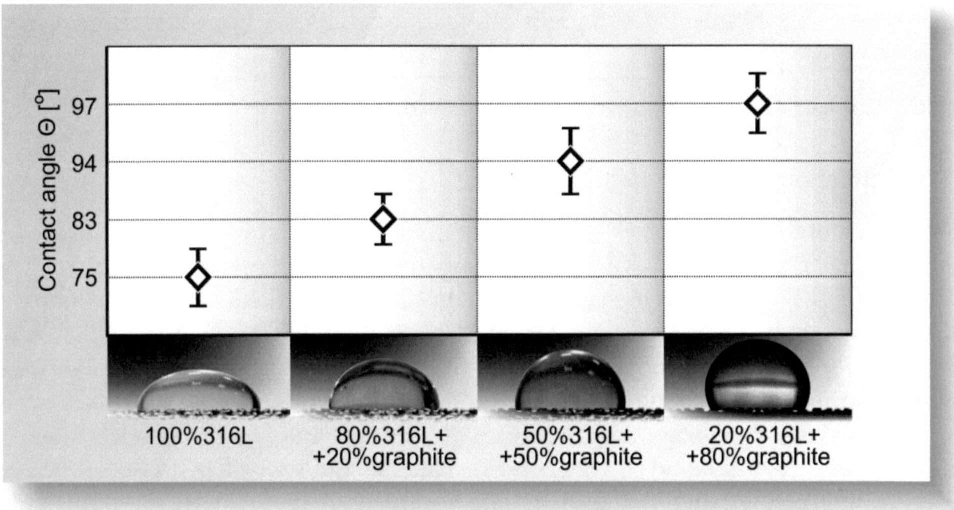

Fig. 13. Contact angle evaluated for graphite-steel composites.

Composites	Contact angle Θ [°]
100% 316L	75 ± 0.15
80% 316L + 20% grafit	83 ± 0.23
50% 316L + 50% grafit	94 ± 0.28
20% 316L + 80% 434L	97 ± 0.32
100% grafit	102 ± 0.37

Table 7. Values of contact angles for graphite-steel composites.

The authors of the study [56] demonstrated that material porosity affects contact angle. Fig. 14 presents the relationship between contact angle and sinter porosity. As can be observed, the relationship of both parameters which characterize the surface is non-linear. The materials whose porosity varies from 10 to 12% are numbered among hydrophobic materials. In the case of graphite-steel composites, with porosity higher than 12%, contact angle is lower than 90°.

5.6 Sinter roughness measurements

The available literature reports that contact resistance and wettability depend on surface geometry [57-58]. If a material is hydrophobic, a drop covers roughness in the surface and smoothens the unevenness (homogeneously) or it only touches the roughness, leaving a space between the drop and the solid (heterogeneously) (see Fig. 15).

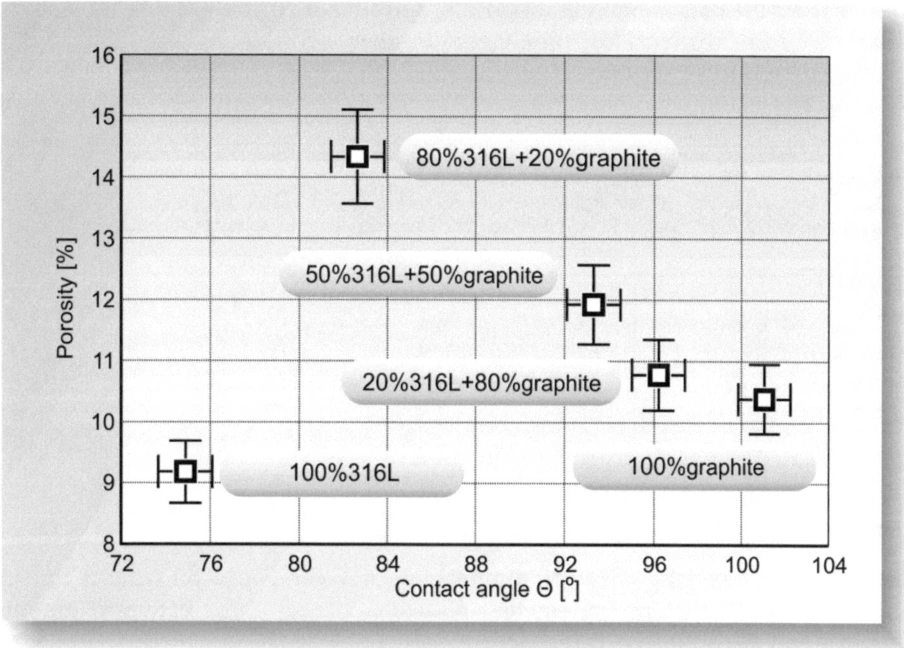

Fig. 14. Effect of porosity on contact angle in graphite-steel composites.

Fig. 15. Behavior of a drop on hydrophobic material depending on surface geometry: (a) homogeneous, (b) heterogeneous.

Table 8 contains the values of height and longitudinal parameters for 316L sintered steel and graphite-steel composites. The substantial impact on surface geometry in sinters is from the presence of graphite. In the case of rough surface, an insignificant contribution of contact surface is observed. It is essential for fuel cells that the surface of the material for these parts is smooth, which is obtained through polishing or covering the surface with a coating [59]. Graphite, as a material with high porosity is subjected to polishing in order to obtain the smooth surface.

As results from the data contained in Table 8, addition of 20% of graphite to the composite considerably increases surface roughness. Further addition of graphite to the steel insignificantly reduces roughness, but it is still higher than roughness in 100% steel sinter 316L. This fact should be closely associated with the porosity revealed for individual composites.

Composites	Parameters of surfach geometry			
	Height feature of profile		Lengthwise feature of profile	
	R_a [μm]	R_z [μm]	S_m [mm]	D_p [%]
100% 316L	4.67	55.7	0.06	25.5
80% 316L + 20% grafit	7.12	76.5	0.06	32.4
50% 316L + 50% grafit	6.67	70.9	0.06	28.6
20% 316L + 80% grafit	6.04	65.4	0.06	24.1
100% grafit	5.46	52.3	0.05	23.6

Table 8. Parameters of surface geometry in graphite-steel composites.

5.7 Measurements of contact resistance in composites

In order to determine the effect of composition of a composite on contact resistance of interfacial contact, the measurements of contact resistance between graphite-steel composites and the diffusion layer were carried out (Fig. 7). An increase in stress value causes the decrease in contact resistance, whereas at high values of pressure force, contact resistance does not change. The values of contact resistance are the lowest for the system of 316L steel sinter - diffusion layer (Fig. 16). Addition of graphite to steel sinter elevates contact resistance by nearly 40 mΩ cm2 in the case of a composite 80% 316L + 20% graphite.

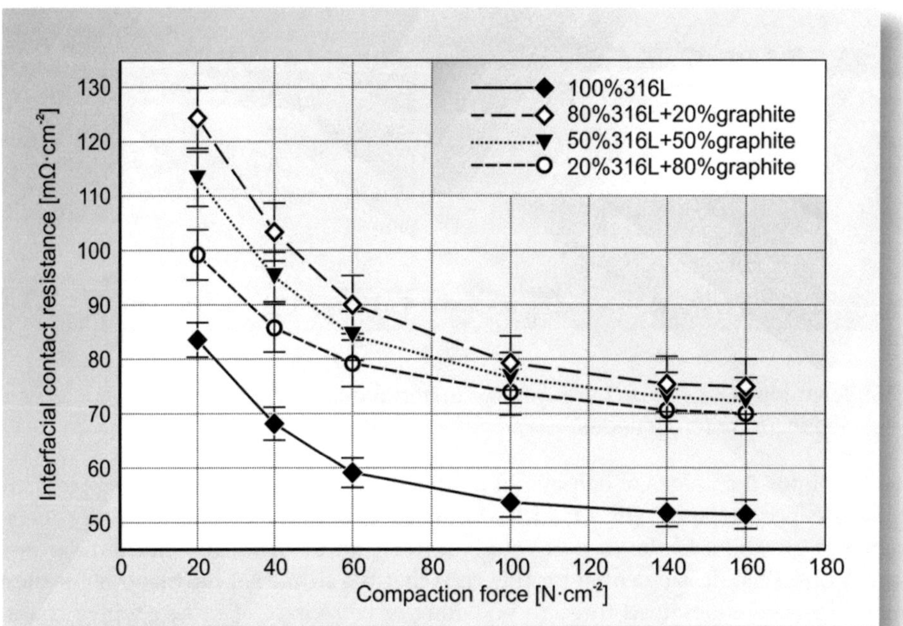

Fig. 16. Surface contact resistance for graphite-steel composites depending on stress.

5.8 Assessment of corrosion resistance in sinters

With regard to application of the used materials, one should determine corrosion resistance in the sinters. Fig. 17 presents the patterns of potentiokinetic curves recorded under conditions of work of fuel cell. The sinter 100% 316L is subjected to passivation both under conditions of cathode operation and under conditions of anode operation. Corrosion potential of the sinter 100% 316L in the analyzed environment does not change whether the solution was saturated with oxygen or hydrogen and amounts to -0.30 V vs. NEK. In the case of composite graphite-steel materials, addition of graphite caused an increase in corrosion resistance of the sinter. As results from the profile of the potentiokinetic curves, value of current density in the anode range is decreased even by two orders of magnitude. The value of corrosion potential in 316L+graphite sinters is insignificantly changed or remains at the same level compared to E_{kor} for the sinter of 100% 316L. It should be noted that the value of corrosion potential for the sinter of 100% graphite is shifted towards positive values and amounts to ca. 0.09 V vs. NEK in the solution saturated with O_2, and ca. -0.02 V vs. NEK in the solution saturated with H_2.

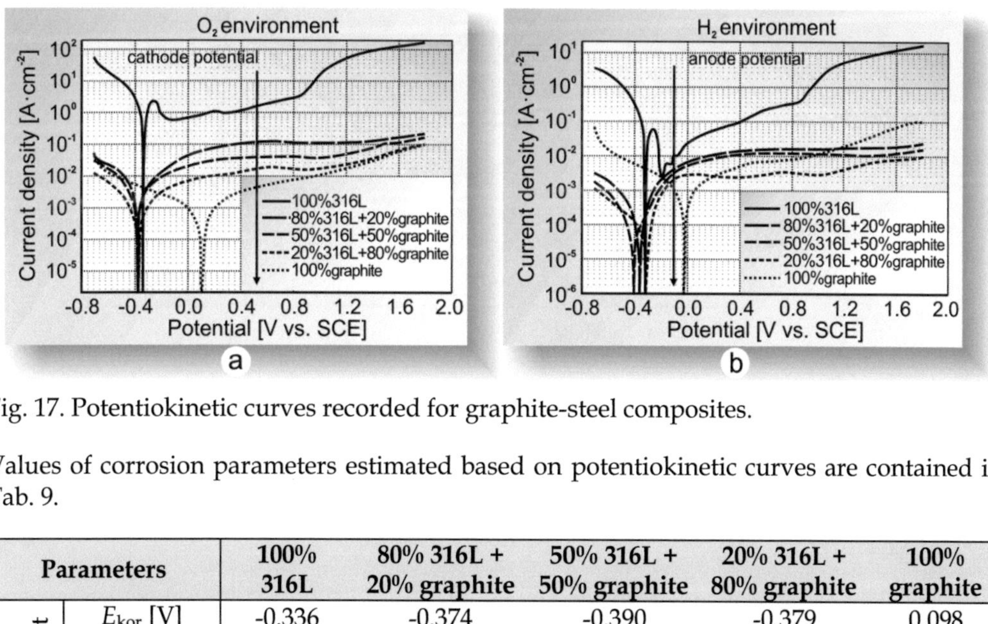

Fig. 17. Potentiokinetic curves recorded for graphite-steel composites.

Values of corrosion parameters estimated based on potentiokinetic curves are contained in Tab. 9.

Parameters		100% 316L	80% 316L + 20% graphite	50% 316L + 50% graphite	20% 316L + 80% graphite	100% graphite
O_2 environment	E_{kor} [V]	-0.336	-0.374	-0.390	-0.379	0.098
	i_{kor} [A cm^{-2}]	$90.7 \cdot 10^{-4}$	$22 \cdot 10^{-4}$	$13 \cdot 10^{-4}$	$8.00 \cdot 10^{-4}$	$5.53 \cdot 10^{-4}$
	i przy 0.6V [A cm^{-2}]	2.263	0.128	0.040	0.017	0.005
	R_p [Ω cm^2]	30.56	15 784.4	25 615.6	61 244.2	102 404.4
H_2 environment	E_{kor} [V]	-0.303	-0.357	-0.406	-0.314	-0.026
	i_{kor} [A cm^{-2}]	$58.0 \cdot 10^{-4}$	$9.42 \cdot 10^{-5}$	$7.76 \cdot 10^{-5}$	$6.12 \cdot 10^{-5}$	$9.42 \cdot 10^{-5}$
	i przy -0.1V [A cm^{-2}]	0.008	0.004	0.030	0.002	1.26
	R_p [Ω cm^2]	687.16	285 641.3	450 340.7	533 957.2	638 985.3

Table 9. Corrosion parameters of graphite - stainless steel composites.

Fig. 18 presents the effect of addition of graphite to sintered steel on polarization resistance for the material in the corrosion environment used in the study. The highest corrosion resistance was found for the sinter of 100% graphite, whereas sinter of 100% of 316L steel, compared to graphite, exhibit nearly 1000 time lower polarization resistance in the environment of H_2 and several thousand times lower in the environment of O_2. The sinters are characterized by higher corrosion resistance in the solution saturated with hydrogen (including the sinter of 100% 316L), compared to the O_2 solution.

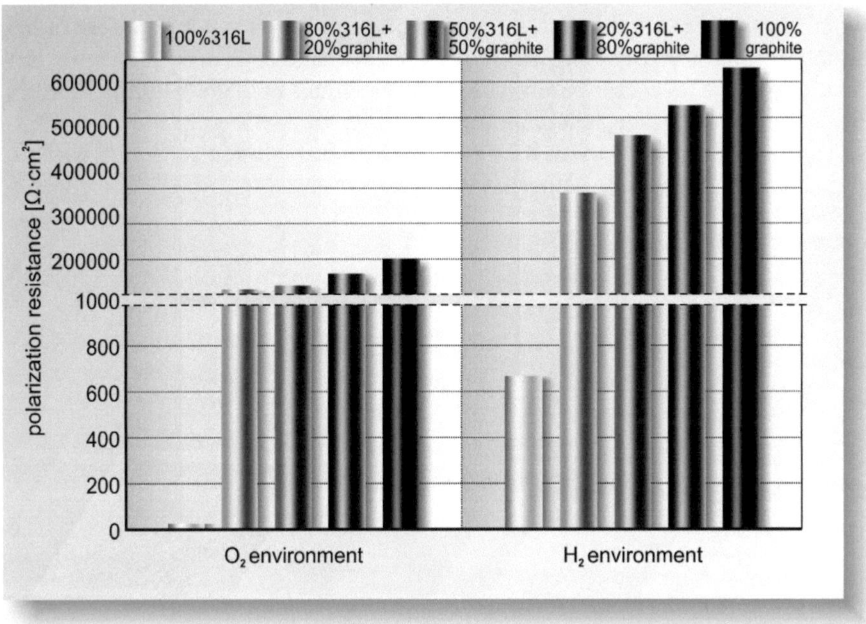

Fig. 18. Change in polarization resistance depending on proportion of graphite in a composite.

6. Conclusions

The main task of bipolar plates in fuel cells is to ensure even distribution of reactants on electrode surface. In order to meet this requirement, solid materials are cut the channels with a variety of shapes. In the case of use of porous material, obtained by means of powder metallurgy, surface unevenness forms the channels which supply and discharge media. From the standpoint of economics, the intended porosity of material is beneficial. With consideration of seeking new solutions and opportunities in the field of material choice in order to facilitate and optimize operation of machines, it seems entirely legitimate to investigate the problems analyzed in the present study.

Based on the investigations, the following conclusions were drawn:

- through modification of stainless steel with addition of graphite, it is possible to obtain materials with desired density (according to *DOE* requirements), porosity, mechanical strength and corrosion resistance;

- percentage of individual components in the proposed material impacts on stereological parameters of composites;
- depending on the ratio of graphite to steel in a composites, the materials with different porosity can be obtained; addition of 20% of graphite significantly elevates composite porosity;
- addition of graphite to steel affects surface energy in the material, and, consequently, the composites with 50% and more of graphite are numbered among a group of non-wettable (hydrophobic) materials;
- proportion of individual components in the sinter affects the height and longitudinal parameters of surface geometry in composites;
- porosity in the composites included in the study, values of surface geometry and grain diameter affect contact resistance between the composites (bipolar plate in fuel cell) and diffusion layer; the lowest value of interfacial resistance was found for 100% 316L material which exhibits the lowest porosity and the lowest values which characterize surface geometry (the smoothest);
- the highest corrosion resistance among the materials included in the study is observed for 100% graphite; in the case of composite graphite-steel materials, addition of graphite caused elevated corrosion resistance in the sinter; addition of 20% of graphite increased the value of polarization resistance (R_p) by several hundred times: from 30.56 Ω cm^2 to 15 784,4 Ω cm^2 in O_2 environment.

7. References

[1] Jayakumar K., Pandiyan S., Rajalakshmi N., Dhathathreyan K.S., Cost-benefit analysis of commercial bipolar plates for PEMFC's, J. Power Sources, 161 (2006) 454-459.

[2] U.S. Department of Energy (DOE) www.energy.gov

[3] Hermann A., Chaudhuri T., Spagnol P., Bipolar plates for PEM fuel cells: A review, Int. J. Hydrogen Energy, 30 (2005) 1297-1302.

[4] Mehta C., Cooper J.S., 2003, Review and analysis of PEM fuel cell design and manufacturing, J. Power Sources, 114 (2003) 32-53.

[5] Lee S.-J., Huang C.-H., Lai J.-J., Chen Y.-P., Corrosion- resistance component for PEM fuel cells, J. Power Sources, 131 (2004) 162-168.

[6] Antepara I., Villarreal I., Rodríguez-Martínez L.M., Lecada N., Castro U., Leresgoiti A., Evaluation of ferritic steels for use as interconnects and porous metal supports in IT-SOFCs, J. Power Sources, 151 (2005) 103-107.

[7] Lee S.-J., Lai J.-J., Huang C.-H., 2005, Stainless steel bipolar plates, J. Power Sources, 145 (2005) 362-368.

[8] Kumar A., Reddy R. G., Materials and design development for bipolar/end plates in fuel cells, J. Power Source, 129 (2004) 62-67.

[9] Mathur R.B., Dhakate S.R., Gupta D.K., Dhami T.L., Aggarwal R.K., Effect of different carbon fillers on the properties of graphite composite bipolar plate, J. Mat. Proc. Tech., 203 (2008) 184-192.

[10] Dhakate S.R., Sharma S., Borah M., Mathur R.B., Dhami T.L., Expanded graphite-based electrically conductive composites as bipolar plate for PEM fuel cell, Int. J. Hydrogen Energy, 33 (2008) 7146-7152.

[11] Yasuda E., Enami T., Hoteida N., Lanticse-Diaz L.J., Tanabe Y., Akatsu T., Carbon alloys- multi-functionalization, Materials Sci. Engineering B, 148 (2008) 7-12.

[12] Chung C.-Y., Chen S.-K., Chiu P.-J., Chang M.-H., Hung T.-T., Ko T.-H., Carbon film-coated 304 stainless steel as PEMFC bipolar plate, J. Power Sources, 176 (2008) 276-281.

[13] Feng K., Cai X., Sun H., Li Z., Chu P.K., Carbon coated stainless steel bipolar plates in polymer electrolyte membrane, Diamond & Related Materials, 19 (2010), 1354-1361.

[14] Fu Y., Lin G., Hou M., Wu B., Shao Z., Yi B., Carbon-based films coated 316L stainless steel as bipolar plate for proton exchange membrane fuel cells, Int. J. Hydrogen Energy, 34 (2009) 405-409.

[15] Andre J., Antoni L., Petit J.-P., Corrosion resistance of stainless steel bipolar plates in a PEMFC environment: A comprehensive study, Int. J. Hydrogen Energy, 35 (2010) 3684-3697.

[16] Kraytsberg A. Auinat M., Ein-Eli Y., Reduced contact resistance of PEM fuel cell's bipolar plates via surface texturing, J. Powers Sources, 164 (2007) 697-703.

[17] Kim J.S., Peelen W.H.A., Hemmes K., Makkus R.C., Effect of alloying elements on the contact resistance and passivation behavior of stainless steel, Corr. Sci., 44 (2002) 635-655.

[18] Hodgson D.R., May B., Adcock P.L., Davies D.P., New lightweight bipolar plate system for polymer electrolyte membrane fuel cells, J. Powers Sources, 96 (2001) 233-235.

[19] Li M.C., Zeng C.L., Luo S.Z., Shen J.N., Lin H.C., Cao C.N., Electrochemical corrosion characteristics of type 316 stainless steel in simulated anode environment for PEMFC, Electrochim. Acta, 48 (2003) 1735-1741.

[20] Geng S., Li Y., Ma Z., Wang L., Wang F., Evaluatioof electrodeposited Fe-Ni Allom on ferritic stainless steel solid oxide fuel Ce;; InterConnect, Journal of Power Sources, 195 (2010) 3256-3260.

[21] Paulauskas I.E., Brady M.P., Meyer III H.M., Buchanan R.A., Walker L.R., Corrosion behavior of CrN, Cr2N and π phase surfaces on nitrided Ni-50Cr for proton exchange membrane fuel cell bipolar plates, Corr. Sci., 48 (2006) 3157-3171.

[22] El-Enim S.A.A., Abdel-Salam O.E., El-Abd H., Amin A.M., New electroplated aluminum bipolar plate for PEM fuel cell, J. Power Sources, 177 (2008) 131-136.

[23] Nikam V.V., Reddy R.G., Corrosion studies of a copper-berylium alloy in a simulated polymer electrolyte membrane fuel cell environment, J. Power Sources, 152 (2005) 146

[24] Nikam V.V., Reddy R.G., Copper alloy bipolar plates for polymer electrolyte membrane fuel cell, Electrochim. Acta, 51 (2006) 6338-6345.

[25] Heinzel A., Mahlendorf F., Niemzig O., Kreuz C., Injection moulded low cost bipolar plates for PEM fuel cells, J. Power Sources, 131 (2004) 35-40.

[26] Radhakrishnan S., Ramanujam B.T.S., Adhikari A., Sivaram S., High-temperature, polymer hybrid composites for bipolar plates: Effect of processing conditions on electrical properties, J. Power Sources, 163 (2007) 702-707.

[27] Makkus R.C., Janssen A.H.H., F. A. de Bruijn, R. K. A. M. Mallant, Stainless steel for cost-competitive bipolar plates in PEMFCs, Fuel Cells Bulletin, 17 (2000) 5-9.

[28] Wang S.-H., Peng J., Lui W.-B., Surface modification and development of titanium bipolar plates for PEM fuel cells, J. Power Sources, 160 (2006) 485-489.

[29] Mennola T., Design and experimental characterization of polymer electrolyte membrane fuel cells, Thesis, Helsinki University of Technology, Espoo, 2000.

[30] Dudek A., Włodarczyk R., Nitkiewicz Z., Structural analysis of sintered materials used for low-temperature fuel cell plates, Materials Science Forum, 638-642 (2010) 536-541.

[31] Wlodarczyk R., Dudek A., Sintering stainless steel as bipolar plate material for polymer electrolyte membrane fuel cell, Steel Research, 81 (2010) 1288-1291.

[32] Wlodarczyk R., Dudek A., Properties and application of sintered stainless steel as interconnectors in fuel cell, Solid State Phenomena, 165 (2010) 231-236.

[33] Dudek A., Wlodarczyk R., Fuel cells as unconventional energy sources, Materials Engineering 2010, Collective monograph; Material and exploitation problems in modern Material Engineering, Monography 6 (2010) 194-204.

[34] Wlodarczyk R., Dudek A., Nitkiewicz Z., Application of austenite sinters as parts in hydrogen fuel cell, Engineering and quality production, Monography, Dnipropetrovsk 2010, 134-149.

[35] Wlodarczyk R., Effect of pH on corrosion of sintered stainless steel used for bipolar plates in polymer exchange membrane fuel cells, 6th International Conference Mechatronic System and Materials, Opole 2010, 219-220.

[36] PN-EN ISO 643, Stal. Mikrograficzne określanie wielkości ziarna.

[37] PN-74/M-04255 Struktura geometryczna powierzchni- Falistość powierzchni-Określenia podstawowe i parametry.

[38] PN-87/M-04251 Struktura geometryczna powierzchni- Chropowatość powierzchni-Wartości liczbowe parametrów.

[39] Davies D.P., Adcock P.L., Turpin M., Rowen S.J., Stainless steel as bipolar plate material for solid polymer fuel cells, Journal of Power Sources, 86 (2000) 237-242.

[40] Davies D.P., Adcock P.L., Turpin M., Rowen S.J., Stainless steel as bipolar plate material for solid polymer fuel cells, Journal of Power Sources, 86 (2000) 237-242.

[41] Zhang L., Liu Y., Song H., Wang S., Zhou Y., Hu S.J., Estimation of contact resistance in proton exchange membrane fuel cells, Journal of Power Sources, 162 (2006) 1165-1171.

[42] Barber M., Sun T.S., Petrach E., Wang X., Zou Q., Contact mechanics approach to determine contact surface area between bipolar plates and current collector in proton exchange membrane fuel cells, Journal of Power Sources, 185 (2008) 1252-1256.

[43] Cheng X., Shi Z., Glass N., Zhang L., Zhang J., Song D., Liu Z.-S., Wang H., Shen J., A review of PEM hydrogen fuel cell contamination: Impacts, mechanisms, and mitigation, J. Power Sources, 165 (2007) 739-756.

[44] Shi M., Anson F.C., Dehydration of protonated Nafion® coatings induced by cation exchange and monitored by quartz crystal microgravimetry, J. Electroanal. Chem., 425 (1997) 117-123.

[45] Inaba M., Kinumoto T., Kiriake M., Umebayashi R., Tasaka A. Ogumi Z., Gas crossover and membrane degradation in polymer electrolyte fuel cells, Electrochimica Acta, 51 (2006) 5746-5753.

[46] André J., Antoni L., Petit J.-P., De Vito E., Montani A., Electrical contact resistance between stainless steel bipolar plate and carbon felt in PEMFC: A comprehensive study, International Journal of Hydrogen Energy, 34 (2009) 3125-3133.

[47] Gülzow E., Schulze M., Gerke U., Bipolar concept for alkaline fuel cells, Journal of Power Sources, 156 (2006) 1-7.

[48] Gamboa S.A., Gonzalez-Rodriguez J.G., Valenzuela E., Campillo B., Sebastian P.J., Reyes-Rojas A., Evaluation of the corrosion resistance of Ni-Co-B coatings in simulated PEMFC environment, Electrochimica Acta, 51 (2006) 4045-4051.

[49] Shores D.A., Deluga G.A., Handbook of fuel cells - fundamentals, technology and applications, New York; Wiley; (2003) 273.

[50] Lee Y.-B., Lee C.-H., Lim D.-S., The electrical and corrosion properties of carbon nanotube coated 304 stainless steel/polymer composite as PEM fuel cell bipolar plates, International Journal of Hydrogen Energy, 34 (2009) 9781-9787.

[51] Borup R.L., Vanderburgh N.E., Design and testing criteria for bipolar plate materials for PEM fuel cell applications, Proces. Mat., Res., Soc., Symp., 393 (1995) 151.

[52] Fleury E., Jayaraj J., Kim Y.C., Seok H.K., Kim K.Y., Kim K.B., Fe-based amorphous alloy as bipolar plates for PEM fuel cell, Journal of Power Sources, 159 (2006) 34-37.

[53] Amin M. A., Ibrahim M.M., Corrosion and corrosion control of mild steel in concentrated H2SO4 solutions by a newly synthesized glycine derivative, Corrosion Science, 53 (2011) 873-885.

[54] Bala H., Korozja materiałów - teoria i praktyka, Wydawnictwo Wydziału Inżynierii Procesowej, Materiałowej i Fizyki Stosowanej, Częstochowa, 2002.

[55] Parthasarathy G., Sreedhar B., Chetty T.R.K., Spectroscopic and X-ray diffraction studiem on fluid deposited rhombohedral graphite from the Eastern Ghats Mobile Belt, India, Current Science, 90 (2006) 995-1000.

[56] Tang Y., Yuan W., Pan M., Wan Z., Feasibility study of porous copper fiber sintered felt: A novel porous flow field in proton exchange membrane fuel cell, Int. J. Hydrogen Energy, 35 (2010) 9661-9677.

[57] Kim D.-K., Lee D.-G., Lee S., Correlation of microstructure and surface roughness of disc drums fabricated by hot forging of an AISI 430F stainless steel, Metallurgical and Materials Transactions 32A, (2001) 1111-1116.

[58] Hakiki N.E., Structural and photoelectrochemical characterization of oxide films formed on AISI 304 stainless steel, Journal of Applied Electrochemistry, 40 (2010) 357-364.

[59] Mahabunphachai S., Cora Ö. N., Koç M., Effect of manufacturing processes on formability and surface topography of proton exchange membrane fuel cell metallic bipolar plates, Journal of Power Sources, 195 (2010) 5269-5277.

10

Oxidation Resistance of Nanocrystalline Alloys

Rajeev Kumar Gupta[1], Nick Birbilis[1] and Jianqiang Zhang[2]
[1]*Department of Materials Engineering, Monash University,*
[2]*School of Materials Science & Engineering, The University of New South Wales,*
Australia

1. Introduction

Nanocrystalline (nc) materials are single or multi-phase polycrystalline solids with a grain size of a few nanometers, typically less than 100 nm. Owing to the very fine grain size, the volume fraction of atoms located at grain boundaries or interfaces increases significantly in nanocrystalline materials [1]. A simple geometrical estimation, where the grains are assumed as spheres or cubes, yields the following values for the volume fraction of the interfaces: 50% for 5 nm grains, 30% for 10 nm grains and about 3% for 100 nm grains [2-5]. These values of interface volume fraction are several orders of magnitude higher than those of conventional microcrystalline materials. Consequently, nanocrystalline materials exhibit properties that are significantly different from and often improved over, their conventional microcrystalline (mc) counterparts. For example, nanocrystalline materials exhibit increased mechanical strength [6-10], enhanced diffusivity [11], improved corrosion resistance (some nanocrystalline materials) [12-20], optical, electrical and magnetic properties [21-24]. Due to their unique properties, nanocrystalline materials have attracted considerable research interests and the field of nanocrystalline materials has now become one of major identifiable activities in materials science and engineering.

Metals exposed to high temperature oxygen-containing environments form oxides. If an oxide scale can form and the oxide is dense and adherent, then this scale can function as a barrier isolating the metal from the external corrosive atmosphere. This oxide scale is called protective oxide scale. On the other hand, if a non-protective oxide scale is formed, oxygen can penetrate through the scale and the oxidation will extend further into the metal substrate, causing a rapid metal recession. Of all oxides, chromia and alumina are two kinds of oxides thermodynamically and kinetically feasible to meet the requirement for resisting high temperature oxidation.

Alloys containing aluminium or chromium can be selectively oxidised to form alumina or chromia. To form an external oxide scale, the concentration of aluminium or chromium should reach a critical value. Nanocrystalline alloys promote this selective oxidation process by reducing the critical value. For example, in the conventional microcrystalline Ni-20Cr-Al alloy system, > 6 wt% Al is required to form a protective Al_2O_3 scale at 1000°C [25]. If the Al content is lower than 6 wt%, complex oxide mixtures consisting of Cr_2O_3, $NiCr_2O_4$ and internal Al_2O_3 form, resulting in high reaction rates and poor oxidation resistance. With a nano-crystalline alloy structure, this value can be substantially reduced to 2 wt% Al, when the grain size is 60 nm [26]. This promotion effect is also evident for the K38G alloy containing 3.5-4 wt% Al and 16% Cr, which forms external Cr_2O_3 scale and internal Al_2O_3 precipitates in the cast form (large grains) but only Al_2O_3 when in the form of sputtered nano-crystalline structure [27-30].

The unique structure and high grain boundary fraction, enhances diffusion of impurities and alloying elements, and changes materials thermodynamic properties [31-33] which are expected to cause a considerable difference in the resistance of nanocrystalline materials to environmental degradation (oxidation) at high temperatures. For practical application of these nanocrystalline materials, an acceptable level of resistance to environmental degradation is required. However, the effect of the nanocrystalline structure on the high-temperature oxidation resistance has attracted only a limited research attention. Oxidation resistance of nanocrystalline Ni-Cr-Al [34-36], Fe-Cr [37-39], and Zr [40-43] based alloys have been mainly investigated and in most of the cases, oxidation resistance has been reported to improve due to nanocrystalline structure.

The properties of oxides (Al_2O_3, SiO_2 or Cr_2O_3) formed during oxidation [44,45,46] also depend upon the grain size of the alloy and nanocrystalline structure alters the properties of oxide. For example, more uniform oxide scale with finer grain size and higher Cr or Al content is formed on the nanocrystalline alloys [34-36]. The oxide scales formed from nanocrystalline materials exhibit enhanced plastic deformation (due to fine grain size of formed oxide), which can release the stresses accumulated in the scales, and therefore the scale spallation tendency is reduced. It was reported that cyclic and long-time oxidation resistance was significantly improved by applying nanocrystalline coatings on type 304 stainless steel [47,48], Ni-Cr-Al [27], Co-Cr-Al [49], and Ni-(Co)-Cr-Al [28-30].

In order to investigate the possible differences in oxidation resistance along with any underlying mechanisms, understanding the nanocrystalline structure of a material is essential. This chapter will therefore first describe the structure of nanocrystalline materials, their thermodynamic properties and the possible effects of changes in the material structure (caused by such fine grain size) that may influence the oxidation resistance of a material.

2. Structure and properties of nanocrystalline alloys

2.1 Dual phase model

The unique properties of nanocrystalline materials are associated with very fine grain size, whereby, depending upon the grain size, interfaces can include up to 50% of the atoms in the material [2-5]. Therefore determination of the structure and associated properties of individual features of a nanocrystalline structure becomes very important. Various models representing the structure of nanocrystalline materials, such as "gas like" model as suggested by Birringer et al. [5] and a "frozen gas like" model suggested later [1,50], are proposed in the literature. However, the structure of nc-materials, in general, may be described as a composition of two components: a crystalline component (CC), which is formed by small equiaxed single crystals each with random crystallographic orientations and the intercrystalline component (IC), which is formed by the interfaces between the crystallites (grain boundaries) and intersection points of these interfaces (triple junctions). The second component may be characterized by the reduced atomic density and inter-atomic spacing deviating from those in the perfect crystal lattice. The IC surrounds the nanometer-sized crystals and forms a network between them [1,21]. As grain size reduces, the IC increases and it may even exceed CC.

Various researchers support the view of the two phase model of nanocrystalline metals. For example, extended x-ray-adsorption fine structure (EXAFS) and Mössbauer spectroscopy of ball-milled iron indicated the presence of two phases as characterized by significantly different atomic arrangements. These different atomic arrangements can be attributed to the presence of interfacial region and crystalline region [51]. Similarly, EXFAS investigation of nanocrystalline Fe and Pd indicated a large reduction in the atomic coordination number, supporting the idea of a very disordered structure at the interfaces [52-56]. Positron-lifetime spectroscopy measurements showed a large density of vacancy-like defects in grain boundaries and relatively large free volume at the triple points arising from misorientation-induced atomic instability of these sites [57]. Elastic relaxation of the interfaces occurred with a very different parameter than conventional coarse grain size polycrystalline materials. Modelling of thermoplastic properties and structure demonstrates that two phase model is an appropriate mean to account for the vibrational density of states and excess energy density in terms of grain boundary [58]. Similar to experimental findings, computer simulation of nanocrystalline iron has shown that grain boundary component in nanocrystalline material is very high and is a strong function of grain size [59,60].

In discussing the structure of nanocrystalline materials further, the following terminology will be used in the text. Three types of grain contacts are possible in a polycrystalline material. They include, a) contact surfaces, b) contact lines and c) contact points. Surfaces of two grains which contact one another are called as contact interfaces. A contact line may represent a common line for three or more adjacent grains. A contact line of three grains is called triple junction. The boundary of grain is its surfaces. Grain boundaries which are seen in the metallographic slides are the section of interfaces by slide plane. A triple point is a section of triple junction by a plane. A detailed description of the terminology used here can be found elsewhere [61].

2.2 Volume fraction of crystalline and intercrystalline regions

Mutschele and Kirchheim [62] proposed following relation to evaluate the volume fraction (C_{ic}) of nanocrystalline materials associated with intercrystalline regions,

$$C_{ic} = 3\delta / d \tag{1}$$

where, δ is the average grain boundary thickness and d is the average diameter of the grains and grains are considered to be cubes. Later, Palumbo et. al. [2] have shown that equation 1 was not suitable for the calculation of volume associated with triple points and to make it more general (to account for the triple points associated with the intercrystalline component) they proposed following relationship for the calculation of the intercrystalline component (V_t^{ic}):

$$V_t^{ic} = 1 - \left[\frac{d-\delta}{d}\right]^3 \tag{2}$$

where, d is the maximum diameter of an inscribed sphere. This yields the following relation for grain boundary volume fraction (V_t^{gb}),

$$V_t^{gb} = \left[\frac{3\delta(d-\delta)^2}{d^3} \right] \tag{3}$$

The volume fraction associated with triple points (V_t^{tp}) is then given by,

$$V_t^{tp} = V_t^{ic} - V_t^{gb} \tag{4}$$

Using above equations and applying a boundary thickness (δ) of 1 nm [2, 62,63], the effect of grain size (*d*), in the range of 2 nm to 1000 nm, on the calculated volume fractions corresponding to intercrystalline regions, grain boundaries, and triple junctions, is shown in Figure 1. The intercrystalline component increases from a value of 0.3% (at a grain size of 1000 nm) to a maximum value of 87.5% at a 2 nm grain size (Figure 1). The volume fractions associated with intercrystalline regions and perfect crystal are equivalent (i.e., 50%) at a grain size of ~ 5 nm. In assessing the individual elements of the intercrystalline fraction, it is noted that the triple junction volume fraction displays greater grain size dependence than that of the grain boundaries. In the range 100 nm to 2 nm, the triple junction volume fraction increases by three orders of magnitude, while the grain boundary volume fraction increases by little over one order of magnitude. In the nanocrystalline range (i.e., d ~ 10 nm), the grain boundary fraction only increases from ~27% at 10 nm, to a maximum value of ~ 44% at 3 nm. Over the same range of grain sizes, the triple junction fraction increases from ~3% to a value of 50%.

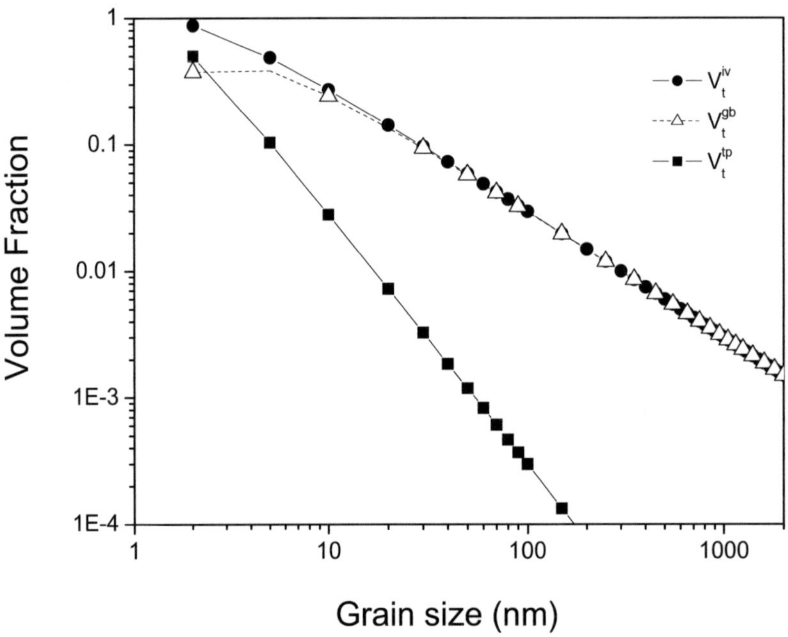

Fig. 1. The effect of grain size on the volume fractions of intercrystalline region, grain boundaries and triple points; calculated from equations 2-5 and assuming the grain boundary thickness to be 1 nm [2].

Based on Figure 1, properties which are influenced by the grain boundary and triple points are expected to be altered significantly when grain size is refined to, or below, 100 nm. The effect of triple points is more pronounced when the grain size is less than 10 nm; consequently material properties with grain size less than 10 nm would be significantly different than those with grain size > 10 nm.

2.3 Thermodynamic properties of nanocrystalline materials

Many researchers have described grain boundaries in nanocrystalline materials as more disordered than those in conventional microcrystalline materials [64-67]. For example, investigations on nanocrystalline Fe have demonstrated that grain boundaries in nanocrystalline Fe differ significantly from the grain boundaries in conventional polycrystalline Fe [64-69]. Thermodynamic properties (specific heat at constant pressure, heat of fusion and stored enthalpy) of ball-milled Fe and other nanocrystalline materials, investigated by Fecht suggested [64-69] that the grain boundaries' energy in nanocrystalline materials to be considerably greater than in the case of equilibrated grain boundaries in microcrystalline materials. In conventional polycrystalline materials, grain boundary energy, as determined by experiments, as well as static and dynamic simulations, is approximately 1 J/m^2, whereas, this value soars to 4 J/m^2 in nanocrystalline materials [70-75].

Significantly different thermodynamic properties of nanocrystalline materials are expected to increase the Gibbs free energy of the materials alloys which can be represented as per following relationship:

$$G= V_t^{gb} .G^{gb}+ V_t^{tp} .G^{tp}+ V_t^{cc} .G^{cc} \tag{5}$$

where, G^{gb}, G^{tp} and G^{cc} are the standard gibbs free energies of grain boundaries, triple points and grains.

Increases in the interfacial energy may lead to a significant increase in the free energy which can be described simply as (neglecting second order contributions due to specific heat differences):

$$\Delta G= \Delta H -T\Delta S \tag{6}$$

Enthalpy difference, ΔH, is shown to be quite higher in nanocrystalline materials than conventional microcrystalline materials. For example it has been shown that ΔH of nanocrystalline Fe increases with decrease in grain size [64-67,76]. Similar behaviour was reported for nanocrystalline copper as well. For example, nanocrystalline Cu, prepared as a powder by vapour deposition followed by compaction releases 300 J/mol at 430K when analysed immediately after compaction and 53 J/mol at 450K when analysed five days after preparation. Such values of enthalpy release have also been confirmed by a study on nanocrystalline Cu prepared by electrodeposition and cold rolling [77-79]. Comparison of these data show that nanocrystalline materials are far from equilibrium, not only because they contain a large amount of interfaces but also because these interfaces are not equilibrated. Therefore, these materials should have high value of free energy which may result in higher reaction rate at the nanocrystalline surfaces.

The total free energy also depends upon entropy term (equation 1), however, evaluation ΔS is not straightforward since there is a little reported in the literature on the entropy contribution from grain boundaries and interfaces for crystals of any size. Although it is expected that this entropy contribution is small and it can also be conceived that non-equilibrated grain boundaries have higher entropy. A value of 0.36 mJ/m^2 K has been estimated for as-prepared nanocrystalline Pt, in contrast to the value for conventional grain boundaries of 0.18 mJ/m^2 K [80]. In fact, the excess entropy per atom sitting in a grain boundary is a substantial part of the entropy of fusion, but the overall entropy per mole of substance sums up to a limited amount even for materials with very small grains. Using this knowledge of enthalpy and entropy, the free energy of nanocrystalline copper has been reported to be higher than that of coarse grain copper [80,81].

2.4 Diffusion in nanocrystalline materials

In general, atomic transport in nanocrystalline materials differs substantially from that in coarse-grained material, due to the crystallite interfaces providing paths of high diffusivity. In conventional microcrystalline materials, crystal volume self-diffusion or substitutional diffusion dominates, at least at temperatures higher than approximately half of the melting temperature. Interface diffusion, in combination with a high fraction of atoms in interfaces, gives rise to modified physical properties of nanocrystalline solids. Furthermore, diffusion processes may control the formation of nanocrystalline materials, for example, by means of crystallization of amorphous precursors, as well as the stability of nanocrystalline materials (relaxation, crystallite growth), their reactivity, corrosion behaviour, or interaction with gases. The relevance of diffusion-controlled processes demands a comprehensive understanding of atomic diffusion in nanocrystalline materials. Detailed discussion of the diffusion processes in the nanocrystalline material is out of the scope. For the readers interest it could be found elsewhere [82-85]. A recent study on nanocrystalline Fe has shown that the diffusion coefficient of Cr in Fe can be enhanced by several orders of magnitude by reducing the grain size to nanometer level [11].

Diffusion in a material can be expressed as the combined effect of diffusion through the grain boundaries and lattice diffusion and can be written as:

$$D_B = fD_{gb} + (1-f)D_b \rightarrow D_B = f\left(D_{gb} - D_b\right) + D_b \qquad (7)$$

where, f is the grain boundary fraction, D_{gb} is the grain boundary diffusion coefficient and D_b is the bulk diffusion coefficient of B in the alloy. Assuming the cubic shape of grains, the grain boundary area fraction (f) can be calculated as per equation (1). Because D_{gb} is much larger than D_b, the effective diffusion coefficient of nanomaterials increases significantly by their high area proportion of grain boundaries.

3. Factors effecting the oxidation behavior of a nanocrystalline alloy

An effective protection of metallic materials against high temperature oxidation is based on the protective oxide scale which acts as diffusion barrier, isolating the material from the aggressive atmosphere. The principle is simple, however its application is complex; to act as a real barrier the oxide film needs to be dense and homogenous and has to cover entire surface of the alloy. Oxide scale should possess mechanical properties as close possible to the base material and most importantly it should be adherent to the substrate even in the presence of large thermal shocks. These parameters largely depend upon the alloy composition and microstructure and can be optimized choosing the right combination of the two. The development of the nanocrystalline structure has provided a large scope of modification of the microstructure and to investigate the effect of nanocrystalline structure on the properties of oxide scale formed and therefore the resultant oxidation resistance. Since nanocrystalline materials are thought to be very reactive due to presence of large fraction of defects, it was supposed that they may possess poor oxidation resistance. Here both the possibilities of improvement and deterioration of oxidation resistance due to a nanocrystalline structure are discussed:

3.1 Deterioration of oxidation resistance caused by nanocrystalline structure

The following are possibilities which may lead to a higher oxidation rate in a nanocrystalline structure:

1. It was described in the previous section that free energy of a nanocrystalline alloy is increased as the atoms residing at the grain boundaries are more reactive than the atoms at grains. This increased free energy of nanocrystalline materials would accelerate the reactions occurring upon them and therefore oxidation rate of nanocrystalline structure is expected to increase, leading to poor resistance to environmental degradation.
2. Increase in oxide nucleation sites and, therefore, formation of oxide scale with comparatively finer grain size through which diffusion of oxygen and metal would be faster because of enhanced diffusion through the grain boundaries. Such phenomenon occurring in Ni-Cr-Al alloy accelerates diffusion of Al through the oxide which facilitates the formation of Al oxide [34] and leads to substantial improvement in oxidation resistance. However, such diffusion in a pure metal may lead to a significant higher oxidation rate if a non-protective oxide scale forms. For example in cause of pure Ni, nanocrystalline structure has reported to increase the oxidation rate because of increased diffusion of Ni and oxygen through the grain boundaries of oxide formed on the metal [86].
3. In the case of alloys where the concentration of solute atoms is lower than a critical value for external oxide scale formation, internal oxidation occurs. The oxidation rate is enhanced because of increased diffusivity of oxygen through grain boundaries leading to severe internal oxidation near the surface. Rapid oxidation occurs for these alloys.

3.2 Improvement in oxidation resistance caused by the nanocrystalline structure

Improvement in oxidation resistance of some engineering alloys where protective oxide scales are formed at high temperatures is noticed in their nanocrystalline forms. Improved oxidation resistance of FeBSi [87], Ni-based alloys [88-93], Zr and its alloys [40-43], Cr-33Nb [94], Fe-Co based alloys [95,96] and Cu-Ni-Cr alloys [97] is reported in their nanocrystalline form (in comparison to their microcrystalline counterparts). The mechanistic role of a nanocrystalline structure leading to the improved oxidation resistance is discussed below:

3.2.1 Enhanced diffusivity and oxidation resistance

Certain alloys can develop a continuous layer of the protective oxide of more reactive alloying elements which forms basis for the development of oxidation resistanct alloys. Such alloys are Fe, Ni, Co based, with Al, Cr or Si as the reactive alloying additives. For example, Iron-chromium alloys (such as stainless steels) are the most commonly employed oxidation resistant materials. It has been established that when time-dependent inward flux of oxygen is less than the time-dependent outward flux of solute (Cr), a continuous layer of Cr-oxide is formed at or very near the surface. Formation of such oxide layer and therefore oxidation resistance of Fe-Cr alloys depend upon the supply of solute from the alloy to alloy/oxide scale. It was established in the literature that a fine grain (~17 μm or less) stainless steel easily developed a uniform layer of Cr_2O_3. For an alloy with grain sizes greater than ~40 μm, this protective layer of Cr_2O_3 was difficult to form due to insufficient grain boundary diffusion and inadequate chromium supply [98-100]. In nanocrystalline materials where grain size is very fine and diffusion coefficients are high, such Cr oxide formation should be facilitated by a large extent. Since enhanced diffusion of alloying elements in the nanocrystalline structure facilitates the formation of protective oxides, therefore alumina, chromia and/or silica forming alloys should have improved oxidation resistance in their nanocrystalline form.

3.2.2 Nucleation sites and lateral growth of passive film

In most alloys, nucleation and growth are the mechanism of oxide scale formation during oxidation [44,45,101-103]. It is widely reported that nucleation of oxide is favoured at the high energy sites, i.e., surface defects in the form of dislocations, grain boundaries, triple points, impurities etc. Since nanocrystalline materials are composed of the large fraction of surface defects therefore they offer a large fraction of closely spaced nucleation sites. During the lateral growth of oxide, these nucleation sites become very important as the presence of closely spaced nucleation sites reduces the lateral distance necessary for the lateral growth of a uniform oxide layer to cover the entire surface.

Lobb and Evans [104] have reported an improvement in the oxidation resistance of fine grained conventional microcrystalline materials as a result of increased grain boundary area fraction as the grain boundaries acts as the preferential nucleation sites for Cr_2O_3. They reported that the oxide film nucleated at the grain boundaries needs to grow laterally for the formation of a uniform oxide layer and reported that the finer the grain size, the better the uniformity of oxide should be. Later it was shown that above a critical grain size, formation of a uniform Cr layer was not possible which could be understood based on the combined role of diffusion and nucleation site densities in the alloys [44,45,98-100]. However, these studies were performed on the material where minimum grain size was a few microns. Nanocrystalline materials offer a huge amount of nucleation sites. Therefore, formation of a homogenous compact layer able to cover whole surface is facilitated, which is expected to result in a significant improvement in the oxidation resistance.

3.2.3 Structure of the oxide scale

Mechanical properties, microstructure, adhesion and growth of oxide scale has been reviewed recently [105,106] and it has been found that oxidation resistance of a metal largely depends upon their physical properties (e.g., crystal size, morphology and crystallographic orientation of oxides formed, lattice mismatch with the base metal, adhesion of oxide layer) of oxide [105,106]. For example, fine grained oxide scales often show a fast creep rate at high temperatures, releasing the stresses accumulated in the scales and, therefore, decrease in scale spallation tendency. This may have important implication in reducing the spallation of oxides at high temperature. Since nanocrystalline materials provide several orders of magnitude more nucleation sites than the microcrystalline materials therefore grain size of the oxide developed on nanocrystalline materials is expected to be finer. This finer grain size of oxide suppresses oxide scale spallation in nanocrystalline form. This effect has been demonstrated in high temperature oxidation tests of several nanocrystalline alloys [27-30,47].

It was also proposed that fine nano-sized oxide structure reduces conductivity which suppresses the transport of the oxidizing species, enhancing oxidation resistance. This proposal was successfully used to explain improved oxidation resistance of Zr and Zr based alloys [107]. However, it is not clear if this model also applies for other alloys.

It is important to note that the above factors can operate both indifferently and in combination [12]. The nature of the influence of nanostructure on the diffusion-assisted corrosion (viz., high temperature oxidation) depends on the role of the predominantly diffusing species in a given alloy. For example, oxidation resistances of an iron-aluminide and an Fe-B-Si alloy in the nanocrystalline state are reported to be superior to that in their microcrystalline state [87,108]. This behaviour is attributed to Al and Si, the well-known protective oxide film formers, being the predominantly diffusing species respectively in the two alloys, and the nanostructure facilitating their diffusion and expedited formation of protective films (of Al/Si oxide). Therefore, it is important to consider all the factors effecting the oxidation resistant, and the net effect of all the parameters would determine the change in the oxidation rate caused by nanocrystalline structure.

4. Critical concentration of solute required for the transformation of internal to external oxidation as a function of grain size

Oxidation of engineering alloys is very complex as the components of the alloys has different affinities for the oxygen, and reacting atoms do not diffuse at the same rates in the oxides or alloy substrates. Various types of oxides can be formed on and in the alloy. Atomic ratios of the elements in the oxide scale may differ significantly from those in the alloy. When oxygen and metal atoms diffuse and react at the surface of an alloy, an external oxide layer is formed on the surface and this is termed as the "external oxidation". For external oxidation, outward flow of metal atoms must exceed the inward flow of oxygen, whereas, when inward flow of oxygen exceeds the outward flow of the metal atoms, oxygen diffuses inside the metal and oxidation takes places within the alloy. This process is termed as "internal oxidation" which leads to catastrophic loss in the material property [44,45,109]. Figures 2 and 3 schematically show internal and external oxidation for alloy A-B under the conditions where only B oxidises and both A and B oxidise, respectively.

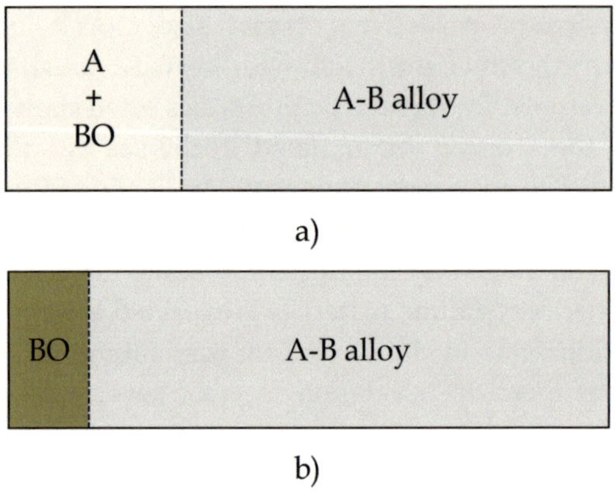

Fig. 2. Wagner's model of the transition from the internal to external oxidation of alloy A-B under the condition where only B can oxidize: a) B is less than the critical amount of B required for the transition and b) B is higher than the critical content of B required for the transition from the internal to external oxidation [44,45, 110-112].

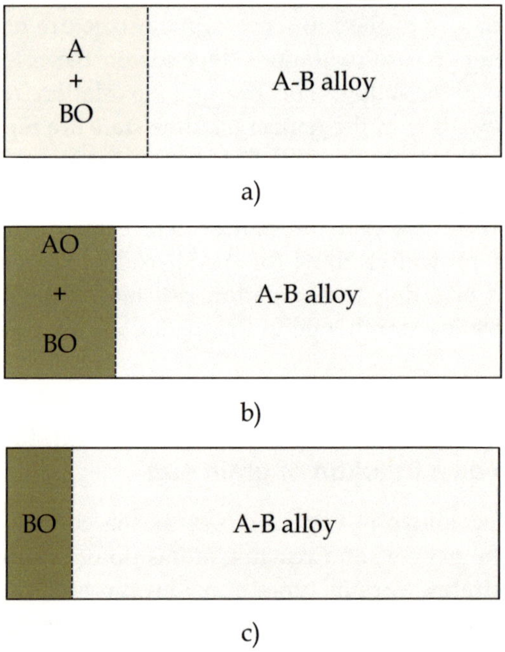

Fig. 3. Wagner's model of the transition from the internal to external oxidation of alloy A-B under the condition where both A and B can oxidize: transition from internal oxide (a) to external composite oxide scale (b) to external exclusive oxide scale (c) [44,45,110-112].

From a kinetic point of view, there is a minimum concentration of oxide former required to ensure that the alloy chromium (or any other oxide former) diffusion flux must be sufficient to outweigh inward oxygen diffusion, thus favouring external scale growth over internal oxide precipitation. The required concentration, $N_M(Crit)$, is found from Wagner's analysis [45,110-112] to be:

$$N_B(Crit1) = \left(\frac{g.\pi}{3} . \frac{V_{AB}}{V_{BO_v}} . \frac{N_O^{(s)} D_O}{D_B}\right)^{1/2} \tag{8}$$

Here $N_O^{(s)} D_O$ is the oxygen permeability in the alloy matrix, V_{AB}, V_{BO_v} are the molar volumes of alloy and oxide, D_B is the diffusion coefficient of chromium or aluminium in the alloy, and g the critical volume fraction of oxide required to form a continuous layer. According to this equation, the critical value for external chromia formation in Ni-Cr at 1000°C is $N_{Cr} = 0.29$ and for alumina formation in Ni-Al at 1200°C is $N_{Al} = 0.11$ [45], both in agreement with the experimental measurements. The requirement for these relatively large concentrations of Cr or Al to form a complete protective scale will, in many cases, change other alloy properties, which limits the applicability of this approach, particularly at lower temperatures.

No sooner has the continuous scale of chromia or alumina become established than the steady-state growth of this oxide starts. For chromia/alumina scale growth to be maintained, an alloy must supply chromium/aluminium by diffusion from its interior at a rate sufficient to balance the rate at which the metal is consumed in forming new oxide. Wagner's diffusional analysis [112] leads to the requirement:

$$N_B(Crit2) = \frac{V_{AB}}{V_{BO_v}} \left(\frac{\pi k_p}{2D_B}\right)^{1/2} \tag{9}$$

where V_{AB}, V_{BO_v} are the molar volumes of alloy and oxide, k_p is the parabolic rate constant for scale thickening:

$$x^2 = 2k_p t \tag{10}$$

with x the scale thickness formed in time, t, and D_B the alloy interdiffusion coefficient. The quantity $N_B(Crit2)$ is the minimum original alloy chromium concentration necessary to supply metal to the alloy-scale interface fast enough to support exclusive chromia scale growth.

To avoid internal oxidation and maintain external chromia/alumina scale, the concentration of aluminium/chromium should be higher than both N_B(Crit1) and N_B(Crit2). In nanocrystalline materials, D_B could be very high [11,12] according to Eqn 6, and therefore both N_B(Crit1) and N_B(Crit2) decrease significantly which will lead to substantial increase in the formation of a protective alumina/chromia external scale, therefore increased oxidation resistance. Assuming the alloy grains are cubic, the area proportion of grain boundary f can be calculated $f=2\delta/d$. Also considering Dgb>>Db, Eqn 7 can be simplified as [113]:

$$N_B(Crit1) = A(D_b + \frac{2\delta}{d} D_{gb})^{-1/2} \tag{11}$$

where

$$A = (\frac{g\pi N_O^{(s)} D_O V_{AB}}{2V_{BO_v}})^{1/2},$$

thus reduction in grain size results in the reduction of N_B(Crit1).

Using Equations (1) and (6), the critical concentration N_B(Crit2) is calculated as:

$$N_B(crit2) = B.\left(D_b + \frac{2\delta}{d}D_{gb} - \frac{2\delta}{d}D_b\right)^{-\frac{1}{2}} = B\left(\frac{\sqrt{d}}{\sqrt{2\delta(D_{gb}-D_b)+d.D_b}}\right) \quad (12)$$

Where,

$$B = \frac{V_{AB}}{V_{BO_v}}\left(\frac{\pi k_p}{2}\right)^{1/2}.$$

At a given temperature and external oxygen partial pressure, N_B (crit1) and N_B (crit2) are a function of only grain size *(d)* of the alloy, diffusion coefficients (D_{gb} and D_b) and grain boundary width (δ) of the alloy.

The critical concentration of B required, where only bulk diffusion can control the oxidation (i.e., the area fraction of grain boundary is very small such that $D=D_b$), can be calculated as:

$$N_{oB}(crit2) = B.(D_b)^{-\frac{1}{2}} \text{ and } N_{oB}(crit1) = A.(D_b)^{-\frac{1}{2}} \quad (13)$$

From equations (10), (11) and (12), for an alloy of grain size, *d*, the ratio *(X)*, critical concentration of B required for external oxidation in material with a grain size *d* to that when the bulk diffusion coefficient dominates, can be given as:

$$X = \frac{N_B(crit1)}{N_{oB}(crit1)} = \frac{N_B(crit2)}{N_{oB}(crit2)} = \sqrt{\frac{d}{2\delta\left(\frac{D_{gb}}{D_b}-1\right)+d}} = \sqrt{\frac{1}{f\left(\frac{D_{gb}}{D_b}-1\right)+1}} \quad (14)$$

Since, the values of D_{gb} and D_b are available from the literature, the value of *X* can be calculated as a function of grain size, time and temperature. Above relationship may be very useful in comparing the change in the critical amount of B (Al or Cr) caused by nanocrystalline structure at a given temperature.

Given that the critical amount of B (N_B(crit1)) required for a material with grain size d_0 is *w*, according to equation (13), the ratio for the critical amount of B required for a grain size of *d* to that for the grain size d_0 can be expressed as *X'*:

$$X' = \frac{N_B}{w} = \sqrt{\frac{d(2\delta(D_{gb}-D_b)+d_0.D_b)}{d_0(2\delta(D_{gb}-D_b)+d.D_b)}} \quad (15)$$

For Fe-Cr alloy, X' is calculated based on available data listed in the Table 1 (extrapolated from the literature [114,115] with d_0 assumed to be 5 micron, and plotted it as a function of grain size in Fig 4. Figure 4 clearly shows that Cr required for such transition decreases substantially when grain size is below 100 nm.

Temperature	Lattice (D_b) m^2/s	GB (D_{gb}) m^2/s
300 °C	1.2×10^{-26}	8.6×10^{-22}
840°C	1.5×10^{-15}	3.7×10^{-12}

Table 1. Diffusion coefficients of Cr in Fe-Cr alloy [114,115]

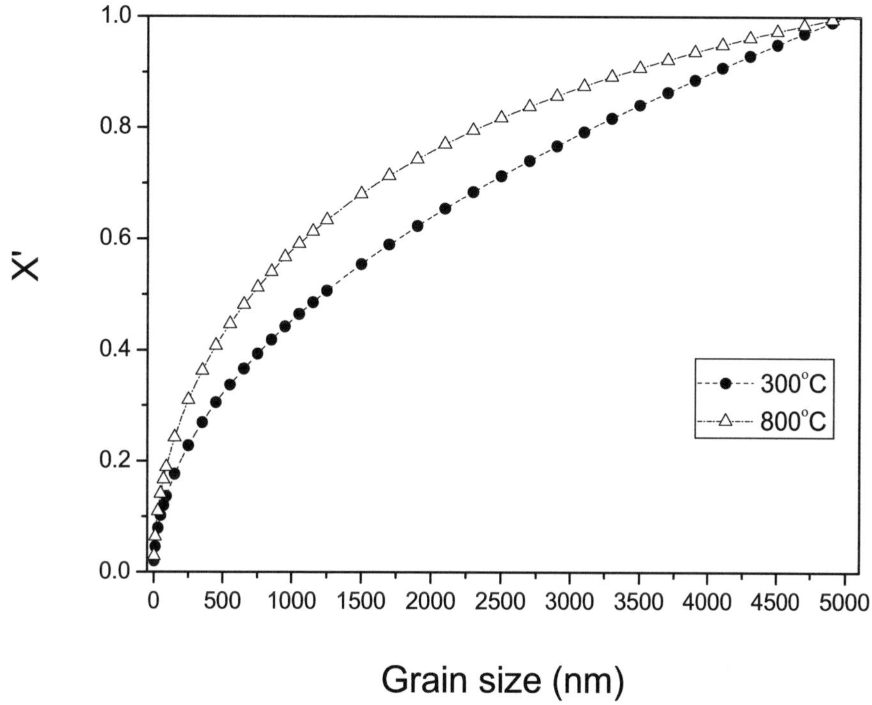

Fig. 4. Plot of the grain size with X′ (ratio of critical amount of Cr required for external passive oxide layer formation in a Fe-Cr alloy of grain size d to critical amount of Cr required for such transition in a Fe-Cr alloy with a grain size of 5 μm) [12,39].

5. Oxidation resistance of nanocrystalline alloys – Some examples

5.1 Fe-Cr alloys

5.1.1 Oxidation resistance of nanocrystalline and microcrystalline Fe-10Cr alloys

The oxidation behaviour of Fe-Cr alloys have been extensively investigated in the literature and reviewed by Wood [102]. Fe-Cr alloys are known for their high oxidation resistance due to formation of a Cr rich oxide layer which largely depends upon the selective oxidation of Cr. As already demonstrated before, a nanocrystalline structure is expected to cause a significant reduction in the Cr concentration required for chromia scale formation. For the validation of this hypothesis various nanocrystalline and microcrystalline Fe-10Cr and Fe-20Cr alloys were prepared by high energy ball milling followed by compaction and sintering which are described elsewhere [116]. The oxidation behaviour of such Fe-Cr alloys was investigated in a temperature range of 300 to 400°C by continuous weight gain experiments. The experimental details can be found elsewhere [12,38,39].

Weight gain curves representing the oxidation kinetics of nanocrystalline and microcrystalline Fe10Cr alloys, in the temperature range of 300-400°C are presented in

Figures 5-7 [12,38,39]. Oxidation kinetics at 300°C shows the microcrystalline alloy to be oxidizing at a considerably greater rate than the nanocrystalline alloy. After 3120 minutes of oxidation, weight gain of microcrystalline Fe-10Cr alloy was found to be nearly seven times greater than that of nanocrystalline alloy of same chemical composition.

Besides the considerably higher weight gain of the microcrystalline alloy, the evolution of oxidation kinetics was also different. Both nanocrystalline and microcrystalline Fe10Cr alloys follow parabolic kinetics for the first 240 minutes of oxidation (as evident in the weight-gain^2 versus time plot in Figures 5b and 6b). However during subsequent oxidation, nanocrystalline Fe10Cr alloy show considerable deviation from the parabolic behaviour whereas, microcrystalline alloy of same chemical composition continued to follow the parabolic kinetics (Figures 5c and 6c). The marked deviation of the nanocrystalline Fe-10Cr alloy from the parabolic behaviour is accounted for the insignificant increase in the weight-gain of this material after the first 240 minutes of oxidation (Figures 5a, 6a, 5b and 6c). This behaviour could be attributed to some critical change in the chemical characteristic of the oxide scale formed on both nanocrystalline and microcrystalline alloys before and after 240 minutes of oxidation which was described by Gupta et al using SIMS [12,38,39] analysis of oxide formed during various period of oxidation.

Oxidation kinetics of nanocrystalline and microcrystalline Fe-10Cr alloys at 350 and 400°C are presented in Figures 6 and 7. The trend of greater oxidation rate of the microcrystalline alloy, as seen at 300°C is also followed at the two higher temperatures. However, the influence of nanocrystalline structure in improving the oxidation resistance was extraordinarily enhanced at these higher temperatures as indicated by the comparative weight gains after 3120 minutes of oxidation: weight gain of microcrystalline Fe-10Cr alloy was found to be 18 times greater than that of the nanocrystalline Fe-10Cr alloy at 350°C, and nearly 17 times greater at 400°C.

A close observation of the data as presented in Figures 5-7 show that both nanocrystalline and microcrystalline Fe-10Cr alloys follow parabolic kinetics ,i.e., (weight grain per unit area)^2 = kt. The rate constants (k) in nanocrystalline alloy changes with time (Table 2). Oxidation kinetics of nanocrystalline Fe-10Cr can be divided in the two stages, each stage characterized by a unique k value (Table 2). Microcrystalline alloy, on the other hand follow a single parabolic rate constant. As presented in the Table 2, k value for microcrystalline (mc) Fe10Cr alloys are more than an order of magnitude greater than either of the k values for nanocrystalline (nc) Fe10Cr alloy at the three temperatures.

	nanocrystalline Fe10Cr	microcrystalline Fe10Cr
300°C	5.65×10^{-13} (1st stage) and 7.42×10^{-14} (2nd stage)	7.74×10^{-12}
350°C	1.04×10^{-12} (1st stage) and 1.7×10^{-13} (2nd stage)	1.46×10^{-10}
400°C	1.34×10^{-12} (1st stage) and 5.69×10^{-13}(2nd stage)	2.53×10^{-10}

Table 2. Parabolic oxidation rate constants (k) values in $g^2cm^{-4}s^{-1}$[12]

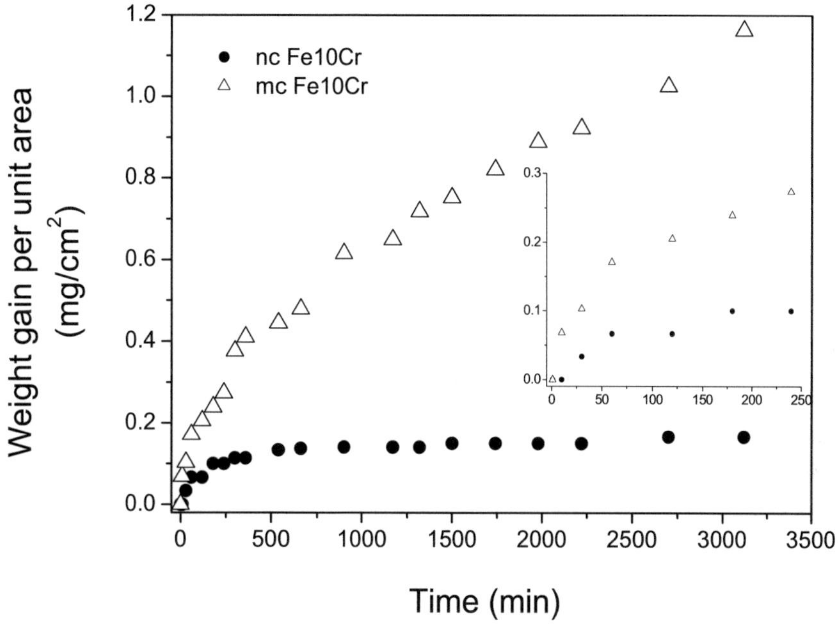

Fig. 5a. Oxidation kinetics of nanocrystalline (nc) and microcrystalline (mc) Fe-10Cr alloys at 300°C as represented by weight-gain vs time plots for 3120 minutes [12,37,39]. Inset shows the zoom of the region representing initial periods of oxidation (up to 240 minutes of oxidation).

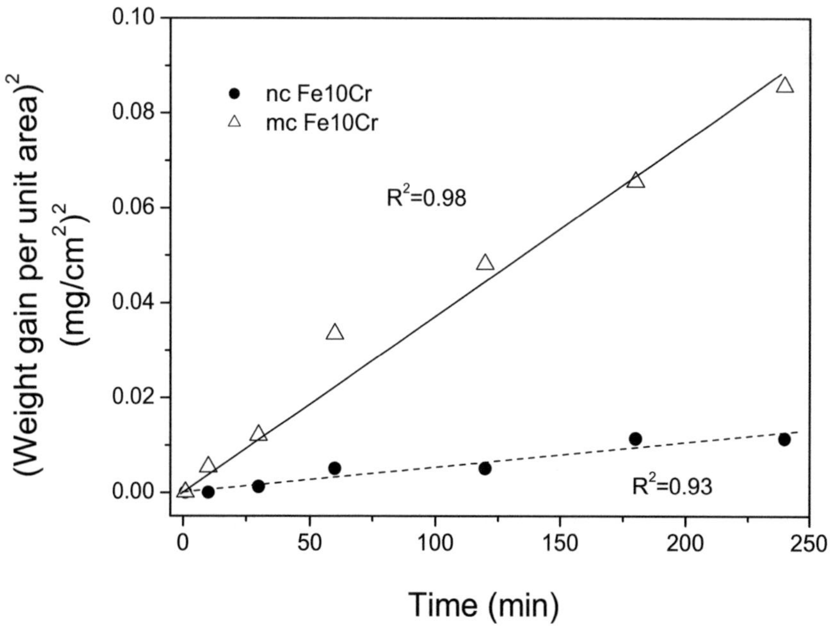

Fig. 5b. Oxidation kinetics of nanocrystalline (nc) and microcrystalline (mc) Fe-10Cr alloys oxidised at 300°C: weight-gain^2 with time (up to 240 minutes) suggesting parabolic kinetics for both mc and nc alloys [12,37,39]

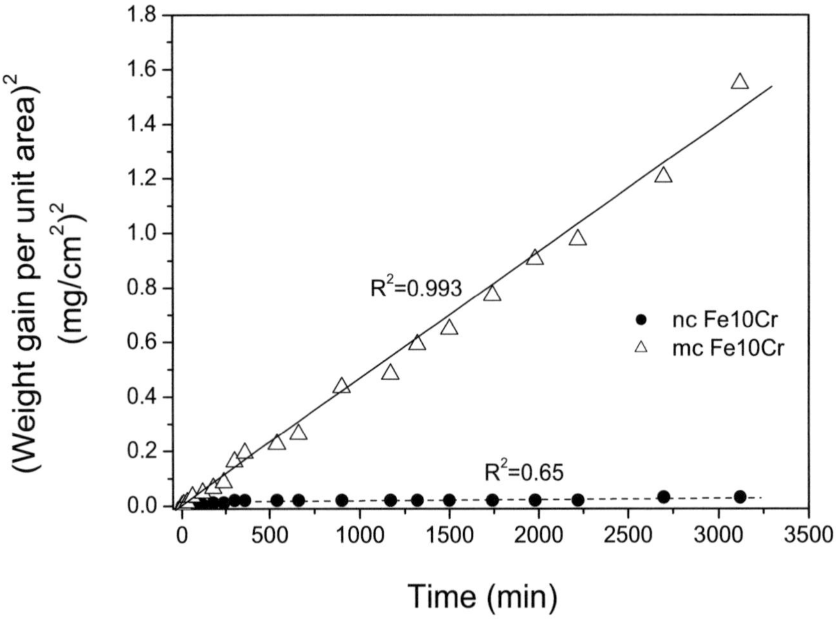

Fig. 5c. Oxidation kinetics of nanocrystalline (nc) and microcrystalline (mc) Fe-10Cr alloys, oxidised at 300°C: weight-gain^2 with time, suggesting parabolic kinetics for mc alloy but departure from parabolic kinetics for nc alloy [12,37,39].

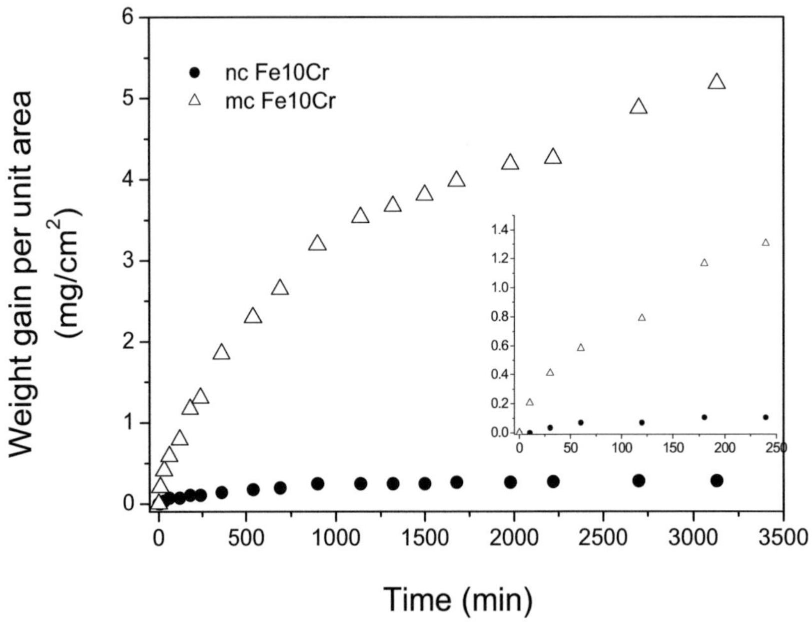

Fig. 6a. Oxidation kinetics (weight-gain vs time plot) of nanocrystalline (nc) and microcrystalline (mc) Fe-10Cr alloys, during oxidation at 350°C for 3120 min in air [12,39]. Inset shows the zoom of the region showing initial periods of oxidation (up to 240 minutes of oxidation).

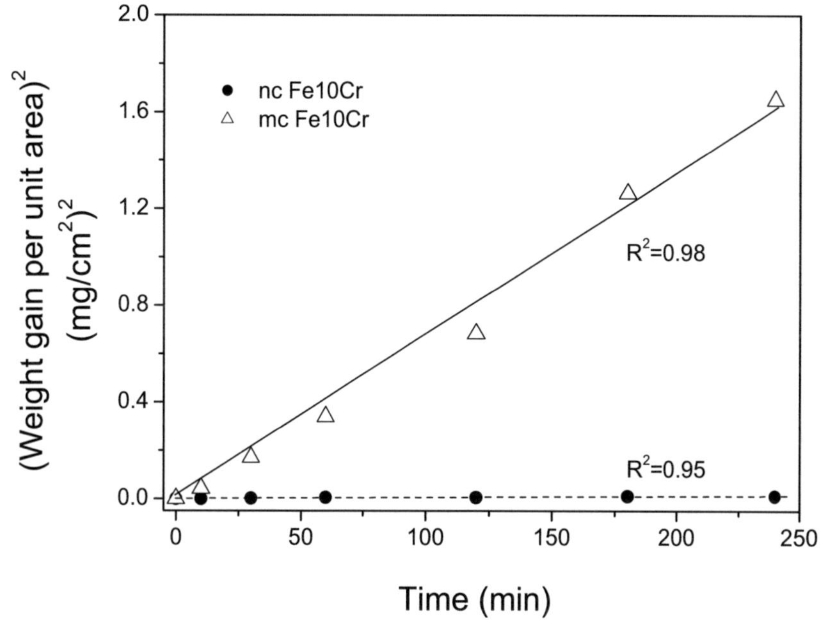

Fig. 6b. Oxidation kinetics of nanocrystalline (nc) and microcrystalline (mc) Fe-10Cr alloys oxidised at 350°C: weight-gain² with time (up to 240 minutes) suggesting parabolic kinetics for both mc and nc alloys [12,39]

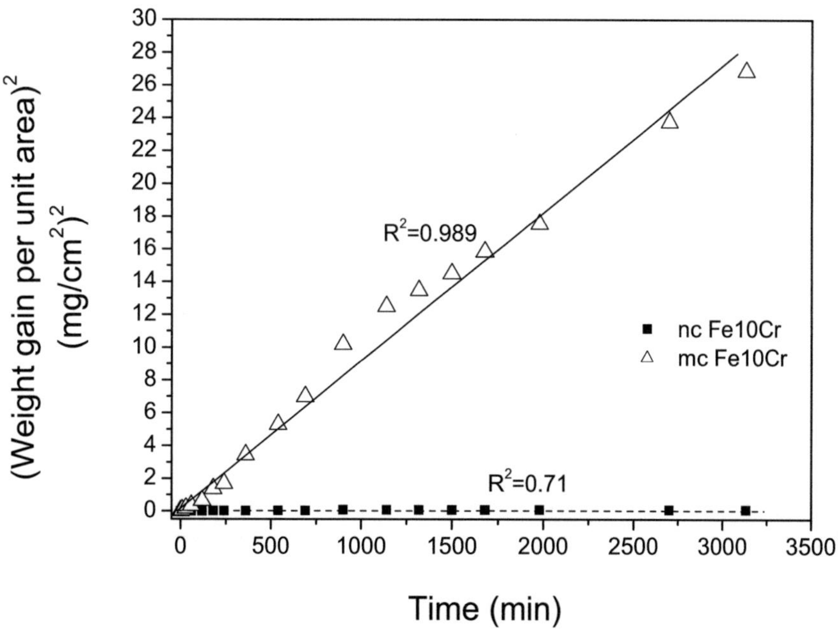

Fig. 6c. Oxidation kinetics of nanocrystalline (nc) and microcrystalline (mc) Fe-10Cr alloys, oxidised at 350°C: weight-gain² with time, suggesting parabolic kinetics for mc alloy but departure from parabolic kinetics for nc alloy [12,39].

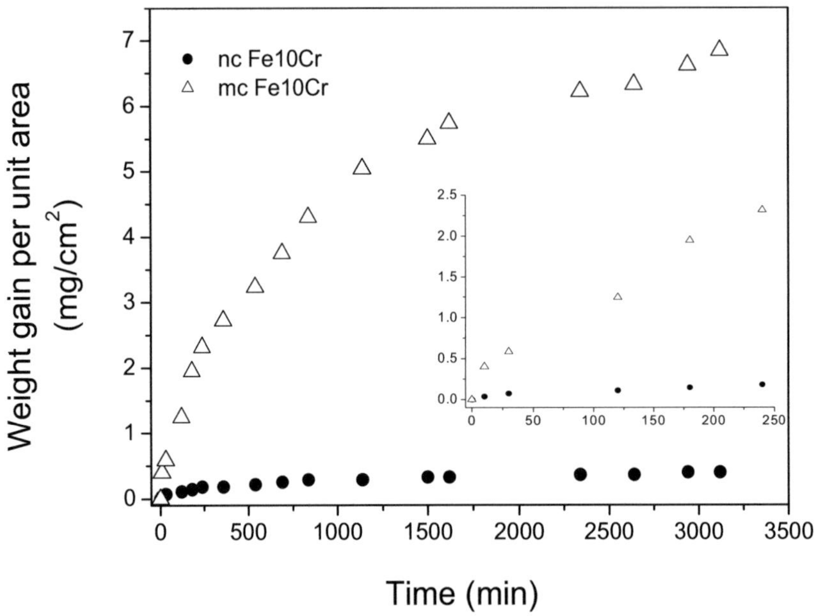

Fig. 7. Oxidation kinetics (weight-gain vs time plot) of nanocrystalline (nc) and microcrystalline (mc) Fe-10Cr alloys, during oxidation at 400°C for 3120 min in air [12,39]. Inset shows a zoom of the region of initial periods of oxidation (up to 240 minutes of oxidation).

5.1.2 SIMS depth profile of oxidized Fe-10Cr alloys

In order to understand the mechanism of improved oxidation resistance of nanocrystalline Fe-10Cr alloy, the composition, including Cr content of the thin oxide films developed on the nanocrystalline and microcrystalline alloys was characterized. The thin oxide films formed over nanocrystalline Fe-10Cr and microcrystalline Fe-10wt%Cr alloys at 300, 350 and 400°C in air were characterised by SIMS depth profiling [12,39].

Oxidation resistance of Fe-Cr alloys was associated the development of a protective layer of Cr_2O_3. Depth profiles for Cr, O and Fe for the nanocrystalline and microcrystalline Fe-10Cr alloys oxidized at the three temperatures for 30, 120 and 3120 min were obtained [12,32,33]. It was found that the oxide film developed on microcrystalline Fe-10Cr alloy is considerably thicker than that on nanocrystalline Fe-10Cr alloy at the three test temperatures [12,39].

The most relevant findings of the SIMS analyses as reported in the literature [12,38,39] are the depth profiles of chromium and their consistency with the trends of oxidation kinetics. The Cr depth profiles obtained after 52 hours of oxidation are presented in the Figures 8-10. At each of the oxidation temperatures, Cr content of the inner layer of nanocrystalline Fe-10Cr alloy was invariably found to be considerably higher than the highest Cr content in the inner layer of microcrystalline Fe-10Cr alloy. This provides an explanation for the greater oxidation resistance of the nanocrystalline Fe-10Cr alloy (as shown in Figures 8-10), since oxidation resistance of Fe-Cr alloys is governed primarily by the Cr content of the thin oxide scale.

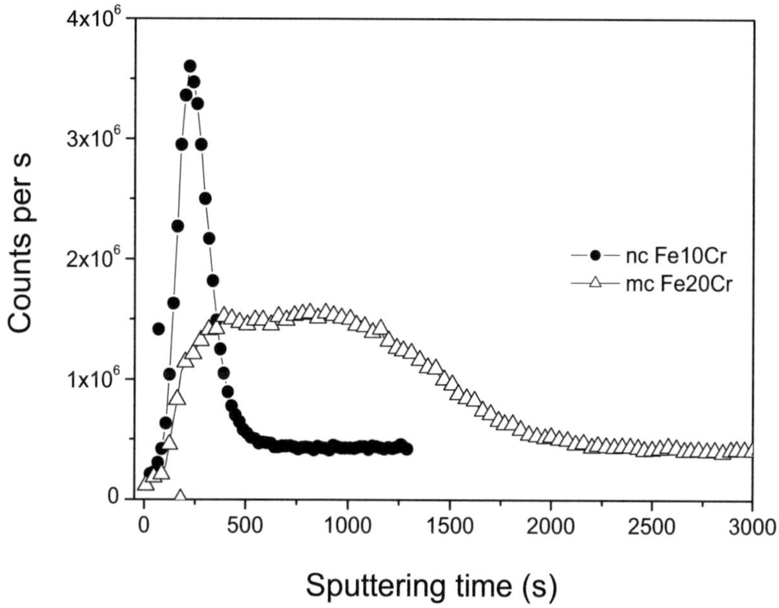

Fig. 8. SIMS depth profiles for Cr the in the oxide scale developed during oxidation of nanocrystalline (nc) and microcrystalline (mc) Fe-10Cr alloys at 300°C for in air for 3120 minutes, using a Cameca ims (5f) dynamic SIMS instrument. SIMS parameters were: Cs^+ ion primary beam (10 nA), depth profiling of craters of 250 μm × 250 μm area [12,39].

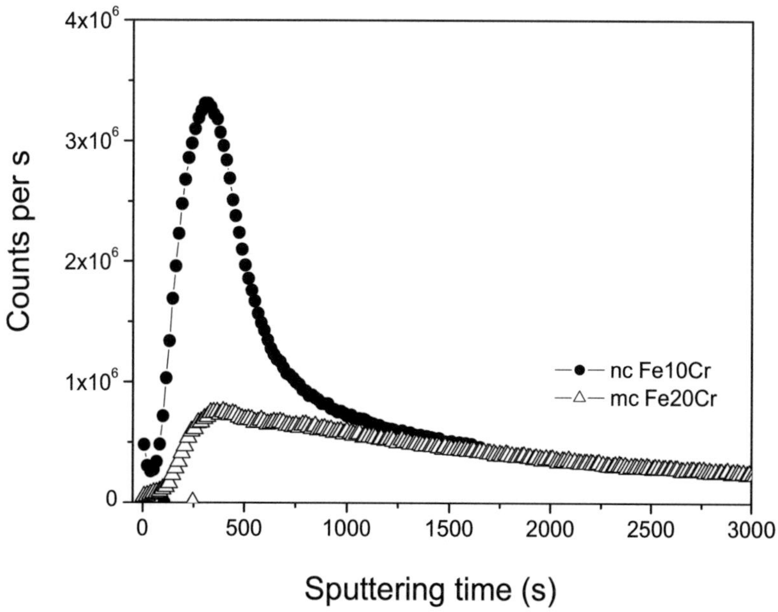

Fig. 9. SIMS depth profiles for Cr the in the oxide scale developed during oxidation of nanocrystalline (nc) and microcrystalline (mc) Fe-10Cr alloys at 350°C for in air for 3120 min, using a Cameca ims (5f) dynamic SIMS instrument. SIMS parameters were: Cs^+ ion primary beam (10 nA), depth profiling of craters of 250 μm × 250 μm area [12,39].

SIMS analysis as carried out in our previous work [12,39] provides a qualitative analysis of Cr enrichment of the surface. Based on such qualitative analysis of Cr content, a Cr_2O_3 oxide layer was proposed to develop in nanocrystalline alloy, whereas, it was proposed that a mixed Fe-Cr oxide layer forms in case of microcrystalline alloy. A Future study quantifying the Cr, Fe and O contents of oxide layer and their oxidation states using techniques such as X-ray photoelectron spectroscopy (XPS) must provide a better understanding of the effect of nanocrystalline structure on the chemical composition of oxide layer.

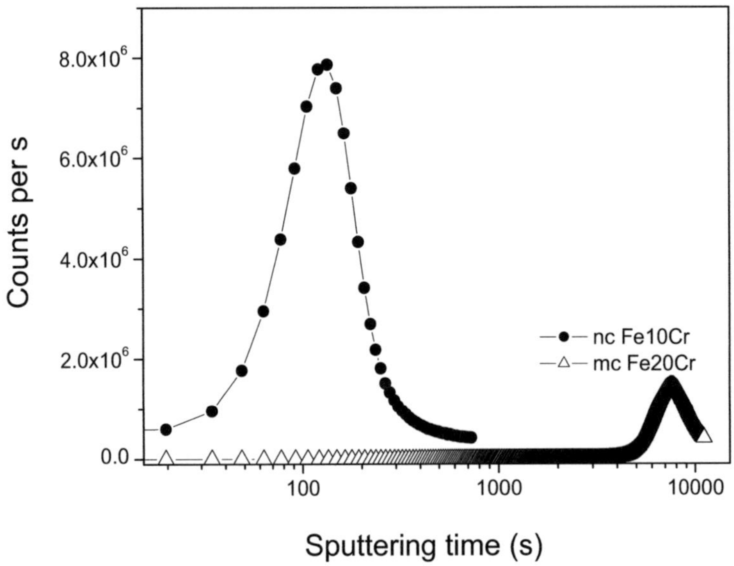

Fig. 10. SIMS depth profiles for Cr the in the oxide scale developed during oxidation of nanocrystalline (nc) and microcrystalline (mc) Fe-10Cr alloys at 400°C for in air for 3120 min, using a Cameca ims (5f) dynamic SIMS instrument. SIMS parameters were: Cs^+ ion primary beam (10 nA), depth profiling of craters of 250 μm × 250 μm area [12,39].

5.1.3 Oxidation resistance of nanocrystalline Fe10Cr versus oxidation of microcrystalline Fe20Cr alloy

For developing an understanding of how the considerably greater oxidation resistance of nanocrystalline Fe-10Cr alloy (in comparison with microcrystalline Fe-10Cr alloy) compares with the resistance of an alloy with much higher Cr content, samples of microcrystalline Fe-20Cr alloys were also oxidized at 350 °C for durations up to 3120 minutes [12,39]. However, what is most relevant to note is that the weight gain at the end of 3120 minutes of oxidation of microcrystalline Fe-20Cr alloy is similar to that of the nanocrystalline Fe-10Cr alloy at 350 °C for same period of time (shown in Figure 11), suggesting the degree of oxidation resistance conferred due to nanocrystalline structure at only 10% chromium to be similar to that of the alloy with 20% chromium but microcrystalline structure. This finding may have wide industrial applications in developing steel with low Cr but very high oxidation resistance as exhibited by Fe20Cr alloy.

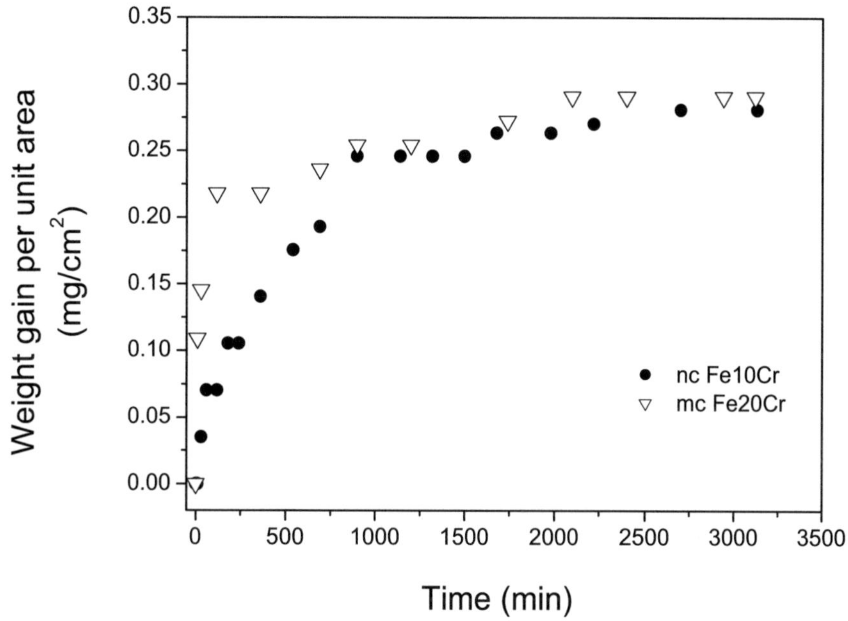

Fig. 11. Comparison of oxidation resistance of nanocrystalline Fe10Cr alloy with that of microcrystalline Fe20Cr alloy at 350°C in air [12,39].

5.2 Ni-Cr-Al based alloys

Oxidation resistance of Ni-Cr-Al based alloys largely depend upon the chemical and physical properties of alumina scale formed during high temperature oxidation. Most of the Ni-Cr-Al alloy contains enough Cr to form external Cr_2O_3 scale but for the application over 1000°C, the chromia scale does not provide any beneficial effect due to the volatilisation problems. Formation of a compact protective Al_2O_3 scale is most efficient in protecting material from the high temperature oxidation. In general, a conventional Ni-20Cr-Al system requires more than 6 wt% of Al to form protective oxide scale which largely depends upon the a) diffusion coefficient of the Al from the bulk to alloy/oxide interface and b) diffusion of Al in the formed oxide scale [45,118].

Since nanocrystalline materials possess significantly higher diffusion coefficient caused by higher fraction of grain boundaries therefore Al required for formation an exclusive Al oxide film can be reduced significantly and nanocrystalline materials should show improved oxidation resistance [88-93,119]. Wang et al [119] were among first researchers to investigate the oxidation resistance of nanocrystalline Ni-Al-Cr alloys and reported a significant improvement in the oxidation resistance of NiCrAl alloys due to nanocrystalline structure. Later, various authors have investigated the oxidation resistance of nanocrystalline NiCrAl alloys with various Al and Cr contents and produced by different methods but in all the cases nanocrystalline structures was reported to enhance the oxidation resistance. Most comprehensive work showing effect of the nanocrystalline structure on NiCrAl alloys was that of Gao et al [88] who reported excellent oxidation behaviour of a nanocrystalline coating of Ni20CrAl alloy over its microcrystalline counterpart. They reported a coating with grain size less than 70 nm may increase the oxidation rate by 4 times. Increment in oxidation resistance due to grain refinement was more pronounced when grain size was below 100 nm. It has been shown than Al content required to prevent external oxidation can be reduced from 6% to only 2% by reducing the grain size to ~ 60 nm [88-93]. These findings may have large industrial implications as it would provide an opportunity to achieve the desired oxidation resistance with lower Al content.

6. Challenges for application of nanocrystalline alloys at high temperature

For high temperature application, a major problem is thermal stability of the nanocrystalline structure. Thermodynamically, such fine structures could not survive for long times at high temperatures because of the large specific surface energy (driving force) for grain growth. An effective way to prevent this grain growth is by introducing stable second-phase particles in the microstructure, where they play a particle pinning effect (Zener effect) on grain growth [120]. Surprisingly, the coarsening kinetics for nanocrystalline alloys prepared by sputtering is rather slow even without apparent second-phase particles in the microstructure. Lou et al. [19] found that after 100 h oxidation in air at 1000°C, the grains of a sputtered K38G nanocrystalline coating were somewhat coarsened, but still rather fine-grained, growing from 20-100 nm to 200-1000 nm. It is not clear if this high stability against grain growth comes from grain boundary segregation of alloy elements (e.g. K38G alloy contains 3.6% Ti, 1.7% Mo, 2.6% W, 1.7% Ta in the Ni-Cr-Al base), which anchor or retard the grain boundary movement (similar to Zener effect) although no particle formation was apparent, or from the sputtered coating structure itself (columnar structure). Unfortunately, no systematic investigation of temperature and alloy composition effects on the thermal stability of these nanocrystalline alloys is available to date, future research should be carried out for high temperature application of nanocrystallline alloys.

Another problem limiting the application of nanocrystalline materials is preparation of nanocrystalline alloys. Currently, the bulk metallic nanomaterials can only be prepared at the laboratory scale, usually by compacting prepared nanocrystalline powders. However, consolidation of the nanopowders into bulk materials needs high temperature and pressure which may considerably coarsen the structure. Because of this difficulty, surface nanocoating has been considered a potential industry application. Nanocrystalline costing are often prepared by chemical vapour deposition (CVD), physical vapour deposition (PVD), electrochemical deposition, electro-spark deposition, and laser and electron beam surface treatment.

7. Future research

All studies reporting excellent oxidation resistance of nanocrystalline NiCrAl alloys were conducted at high temperatures. However, the grain growth behaviour of the alloys at these elevated temperatures was not determined and therefore many questions remain unanswered. Detailed studies investigating the effect of such fine grain size on the oxide scale formation along with grain growth of material will be helpful in understanding the underlying mechanism of the improved oxidation resistance.

Oxidation behaviour of Fe-Cr alloys as described in the section 5.1 was investigated at moderate temperatures [12,37-39]. Choice of moderate temperature was motivated by: 1) a very high difference in the grain boundary and lattice diffusion coefficient values at moderate temperatures and 2) higher grain growth at elevated temperatures. Grain growth of the nanocrystalline materials at high temperatures limits their use for high temperature applications. However, it was shown recently that addition of Zr to Fe-Cr based alloys prevents grain growth of these materials [120,121] and therefore such alloys with small amount of Zr (i.e., Fe-Cr-Zr alloys) will be ideal for investigation of oxidation resistance in the temperature range of 600-800°C.

Grain size of Fe-Cr alloys (used for the investigation of the effect of nanocrystalline structure on oxidation resistance) was limited to 54 (±4) nm which could be further decreased with the recent advancements in the sample preparation techniques such as one recently developed by Gupta et al [122] where an artefact free FeCrNi alloy with a grain size less than 10 nm was produced by in-situ consolidation technique. Further investigations on such alloys with grain size below 10 nm will demonstrate pronounced effect of triple points and grain boundaries and it may be possible to develop stainless steels with further improved oxidation resistance but less Cr content.

Improved oxidation resistance of nanocrystalline Fe-Cr or Ni-Cr-Al alloys have been attributed to the greater Cr and/or Al enrichment of the oxide scale (i.e., change in chemical composition of the oxide scale) due to faster diffusion of Cr and/or Al. However, physical properties of oxide scale, which are very important in determining the oxidation resistance of an alloy, have attracted only a little research attention. Investigation of the physical properties (grain size, morphology, crystallographic details etc.) of the oxide scale formed on the nanocrystalline alloys will further help in understanding the effect of nanocrystalline structure on the oxidation resistance of an alloy.

8. Concluding remarks

Nanocrystalline materials are being investigated due to their unique properties. More generally, development of materials resistant to environmental degradation is not the main focus of nanocrystalline metals research to date, but it seems possible that nanocrystalline metallic materials may lead to a substantial increment in oxidation resistance; caused by promoted oxide scale formation, improved adherence and reduced spallation tendency of the oxide scale. Nanocrystalline Fe-Cr and M-Cr-Al alloys have demonstrated improved oxidation resistance and present potential to be used for high temperature applications in future. More fundamental investigations are required to fully characterise the oxidation phenomenon and underlying principles for nanocrystalline materials.

9. References

[1] H. Gleiter, Progress in Materials Science, 33 (1989) 223.

[2] G. Palumbo, S. J. Thorpe, K. T. Aust, Scripta Metall. Mater. 24 (1990) 2347.

[3] C. C. Koch, K.M. Yousef, R. O. Scattergood, K. L. Murty, Advanced Eng. Mater. 7 (2005) 787

[4] R. W. Siegel, NanoStructured Materials 4 (1994) 121

[5] R. Birringer, H. Gleiter, H. P. Klein, P. Marquardt, Phys. Lett. A 102 (1984) 365.

[6] M. A. Meyers, A. Mishra, D. J. Benson, Progress in Materials Science 51 (2006) 427.

[7] T. Zhu, J. Li, Progress in Materials Science 55 (2010) 710.
[8] C. C. Koch, Encyclopedia of Materials: Science and Technology (2001) 5901.
[9] C. C. Koch, Scripta Materialia, 49 (2003) 657.
[10] S. Cheng, E. Ma, Y. M. Wang, L. J. Kecskes, K. M. Youssef, C. C. Koch, U. P. Trociewitz, K. Han, Acta Materialia 53 (2005) 1521.
[11] Z. B. Wang, N. R. Tao, W. P. Tong, J. Lu, K. Lu, Acta Materilia 51 (2003), 4319.
[12] R. K. Gupta, PhD thesis, Synthesis and corrosion behaviour of nanocrystalline Fe-Cr alloys, Monash University (2010).
[13] K. D. Ralston, N. Birbilis, Corrosion 66 (7) (2010) 075005 - 1
[14] G. Meng, Y. Li, F. Wang, Electrochimica Acta 51 (2006) 4277.
[15] C. T. Kwok, F. T. Cheng, H. C. Man, W. H. Ding, Materials Letters 60 (2006) 2419.
[16] Y. Li, F. Wang, Electrochimica Acta 51 (2006) 4426.
[17] X. Y. Wang, D. Y. Li, Electrochimica Acta 47 (2002) 3939.
[18] Sh. Hassani, K. Raeissi, M. Azzi, D. Li, M. A. Golozar, J. A. Szpunar, Corrosion Science 51 (2009) 2371.
[19] Kh. M. S. Youssef, C. C. Koch, P.S. Fedkiw, Corrosion Science 46 (2004) 51.
[20] L. Wang, Y. Lin, Z. Zeng, W. Liu, Q. Xue, L. Hu, J. Zhang, Electrochimica Acta 52 (2007) 4342.
[21] H. Gleiter, Acta Mater 48 (2000) 1.
[22] J. Y. Fan, X. L. Wu, K. Paul, K. Chu, Progress in Materials Science 51 (2006) 983.
[23] T. Sourmail, Progress in Materials Science 50 (2005) 816.
[24] M. E. McHenry, M. A. Willard, D. E. Laughlin, Progress in Materials Science 44 (1999) 291.
[25] G.R. Wallwork, A.Z. Hed, Oxidation of Metals, 3 (1971) 171.
[26] G. Chen, H. Lou, Surface and Coatings Technology 123 (2000) 92.
[27] Z. Liu, W. Gao, K.L. Dahm, F. Wang, Acta Materialia 46 (1998) 1691.
[28] L. Hanyi, W. Fuhui, X. Bangjie, Z. Lixin, Oxidation of Metals, 38 (1992) 299.
[29] H. Lou, S. Zhu, F. Wang, Oxidation of Metals, 43 (1995) 317.
[30] H. Lou, F. Wang, S. Zhu, B. Xia, L. Zhang, Surface and Coatings Technology, 63 (1994) 105.
[31] W. Xu, X. Song, N. Lu, C. Huang , Acta Materialia 58, (2010) 396.
[32] C. X. wang, C. W. Yang, Materials Science and Engineering R 49 (2005) 157.
[33] P. Keblinski, S.R. Phillpot, D. Wolf, H. Gleiter, Physics Letters A 226 (1997) 205.
[34] G. Chen, H. Lou, NanoStructured Materials, 11 (1999) 637
[35] G. Chen, H. Lou, Surface and Coatings Technology 123 (2000) 92.
[36] G. Chen, H. Lou, Corrosion Science 42 (2000) 1185.
[37] R. K. Singh Raman, R. K. Gupta, Corrosion science 51 (2009) 316.
[38] R. K. Gupta, R. K. Singh Raman, C. C. Koch, Journal of Materials Science 45 (2010) 4884.
[39] R. K. Singh Raman, R. K. Gupta ,C. C. Koch, Philosophical Magazine 90 (2010) 4884.
[40] U. Köster, L. Jastrow, Materials Science and Engineering A 449-451 (2007) 57.
[41] U. Köster, D. Zander, Triwikantoro, A. Rüdiger, L. Jastrow, Scripta Materialia 44 (2001) 1649.
[42] Triwikantoro, D. Toma, M. Meuris, U. Köster , Journal of Non-Crystalline Solids 250-252 (1999) 719.
[43] K. Mondal, U.K. Chatterjee, B.S. Murty, Journal of Non-Crystalline Solids, 334-335 (2004) 544.
[44] P. Kofstad, High Temperature Corrosion, Elsevier Applied Scicnece, New York (1988).
[45] D.J. Young, *High Temperature Oxidation and Corrosion of Metals,* Elsevier, Amsterdam (2008).

[46] M.K. Hossain, Corrosion Science 19 (1979) 1031.
[47] X. Peng, J. Yan, Y. Zhou, F. Wang, Acta Materialia 53 (2005) 5079.
[48] M.D. Merz, Metallurgical and Materials Transactions A, 10 (1979) 71.
[49] F. Wang, H. Lou, W. Wu, Vacuum, 43 (1992) 752.
[50] X. Zhu, R. Birringer, U. Herr, H. Gleiter, Phys. Rev. B 35 (1987) 9085.
[51] L. D. Bianco, A. Hernando, E. Bonetti, E. Navarro, Phys. Rev. B 56 (1997) 8894.
[52] T. Haubold, R. Birringer, B. Lengeler, H. Gleiter, Phys. Lett. A 135 (1989) 461.
[53] T. Haubold, W. Krauss, H. Gleiter, Philos. Mag. Lett 63 (1991) 245.
[54] A. DiCicco, M. Berrettoni, S. Stizza, E. Bonetti, G. Cocco Phys. Rev. B 12 (1994) 386.
[55] A.D. Cicco, M. Berrettoni, S. Stizza, E. Bonetti, Physica B 208-209 (1995) 547.
[56] R. K. Islamgaliev, R. Kuzel, E.D. Obraztsova, J. Burianek, F. Chmelik, R.Z. Valiev, Mater Sci and Engineering A 249 (1998) 152.
[57] H. E. Schaefer, R. Wurschum, R. Birringer, H. Gleiter, Phys. Rev. B 38 (1988) 9545.
[58] J. Wang, D. Wolf, S. R. Phillippot, H. Gleiter, Philos. Mag. A 73 (1996) 517
[59] D. Chen, Mater Sci and Engineering A 190 (1995) 193.
[60] D. Chen, Materials Letters 21 (1994) 405.
[61] R.E. Read-Hill, Physical Metallurgy Principles, Litton Educational Publishing, Inc. 1973
[62] T. Mutschele, R. Kirchheim, Scripta materialia 21 (1987) 1101.
[63] R. Kirchheim, T. Mutschele, W. Kieninger, H. Gleiter, R. Birringer, T. D. Koble, Mater Sci and Engineering A 99 (1988) 457.
[64] H. J. Fecht, NanoStructured Materials 1 (1992) 125.
[65] H. J. Fecht, Phys. Rev. Lett 65 (1990) 610.
[66] H. J. Fecht, E. Hellstern, Z. Fu, W.L Johnson, Adv. Powd. Met 2 (1989) 111.
[67] H. J. Fecht, Materials Science and Engineering A 179-180 (1994) 491.
[68] E. Hellstern E, H. J. Fecht, Z. Fu Z, W. L. Johnson, J. appl. Phys 65 (1989) 305.
[69] G. Wallner, E. Jorra, H. Franz, J. Peisl, R. Birringer, H. Gleiter, T. Haubold. W. Petry, MRS Symp. Proc 132 (1989) 149.
[70] J. Horvath, R. Birringer, H. Gleiter, Solid State Comm 62 (1987) 319.
[71] R. Birringer, P. Zimmer, Acta Materialia 57 (2009) 1703.
[72] A. Seeger, G. S. Acta metall. 1959;7:495.
[73] D. A. Smith, V. Vitek, R.C. Pond, Acta metal 24 (1977) 475.
[74] H. J.Fros, M. F. Ashby, F. Spaepen, Scripta matallurgica 14 (1980) 1051.
[75] D. Wolf, Philos. Mag. B 69 (1989) 667.
[76] Y. H. Zhao, H. W. Sheng, K. Lu, Acta Mater 49 (2001) 265.
[77] A. Lucci, G. Riontino, M. C. Tabasso, V. G. Tamanini, Acta Metall. 26 (1978) 615.
[78] B. Gu¨ nther, A. Kupmann, H. D. Kunze, 27 (1992) 833.
[79] L. Lu, M. L. Sui, K. Lu, Acta Mater. 49 (2001) 4127
[80] A. Tscho¨ pe, R. Birringer, Acta Metall. 41 (1993) 2791.
[81] A.T. Dinsdale, Calphad 15 (1991) 317.
[82] Y. K. Huang, A. A. Menovsky, F. R. Boer, NanoStructured Materials Materials 2 (1992) 587.
[83] R. Würschum, S. Herth, U. Brossmann, Advanced Engineering Materials, 5 (2003) 365
[84] I.V. Belova, G.E. Murch, J. of physics and chemistry of solids 64 (2003) 873
[85] H. Mehrer, Diffusion in Nanocrystalline Materials, springer Series in Solid-State Sciences (2007)
[86] G. Shujiang , F. Wang, S. Zhang, Surface and Coatings Technology 167 (2003) 212.
[87] H. Y. Tong, F. G. Shi , E. J. Lavernia, Scripta Metall Mater 32 (1995) 511.
[88] W. Gao, Z. Liu, Z. Li, Advanced Materials, 13 (2001) 1001.

[89] G. Chen, H. Lou, Scripta materialia 43 (2000) 119.
[90] G. F. Chen, H. Y. Lou, Materials Letters 45 (2000) 286.
[91] L. Liu, F. Wang, Materials Letters 62 (2008) 4081.
[92] F. Wang, X. Tian, Q. Li, L. Li, X. Peng, Thin and Solid Films 516 (2008) 5740.
[93] G. Cao, L. Geng, Z. Zheng, M. Naka, Intermetallics 15 (2007) 1672.
[94] H. Z. Zheng, S. Q. Lu, X. J. Dong, D. L. Quyang, Materials Science and Engineering A 496 (2008) 524.
[95] J. E. May, S. E. Kuri, P. A. P Nascente, Mater Sci and Engineering A 428 (2006) 290.
[96] J. E. May, G. Galerie, T. P. Busquim, S. E. Kuri, Materials and Corrosion 58 (2008) 87
[97] Y. Niu, Z. Q. Cao, F. G, Farne, G. Randi, C. L. Wang, Corrosion Science 45 (2003) 1125.
[98] S. Leistikow, I. Wolf and H.J. Grabke, Werkst. Korros., 38 (1987) 556.
[99] Y. Shida, N. Ohtsuka, J. Muriama, N. Fujino and H. Fujikawa, Proc. JIMS-3, High Temp. Corros., Trans. Jap. Inst. Met., 1983, 631.
[100] H. J. Grabke, E. M. Muller-Lorenz, S. Strauss, E. Pippel, J. Woltersdorf, Oxidation of Metals, 50 (1998) 314.
[101] M. Martin, N. Lakshmi, U. Koops, H.-I. Yoo, Z. Phys. Chem. 221 (2007) 1449.
[102] G. C. Wood, Oxidation of Metals, 2 (1970)
[103] V. R. Howes, Corrosion Science, 7 (1967) 735
[104] R. Lobb, H. Evans, Metal Science 26 (1981).
[105] M. Schutz, Oxidation of Metals, 44, (1995), 29
[106] H. Hindam, D. P. Whittle, Oxidaiton of Metals, 18 (1982) 245
[107] X. Y. Zhang, M.H. Shi, C. Li, N. F. Liu, Y.M. Wei, Mater Sci and Engineering A 448 (2007) 259.
[108] O. E. Kedim, S. Paris, C. Phigini, F. Bernard, E. Gaffet, Z. A. Munir, Mater Sci and Engineering A 369 (2004) 49.
[109] W. W. Smeltzer, D. P. Whittle, J. Electrochem. Soc. 125 (1978) 1116.
[110] C. J. Wagner, J. Electrochem. Soc. 99 (1952) 369.
[111] C. J. Wagner, J. Electrochem. Soc. 103 (1956) 571.
[112] C. J. Wagner, J. Electrochem. Soc. 63 (1959) 772.
[113] X. Peng, F. Wang, Oxidation - resistant nanocrystalline coatings, Development in high-temperature corrosion and protection of materials, W. Gao, Z. Li (Eds.), CRC Press, 2008.
[114] I. Kaur, W. Gust, L. Kozma, Handbook of grain and interphase boundary diffusion data: Stuttgart: Zigler Press, (1989), 523.
[115] A. W. Bowen, G .M. Leak, Metall. Trans. 1 (1970), 1695.
[116] R. Gupta, R. K. Singh Raman, C. C. Koch, Materials Science and Engineering A 494 (2008) 253.
[117] J. G. Goedjen, D. A. Shores, Oxidation of Metals 37 (1992) 125
[118] F. Wang, Oxidation of Metals 47 (1997) 247.
[119] F.J. Humphreys, M. Hatherly, Recrystallization and related annealing phenomena, Pergamon, 1996.
[120] R. K. Gupta, R. K. Singh Raman, C. C. Koch, TMS 2008 Annual Meeting 1 (2008) 151-157.
[121] K.S. Darling, R.N. Chan, P.Z. Wong, J. E. Semones, R.O. Scattergood, C.C. Koch, Scripta Materlialia 59 (2008) 530.
[122] R. K. Gupta, K. S. Darling, R. K. Singh Raman, K. R. Ravi, C. C. Koch, B. S. Murty, R. O. Scattergood, Journal of Materials Science (2011) DOI 10.1007/s10853-011-5986-6.

11

Corrosion Behavior of Stainless Steels Modified by Cerium Oxides Layers

Emilia Stoyanova and Dimitar Stoychev
Institute of Physical Chemistry, Bulgarian Academy of Sciences
Bulgaria

1. Introduction

The modifying of the surface, which involves altering only the surface layers of a material, is becoming increasingly important with the aim to enhance the corrosion resistance of many kinds of materials. The advantage of this approach lies in the fact that the natural physical and mechanical properties of the material are retained, while at the same time the corrosion resistance is increased. It is well known that electroplated zinc coating is employed as active galvanic protection for low and middle-content alloyed steels (Almeida et al.,1998; Hagans & Hass, 1994; Kudryavtsev, 1979; Lainer, 1984; Zaki, 1988). However, zinc is a highly reactive element, and therefore high corrosion rates of this coating are observed in cases of indoor and outdoor exposures. For this reason a post-treatment is needed to increase the lifetime of zinc coatings. This kind of treatment is applied in the current industrial practice to prolong the lifetime of zinc coatings and the steel substrates, respectively. This treatment consists of immersion in a chemical bath, which forms a conversion layer over the plated zinc. The so formed layer is a dielectric passive film with high corrosion resistance and it is also a better surface for paint adherence (Zaki, 1988). The main problem with the traditionally applied post-treatment procedures is the presence of Cr^{6+} salts that are considered to be carcinogenic substances, which are known to be very harmful to human health and environment (Schafer & Stock, 2005) and whose use is forbidden by European regulations (Hagans & Hass, 1994).

Molybdates, tungstates, permanganates and vanadates, including chromium-like components, were the first chemical elements to be tested as hexavalent chromium substitutes (Almeida et al., 1998a, 1998b; Korobov et al, 1998; Schafer & Stock, 2005; Wilcox & Gabe, 1987; Wilcox et al., 1988). Recently many alternative coatings have been developed, based on zirconium and titanium salts (Barbucci et al., 1998; Hinton, 1991), cobalt salts (Barbucci et al., 1998; Gonzalez et al., 2001) and organic conductive polymers (Gonzalez et al., 2001; Hosseini et al., 2006). The use of salts of rare-earth metals as the main component in the electrolytes, developed for the formation of cerium, lanthanum and other oxide protective films is also a very promising alternative to the chromate films and it is one of the advanced contemporary methods for corrosion protection of metals and alloys (Bethencourt et al., 1998; Crossland et al., 1998; Davenport et al., 1991; Fahrenholtz et al., 2002; Forsyth et al., 2002; Hinton, 1983, 1992; Hosseini et al., 2007; Liu & Li , 2000; Montemor et al., 2002; Montemor & Ferreira, 2008; Pardo et al., 2006; Wang et al., 1997). However, some aspects of the preparation and of the corrosion behavior of these coatings are not quite clear yet and their practical utilization is still uncertain. In order to find an attractive alternative to Cr^{6+} conversion coating, several treatment procedures that should manifest both efficient anti-corrosive behavior as well as an optimal benefit/cost ratio, and mainly insignificant environmental impact, have yet to be developed. It has been found out that cerium species can be successfully applied to protect zinc from corrosion (Aramaki, 2001, 2002; Arenas et al., 2003, 2004; Ferreira et al., 2004; Otero et al., 1996, 1998; Wang et al., 2004; Virtanen et al., 1997), aluminum and aluminum containing alloys (Aldykiewicz et al., 1995; Amelinckx et al., 2006; Arnott et al., 1989; Davenport et al., 1991; Mansfeld et al., 1989,1991, 1995; Pardo et al., 2006; Zheludkevich et al., 2006; Di Maggio et al., 1997; Lukanova et al.. 2008), stainless steels (Breslin et al., 1997; Lu & Ives, 1993, 1995), magnesium containing alloys (Arenas et al., 2002; Liu et al., 2001) even SiC/Al metal matrix composites. All these can be used to reduce the rate of general corrosion, pitting and crevice corrosion as well as stress corrosion (Breslin et al., 1997; Lu & Ives, 1993, 1995). The oxide films of rare-earth elements and refractory compounds can be formed mostly by means of chemical or electrochemical methods (Amelinckx et al., 2006a, 2006b; Avramova et al., 2005; Balasubramanian et al., 1999; Di Maggio et al., 1997; Marinova et al., 2006; Montemor et al., 2001, 2002; Schmidt et al., 1997; Stefanov et al., 2000a, 2000b, 2004a, 2004b; Stoychev et al., 2000, 2003, 2004; Tsanev et al., 2008; Tyuliev et al., 2002; Valov et al., 2002; Zheludkevich et al., 2005). It is supposed that cerium oxide/ hydroxide formation is the main reason for the corrosion protection property of cerium compounds. In spite of the growing number of investigations during the last years, focused on the mechanisms via which the oxides of rare-earth metals (mainly cerium oxides) lead to improvement of the corrosion stability of the systems "oxide(s)/protected metal", still a series of issues remain problematic. The first hypotheses in this respect have been put forward by Hinton (Hinton & Wilson, 1989; Hinton, 1992). He supposed in his early works that the cathodic reactions (reduction of oxygen and evolution of hydrogen) lead to alkalization of the near-to-the-electrode layer, which in its turn results in precipitation of the oxide of the rare-earth element, respectively in formation of protective film on the electrode surface.

The modern technologies for surface treatment, aimed at modifying the surface composition and structure of metals and alloys, including stainless steels, are becoming more and more important instruments for improving their stability to corrosion and for attributing the desired outside appearance and/or functional properties. Wang and coworkers (Wang et al., 2004) have studied the corrosion resistance of stainless steel SS304 after immersion treatment in electrolytes, containing Ce^{3+} ions, $KMnO_4$ and sulfuric acid. The obtained experimental results prove the considerable increase in the corrosion stability of steel in 3.5% NaCl solution. The corrosion potential of the steel, treated by immersion, has more positive values than that of the non-treated steel, while the potential of pitting formation is also shifted in the positive direction, which is the criterion for weakening the tendency of the studied steel to undergo pitting formation. It has also been observed that the values of the current of complete passivation are decreased by one order of magnitude. As far as the cathodic reaction is concerned, the cerium conversion coatings blocked the matrix steel, which caused the reduction of the oxygen and protons to take place at a higher over-potential and the cathodic reaction was inhibited. The analysis of the chemical state of cerium in the conversion film indicated that the prevailing amount of cerium is in trivalent state. Aldykiewicz and coworkers (Aldykiewicz et al, 1996) have investigated the influence of cerium oxide, deposited on aluminum alloys and they ascertained that the trivalent cerium is oxidized to tetravalent by the oxygen dissolved in the electrolyte, which leads to precipitation/formation of non-soluble CeO_2 on the cathodic sections of the electrode surface. Montemor and coworkers (Montemor et al., 2001, 2002) studied the effect of the composition of the electrolytes (based on $Ce(NO_3)_3$ and the regime of preparing conversion layers on galvanized (zinc coated) steel. The increase in the thickness of the cerium conversion films in the process of formation leads to their enrichment in Ce^{4+}. According to the same authors the conversion films, formed in $La(NO_3)_3$, are more efficient in view of anticorrosion protection, compared to those formed in electrolytes, containing $Ce(NO_3)_3$ and $Y(NO_3)_3$ (Montemor et al, 2002). The mechanism, involved in such a process of reducing the corrosion rate of the substrate, may be related to precipitation of cerium oxides and hydroxides in the vicinity of the anodic areas. These precipitates reduce the cathodic activity and hinder the transfer of electrons from the anodic to the cathodic spots.

The mechanism of zinc corrosion inhibition, when zinc is treated in solutions of $Ce(NO)_3$, has been studied by Aramaki (Aramaki, 2001a, 2001b, 2002a; 2002b). He established the formation of hydrated or hydroxylated Ce-rich layer. This process, in its turn, leads to the formation of Ce_2O_3 on the electrode/the protected surface, respectively to inhibition of the cathodic reactions of the corrosion process in solutions of NaCl. The work by Lu and Ives (Lu & Ives, 1995) has been extended further to study the effect of cerium salt solution treatment. Rotating disk assemblies were employed to monitor the cathodic electrode process and its inhibition by cerium salt treatment on austenitic stainless steels in a solution simulating sea water. The reduction of oxygen and hydrogen cations on both kinds of non-treated steels has been shown to be controlled by mass transfer processes in the solution. Cerium treatment effectively inhibits the cathodic reduction of oxygen, which is controlled primarily by charge transfer on the electrode. The over-potential for cathodic reduction of hydrogen cations is increased after the cerium treatment and the electrode reaction is controlled both by the mass transfer process in solution and by the charge transfer on the electrode. As a result of inhibition of the electrode processes, cerium improves the localized corrosion resistance, and in particular the crevice corrosion resistance, of stainless steels.

The electrochemical behavior of stainless steels - SS304 and 316L, following various cerium and cerium/molybdenum prereatment steps, was studied aiming at gaining more information on the process, by which cerium and molybdenum can modify the properties of passive film formed on stainless steels (Breslin et al, 1997). The coatings were analyzed by electrochemical impedance spectroscopy and X-ray photoelectron spectroscopy in order to identify the cerium species, which play the main role in the promotion of the passivation behavior. The pre-treatment step, denoted as $Ce(CH_3CO_2)_3$ and $CeCl_3$, involved immersion treatment of the electrodes in the $Ce(CH_3CO_2)_3$ solution for 1 h and then in the $CeCl_3$ solution for one additional hour at approximately 92°C. Regardless of the nature of the cerium salt, no changes in the rate of the cathodic reduction reaction could be observed. The increase in the corrosion potential (E_{corr}). is mainly due to the decrease in the passive current density, suggesting that treatment in cerium solutions does not affect the rate of the cathodic reaction, but rather reduces the rate of the passive film dissolution. However a significant lowering of the oxygen reduction current could be observed, following the electrodeposition of small amount of cerium onto the electrode surface. Thus, it appears that efficient formation of cerium hydroxide/oxide does not occur upon immersion of electrodes at elevated temperatures in cerium solutions. It was possible to observe a yellow colored film indicative of cerium in the 4^+ oxidation state on the surface of the stainless steel, following a 24-h immersion time interval in the $Ce(NO)_3$ solution at room temperature. Breslin and coworkers (Breslin et al, 1997) proved that the treatment of SS304 in cerium-salt solutions gave rise to an increase in the value of the pitting potential E_{pit}, with the greatest increase resulting from immersion in $CeCl_3$ at 90-95°C for 30 min, followed by immersion in $Ce(NO)_3$ solution at 90-95°C for additional 60 min time interval. The enhanced resistance to the onset of pitting, according to these authors, could be due to the dissolution of surface MnS inclusions during the immersion in the chloride-containing solution and possibly chromium enrichment of the passive film during treatment in the sodium nitrate solution, which is highly oxidizing. The presence of cerium in the solution seemed to have only a minor effect on E_{pit}. The survey of the various mechanisms, proposed in the current literature, indicates that the role of rare-earth elements as inhibitors of corrosion and as protective coatings is not completely elucidated. It is accepted that their presence leads to improvement of the corrosion stability of metals and alloys and therefore they are a promising alternative, in conformity with the requirements for protection of the environment prohibiting the conventional Cr^{6+} conversion treatment.

At the same time it is known that thin films of Ce_2O_3-CeO_2, have also an important functional designation for the manufacture of catalytic converters, where ceria is widely used in such kind of catalytic processes as a reducible oxide support material in emission control catalysis for the purification of exhaust gases from various combustion systems (Trovarelli, 1996). In the so called "three-way automotive catalysis", for example, the reducibility of ceria contributes to oxygen storage/release capability, which plays an important role in the oxidation of CO and hydrocarbons catalyzed on the surface of precious metal particles (Bunluesin et al, 1997). It is because of their specific interactions with oxygen that the cerium oxides are included in the support layers (Al_2O_3, ZrO_2, etc.) of the proper catalytically active components of the converters (noble metals like Pt, Ro, Pd and others) and they participate directly in the decontamination of exhaust gases (reduction of NO_x, oxidation of CO and hydrocarbons, etc.) originating from internal combustion engines (Mcnamara, 2000). In this connection it is important to point out that during the process of operation the main construction elements of the catalytic converters, which are made of stainless steel (Lox et al, 1995; Nonnenmann, 1989) (for example steel OC 404), are subjected simultaneously to over-heating and at the same time to the aggressive action of the nitrogen oxides, being liberated in the course of the processes of combustion, of sulfur oxides, of water vapor and incompletely oxidized hydrocarbons etc. (respectively resulting in formation of HNO_3, H_2SO_4 etc). In this respect and in the light of the data available in the literature about the protective action of the cerium oxides and hydroxides, it is essential to know what is the intimate mechanism of their anti-corrosion action and to what extent they could contribute, in particular, to the prolongation of the exploitation life-time of the catalytic converters, made of stainless steel.

Our studies on the protective effect of mixed Ce_2O_3-CeO_2 films electrochemically deposited on stainless steel OC 404 (SS) in model media of 0.1N HNO_3 and 0.1N H_2SO_4 (Nikolova et al., 2006a, 2006b, 2008; Stoyanova et al., 2006a, 2006b, 2010), have shown that these films in their nature are in fact cathodic coatings. The influence of the change in the concentration of

Ce in the oxide films has been studied in regard to the corrosion potential of the steel in the same corrosion medium. Thereupon it was found out that the increase in the surface concentration of Ce in the oxide films results in a gradual shift of the corrosion potential of steel in the positive direction - from the zone characteristic of anodic dissolution to the zone of deep passivity - defined by the anodic potentiodynamic curve. Moreover it has been proved that there occurs a cathodic reaction of reduction of the electrochemically active CeO_2 - one of the components of the electrodeposited mixed Ce_2O_3-CeO_2 film.

The present work discusses a hypothesis, aimed at elucidation of the question: how the change in the surface concentration of Ce in the mixed Ce_2O_3-CeO_2 oxide film electrodeposited on OC 404 steel influences the processes of anodic passivity of the studied steel, respectively the values of the potentials of complete passivation and the potentials of pitting formation, as well as the current density in passive state, determining the corrosion behavior of the steel under consideration. As far as the oxide of Ce^{3+} (i.e. Ce_2O_3) is chemically unstable and it is dissolved in sulfuric acid medium (Achmetov, 1988) the investigations carried out in ref.(Guergova et al, 2011) established an effective inhibitory action of the cerium ions (Ce^{3+}, Ce^{4+}), passing over from the system Ce_2O_3-CeO_2/SS into the corrosion medium. A possible inhibitory interaction has been supposed to occur on the surface of the steel.

2. Experimental

2.1 Specimen preparation and structure characterization

The stainless steel (SS) samples (SS type OC 404 containing 20% Cr, 5.0% Al, 0.02% C, the rest being Fe) were 10x10 mm plates of steel foil, 50 μm thick. The deposition of the films was carried out in a working electrolyte consisting of absolute ethanol saturated with 2.3 M LiCl and 0.3 M $CeCl_3x7H_2O$ salts. The cathodic deposition was performed in a galvanostatic regime at current density of $0.1 mA.cm^{-2}$. The deposition time interval was 60 min. Platinum coated titanium mesh was used as counter electrode (anode). It was situated symmetrically around the working electrode and its surface was chosen specially to ensure a low anode polarization, which hindered Cl^- oxidation. Because of the relatively low equivalent conductance of the working electrolyte (χ - 1.10^{-2} Ω^{-1} cm^{-1}), it becomes warmed up during the electrolysis. For this reason, the electrochemical measurements were carried out in a specially constructed electrochemical cell. The cell was kept at a constant temperature of 5-7°C by circulation of cooling water. The obtained CeO_2-Ce_2O_3 coatings had a thickness of 1μm (Avramova et al., 2005; Stefanov et al., 2004). The system CeO_2-Ce_2O_3/SS was investigated prior to and after thermal treatment (t.t.) at 450°C for 2 h in air. The model aggressive solution (0.1N H_2SO_4) was prepared by dilution of analytical grade 98% H_2SO_4 ("Merck") with distilled water. In order to evaluate the inhibitory effect of lanthanide salt, variable concentrations of $Ce(SO_4)_2.4H_2O$ from 0.1 to 1500 ppm were added to 0.1N H_2SO_4.

The morphology and structure of the samples was examined by scanning electron microscopy using a JEOL JSM 6390 electron microscope (Japan) equipped with ultrahigh resolution scanning system (ASID-3D) in regimes of secondary electron image (SEI) and back scattered electrons (BEC) image. The pressure was of the order of 10^{-4} Pa.

2.2 Chemical characterization

The chemical composition and state of the surfaces being formed was studied using X-ray photoelectron spectroscopy (XPS). The XPS studies were performed on an Escalab MkII system (England) with Al K_{α} radiation ($h\nu$ = 1486.6 eV) and total instrumental resolution of ~ 1 eV. The pressure in the chamber was 10^{-8} Pa. The binding energy (BE) was referred to the C1s line (of the adventitious carbon) at 285.0 eV. The element concentrations were evaluated from the integrated peak areas after Shirley-type of linear background subtraction using theoretical Scofield`s photoionization cross-sections.

2.3 Electrochemical (corrosion) characterization

The electrochemical behaviour of the samples (plates 10 x 10 x 0.05mm) was studied in a standard three-electrode thermostated cell (100 ml volume). The model corrosion medium was 0.1 N H_2SO_4 ("p.a." Merck) after deaeration with additionally purified argon at 25°C. A counter electrode, representing a platinum plate (10x10x0.6mm), and a mercury/mercurous sulfate reference electrode (MSE), ($E_{Hg/Hg2SO4}$= +0.642V versus SHE) were used. All potentials in the text are related to MSE. The anodic and cathodic polarization curves were obtained using a 273 EG&G potentiostat/galvanostat and computer-aided processing of the results according to an "Echem" programme, with a potential sweeping rate of 10 mV/s within a potential range from –1.500 to +1.500V. The recording of the potentiodynamic curves was carried out starting from the stationary corrosion potential (E_{st}), measured in the absence of external current (at open circuit) in the anode and cathode directions. We used a separate electrode for each recorded curve. The stationary corrosion potential of the samples under investigation was determined by direct measurement of the function "E_{st}–time" at open circuit (with respect to the same reference electrode) after immersing the samples in 0.1 N H_2SO_4 in the absence and in the presence of Ce^{4+}. The E_{st} was established after a sufficiently long time interval - from few minutes to several decades of minutes until the moment, when the corrosion potential change did not exceed 1-3mV for 5 min.

3. Theoretical background

It is known that one of the basic approaches to promote the corrosion resistance of alloys is to enhance their passivity. It has been established that upon introducing a new component with higher inclination to passive state into the metal or into the alloy, it transfers this property to the main metal or to the alloy. The formation of a system, which is more stable to corrosion (i.e more easily passivated system) could be achieved through promoting the effectiveness of the cathodic process. At first glance this is a self-contradictory statement, however it can be easily explained in the following way. It can be seen from the corrosion diagram, represented in Fig. 1, that if the anodic potentiodynamic curve of the alloy remains one and the same, the rate of corrosion can be changed considerably at the expense of the changing effectiveness of the occurring cathodic process. It is important to note that in the case of non-passivating systems (i.e. systems characterized by anodic behavior until point B of the anodic curve) the corrosion rate is always increasing with the increase in the cathodic efficiency (for example the transition from cathodic curve K_1 to K_2 in Fig.1). In the cases when the corrosion systems are passivating ones and the anodic polarization curve is not a monotonous dependence between the potential and the current and when there exists a region of potentials somewhere between the passivation potential (E_p) and the potential of transpassivity (E_t) (or the potential of pitting formation (E_{pit})), in which the increase of the effectiveness of the cathodic process is leading not to enhancement but rather to abatement of the corrosion rate (for example in the course of the transition from one cathodic process K_2 to another one K_3) one can observe a system more stable to corrosion (easily passivating system). Obviously in this case of minimal corrosion currents there will appear a cross-point between the anodic and the cathodic curves of the corrosion diagram within the zone of stable passive state. Under these conditions it is quite possible that a smaller corrosion current is corresponding to a more efficient cathodic process in comparison to the system displaying a lower cathodic efficiency (if we exclude the conditions where the potential of the system is reaching the value of the potential of transpassivation and the potential of pitting formation - K_4). Therefore one can conclude that during the occurring of an efficient cathodic process the system will pass over spontaneously to a stable passive state and it will be corroding at a much lower rate, corresponding to the current of complete passivation. The stationary corrosion potential of such a system will be more positive than the potential of complete passivation (E_{cp}) and at the same time more negative than the potential of break through the passive film and the potential of transpassivation. In this way the rate of corrosion can be reduced to a considerable extent by the correct use of the phenomenon "passivation".

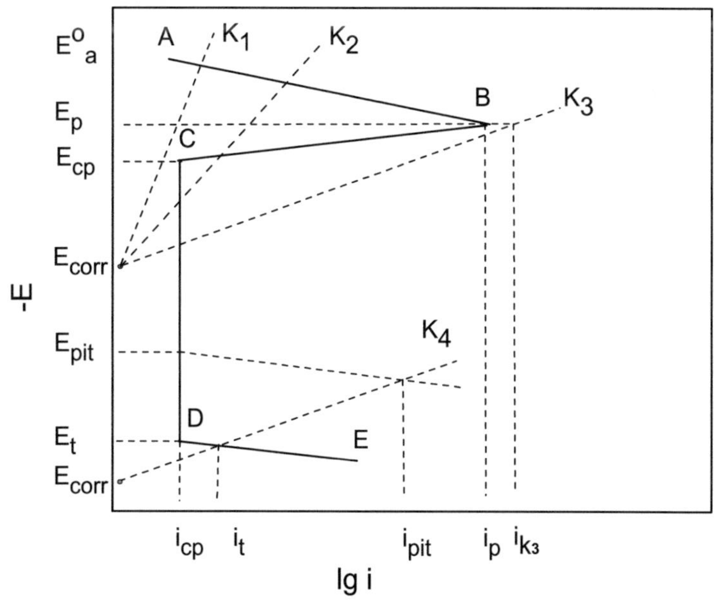

Fig. 1. Schematic polarization diagram explaining the action of the effective cathodic coatings on the steel corrosion: i_p, i_{cp}., i_{pit}, i_t- respectively currents of initial passivation, complete passivation, pitting formation and corrosion in transpassive state.

So, in order to create a system stable to corrosion and to decrease the rate of corrosion, it is necessary to find a way to promote the cathodic effectiveness (for example as it is in this specific case of investigations, carried out by us, to modify the steel surface with CeO_2-Ce_2O_3

oxides as cathodic coating). This theoretical approach has been used in this work, with a view to explain the obtained results, in view of stabilization and restoration of the passive state of OC 404 steel as a consequence of electrochemical formation of the surface CeO_2-Ce_2O_3 layers. We agree with the assumption that for a similar type of modified systems there exists only one passive state of the system (even without applying any external anodic current). Or, in other words, the result is a spontaneously self-passivating system and if in some way it is being led away from its passive state (for example in the case of cathodic polarization or by exerting a mechanical impact), after the termination of the external effect, again the system will pass over to its passive state.

As a matter of fact the increase in the effectiveness of the cathodic process is connected with the character of the cathodic process. The dilemma is whether the promotion of the cathodic efficiency is due only to the process of hydrogen depolarization, on the cerium oxide cathodic coating (in case of steel corrosion in acidic medium) or it is possible that there exists another cathodic process, owing to the oxidative properties of the electrochemically active CeO_2.

Figure 2 represents an example of a corrosion diagram, illustrating the changes in the behavior of the corrosion system upon increasing the surface concentration of the effective cathodic coating (for example in the case of modifying the steel surface with cerium oxides). It follows from the diagram that the shift in the corrosion potential of the system is associated with the increase in the concentration of the cathodic depolarizer (the cerium oxides), which will facilitate the transition from active state into passive state of the system, under the conditions of disturbed passivity. At concentration of the cerium oxides, corresponding to the cathodic curve C_1 (Fig. 2), the rate of dissolution of the steel will

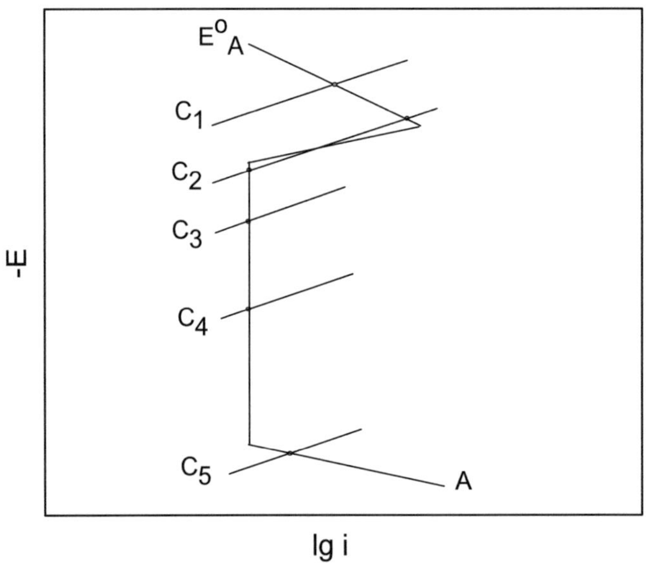

Fig. 2. Schematic polarization diagram illustraiting the influence of the effective cathodic coating on the steel corrosion, respective cathodic curves : c_1, c_2, c_3, c_4, c_5 in case of increasing the surface concentration of cerium oxides.

become commensurable with that of the pure steel. At concentration of the cerium oxides, corresponding to the cathodic curve C_2 it is possible to establish two corrosion potentials of the steel, located respectively in the passive and active regions of dissolution of the anodic potentiodynamic curve. At the higher concentrations of the cerium oxides, represented in the corrosion diagram by means of the respective cathodic potentiodynamic curves of the system cerium oxides/steel - C_3 and C_4, the steel is characterized by a stable passive state and under these conditions the rate of corrosion of the steel is no longer dependent on the surface concentration of the cerium oxides. The influence of the further increase in the concentration of the cerium oxides can be illustrated through the corrosion diagram by means of the cathodic curve C_5 – the rate of steel corrosion will grow up due to the fact that the stationary corrosion potential of the steel will be shifted to the region of transpassivity (Tomashov & Chernova, 1963; 1993). The experimental results, obtained by us, confirm these theoretical concepts.

4. Experimental results and discussion

4.1 Potentiodynamic polarization studies

Figure 3 shows a typical experimentally obtained corrosion diagram E-lg i, illustrating the kinetics of the cathodic and anodic processes on the studied steel in the absence of electrochemically deposited cerium oxides film (the curves 2) and after the deposition of thin oxide films with different surface concentrations of Ce (curves 3-5).

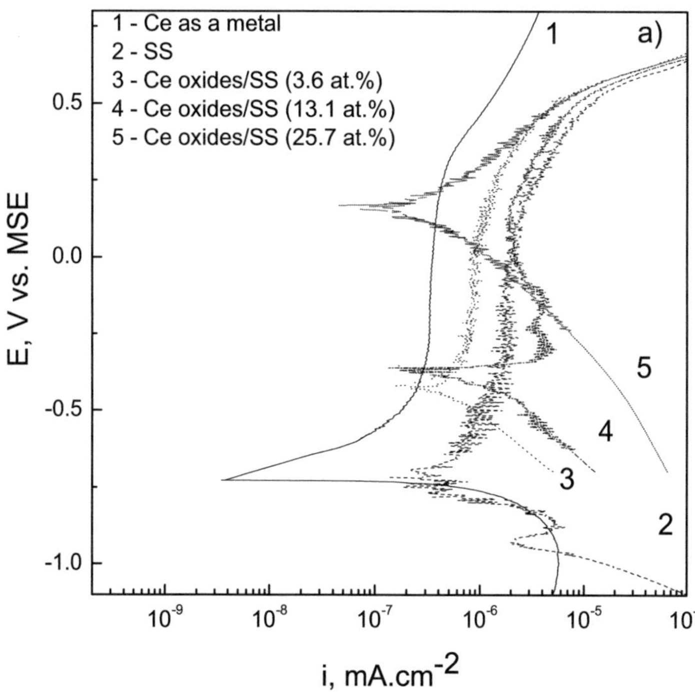

Fig. 3a. Potentiodynamic E–lgi curves for Ce (1), for SS (2) and for the systems CeO_2-Ce_2O_3/SS containing different concentrations of Ce (3-5), obtained in 0.1 N H_2SO_4.

The plotting of the model polarization curves enabled us to follow the changes in E_{corr}, estimated on the basis of the cross-point of the anodic polarization curve of the studied steel (SS) with the cathodic polarization curves of the studied systems, having different surface concentrations of Ce (ranging from 4% up to 30%). Such an approach of considering the partial polarization curves allows us to make the connection between the occurring anodic and cathodic corrosion processes, localizing the cathodic reaction on the rich in cerium zones on the electrode surface. For the sake of comparison the figure represents also the anodic and cathodic potentiodynamic curves of the metallic Ce (curves1). It follows from Fig. 3a that with the increase in the surface concentration of Ce (curves 3-5) the values of the corrosion potential E_{corr} are shifted strongly in the positive direction - from ~ -0.900V (for the non-coated with cerium oxides steel) up to ~ + 0.160 V. Obviously for the non-coated with CeO_2-Ce_2O_3 steel surface the cathodic depolarizing reaction, occurring at voltage ~ -0.900V, is connected with the evolution of hydrogen. As far as the respective cathodic branches of the potentiodynamic curves of the system Ce_2O_3-CeO_2/ SS are concerned, they are also shifted strongly in positive direction in the zone of passivity of the investigated steel.

The slope of the cathodic Taffel's curves grows up considerably from 0.250V up to 0.319 V with the increase in the surface concentration of cerium (Table 1). The change in the slope of these curves confirms the supposition about the occurring of another cathodic reaction, different from that of hydrogen evolution on the heterogeneous electrode surface.

	SS			
Samples	**E_{corr}, V**	**i_{corr}, mA.cm^{-2}**	**$i_{c.p,}$ mA.cm^{-2}**	**b, V**
SS non-covered with cerium oxides	-0.890	2.24 x 10^{-7}	1.89 x 10^{-6}	0.107
SS covered with 3.6 at. % cerium oxides	- 0.432	2.85 x 10^{-7}	9.10 x 10^{-7}	0.250
SS covered with 13.1 at. % cerium oxides	- 0.371	2.93 x 10^{-7}	3.63 x 10^{-6}	0.276
SS covered with 25.7 at. % cerium oxides	+0.161	1.36 x 10^{-7}	-	0.319
	$SS_{t.t.}$			
Samples	**E_{corr}, V**	**i_{corr}, mA.cm^{-2}**	**$i_{c.p.}$, mA.cm^{-2}**	**b, V**
SS non-covered with cerium oxides	- 0.975	1.74 x 10^{-7}	3.21 x 10^{-6}	0.074
SS covered with 4.2 at. % cerium oxides	- 0.486	7.96 x 10^{-7}	2.96 x 10^{-6}	0.151
SS covered with 20.7 at. % cerium oxides	- 0.269	7.98 x 10^{-8}	1.33 x 10^{-6}	0.176
SS covered with 29.6 at. % cerium oxides	+ 0.090	8.19 x 10^{-7}	-	0.304

Table 1. Electrochemical characteristics of coated steel before and after thermal treatment compared to bare steel.

The strong shifting of E_{corr} of the steel surface, covered with cerium oxides in the positive direction depending on the surface concentration of cerium could also be associated with the occurrence of another cathodic process. The values of the corrosion potential in the presence of Ce are more positive than the Flade potential and more negative than the potential of transpassivity of the steel under consideration. Therefore we can conclude that even at surface concentration of Ce about 4% the corrosion potential of the steel is shifted in positive direction reaching potentials more positive than the potential of complete passivation. The improvement of the corrosion stability of the steel as a result of the action of the effective cathodic coating is expressed in the stabilization of the passive state of the steel. One can conclude from Figure 3a that the steel samples with electrochemically deposited cerium oxide film will corrode under the conditions of passivity. Thereupon with the increase in the surface concentration of Ce from 0 up to 3.6 at % a tendency is observed – a decrease in the currents of complete passivation ($i_{c.p.}$,see Table 1). At 13.1%concentration of the cerium oxides the current of complete passivation is of the same order with that of the non-covered steel, while at 25,7% concentration as a result of the strong shifting of E_{corr} in positive direction and its approaching the values within the zone of potentials of pitting formation and transpassivity the anodic potentiodynamic curve is not characterized by a well expressed region of passivity. The change in the currents of complete passivation depends to a considerable extent on the composition of the passive film on the steel. Therefore for the system Ce_2O_3-CeO_2/SS we can assume that it will pass over spontaneously into a stable passive state and that it will be dissolved at a much lower rate of corrosion, corresponding to the values of the anodic currents in the passive state (Fig. 3a).

If in one way or another the system is artificially taken out of its state of passivity (for example by means of cathodic polarization or by some mechanical impact), after discontinuing the external impact, it will restore again its passive state, i.e. what we obtain is a spontaneously self-passivating system.

A similar effect, expressed to an even greater extent, is also observed with the samples of thermally treated system Ce_2O_3-CeO_2/$SS_{t.t}$ (Fig. 3b), in which case it was established that as a result of disruption of the integrity of the oxide film the passive state of the steel is disturbed (Guergova et al., 2008) and conditions are created to increase the rates of the total and the local corrosion. The presence of electrochemically deposited cerium oxide film (in a way analogous to that for the samples of non-treated thermally system Ce_2O_3-CeO_2/SS) shifts strongly the corrosion potential of the system in positive direction (see curves 3–5). This effect determines the restoration of the passive state of the steel, disturbed as a result of its thermal treatment. Upon increasing the surface concentration of the cerium one can observe not only shifting of the corrosion potential of the samples in the positive direction, but also a tendency of decrease in the currents of complete passivation. An exception in this respect is observed at very high concentrations of the cerium oxides ($\geq$ ~29%). Obviously in these cases the corrosion potential of the system CeO_2-Ce_2O_3/$SS_{t.t}$, which is still in the process of being established, starts approaching the value of the reversible redox potential of the couple Ce^{4+}/Ce^{3+}, whereupon the reaction of oxidation of Ce^3 to Ce^{4+} is taking place. As a result of this the character of the anodic curve will be changed (Fig. 3b, curve 5) and the determination of the current of complete passivation of the steel based on this curve would be incorrect.

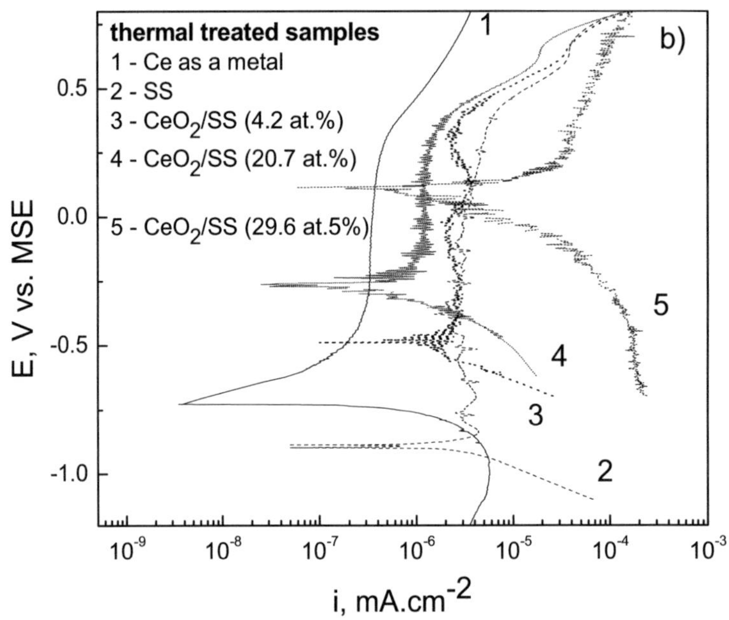

Fig. 3b. Potentiodynamic E–lgi curves for Ce(1), for $SS_{t.t.}$ (2) and for the systems CeO_2-$Ce_2O_3/SS_{t.t.}$ containing different concentrations of Ce (3-5), obtained in 0.1 N H_2SO_4.

4.2 Chronopotentiometric investigation

Fig. 4a and 4b illustrate the altering of the stationary corrosion potentials in the case of open circuit (open circuit potentials) with the SS and $SS_{t.t.}$ samples and with the systems Ce_2O_3-CeO_2/SS and Ce_2O_3-$CeO_2/SS_{t.t.}$ The juxtaposition of the values of the stationary corrosion potentials with the anodic potentiodynamic curves of SS and $SS_{t.t.}$ shows that in the presence of cerium oxide film on the surface of the steel one can observe a strongly manifested tendency to self-assivation. In the cases of non-thermally treated steel its high corrosion resistance and its ability to passivate itself is connected also with the high content of Cr, while the role of the cerium oxides is reduced to promoting the passivation ability and stabilization of its passive state in weakly acidic medium (Stoyanova et al, 2006). In the case of thermally treated steel, however, due to the cracking of the surface passive film, as a result of the thermal treatment its stationary corrosion potential reaches values ($E_{st.}$= -0.975V), characteristic of the corrosion in the active state (Fig. 4b). The disrupted passive state is also a prerequisite for the development of local corrosion in the active anodic sections – pitting and/or inter-crystalline, which is characteristic for this type of steel. It is also seen in Fig. 4b that the electrochemically formed cerium oxide films on the surface of the steel samples lead to strong shifting of the stationary corrosion potential of the steel in the positive direction – to potentials more positive than the potential of complete passivation and more negative than the potential of transpassivity. The established experimental facts unanimously indicate that the electrochemically deposited oxide films on the surface of the steel lead to restoration of its passive state, due to promoted ability of the system to passivate itself under the conditions of the real corrosion process.

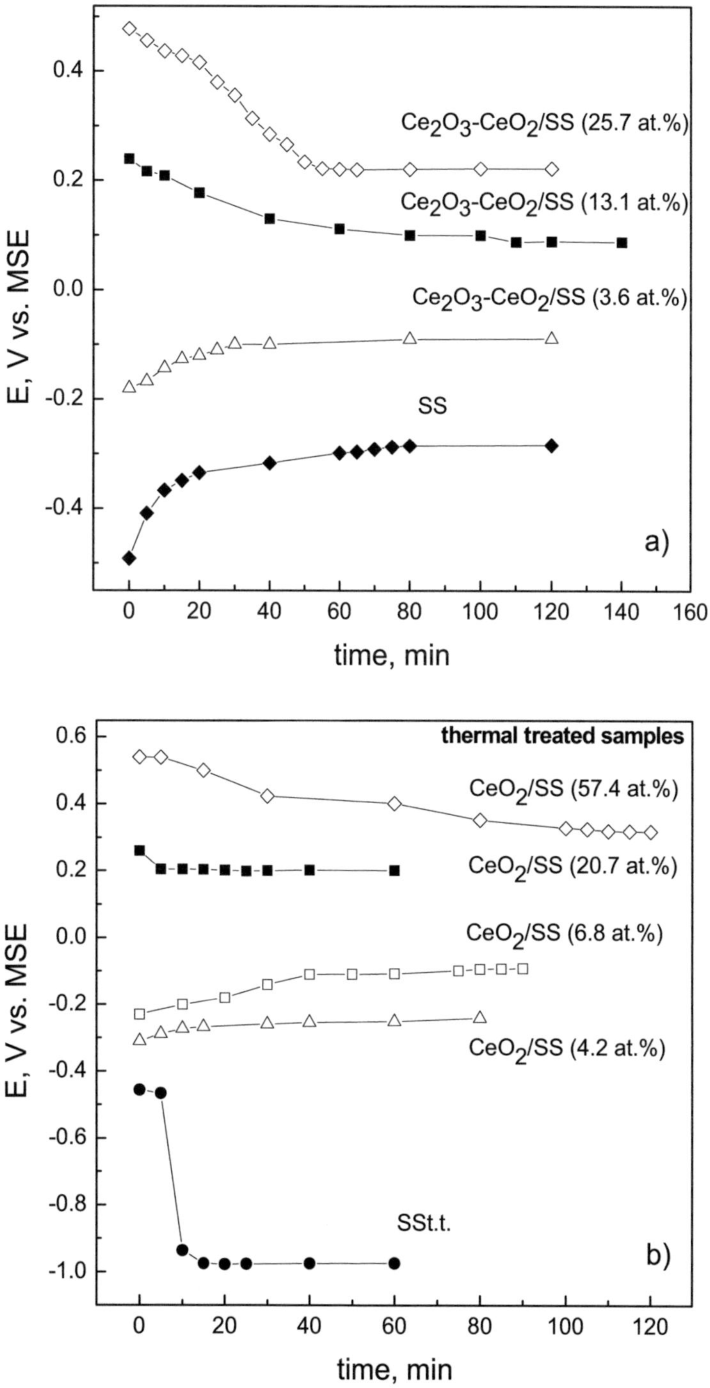

Fig. 4. Open circuit potential vs. time curves for SS (a) and $SS_{t.t.}$(b), as well as for the systems CeO_2-Ce_2O_3/SS (a) and CeO_2-Ce_2O_3/$SS_{t.t.}$(b), containing different concentrations of Ce, obtained in 0.1 N H_2SO_4.

The strong shifting of E_{corr} of the steel sample, modified with cerium oxide film, in the positive direction, depending on the surface concentration of cerium, can be attributed to the occurrence of another cathodic process in addition to the reaction of hydrogen evolution. We can assume that the effective cathodic sections of CeO_2-Ce_2O_3 will participate in the occurring cathodic depolarizing reaction according to the equations (1 and 2) given below:

$$CeO_2 + 2H^+ = Ce(OH)_2^{2+} \quad (1)$$

$$2Ce(OH)_2^{2+} + 2e^- = Ce_2O_3 + H_2O + 2H^+ \quad (2)$$

The above indicated reactions occur as a result of the extraordinary oxidation-reduction capability of the couple CeO_2-Ce_2O_3. The occurring of these reactions means that in the course of the corrosion process the surface film will be changing, enriching itself in Ce_2O_3. On the other side, the reactions (3) and (4) will also take place on the anodic sections of the steel surface and the latter one will lead to passivation:

$$Me \rightarrow Me^{n+} + ne \quad (3)$$

$$Me + H_2O \rightarrow Me_2O_n + ne \quad (4)$$

Taking into account the fact that the oxides of Ce^{3+} of the type Ce_2O_3 are soluble in acids, the reaction reported in (Achmetov, 1988) will also occur:

$$Ce_2O_3 + 6H^+ = 2Ce^{3+} + 3H_2O \quad (5)$$

As well as the respective conjugated reaction of oxidation:

$$Ce^{3+} + 2H_2O = Ce(OH)_2^{2+} + 2H^+ + e^- \quad (6)$$

Obviously, the cathodic reaction of reduction of CeO_2, which occurs at the corrosion potentials, established for the systems CeO_2-$Ce_2O_3/SS_{t.t.}$, is the main reason for restoring and preserving the passive state of the thermally treated steel samples (in accordance with equation 4) during their corrosion in solutions of sulfuric acid.

4.3 The inhibiting effect of cerium ions

In connection with the above statements a next step has been made in the investigations, namely studying the influence of the Ce^{3+} and Ce^{4+} ions as components of the corrosion medium (0.1N H_2SO_4) on the anodic behavior of stainless steel. These investigations were provoked by the observed occurrence of cathodic depolarization reaction of Ce^{4+} (CeO_2) reduction, as a result of which the surface concentration of cerium is decreasing and theoretically it should approach zero value (Stoyanova et al., 2010). For this purpose an inverse experiment was carried out at different concentrations of Ce^{4+} ions in the corrosion medium we monitored the changes in the stationary corrosion potential of the thermally treated steel by the chronopotentiometric method. The aim of this experiment was to prove the occurrence of a reversible reaction of reduction of Ce^{4+}: $Ce^{4+} - e \leftrightarrow Ce^{3+}$, (instead of the reaction of hydrogen depolarization), which in its turn creates also the option to form a film (chemically insoluble) of cerium hydroxides/oxides on the active sections of the steel surface.

Figure 5 illustrates the analogous E-τ dependences at open circuit, obtained upon immersion of SS_{tt} in 0.1 N H_2SO_4 solution, to which various concentrations of Ce^{4+} ions have been added. It was important to find out what is the influence of cerium ions on the corrosion behavior of the samples of thermally treated steel, when the character of the corrosion process is changed as a consequence of the thermal treatment of the steel. It should be reminded at this point that the stationary corrosion potentials (E_{st}) of SS and $SS_{t.t.}$in 0.1N H_2SO_4 solution have values for the non-thermally treated steel E_{st}= ~ -0.300 V, while for the thermally treated steel this value is E_{st}= ~ -0.980 V (Fig. 4)). The registered negative values of E_{st} in the absence of cerium ions for the $SS_{t.t.}$ samples, in our opinion, are connected with the strong cracking of the natural passive film on the surface of SS (Fig. 6). The most probable reason for this loss of the "stainless character" of the steel surface are the revealed sections, determining a several times higher concentration of iron-containing agglomerates in the surface layer (Table 2).

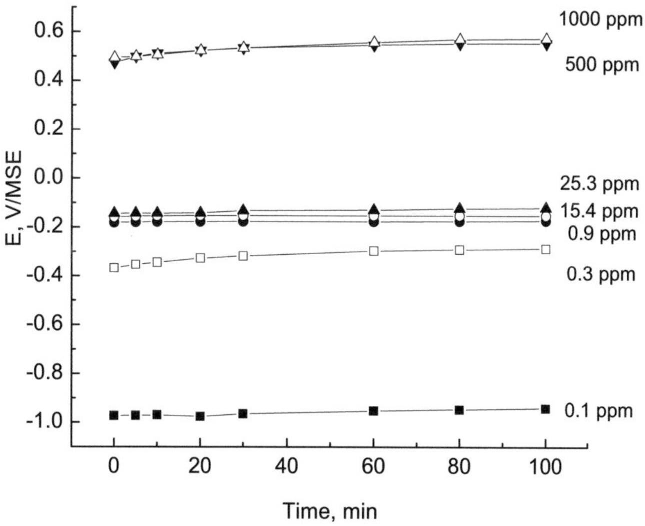

Fig. 5. Evolution of the open-circuit potential for $SS_{t.t.}$ at different concentrations of Ce^{4+} in 0.1 N H_2SO_4 solution.

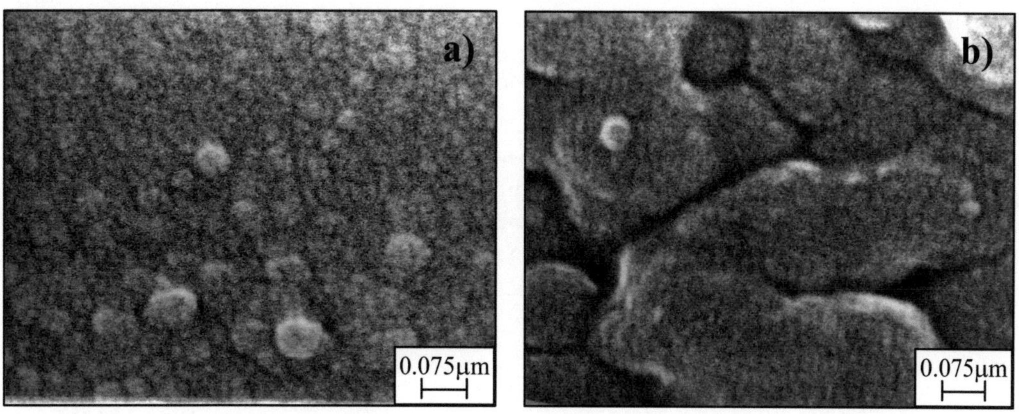

Fig. 6. SEM images of stainless steel before (a) and after thermally treatment (b).

Samples	O, at. %	Fe, at. %	Cr, at. %	Al, at. %	Ce, at. %	Cr/Fe, %	Al/Fe, %	Al/Cr, %	E_{st}, V
SS	58.3	3.1	3.4	35.2	-	1.09	11.35	10.35	_
SS 50h in 0.1N H_2SO_4	66.8	2.7	3.9	26.6	-	1.44	9.85	6.82	-0.209
$SS_{t.t.}$	64.9	7.2	7.0	20.9	-	0.97	2.90	2.99	_
$SS_{t.t.}$ 50h in 0.1N H_2SO_4	65.4	9.1	12.2	13.3	-	1.34	1.46	1.09	-0.300

Table 2. Distribution of the elements (in at. %) on the surface SS and $SS_{t.t.}$ before and after 50h immersion in 0.1N H_2SO_4.

For this reason this cycle of investigations has been carried out using samples of thermally treated steel, since the thermal treatment leads to change in the character of the corrosion process of steel. It should be taken into account that such kind of excessive heat treatment happen to take place both in the formation of catalytic converters, as well as in the course of their operation. In the latter case in the presence of cerium ions in the corrosion medium (Fig. 5), one observes a strong shifting of E_{st} in positive direction (from -0.942 V to -0.286 V), even at relatively low concentrations of Ce^{4+} (0.3 ppm) in the corrosion medium. The further increase in the concentration of Ce^{4+} ions (from 0.3 to 0.9 ppm) results in insignificant changes in E_{st}. Thereupon for SS_{tt} this shift jumps from -0.942 V (at Ce^{4+} ions concentration 0.1 ppm) up to -0.175 V (at Ce^{4+} ions concentration ~ 0.9 ppm). In the consecutive 20 – 30 fold increase in the concentration of Ce^{4+} ions (15-25 ppm) a preservation of the E_{st} value is observed, whereupon it manifests values ~ -0.150 - -0.120 V. The consecutive 20-50 fold increase in the concentration of Ce^{4+} ions (500-1000 ppm) leads to strong shifting of E_{st} in positive direction reaching values of about +0.510 - +0.570V.

These results prove that in the case of samples of thermally treated steel non-coated with Ce_2O_3-CeO_2 one observes analogous changes in the stationary corrosion potential of the steel electrode, which have already been registered for the system Ce_2O_3-$CeO_2/SS_{t.t.}$. The juxtaposition of the above-mentioned changes in E_{st} at open circuit (conditions of self-dissolution) with the characteristic zones (corrosion potential, Flade potential, zone of passivity, transpassivity region), defined by the cathodic and anodic potentiodynamic E-lgi polarization curves (conditions of external cathodic and anodic polarization) for SS_{tt} (Fig. 7.) in 0.1 N H_2SO_4 solution not-containing Ce^{4+}, shows the following. The addition of cerium ions causes shifting and establishing stationary corrosion potential (Fig. 7) in the zone of passivity of the steel. Evidently, this effect will lead also to improvement of the passivation ability, respectively to improvement of the stability to corrosion, of the steel in sulfuric acid medium, which is of great importance for the specific case of thermally treated steel, when the inhibitory action of the Ce^{4+} ions eliminates the negative influence of the cracking of the natural passive film on the steel.

The recovery of the passive state of SS_{tt}, characterized by disrupted passive film, in our opinion, is brought about also by some other reasons. It is caused by the flow of internal cathodic current (instead of external anodic current), which is determined by the occurring of a reduction reaction $Ce^{4+} \leftrightarrow Ce^{3+}$ in the redox system Ce^{4+}/Ce^{3+}. Therefore the Ce^{4+} ions, as component of the corrosion medium, are acting as inhibitor, exerting an oxidative effect. It follows from (Fig. 7), that the increase in the concentration of the inhibitor in the corrosion environment leads to a substantial decrease in the corrosion current - from 1.10^{-6} (at inhibitor concentration of 0.1 ppm) - to $1.\ 10^{-8}$ $A.cm^{-2}$ (at inhibitor concentration ~0.9 ppm). What is making impression is the fact that the further increase in the concentration of the Ce^{4+} ions in the corrosion medium from ~0.9 ppm (which could be accepted as "critical") up to 1000 ppm influences to a smaller extent the rate of corrosion. It is necessary also to point out that with the increase in the concentration of cerium ions the corrosion potential is shifted in positive direction, whereupon its values remain more positive than the potential of complete passivation and more negative than the potential of depassivation of the steel within the entire interval of studied concentrations – an effect analogous to the one already established for the systems Ce_2O_3-$CeO_2/SS_{t.t.}$in 0.1 N H_2SO_4 solution.

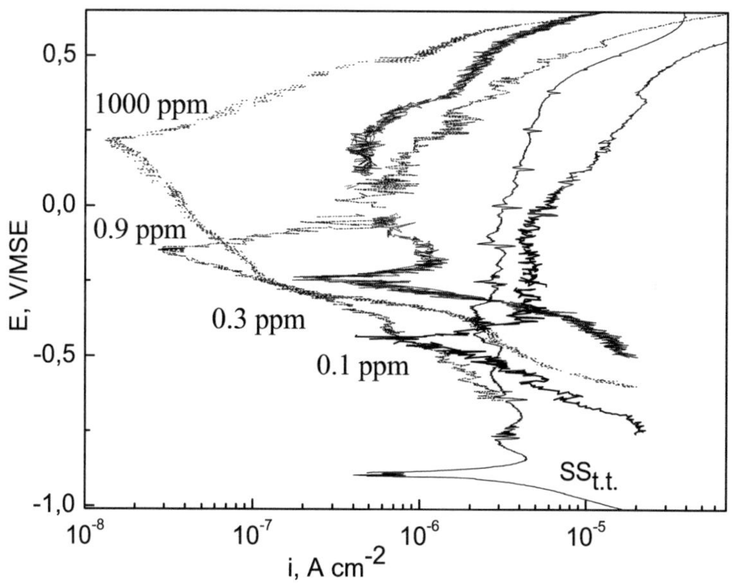

Fig. 7. Potentiodynamic E-lgi curves of $SS_{t.t.}$ at different concentrations of Ce^{4+} in 0.1 N H_2SO_4 solution.

Table 3 lists the electrochemical kinetic parameters: corrosion current density (i_{corr}), corrosion potential ($E_{corr.}$) and degree of inhibition efficiency (Z, %), characterizing the corrosion process in the presence and in the absence of cerium ions, determined on the basis of the results represented in Fig. 7. The degree of inhibition efficiency has been calculated on the basis of the equation:

$$Z\% = (i^o_{corr.} - i_{corr.}) / i^o_{corr.} x\ 100 \quad (7)$$

where i^o_{corr} and i_{corr} are the values of the corrosion current density in the absence and in the presence of cerium ions.

Samples	E, V	i_{corr}, A m^{-2}	Z, %
	SS after thermal treatment		
$SS_{t.t.}$	-0.900	1.5 x 10^{-6}	-
with 0.1 ppm	-0.435	5.8 x 10^{-7}	37.8
with 0.3 ppm	-0.238	2.2 x 10^{-7}	86.8
with 0.9 ppm	-0.156	3.2 x 10^{-8}	98.2
with 1000 ppm	0.212	1.6 x 10^{-8}	99.4

Table 3. Electrochemical parameters characterizing corrosion behaviour of $SS_{t.t.}$.

It is seen from the table that upon increasing the concentration of Ce^{4+} ions in the corrosion medium the degree of protection reaches values up to 99% for the samples of thermally treated steel. The obtained data about the promotion in the efficiency of the inhibiting action with the increase in the concentration of Ce^{4+} in the corrosion medium for the thermally treated steel (Table 3) supposes an interconnection between the inhibitor concentration and the degree of surface coverage, Q following the equation (8):

$$Q = (i^{o}_{corr} - i_{corr}) / i^{o}_{corr} \quad (8)$$

where i^{o}_{corr} and i_{corr} are respectively the corrosion current density, obtained by extrapolation of anodic and cathodic potentiodynamic curves in the absence and in the presence of various concentrations of the inhibitor in the corrosion medium. On the basis of the obtained data about the fraction of surface coverage of steel electrode as a function of the concentration of the inhibitor, one can accept that the adsorption process obeys Langmuir's isotherm. According to this isotherm the interconnection between the fraction of surface coverage and the concentration of the inhibitor is the following:

$$Q = KC.(1+KC), \text{ and respectively:} \quad (9)$$

$$C/Q = 1/K + C \quad (10)$$

where K is the adsorption constant and C is the concentration of the inhibitor. The dependence C/Q as a function of C for the thermally treated steel is represented in Fig. 8. It is seen that the experimental data describe a linear dependence, whereupon the coefficient of the linear regression and the slope of the straight line of this dependence approach a value of 1, which proves the validity of Langmuir's isotherm in our case.

The constant K in the equation (9) is connected with the standard free energy of adsorption (Δ G) in accordance with the equation:

$$K = (1/55.5) \exp(-\Delta DG^{o}_{ads} / RT) \quad (11)$$

The value of K, determined graphically based on the plot of the dependence C/Q as a function of C, is $44,6x10^{6}M^{-1}$, while the value of $(-\Delta G^{o}_{ads})$ amounts to 10.35 к call.mol^{-1}.The relatively low value of ΔG^{o}_{ads}, is indicative of electrostatic forces of interaction between the ions of the inhibitor and the steel surface. Or in other words the interaction of the inhibitor with the surface of the thermally treated steel has physical nature. On the basis of the obtained electrochemical corrosion data from the potentiodynamic curves in the presence and in the absence of inhibitor and judging from the measurements of the stationary corrosion potential of the steel, depending on the concentration of the inhibitor at open circuit we could suppose that under the conditions of internal anodic polarization in the presence of inhibitor the nature of the passive layers remains the same as in the case of external anodic polarization. As far as we can judge the specific action of the inhibitor is manifested in the formation of an adsorption layer, which is transformed into bulk phase, on the active anodic sections of the surface of the metal. In view of the XPS analyses (Guergova, 2011) after 500-hour interval of staying of the thermally treated steel in the aggressive medium in the presence of Ce^{4+} ions (25 ppm), on the surface of the studied film in the region of the Ce3d XPS band one can observe the appearance of a certain amount of cerium (1.5 at. %) in the form of Ce_2O_3. The cerium is most probably incorporated into the surface film as a result of the stay of the steel sample in the inhibited corrosion medium, which leads to its modifying as a consequence of the formation of mixed oxides of the type of cerium aluminates and chromates (Burroughs et al., 1976; Hoang et al., 1993). In support of such hypotheses comes the absence of visible corrosion damages on the surface of $SS_{t.t.}$ exposed for 500 h in 0.1N H_2SO_4 solution in the presence of Ce^{4+} ions (Fig. 9). Of course, from purely electrochemical point of view, the ability of the inhibitor to define strongly positive oxidation-reduction potential of the steel is connected with the proceeding of reduction of Ce^{4+} into Ce^{3+}. In order to investigate the kinetics of reduction of the Ce^{4+} ions to Ce^{3+}, in the region of potentials, characteristic of the passive state of the steel under consideration, we plotted the anodic and the cathodic potentiodynamic curves, characterizing the behavior of the oxidative-reductive couple Ce^{4+}/Ce^{3+} at various concentrations of Ce^{4+} in 0.1 N H_2SO_4 solution, on an inert support of platinum (Fig. 10). Such an approach (Tomashov & Chernova, 1965), to our mind, enables the complete elucidation of the mechanism of inhibitory action of the cerium ions. It allows direct juxtaposition of the changes in the values of the corrosion potentials (respectively the corrosion currents) of the steel in their presence with the values of the reversible redox potentials (respectively the exchange currents) of the couple Ce^{4+}/ Ce^{3+} at comparable concentration levels.

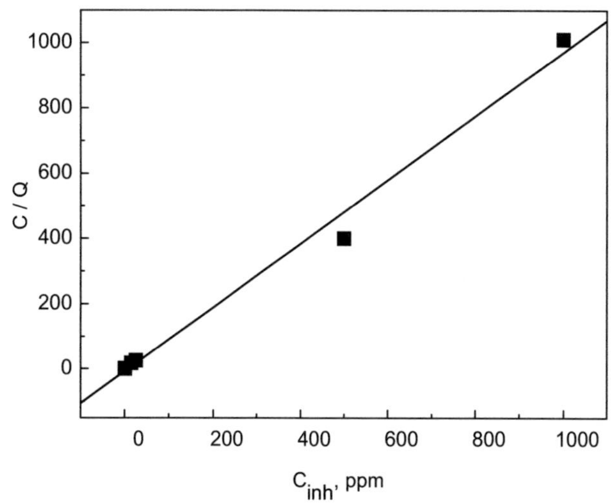

Fig. 8. Langmuir adsorption plots for $SS_{t.t.}$ in 0.1 N H_2SO_4 solution at different concentrations of Ce^{4+} ions.

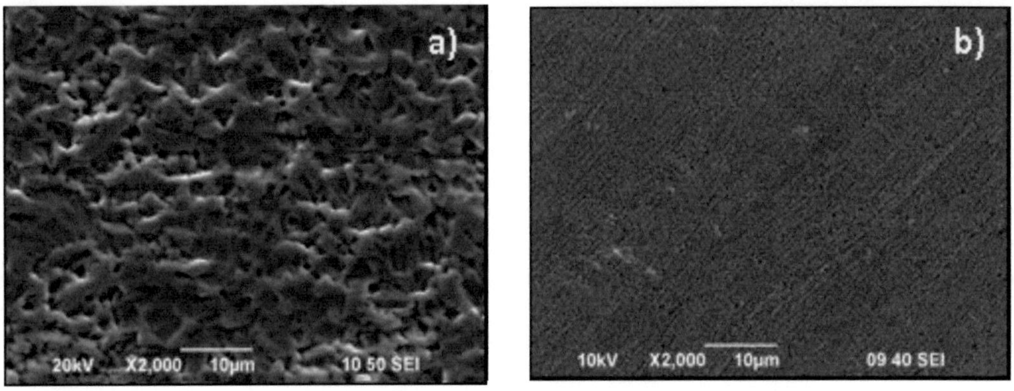

Fig. 9. SEM images on thermally treated stainless steel after 500 h immersion in 0.1 N H_2SO_4 without Ce^{4+}(a) and in the same media with 25 ppm Ce^{4+} (b).

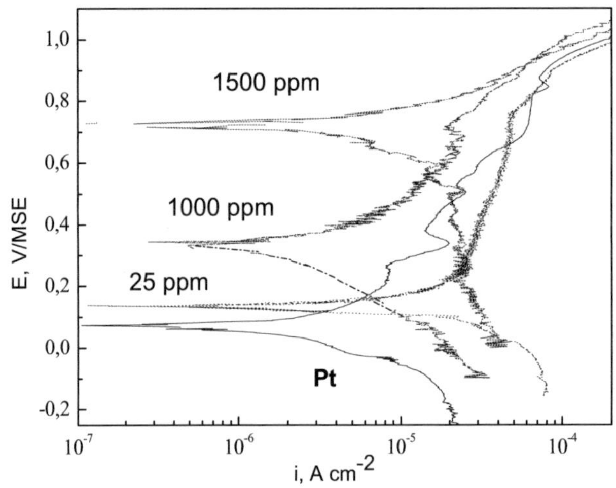

Fig. 10. Potentiodynamic E-lgi curves of Pt at different concentrations of Ce^{4+} in 0.1 N H_2SO_4 solution.

The comparison of the obtained results gives evidence that with the increase in the concentration of the cerium ions in 0.1 N H_2SO_4 solution the equilibrium oxidation-reduction potential of the system Ce^{4+}/ Ce^{3+} is shifted in positive direction (Fig.10 and Table 4), in correspondence with the equation of Nernst, whereupon at all the studied concentrations it is located in the zone of potentials, characteristic of the passive state of steel (Fig. 7). Thereupon the corrosion potentials of the steel are more negative than the equilibrium oxidation-reduction potentials of the system Ce^{4+}/Ce^{3+}. At the same time, the juxtaposition of the corrosion currents for the steel in the presence of cerium ions with the exchange currents for the system Ce^{4+}/Ce^{3+}on Pt, at comparable concentration levels of the cerium ions, shows that they have quite close values.

The so obtained data give us the reason to classify the studied oxidation-reduction couple as an inhibitor having an oxidative effect, which does not influence directly the kinetics of the anodic process. Its action is expressed in its participation in the depolarizing reaction of the corrosion process (i.e. oxidative depolarization), respectively in the establishment of oxidative-reductive potential of the medium more positive than the potential of complete passivation of the steel. This effect, in its turn, defines the value of the stationary corrosion potential of the steel to be more positive than the potential of complete passivation.

Samples	E_o, V	i_o, A cm^{-2}
Pt metal	0.070	2.17 x 10^{-7}
Pt with 0.3 ppm	0.107	8.99 x 10^{-6}
Pt with 25 ppm	0.135	6.35 x 10^{-7}
Pt with 1000 ppm	0.331	6.08 x 10^{-7}
Pt with 1500 ppm	0.722	7.67 x 10^{-7}

Table 4. Reversible redox potentials E_o, and equilibrium currents io, of the system Ce^{4+} / Ce^{3+} on Pt at different concentrations of the Ce^{4+} in the corrosion medium.

In order to prove the integral nature of cerium oxides films as efficient cathodic coating, involved directly in the corrosion process and the role of cerium ions as inhibitors possessing oxidative effect, participating also directly in the corrosion process and leading to the formation of phase layer of cerium oxides on the active cathodic sections of the steel surface, we compared the dependences E-lgi for the systems Ce_2O_3-CeO_2/$SS_{t.t.}$ as well as for the system Ce^{4+}/Ce^{3+}/Pt. - Fig. 11. It is seen in Fig. 11 that the corrosion currents of the systems Ce_2O_3-CeO_2/$SS_{t.t.}$are close in value to the exchange current of the oxidation-reduction current of the couple Ce^{4+}/Ce^{3+} on Pt. The differences between the corrosion potential of the systems Ce_2O_3-CeO_2/$SS_{t.t}$ and of the equilibrium oxidation-reduction potential of the couple Ce^{4+}/Ce^{3+} can be explained by the discrepancies in the surface and bulk phase concentrations of the components.

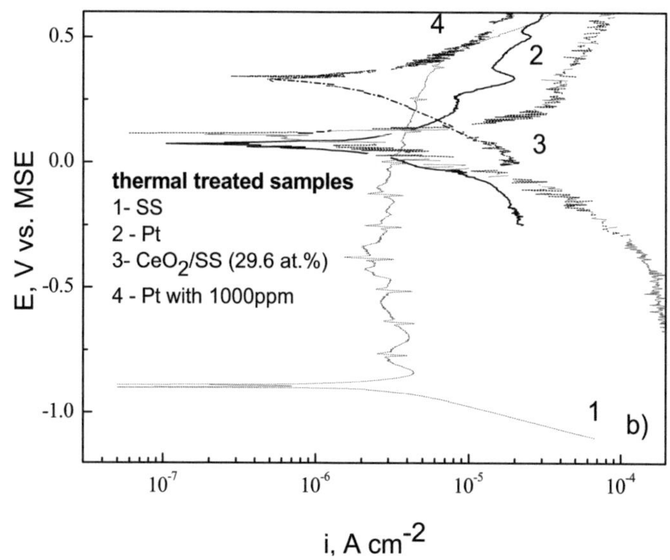

Fig. 11. Potentiodynamic E–lgi curves for $SS_{t.t}$ (1)., for Pt (2) and for the systems CeO_2-Ce_2O_3/Pt (3); Ce^{4+}/Ce^{3+}/Pt (4); obtained in 0.1 N H_2SO_4 solution.

4.4 XPS and SEM results

In confirmation of these results and the conclusions come also the data of the XPS analyses of the samples, having electrochemically deposited cerium oxide films, characterizing the changes in the chemical state and in the composition of the surface film, depending on the time interval of the immersion stay of the samples in $0.1NH_2SO_4$ solution (Stoyanova et al., 2010). Table 5 represents the results for the sample with surface concentration of electrochemically deposited cerium oxide layer 45.1 at.% .It is seen that after 1000 hours of exposure to the corrosion medium the surface concentration of cerium is decreased from 45.1aт.% down to 0.2 at.%. This result is convincing evidence for the occurring of depolarizing reaction involving the participation of the rich in CeO_2 sections of the surface, acting as effective cathodes, in accordance with the equations (1-4). It becomes evident that the presence of cerium oxide film determines the establishment of more positive stationary corrosion potential of the system, due to the proceeding of the reactions 1- 6, the surface passive film will become modified, whereupon its composition, respectively the ratio Cr/Fe, will become different.

Time of exposure, h	E_{st}, V	C, at. %	O, at. %	Al, at. %	Fe, at. %	Cr, at. %	Ce, at. %
as deposited	0.151	15.6	37.6	1.0	0.1	0.6	45.1
18	0.169	50.1	35.7	1.6	0.5	0.4	11.7
200	0.249	64.4	31.5	0	0	0	4.1
250	0.236	63.7	31.5	0	0	0.2	4.6
400	0.239	68.5	27.4	0	0	0	4.1
1000	0.060	31.4	42.9	19.9	0.7	4.9	0.2

Table 5. Concentration of the elements (in at. %) on the surface layers of the system CeO_2-Ce_2O_3/SS after thermal treatment and after corrosion test in 0.1 N H_2SO_4.

In this cycle of experimental runs, using the XPS method, the changes were monitored, occurring in the chemical composition of the passive film of the system CeO_x /SS, during prolonged exposure of the samples in 0.1 N H_2SO_4 solutions (Table 5). The analyses were carried out after the 18th, 200th, 250th , 400th , and 1000th hour - time intervals of exposure. Within the interval 200-400 hours E_{st} remains practically the same, while after 1000 hours of exposure it is shifted strongly in the negative direction, reaching a value of about ~ +0.060 V. To obtain further information about the influence of ceria on the corrosion behaviour of as-deposited sample we analyzed in depth the Ce3d and O1s XPS spectra. As it has already been discussed in our previous papers (Nikolova et al., 2006; Stoyanova et al, 2006), the Ce3d spectrum is a complex one, due to the fact that the peak is spin-orbital split into a doublet, each doublet showing extra structure due to the effect of the final state. There are 8 peaks assignments in the spectra labelled according to Burroughs (Burroughs et al., 1976), where the peaks V, V^{II}, V^{III} and U, U^{II}, U^{III} refer to the 3d5/2 and 3d3/2 respectively and they are characteristic of Ce(IV) 3d final states. The peaks labelled as V^{I} and U^{I} refer to 3d5/2 and 3d3/2 they are characteristic of Ce(III)3d final state (Fig.12).The literature data make it obvious that the chemical state of Ce could be evaluated based on the percentage of the area of the u''' peak, located at 916.8 eV with respect to the total Ce3d area. So if the percentage of the u''' peak related to the total Ce3d area varies from 0 to 14%, then the Ce^{4+} percentage related to the total amount of Ce varies from 0 to 100%. In our case the u'''% amounts change as a result of dipping the as-deposited sample into 0.1 N H_2SO_4 solution, so we observed also change in the percentage of Ce^{4+} and Ce^{3+} on the surface (Table 6). The obtained O1s X-ray photoelectron spectra, recorded after different time intervals of exposure, are quite complicated. These spectra had to undergo de-convolution procedure to analyze the contribution of the separate components in them.

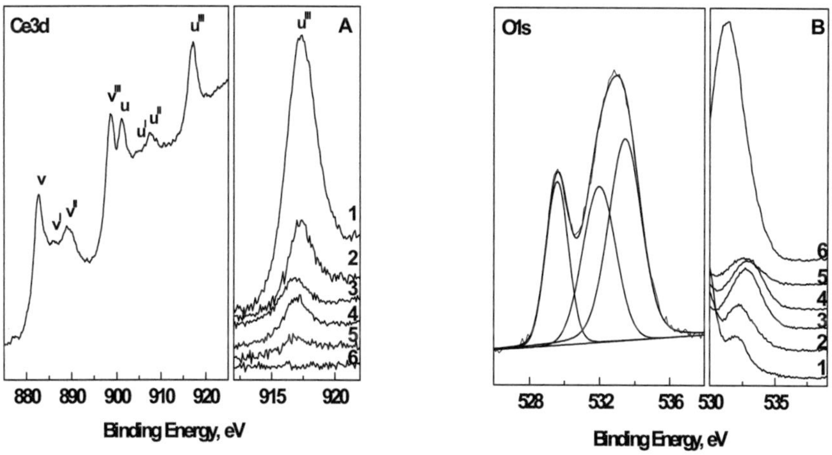

Fig. 12. Ce3d and O1s XPS spectra of CeO_2-Ce_2O_3/SS sample after 250 h exposition to 0.1 N H_2SO_4, Panel A and B correspond to high BE and high BE portion of Ce3d and O1s spectral regions taken at different step of corrosion test. (1) as-deposited; (2) after 18 h ; (3) after 200 h; (4) after 250 h; (5) after 400 h; (6) after 1000 h exposition to 0.1 N H_2SO_4 solution.

The chemical composition of the surface film of the CeO_2-Ce_2O_3/$SS_{t.t.}$ systems under consideration is shown in Table 5. It is important to note that on the surface of the 'as-deposited" samples, covered with cerium oxide film, we detected also the presence of iron, chromium and aluminum in addition to the cerium. The latter is mainly in the valence state Ce^{4+} i.e. in the form of CeO_2. After 18 hours of exposure to 0.1 N H_2SO_4 solution the chemical composition on the surface of the system has changed. The surface concentration of cerium drops down from 45.1 to 11.7 at. %, whereupon Ce^{3+} appears in the form of $Ce(OH)_3$ and Ce_2O_3. After continuing the exposure further (200-400 hours) one observes evolution of the spectra with respect to cerium and oxygen (Fig.12). The main peak in the spectrum of oxygen, having binding energy of about ~529.5 eV, is attributed to the presence of O^{2-} ions, which exist basically as Ce-O bonds in the crystal lattice of the cerium oxide being formed. The second peak, located at 531.7 eV, is associated with the existence of OH^- groups on the surface, while the presence of a peak at 533 eV shows that there is adsorbed water on the surface of the studied passive films (533 eV) (Hoang et al., 1993; Paparazzo, 1990). It can be seen in the spectra that after exposure of the samples in corrosion medium for 200-400 hours the respectively detected high-energy peak in the spectrum of oxygen for these samples is growing up initially, while afterwards it decreases its intensity. This effect, in our opinion, is owing to consecutive enrichment and then impoverishment of the surface layer of the film in OH^- groups i.e. adsorbed water molecules (Fig.12, Table 6). The obtained results gives us the reason to draw the conclusion that the surface passive film under these conditions at this stage consists of CeO_2, Ce_2O_3, $Ce(OH)_4$ as well as $CeO(OH)_2$ and $Ce(OH)_3$, whose existence has been ascertained also by some other authors (Huang et al., 2008).

Time of exposure, h	O1s, eV	Percentage of oxygen contribution to the total	Bonds	Ce3d, eV	Percentage of Ce^{4+} to the total Ce
as deposited	529.2	59	Ce-O	883.0	100
	531.4	41	Ce-OH		
18	529.2	43	Ce-O	882.9	85.35
	531.5	42.5	Ce-OH		
	533.2	14.5	Others		
200	529.4	21.4	Ce-O	882.4	76
	531.9	33.6	Ce-OH		
	533.3	45	Others		
250	529.5	23	Ce-O	882.5	76
	531.9	32	Ce-OH		
	533.4	45	Others		
400	529.5	23	Ce-O	882.5	76
	532.1	45	Ce-OH		
	533.4	32	Others		
1000	531.2	100	OH	882.0	-

Table 6. Calculated contribution of oxygen and percentage of Ce^{4+}, depending of the exposure time in 0.1 N H_2SO_4. Types of the chemical bonds and values of binding energy.

After 1000 hours of exposure the quantity of cerium is drastically decreased, as a consequence of the occurring reactions 1-4 and only some insignificant amounts of cerium have been registered in the valence state Ce^{3+}, i.e. in the form of Ce_2O_3. Only a single peak has been detected in the spectrum of oxygen, having a binding energy of 531,2 eV. Chromium, aluminum and iron have also been detected (Table 5). On the basis of the values of their binding energies (Table 5), including also the location of the O1s peak (Fig.12), we can also conclude that they exist in the form of oxides and hydroxides: Cr_2O_3, $Cr(OH)_3$, Fe_2O_3, FeOOH, Al_2O_3 and $Al(OH)_3$. The high concentration of carbon registered in the surface film is owing to the considerable amount of carbonates adsorbed during the thermal treatment in a high-temperature oven.

In support of the conclusions, drawn on the basis of the above results, evidence is also given by the direct SEM observations carried out. It follows from the electron microscopic studies of the samples, exposed to the corrosion medium, that the disruption of the passive state of steel at the initial stages (until the 50th hour) leads to appearance of local corrosion and its propagation and spreading further to give total corrosion (Fig.13 a,b). For the sake of comparison Fig. 13 shows the same surface in the presence of cerium oxide coating after 50 hours of exposure to the corrosion medium. In this case no corrosion damages can be observed on the surface not only after the 50th hour, but even after 1000 hours of exposure (Fig.13) to the corrosion medium, as a result of its modifying, already discussed above.

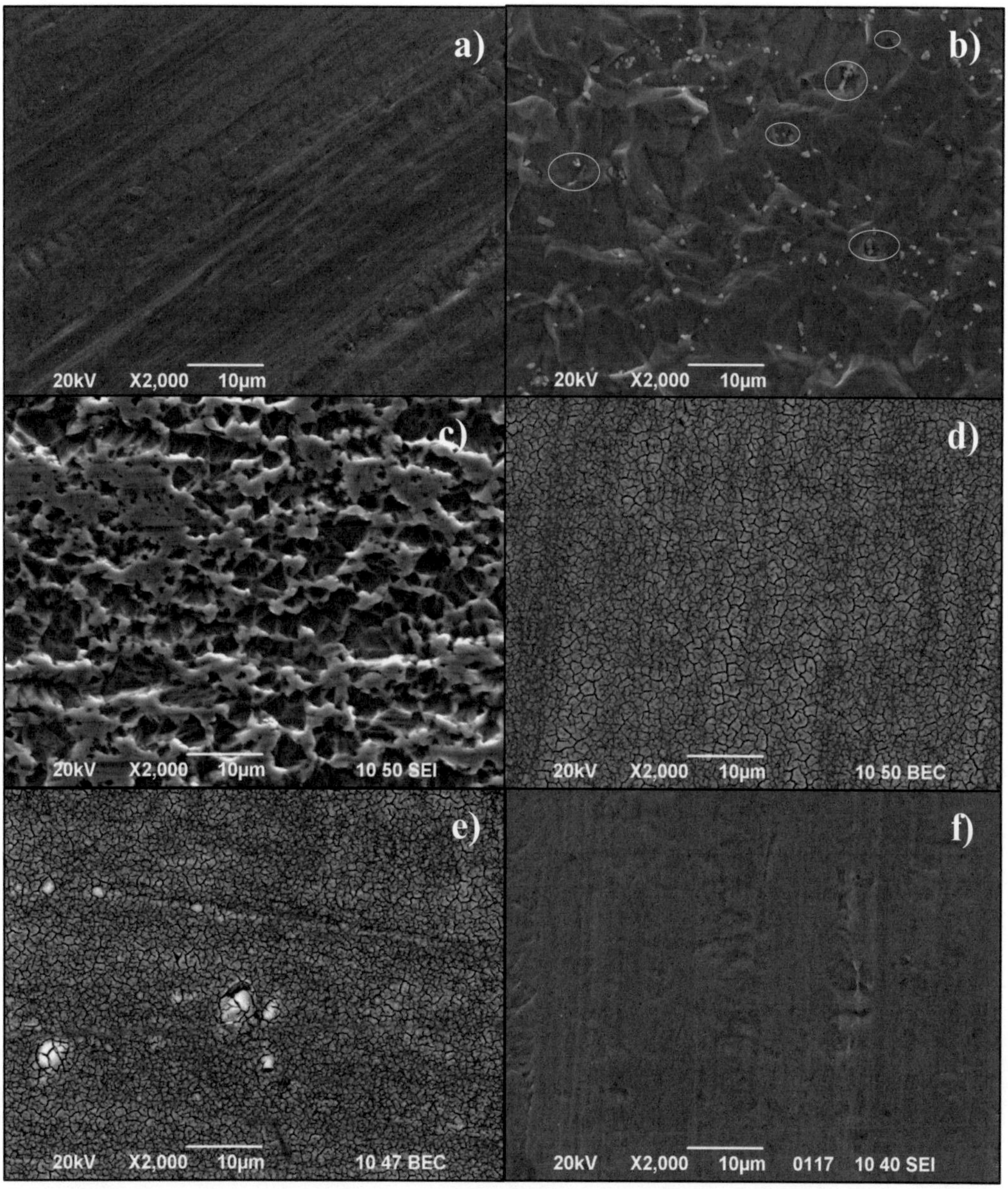

Fig. 13. SEM images for the samples: a) $SS_{t.t.}$; b) $SS_{t.t.}$ after 50 h exposition in 0.1 N H_2SO_4 (O – areas of local corrosion); c) $SS_{t.t.}$ after 1000 h exposition; d) CeO_2-$Ce_2O_3/SS_{t.t.}$; e) CeO_2-$Ce_2O_3/SS_{t.t.}$ after 50 h exposition in 0.1 N H_2SO_4; f) CeO_2-$Ce_2O_3/SS_{t.t.}$ after 1000 h exposition.

5. Conclusions

The above discussed results elucidate the mechanism of action of cerium oxide coatings as effective cathodes and of cerium ions (when they are present as a component of the corrosion medium) – as inhibitor having oxidative action, leading to improvement of the corrosion stability of stainless steels. They explain the improved ability of the steel to undergo passivation, respectively to recover its passive state in cases of disruption of its surface passive film – especially in this specific case, studied by us, i.e. disruption as a result of thermal treatment.

This effect is associated with a strong shifting of the stationary corrosion potential of the steel in positive direction, moving over from potentials, characteristic of corrosion in active state to potentials, located inside the zone of passivity. In this respect, another basic purpose of the investigations was the elucidation of the mechanism of action of the cerium oxides film and in particular collecting experimental evidence for the supposition about the occurring of an efficient depolarization reaction of CeO_2 reduction, resulting in a state of passivity, instead of hydrogen depolarization reaction. For this purpose we considered also the decrease in the surface concentration of Ce in the passive layer under the conditions of the actual corrosion process (self-dissolution) of the stainless steel, by means of XPS, EDS and ICP-AES analyses. A decrease in the surface concentration of CeO_2 (Ce^{4+}) has been observed, which is known to be chemically inert in acidic media. The obtained results prove the occurrence of an effective cathodic process of Ce^{4+} reduction into Ce^{3+} in the surface oxide film. These results elucidate in details the corrosion behavior of the system Ce_2O_3-CeO_2/steel in 0.1N H_2SO_4. They explain the improved ability of self-passivation, respectively the restoration of the passive state of the thermally treated steel in the presence of cerium oxides as components of the modified passive surface film.

It has also been shown that the couple Ce^{4+}/Ce^{3+}, as component (inhibitor) of the corrosion medium, displays analogous action. It is expressed in the occurrence of a reaction of reduction of the oxidative component of the medium - Ce^{4+}, leading to consumption of the electrons removed in the course of oxidation, respectively passivation, of the active anodic sections.

6. Acknowledgements

The authors gratefully acknowledge financial support by the Bulgarian National Science Fund under Contract DO 02-242/TK 01-185.

7. References

Achmetov, N.S., (1988), *General and Inorganic Chemistry* (in Russian, Second edition), Publishing House "Vyshaya Shkola", Moscow

Aldykiewicz Jr A.J., Davenport A.J., Isaacs H.S., (1995).Investigation of cerium as a cathodic inhibitor for aluminum-copper alloys. *Journal of the Electrochemical Society*, Vol. 142, No.10, pp. 3342-3350. ISSN: 0013-4651

Aldykiewicz Jr A.J., Davenport A.J. Isaacs H. S. (1996). Studies of the formation of cerium-rich protective films using X-ray absorption near-edge spectroscopy and rotating disk electrode methods. *Journal of the Electrochemical Society* Vol.143 No1, pp. 147-154, ISSN: 0013-4686

Almeida, E., Diamantino, T.C, Figueiredo M.O., Sa C. (1998). Oxidizing alternative species to chromium VI in zinc - galvanized steel surface treatment. Part I-A morphological and chemical study. *Surface and Coating Technology*, Vol. 106, pp. 8-17. ISSN: 0257-8972

Almeida, E., Fedrizzi, L., Diamantino, T.C. (1998). Oxidizing alternative species to chromium VI in zinc - galvanized steel surface treatment Part 2-An electrochemical study. *Surface and Coating Technology*, Vol. 105, pp. 97-101, ISSN: 0257-8972

Amelinckx, L., Kamrunnahar, M., Chou P, Macdonald D., (2006). Figure of merit for the quality of ZrO_2 coatings on stainless steel and nickel-based alloy surfaces. *Corrosion Science*, Vol. 48, (April, 2006) pp. 3646-3667, ISSN: 0010-938X

Aramaki, K. (2001). The inhibition effects of cation inhibitors on corrosion of zinc in aerated 0.5M NaCl. *Corrosion Science*, Vol. 43, (June, 2001), pp.1573-1588, ISSN: 0010-938X

Aramaki K. (2001). Treatment of zinc surface with cerium(III) nitrate to prevent zinc corrosion in aerated 0.5 M NaCl, *Corrosion Science*, Vol. 43, No.11 pp. 1201-1215, ISSN: 0010-938X

Aramaki, K. (2002). Preparation of chromate-free, self-healing polymer films containing sodium silicate on zinc pretreated in a cerium (III) nitrate solution for preventing zinc corrosion at scratches in 0.1 M NaCl. *Corrosion Science*, Vol. 44 (August 2001), pp. 1375-1389, ISSN: 0010-938X

Aramaki, K. (2002). Cerium (III) chloride and sodium octylthiopropionate as an effective inhibitor mixture for zinc in 0.1M NaCl. *Corrosion Science*, Vol.. 44, (June 2001), pp. 1361-1374, ISSN: 0010-938X

Aramaki, K.,(2002). Self-healing protective films prepared on zinc by treatment with cerium (III) nitrate and sodium phosphate. *Corrosion Science*, Vol. 44, No.11, pp.2621-2634. ISSN: 0010-938X

Arenas, M.A., Conde A., de Damborenea, J., (2002). Cerium: a suitable green corrosion inhibitor for tinplate. *Corrosion Science*, Vol.44, No.3, pp. 511-520, ISSN: 0010-938X

Arenas, M.A., de Damborenea, J.,(2003). Growth mechanism of cerium layers on galvanised steel. *Electrochimica Acta*, Vol. 48, No. 24, (October, 2003) pp. 3693-3698, ISSN: 0013-4686

Arenas, M.A, Garcia, I., de Damborenea, J., (2004). X-ray photoelectron spectroscopy study of the corrosion behaviour of galvanised steel implanted with rare earths. *Corrosion Science*, Vol. 46, No. 4, pp. 1033-1049, ISSN: 0010-938X

Arnott, D.R., Hinton, B.R.W & Ryan N.E, (1989). Cationic-film-forming inhibitors for the protection of the AA 7075 aluminum alloy against corrosion in aqueous chloride solution. *Corrosion*, Vol. 45, No. 1, pp. 12-18

Avramova, I., Stefanov, P., Nicolova, D., Stoychev, D., Marinova, T., (2005). Characterization of nanocomposite CeO_2-Al_2O_3 coatings electrodeposited on stainless steel. *Composites Science and Technology*, Vol. 65, pp. 1663-1667 ISSN: 0266 - 3538

Avramova, I., Stoychev, D., Marinova, T., (2006). Characterization of thin CeO_2-ZrO_2-Y_2O_3 films electrochemical deposited on stainless steel. *Applied Surface Science*, Vol. 235, pp. 1365-1370, ISSN: 0169-4332

Balasubramanian, M., Melendres, C.A., Mansour, A. N., (1999). An X-ray absorption study of the local structure of cerium in electrochemically deposited thin films. *Thin Solid Films*, Vol. 347, No. 1-2, (June 1999), pp. 178-183, ISSN: 0040-6090

Barbucci A., Delucchi M, Cerisola G. (1998), Study of chromate free pretreatments and primers for the protection of galvanized steels, *Progress In Organic Coatings*, Vol. 33, No. 2, pp. 131-138, ISSN: 0300-9440

Bethencourt, M., Botana, F.J., Calvino, J.J, Marcos, M., RodrÍguez-Chacón, M. A., (1998). Lanthanide compounds as environmentally-friendly corrosion inhibitors of aluminium alloys : a review. *Corrosion Science* . Vol. 40, pp. 1803-1819, ISSN: 0010-938X

Breslin, C.B, Chen, C., Mansfeld, F., (1997). The electrochemical behavior of stainless steels following surface modification in cerium containing solutions. *Corrosion Science*, Vol. 39, No. 6, pp. 1061-1073 ISSN: 0010-938X

Bunluesin, T., Gorte, R.G., Graham, G.W., (1997). CO oxidation for the characterization of reducibility in oxygen storage components of three-way automotive catalysts. *Applied Catalysis B: Environmental,* Vol. 14, No. 1-2, (December 1997), pp. 105-115 ISSN: 0926 - 3373

Burroughs, P. Hamnett, A., Orchard A.F., Thornton G. J., (1976). Satellite structure in the X-ray photoelectron spectra of some binary and mixed oxides of lanthanum and cerium. *Journal of the Chemical Society, Dalton Transactions,* Vol. 17, pp. 1686-1698, ISSN: 1472 - 7773

Crossland, A.C., Thompson, G.E., Skeldon, P., Smith G. C. (1998) Anodic oxidation of Al-Ce alloys and inhibitive behaviour of cerium species. *Corrosion Science*, Vol.40, pp.871-885, ISSN: 0010-938X.

Davenport, A.J., Isaacs, H.S., Kendig, M.W. (1991). XANES investigation of the role of cerium compounds as corrosion inhibitors for aluminium. *Corrosion Science,* Vol. 32, No.5-6, pp. 653-663, ISSN: 0010-938X.

Di Maggio, R., Rossi, S., Fedrizzi, L., Scardi P. (1997). ZrO_2-CeO_2 films as protective coatings against dry and wet corrosion of metallic alloys. *Surface and Coatings Technology*, Vol. 89,No. 3, pp.292-298, ISSN:0257-8972.

Fahrenholtz, W.G., O'Keefe, M.J., Zhou, H., Grant, J.T., (2002). Characterization of cerium – based conversion coatings for corrosion protection of aluminum alloys. *Surface and Coatings Technology*, Vol. 155, No.2-3, pp.208-213, ISSN: 0257-8972

Ferreira, M.G., Duarte, R.J., Montemor, M. F., Simoes A.M., (2004). Silanes and rare earth salts as chromate replacers for pretreatments on galvanized steel. *Electrochimica Acta*, Vol. 49, No. 17-18, pp. 2927-2935, ISSN: 0013-4686

Forsyth, M., Forsyth, C.M., Wilson, K., Behrsing, T., Deacon, G.B., (2002). ART characterization of synergistic corrosion inhibition of mild steel surfaces by cerium salicylate. *Corrossion Science* Vol. 44, No. 11, (November 2002) pp.2651-2656, ISSN: 0010-938X

Gonzalez, S., Gil, M.A., Hemandez, J.O., Fox, V., Souto, R. M. (2001). Resistance to corrosion of galvanized steel covered with an epoxy - polyamide primer coating, *Progress In Organic Coatings,* Vol. 41, No. 1-3, pp. 167-170, ISSN: 0300-9440

Guergova, D., Stoyanova, E., Stoychev, D., Atanasova, G., Avramova, I., Stefanov, P. (2008). Influence of calcination of SS OC 4004 with alumina or ceria layers on their passive state in different acid media. *Bulgarian Chemical Communications*, Vol. 40, No. 3, pp227-232

Guergova, D., Stoyanova, E., Stoychev, D., Avramova, I., Stefanov, P., (2011). Investigation of the inhibiting effect of cerium ions on the corrosion behaviour of OC404 stainless steel in sulfuric acid medium. *Open Corrosion Journal* in press, ISSN: 1876-5033

Hagans, P.L, Hass, C.M., (1994) *ASM Handbook Volume 05: Surface Engineering*, Vol.5, pp. 405-411, ISBN: 978-087170-384-2

Hinton, B.R.W., (1991). Corrosion prevention and chromates. *Metal Finishing* Vol. 89, pp. 55-61, ISSN: 0026-0576

Hinton, B.R.W., Wilson, L. (1989). The corrosion inhibition of zinc with cerous chloride,*Corrosion Science,* Vol. 29, No. 8, 1989, pp. 967-975, 977-985, ISSN: 0010-938X

Hinton, B.R.W., (1992). Corrosion inhibition with rare earth metal salts, *Journal of Alloys and Compounds,* Vol. 180, No. 1-2, (25 March 1992), pp. 15-25, ISSN: 0925-8388

Hoang, M., Hughes, A.E., Turney, T.W., (1993). An XPS study of Ru-promotion for Co/CeO_2 Fischer-Tropsch catalyst. *Applied Surface Science.* Vol. 72, No. 1, pp. 55-65, ISSN: 0169 - 4332

Hosseini M.G., Sabouri, M., Shahrabi, T. (2006). Comparison between polyaniline - phosphate and polypyrrole- phosphate composite coatings for mild steel corrosion protection. *Material and Corrosion*, Vol. 57, pp. 447 - 453, ISSN: 0947-5117

Hosseini, M.G., Ashassi-Sorkhabi, H., Ghiasvand, H.A.Y., (2007). Corrosion protection of electro-galvanized steel by green conversion coatings. *Journal of Rare Earths,* Vol. 25, pp. 537-243, ISSN: 1002-0721

Huang Xingqiao, Ning Li, Huiyong Wang, Hanxiao Sun, Shanshan Sun, Jian Zheng, (2008).Electrodeposited cerium film as chromate replacement for tin plate, *Thin Solid Films,* Vol.516, No. 6, (January 2008) pp. 1037-1043, ISSN: 0040 - 6090

Korobov, V.I., Loshkarev, Y.M., Kozhura, O.V. (1998). Cathodic treatment of galvanic zinc coatings in solutions of molybdates, *Russian Journal of Electrochemistry,* Vol. 33, pp. 55-62, ISSN: 1023 -1935

Kudryavtsev, N.T. (1979). *Electrolytic Metal Coatings,* In Russian, Publ. House "Khimiya", Moscow

Lainer, V.I. (1984) *Protective Metal Coatings,* In Russian, Publ. House "Metalurgia", Moscow

Liu, R., Li, D.Y., (2000). Effects of yttrium and cerium additives in lubricants on corrosive wear of stainless steel 304 and Al alloy 6061. *Journal of Materials Science.* Vol. 35, No. 3, pp. 633-641 ISSN: 0022-2461

Liu, H., Yang, J., Liang, H.-H., Zhuang, J.-H., Zhou W.-F., (2001). Effect of cerium on the anodic corrosion of Pb-Ca-Sn alloy in sulfuric acid solution. *Journal of Power Sources,* Vol. 93, No. 1-2, pp.230-233, ISSN: 0378-7753

Lox, E.S., Engler, B.N., (1995) in: A. Frennet, J.M. Bastin (Eds.), Catalysis and Automotive Pollution Control III, Elsevier, Amsterdam, 1995, p.1559 (Chapter: Enviromental Catalyis - Mobile sources)

Lu, Y.C., Ives, M.B., (1993). The improvement of the localized corrosion resistance of stainless steel by cerium. *Corrosion Science,* Vol. 34, No. 11, pp. 1773-1781, ISSN: 0010-938X

Lu, Y.C., Ives, M.B.(1995). Chemical treatment with cerium to improve the crevice corrosion resistance of austenitic stainless steels. *Corrosion Science,* Vol. 37, No. 1 pp.145-155, ISSN: 0010-938X.

Lukanova, R., Stoyanova, E., Damyanov, M., Stoychev, D.(2008). Formation of protective films on Al in electrolytes containing no Cr^{6+}ions. *Bulgarian Chemical Communications*, Vol.40, No.3, pp.340-347.

Mansfeld, F., Lin, S. and Shin, H., (1989). Corrosion protection of Al alloys and Al-based metal matrix composites by chemical passivation, *Corrosion*, Vol. 45 No8, pp. 615-630.

Mansfeld, F., Wang. V., Shih, H., (1991).Development of 'stainless aluminum', *Journal of the Electrochemical Society*, Vol. 138, No.12, pp. L74-L75. ISSN: 0013-4651.

Mansfeld. F., Wang. Y, (1995). Development of "stainless" aluminium alloys by surface modification. *Materials Science and Enginerring A*, Vol.198, No.12 pp.51-61, ISSN: 0921-5093.

Marinova, T., Tsanev, A., Stoychev, D. (2006). Characterisation of Mixed Yttria and Zirconia Thin Films. *Materials Science and Engineering B*, Vol.130, No. 1-3, pp. 1-4. ISSN: 0921-51-07.

Mcnamara, J.M. (2000). Health effects of vehicle emissions a review from the second international conference, *Platinum Metals Review*, Vol.44, No.2, pp. 71-73.

Montemor, M.F., Simoes, A.M., Ferreira, M.G.S.(2001). Composition and behaviour of cerium films on galvanized steel, *Progress in Organic Coatings*, Vol.43, No.4, 274-281, ISSN: 0300-9440.

Montemor M.F, Simoes, A.M.., Ferreira, M.G.S., (2002). Composition and corrosion behaviour of galvanized steel treated with rare – earth salts: the effect of the cation. *Progress in Organic Coatings*, Vol.44, No2, pp.111-120. ISSN: 0300-9440.

Montemor, M.F., Ferreira, M.G.S., (2008). Analytical characterization of silane films modified with cerium activated nanoparticles and its relation with the corrosion protection of galvanized stell substrates, *Progress in Organic Coatings*, Vol.63, No.3 (October 2008) pp.330-337. ISSN: 0300-9440.

Nonnenmann, M.,(1989). New high-performance gas flow equalizing metal supports for exhaust gas catalysts, *Automobilitech. Z.*, Vol.No. 4, pp.185-192.

Nikolova, D. Stoyanova, E., Stoychev, D,. Stefanov, P., Marinova, Ts (2006). Anodic behaviour of stainless steel covered with an electrochemically deposited Ce_2O_3-CeO_2 film. *Surface & Coatings Technology*,Vol. 201, pp. 1559 - 1567, ISSN: 0257-8972.

Nikolova, D., Stoyanova, E., Stoychev, D., Avramova, I., Stefanov, P. (2006). Stability of the passive state of stainless steel OC 4004 in sulphuric acid solutions improved by additionally electrodeposited oxide layers, Book of papers of the International International Workshop *Nanostructured Materials in Electroplating*, Sandanski, Bulgaria, March, 2006, pp.127-131 Eds.D. Stoychev, E.Valova, I. Krastev, N. Atanassov, March 2006, Sandanski, Bulgaria

Nickolova, D., Stoyanova, E., Stoychev, D. P., Stefanov, P., Avramova I., (2008). Protective effect of alumina and ceria oxide layers electrodeposited on stainless steel in sulfuric acid media. *Surface & Coatings Technology*, Vol. 202, pp. 1876-1888, ISSN: 0257-8972

Otero, E., Pardo, A., Saenz, E., Utrilla, M.V, Hierro, P., (1996). A Study of the influence of nitric acid concentration on the corrosion resistance of sintered austenitic stainless steel. *Corrosion Science*, Vol. 38, No. 9, pp. 1485-1493, ISSN: 0010-938X

Otero, E., Pardo, A., Utrilla, M.V., Saenz, E., Alvarez, J.F. (1998). Corrosion behavior of 304I and 316I stainless steels prepared by powder metallurgy in the presence of sulfuric and phosphoric acid. *Corrosion Science*, Vol. 40, No. 8, pp. 1421-1434, ISSN: 0010-938X

Paparazzo, E., (1990). Surface XPS studies of damage induced by X-ray irradiation on CeO_2 surfaces. *Surface Science* Vol. 234, No.1-2, pp. L253-L25, ISSN: 0039 - 6028.

Pardo, A., Merino, M.C., Arrabal, R., Viejo, F., Carboneras, M., Munoz, J.A., (2006). A surface characterization of cerium layers on galvanised steel. *Corrosion Science*, Vol. 48, pp. 3035-3048, ISSN: 0010 - 938X

Pardo, A., Merino, M., Arrabal, C. R., Merino, S., Viejo, F., Carboneras, M. (2006). Effect of Ce surface treatments on corrosion resistance of A3xx.x/SiCp composites in salt fog. *Surface and Coatings Technology*, Vol. 200, No. 9, pp. 2938-2947, ISSN: 0257-8972

Schafer, H., Stock, H.R. (2005). Improving the corrosion protection of aluminum alloy using reactive magnetron sputtering, *Corrosion Science*, Vol. 47, (June 2004) pp.953-964, ISSN: 0010-938X

Schmidt, H., Langenfeld, S., Naß, R., (1997). A new corrosion protection coating system for pressure-cast aluminium automotive parts. *Materials and Design* , Vol. 18, No. 4-6, pp. 309-313, ISSN: 0261 -3069

Stefanov, P., Stoychev, D, Valov, I., Kakanakova-Georgieva, A., Marinova, T. (2000) Electrochemical deposition of thin zirconia films on stainless steels 316 L. *Materials Chemistry and Physics*, Vol. 65, pp.222-225, ISSN: 0254 - 0584

Stefanov, P., Stoychev, D, Stoycheva, M., Ikonomov, J., Marinova, T., (2000). XPS and SEM characterisation of zirconia thin films prepared by electrochemical deposition. *Surface and Interface Analysis* , Vol. 30, pp. 628-631, ISSN: 1096 - 9918

Stefanov, P., Stoychev, D., Atanasova, G., Marinova, T. (2004). Electrochemical deposition of CeO_2 on ZrO_2 and Al_2O_3 thin films formed on stainless steel. *Surface and Coatings Technology*, Vol. 180-181, pp. 446 – 449, ISSN: 0257-8972

Stefanov, P., Stoychev, D., Aleksandrova, A., Nicolova, D., Atanasova, G., Marinova, T., (2004). Compositional and structural characterization of alumina coatings deposited electrochemically on stainless steel. *Applied Surface Sciences*, Vol. 235, pp. 80-85 ISSN: 0169 - 4332

Stoyanova, E., Nikolova, D., Stoychev, D., Electrochemical behaviour of stainless steel OC4004 with modified passive film in nitric and sulphuric acids, Book of Papers of the International Workshop *Nanostructured Materials in Electroplating* pp. 122-126, Eds. D. Stoychev, E.Valova, I. Krastev, N. Atanassov, March 2006, Sandanski, Bulgaria

Stoyanova, E., Nikolova, D., Stoychev, D., Stefanov, P., Marinova, Ts., (2006). Effect of Al and Ce oxide layers electrodeposited on OC 4004 stainless steel on its corrosion characteristics in acid media. *Corrosion Science,* Vol.48,pp.4037-4052,ISSN:0010-938X

Stoyanova, E., Guergova, D., Stoychev, D., Avramova, I., Stefanov, P., (2010). Passivity of OC404 steel modified electrochemically with CeO_2 - Ce_2O_3 oxide layers in sulfuric acid media. *Electrochimica Acta,* Vol. 55, No. 5, pp. 1725 – 1732, ISSN: 0013 - 4686

Stoychev, D., Ikonomov, I., Robinson, K., Stoycheva, M., Marinova, T., (2000). Surface modification of porous zirconia layers by electrochemical deposition of small amounts of Cu, Co and Co+Cu. *Surface and Interface Analysis,* Vol. 30, pp. 69-73, ISSN: 1096 - 9918

Stoychev, D., Valov, I., Stefanov, P., Atanasova, G., Stoycheva, M., Marinova, T., (2003). Electrochemical growth of thin La_2O_3 films on oxide and metal surfaces, *Materials Science and Engineering* C, Vol. 23, No. 1-2, pp. 123-128 ISSN: 0928 - 4931

Stoychev, D., Stefanov, P., Nikolova, D., Aleksandrova, A., Atanasova, G., Marinova, T., (2004). Preparation of Al_2O_3 thin films on stainless steel by electrochemical deposition. *Surface and Coatings Technology*, Vol.180-181, pp.441-445, ISSN:0257-8972

Tomashov, N.D., Chernova, G.P., (1965). *Passivity and Protection of Metals from Corrosion*, (in Russian) Publ. House "Nauka", Moscow

Tomashov, N.D., Chernova, G.P., (1993). *Theory of Corrosion and Corrosion-stable Materials*, (in Russion) Publ. House "Metallurgia", Moscow

Trovarelli A., (1996). Catalytic properties of ceria and CeO_2-containing materials, *Catalysis Reviews - Science and Engineering* , Vol. 38, No. 4, pp. 439-520, ISSN: 0161 - 4940

Tsanev, A., Iliev, P., Petrov, K., Stefanov, P., Stoychev, D., (2008). Electrocatalytical activity of electrodeposited Zr-Ce-Y/Ni and Co/Zr-Ce-Y/Ni oxide systems at evolution of hydrogen and oxygen. Bulgarian Chemical Communications, Vol. 40, No. 3, pp. 348-354

Tyuliev, G., Panayotov, D., Avramova, I., Stoychev, D., Marinova T., (2002). Thin-film coating Cu-Co oxide catalyst on lanthana/zirconia films electrodeposited on stainless steel. *Materials Science and Engineering C*, Vol.23 No1-2, pp.117-121, ISSN: 0928 - 4931

Valov, I., Stoychev, D., Marinova, T., (2002). Study of the kinetics of processes during electrochemical deposition of zirconia from nonaqueous electrolytes. *Electrochimica Acta*, Vol. 47, No. 28, pp. 4419-4431, ISSN: 0013-4686

Virtanen, S., Ives, M., Sproule, G., Schmuki, P., Graham, M. (1997). A surface analytical and electrochemical study on the role of cerium in the chemical surface treatment of stainless steels. *Corrosion Science*, Vol. 39, No. 10-11, pp. 1897-1913, ISSN: 0010-938X

Wang, K.L., Zhang, Q.B., Sun, M.L., Zhu, Y.M. (1997). Effect of laser surface cladding of ceria on the corrosion of nikel –based alloys. *Surface and Coatings Technology*, Vol. 96, No. 2-3, pp.267-271, ISSN: 0257-8972

Wang, Ch., Jiang, F., Wang, F. (2004). The characterization and corrosion resistance of cerium chemical conversion coatings for 304 stainless steel. *Corrosion Science*, Vol. 46, No. 1, pp. 75-89, ISSN: 0010-938X

Wilcox, G.D., Gabe D.R. (1987). Passivation studies using group VIA anions. V. Cathodic treatment of zinc, *British Corrosion Journal.*, Vol. 22, pp. 254-256, ISSN: 0007-0599

Wilcox, G.D., Gabe, D.R., Warwick, M.E. (1988). The development of passivation coatings by cathodic reduction in sodium molybdate solutions. *Corrosion Science*, Vol. 28, No. 6, pp. 577-585, ISSN: 0010-938X

Zaki N. (1988). Chromate conversion coating for zinc. *Metal Finishing*, Vol. 86, pp.75-83 ISSN: 0026-0576

Zheludkevich, M.L., Serra, R., Montemor, M. F., Fereira, M.G. (2005). Oxide nanoparticle reservoirs for storage and prolonged release of the corrosion inhibitors. *Electrochemistry Communications*, Vol. 7, No. 8, pp. 836-840, ISSN: 1388 - 2481

Zheludkevich, M.L., Serra, R., Montemor, M.F., Ferreira, M.G., (2006) Corrosion protective properties on nanostructured sol-gel hybrid coatings to AA2024-T3. *Surface and Coatings Technology*, Vol. 200, No. 9, pp. 3084-3094, ISSN: 0257-8972

12

Corrosion of Metal – Oxide Systems

Ramesh K. Guduru and Pravansu S. Mohanty
University of Michigan, Dearborn, Michigan
USA

1. Introduction

Corrosion of materials occurs because of several factors; for example the application environment, operational conditions, presence of non-equilibrium phases, failure of the protective phases or layers in the materials, etc. In addition to the electro-chemical phenomena occurring in the corrosion process, operational conditions, such as temperature could influence the corrosion rates to different degrees depending on the materials involved. The effect of temperature is known to be severe on the corrosion phenomenon due to the dependence of corrosion rates on diffusion of materials. From the materials perspective, presence of non-equilibrium phases or second phases and their thermodynamic stability, microstructures, properties, and protective layers could affect the corrosion rates. Usually oxide systems are known for their protective behavior because of their stability and hindrance to the diffusion of different ionic species. Understanding their stability and role in prevention or slowing down of corrosion rates is, therefore, very important for engineers to design new material systems with desired properties and structures for corrosion resistant applications. Although metallic alloys with oxide second phase are extensively used in high temperature applications for creep resistance, literature suggests that addition of different kinds of oxide particles could help control the corrosion properties. In this chapter, an overview will be given on the corrosion behavior of different oxide systems and their role in corrosion resistant applications of the oxide particle embedded metallic systems in different environments, including low and high temperature applications.

2. Corrosion process and inhibition

Corrosion is a continuous degradation process of a material. As shown in figure 1, the corrosion of a given material system can take place because of two external major components, namely the environment or the electrochemical system (eg: atmosphere, acid or corrosive media), and operating conditions shown by arrows (eg: stress or pressure, erosion and temperature etc.). The process of electrochemical corrosion occurs in multiple steps, where the ions are involved with a media for ionic motion, and at the same time the material involved should be conductive enough to participate in the electron transfer for a mutual charge transfer process due to the ionic motion. During the process of corrosion, the materials can undergo changes into a new form of the material which could be protective or reactive in further process. The driving force for the corrosion is usually the thermodynamic instability of a given material system in the superimposed surroundings and working conditions.

There are different types of corrosion that can take place on a material system and they could be uniform type or localized type. In uniform corrosion, as the name suggests, corrosion takes place all over the surface. On the other hand, localized corrosion can be several kinds, such as galvanic corrosion, pitting corrosion, selective attack, stray current corrosion, microbial corrosion, intergranular corrosion, crevice corrosion, thermo galvanic corrosion, corrosion due to fatigue, fretting corrosion, stress corrosion, hydrogen damage etc. (Jones, 1992). For more details on each process, the reader is suggested to refer to any review articles or books on corrosion science.

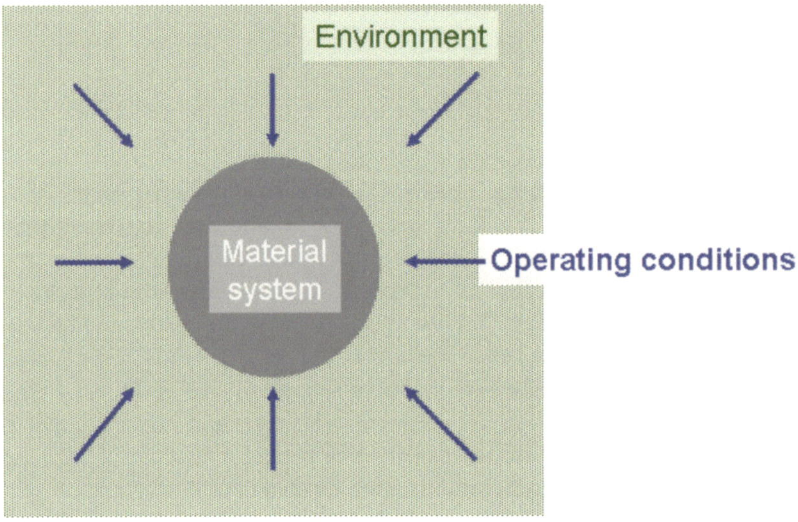

Fig. 1. Schematic for corrosion of a material system with the components involved in the corrosion process.

Different factors contribute to the corrosion under various situations during the service. For example, the components used in hot sections of gas turbine engines and hypersonic vehicles operate under extremely oxidizing, erosive and high temperature conditions, where a combination of high temperature mechanical strength along with excellent oxidation and corrosion resistance are required. In the applications related to marine and aircraft propulsion systems, quite corrosive and erosive environments exist around the components under different operational temperatures with cyclic nature. Therefore depending on the application, the surroundings and operational conditions vary; and usually high strength and high temperature protective coatings are used to meet the requirements of such harsh operating conditions.

There are different approaches adopted to reduce or slow down the corrosion of a material system depending on the type of application and corrosive environment. The simplest and preferred approach among all of the methods is through the application of protective or non-reactive phases over the material system in the form of coatings, which keeps the material from exposure to the surroundings. A classic example for an oxide layer assisted corrosion resistant alloy is stainless steel, in which the alloying element chromium (Cr) forms an impervious stable oxide layer (Cr_2O_3, also called chromia) along the grain boundaries and surface, as shown in the schematic in Fig. 2. Usually grain boundaries are prone to corrosion attack because of defects and high energy sites unless they are protected via passivation. Although, chromia cannot be used alone because of brittleness, chromium enhances passivity when alloyed with other metals and alloying elements in stainless steel.

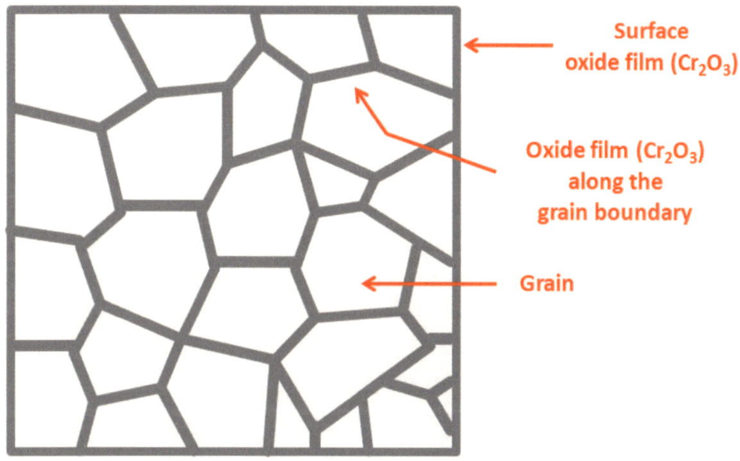

Fig. 2. Schematic for a thin passive layer of chromia (Cr_2O_3) along the grain boundaries as well as on the surface of stainless steel for corrosion protection.

As another example, Aluminum (Al) and Al - alloys could be discussed. However, the corrosion resistance of Al and its alloys can be attributed to the formation of passive oxide layer on their surfaces alone (Jones, 1992), as shown in the schematic in Fig. 3.

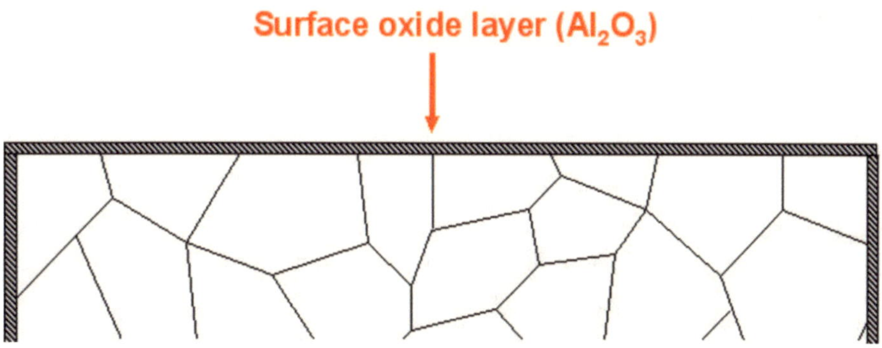

Fig. 3. Schematic for a thin passive layer of alumina (Al_2O_3) on the surface of the Aluminum and/or Aluminum alloys for corrosion protection.

There are a number of oxide systems as protective coatings, as well as dispersoids, demonstrating superior performance in terms of corrosion and other properties, which will be reviewed later in this chapter. However, in cyclic operating conditions with temperature fluctuations and wear conditions the oxide layers may not be suitable; as they can break down due to mismatch of the thermal coefficient of expansion (CTE) with underneath phases, or due to wear, or combination of both, and thereby lead to localized pitting, crevice corrosion, etc., of the underlying substrate. In addition, high temperatures can enhance the diffusion rates. To this end, protective coatings with oxide particle embedded systems are found to be more useful. Most of the high temperature coatings and oxide dispersion strengthened (ODS) alloys are embedded with highly stable oxide phases, which can provide mechanical stability as well as enhance the corrosion and oxidation resistance. In addition to the oxide dispersoids, ODS alloys employ alloying elements (eg: Cr, Al etc.) in such a way that the oxide layers are formed on the surfaces as well as at the grain boundaries at high temperatures during the operation, which then act as protective layers from the corrosion point of view. The oxide dispersoids in the ODS alloys can provide mechanical stability with improved creep resistance.

Here, we will touch base on the corrosion phenomenon of oxide layer and oxide particle assisted corrosion behavior of metallic materials at low and high temperature applications with a brief review, and a case study will be presented on the corrosion phenomenon of oxide particle embedded high temperature composite coatings developed by thermal spray technique.

3. Oxide particle embedded metallic systems

Metal - oxide dispersed systems are well known for excellent mechanical properties because of high strength of the reinforcing ceramic oxide phases. Dispersion of hard oxide particles also enhances the surface properties, such as hardness and wear resistance, which are critical for tribological applications. The oxide particles also improve high temperature creep strength of the metallic materials by acting as obstacles to dislocations, reducing the deformation along the grain boundaries due to the diffusion processes or grain boundary rolling mechanism by pinning the grain boundaries. Thus the oxide particle embedded metallic systems have a vital role in many applications. Their processing is usually done in many routes depending on the type of application as well as the amount of material required. Here, we will briefly go through some of these techniques to introduce the reader to different processing routes. However, for more information one can refer to the literature and review articles on composite processing techniques. Following is the list of a few approaches usually employed to develop the metal - oxide composite systems (Kainer, 2006).

Powder processing route: In this approach, metal and oxide powders are blended together using different methods (eg: ball milling or mechanical alloying) and then compacted and sintered or consolidated into required shapes or bulk solids.

Melting route: There are different number of processes fall under this category that involve molten metals. This route is usually applied to low melting metal matrix composites, in which the metal ingots or pieces are melted and then the oxide particles are dispersed in the molten metal prior to solidification.

Electrodeposition: The oxide particles are suspended in an electrolyte which helps develop the matrix coating. Suspension of oxide particles along with continuous stirring in the electrolyte can embed the particles in the metal matrix during electrodeposition process.

Vapor deposition: Physical vapor deposition techniques (eg: electron beam evaporation, sputtering etc.) can be used to develop composite coatings using multiple targets in co-deposition approach with intermittent reactive deposition process.

Spray deposition: Different number of processes have evolved in this category in which a stream of molten metal droplets are deposited on a substrate to build the matrix layer; and for composites, the oxide particles are co-sprayed to embed them in the matrix layers.

Reactive formation: In this approach, selective oxidation of certain phases in the bulk structures with exothermic reactions results in the in-situ formation of composites.

As listed above there are several approaches available for processing metal - oxide systems, and their corrosion properties are going to be dependent on the processing technique employed too. For example, the processing defects like porosity, improper bonding between the matrix and the oxide dispersoids, and their interfacial properties can influence the corrosion behavior quite extensively. Wetting of the oxide particles becomes a critical factor in some of the processing approaches to deal with the particle - matrix bonding. Fig. 4 shows a schematic for interfacial bonding of the second phase particles with matrix along the grain boundaries and triple junctions. In addition, high temperatures in some of the processing techniques may cause an interfacial reaction between the metal matrix and the dispersed second phase particles, thereby the interfacial stability and its properties play an increasingly important role in the corrosion. It is also possible that the interfaces could become prone to corrosion attack by providing preferential sites. In spray deposition approach splat boundaries, porosity, and distribution of the oxide particles may play an important role in deciding the corrosion properties. Added to that, the microstructures of the composites could also vary from process to process. The effect of some of these parameters on the corrosion of different metal - oxide systems is discussed in brief in the following sections.

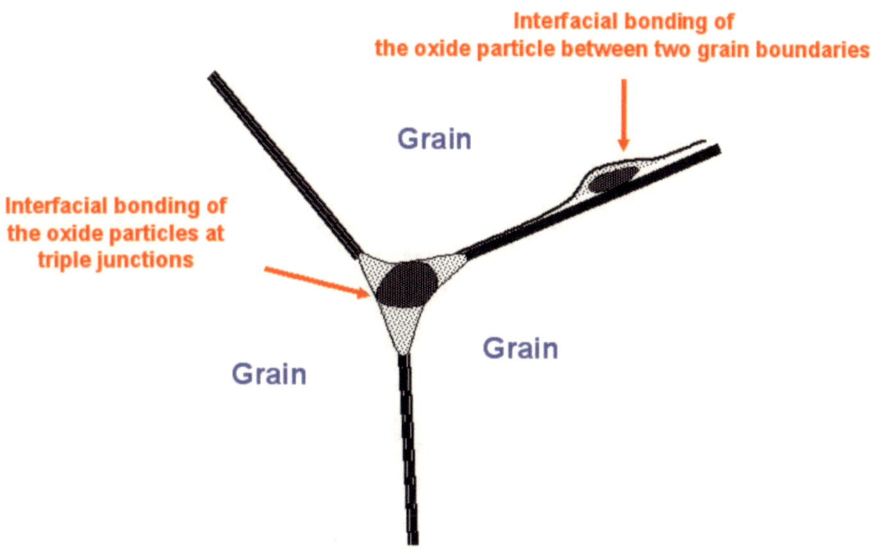

Fig. 4. Schematic for interfacial bonding of second phase particles at grain boundaries and triple junctions.

4. Overview of corrosion phenomena of metal – Oxide systems

Although, the metal matrix composites are well suited for mechanical, tribological and high temperature applications, it is to be clearly noted that their corrosion aspects could be considerably different, as well as complex, compared to the monolithic metallic systems. Corrosion of metal matrix composites could arise due to different reasons, such as electrochemical and chemical interaction between the constituent phases, microstructural effects, and possibly from processing related issues too (Cramer & Covino, 2005). Usually, composites have higher tendency to corrode because of the multiphase structure with metal matrix. If the second phase structure is conductive a galvanic cell can be formed within the system. For example, metallic composites reinforced with graphite or semi-conducting silicon carbide could undergo severe corrosion compared to the pure metals. Galvanic corrosion is not a problem if the second phase dispersoids are insulating, for example, oxide particles.

It is also very important that the second phase particles be uniformly distributed in the metal matrix. The effect of oxide particle size, volume fraction and their pretreatment can also influence the corrosion phenomenon. The other important factors that can contribute to the corrosion are surface morphology, porosity, stresses, bonding, defects at the matrix and dispersoid interfaces, crystallographic structure, and the type of oxide phase dispersed. For example, bonding between Al and Al_2O_3 (alumina) plays a crucial role in the corrosion of Al - Al_2O_3 composites. Usually the corrosion rate of the composites is measured by weight loss and the corrosion studies conducted on Al - Al_2O_3 composites in NaCl solution for prolonged periods showed considerable weight loss due to pits or microcrevice formation in the matrix near the particle-matrix interfaces, as well as from the particle dropout. The corrosion via pit initiation and propagation was determined to be due to the weak spots in the air-formed Al_2O_3 film because of the discontinuities and the second phase particles (Nunes & Ramanathan, 1995). In the case of 6061-T6 alloy mixed with 10 vol% Al_2O_3, poor corrosion resistance was reported to be due to poor bonding at the matrix and oxide particle interfaces (Bertolini et al., 1999). The Al alloys AA 6061 and AA 2014 embedded with Al_2O_3 particles exhibited stress-corrosion cracking when subjected to three-point beam bending along with alternate or continuous exposure to NaCl solution (Monticelli et al., 1997). Although addition of Al_2O_3 may seem to be detrimental in terms of corrosion resistance of Al alloys, with the combination of wear and corrosive conditions, the corrosion resistance of 6061 and 7075 Al alloys was observed to improve with the Al_2O_3 second phase dispersion (Fang et al., 1999; Varma & Vasquez, 2003) along with the enhancement of wear resistance. In marine biological applications, the microbial corrosion was also reported to occur in the Al - Al_2O_3 composites due to biofilm formation at the interfaces of Al and Al_2O_3 particles (Vaidya et al., 1997). In environmental and marine biological applications, the protective chromium oxides are not very benign because of their toxicity and as a result usage of chromia coatings is restricted. However, different rare earth oxides were proposed as alternatives for protection of Al alloys because of their cathodic inhibition properties (Aramaki, 2001; Hamdy et al., 2006; Hinton et al., 1986, 1987; Lin et al., 1992). Usually rare earth oxides are very useful for aerospace applications because of their high temperature oxidation resistance. According to Hamdy et al., (2006) CeO_2 (ceria) treated Al alloys exhibited improved corrosion resistance due to oxide layer thickening. Muhamed & Shibli (2007) also showed improved corrosion performance of Al – CeO_2 composites, but it was not in proportion to the amount of CeO_2 incorporated.

Presence of rare earth oxides was proved to enhance the corrosion resistance of Ni composites also. It was reported that the Ni matrix reinforced with micron CeO_2 particles possessed good corrosion resistance compared to Ni - ZrO_2, Ni - partially stabilized ZrO_2 (PSZ), and pure Ni coatings (Qu et al., 2006). Although the corrosion process usually proceeds along the grain boundaries, in the case of Ni - CeO_2 composites the corrosion path was observed to be preferentially along the Ni/CeO_2 interfaces, instead of Ni grain boundaries. Along with that, higher corrosion resistance of CeO_2 was also observed to enhance the corrosion resistance of Ni/CeO_2 interface. Also, codeposition of CeO_2 particles induced the formation of small equiaxed Ni grains, which resulted in the corrosion along less straight paths and thus lowering the corrosion rates in Ni - CeO_2 composites (Aruna et al., 2006). It is also considered that when CeO_2 nanosized particles are embedded in the nickel matrix, the corrosion path is more seriously distorted as compared to micro-sized particles, which is favorable for corrosion resistance. In fact, the fine grain structure arising from the co-electrodeposition of CeO_2 nanoparticles also promotes good corrosion resistance as compared to coarse grain structure (Qu et al., 2006).

Aruna et al. (2009) showed enhanced performance of wear and corrosion characteristics of Ni based composite coatings by embedding with alumina yttria doped cubic zirconia (AZY, $(1-x)Al_2O_3$-8 mol% yttria stabilized $xZrO_2$ (x = 10 wt%)) particles. The higher Warburg resistance of Ni - AZY and enhanced corrosion resistance was attributed to possible difference in mass transport phenomena in the Ni -AZY composites compared to the pure Ni with increased resistance of Ni grain boundaries in presence of AZY particles and thereby hindered the diffusion of chloride ions (Aruna et al., 2009).

In other examples, Li et al. (2005) demonstrated the effect of the type of oxide particles dispersed on the corrosion behavior of Ni composites. Li et al. (2005) developed nanocomposite coatings consisting of TiO_2 in the form of anatase and rutile in Ni matrix via electrochemical deposition technique, and showed improved corrosion properties of Ni - TiO_2 composites compared to the pure Ni; however, the improvement in corrosion resistance was predominant in the case of anatase dispersed Ni composites. Improved corrosion resistance of Ni – TiO_2 composites was attributed to the inhabitant behavior of TiO_2 particles at the grain boundaries and triple junctions, which are the usual sites for corrosion attack. With an increase in the amount of TiO_2, a decrease in the corrosion rates was also demonstrated because of the increased number of inhabited sites, which reduce penetration of the corrosive solution into the composite coatings. On the other hand, Ni - Al_2O_3 composite coatings (Erler et al., 2003) reported to show poor corrosion resistance compared to the monolithic Ni. Szczygieł and Kołodziej (2005) indicated that the lower corrosion resistance of Ni - Al_2O_3 could be due to poor bonding between the oxide particles and the matrix, which can increase the possibility of dissolution of loosely held Al_2O_3 (alumina) particles at high potentials and result in more nickel exposure to the electrolyte for corrosion attack. In another study by Aruna et al. (2011) the corrosion properties of Al_2O_3 embedded Ni composites showed the oxide phase dependent corrosion performance. Their studies indicated that the corrosion resistance of Ni - α Al_2O_3 was better than the corrosion resistance of Ni – γ Al_2O_3 as well as the Ni – α and γ Al_2O_3 mixture; however, the reason for such behavior was not explained.

At high-temperatures the corrosion failure of a material system results from failure of its protective oxide scale. Different researchers have proved that addition of a small amount of reactive elements (such as Y, Ce, La, and Hf), or their oxides, improves the oxidation resistance of some high temperature alloys by decreasing the growth rate of the oxide and increasing the adherence of the oxide scale to the underlying alloys (Peng et al., 1995). Addition of Y_2O_3, CeO_2, ThO_2, La_2O_3 and Al_2O_3 to Ni - Cr alloys, and Y_2O_3, HfO_2, ZrO_2 and TiO_2 to Co - Cr alloys may promote the formation of Cr_2O_3 protective oxide scale as well as increase its adherence to the ODS alloy system very effectively (Michels, 1976; Stringer & Wright 1972; Stringer et al., 1972; Whittle et al., 1977; Wright et al., 1975). In Ni - 20Cr - Y_2O_3 ODS alloy coatings, presence of Y_2O_3 was observed to promote the formation of Cr_2O_3 scale and thereby the improvement in scale spallation (Lianga et al., 2004). Stringer et al. (1972) proposed that the dispersed oxide particles act as heterogeneous nucleation sites for Cr_2O_3 grains and reduce the internuclear distance for the Cr_2O_3 scale formation, which will allow rapid formation of a continuous Cr oxide film with a finer grain size. The oxide layer with fine grain size can then easily release the thermal stress and therefore prevent crack propagation. Extensive experimental results and detailed mechanistic studies have indicated that the effects of dispersed oxides seem to be independent of the choice of the oxides, as long as they are not less stable than Cr_2O_3 (Lang et al., 1991). According to this mechanism, dispersion of above mentioned oxides expected to be most effective in enhancing Cr_2O_3 scale formation and thus lead to improved resistance to hot corrosion most effectively. According to He and Stott (1996) a short-circuited diffusion of Cr reduced the concentration of Cr in the alloy and thereby facilitated formation of Cr_2O_3 in Ni - 10Cr alloy with presence of Al_2O_3 and Y_2O_3 particles. Quadakkers et al. (1989) reported that Y_2O_3 incorporation in ODS alloys retarded the diffusion of Cr because of prevailing anionic diffusion over cationic diffusion. This mechanism was also supported by Ikeda et al. (1993), who also confirmed that the adhesion of Al_2O_3 could be promoted by the dispersed Y_2O_3 phase in ODS alloys.

According to Carl Lowell et al. (1982), the oxidation and corrosion resistance of ODS alloys was superior compared to the superalloys. However, different corrosion behavior among different ODS alloys, for example Ni based (NiCrAl) and Fe based (FeCrAl) ODS alloys, was attributed to the CTE mismatch and therefore the spallation resistance. Usually lower CTE mismatch between the ferritic ODS alloys and protective alumina film helps reduce the amount of stresses in the oxide during thermal cycling and thereby considerably less, or no, spalling. In contrast, the high CTE of Ni - based ODS alloys directly leads to spalling during cycling from 1100 ^{0}C to room temperature. Similarly, better oxidation and hot salt corrosion behavior is expected for Fe - based ODS alloys compared to the Ni - based ODS alloys. Therefore, it is apparent that the corrosion behavior of ODS alloys is highly dependent on the protective oxide layers formed during the high temperatures compared to the oxide particles embedded within the alloys, unlike the metal – oxide composites; however, formation of a uniform protective oxide scale could be dependent on the embedded oxide particles in the metal matrix. Thus, presence of oxide particles in a metal matrix can directly, as well as indirectly, help enhance the corrosion properties of different alloys and composite systems.

5. Case study on high temperature coatings developed by spray deposition

This case study presents synthesis and characterization of oxide particle embedded high temperature coatings developed by thermal spray technique, which is one of the processing routes discussed in the Section - 3, for boiler coating applications.

As discussed in the earlier sections, high temperature coatings are ubiquitous to industrial power generation, marine applications, and aircraft propulsion systems. Most high temperature coatings operate under extremely harsh conditions with conflicting operational requirements. For instance, coatings used in power plant boilers need to ensure an effective protection against high temperature corrosion under oxidizing, sulfidizing, carburizing environments and erosion from fly ash, as well as having a high thermal conductivity to exchange heat in order to provide an effective and economical maintenance. Further, to avoid premature failure, as discussed in the previously discussed overview section, the high temperature coatings also require good adhesion to the substrate, minimal mismatch in CTE between the coating and the substrate material, good thermal fatigue, and creep resistance (Bose, 2007; Patnaik, 1989; Uusitalo et al., 2004; Yoshiba, 1993; Yu et al., 2002).

Most commercial coating systems do not meet all the required attributes for a given environment. For example, NiCr (55/45 wt.%) alloy is usually recommended for erosion–corrosion protection for boiler tubes in power generation applications (Higuera, 1997; Martinez-Villafan et al., 1998; Meadowcroft, 1987; Stack et al., 1995). Weld overlay coatings of Alloy 625(Ni-21Cr-9Mo-3.5Nb) have also been used for this application. When nickel is alloyed with chromium (>15wt%), Cr oxidizes to Cr_2O_3, which could make it suitable for use up to about 1200°C (Goward, 1986), although in practice its use is limited to temperatures below about 800°C. The efficacy of NiCr coatings deteriorates severely when molten ash deposits consisting of sodium-potassium-iron tri-sulfates $(Na,K)_3Fe(SO_4)_3$ are present. Further, higher Cr content also reduces the creep resistance of NiCr alloys. Particularly, this issue becomes magnified in the case of thermal spray coatings. In addition to the grain boundaries, presence of splat boundaries, an inherent feature in thermal sprayed coatings also contributes to poor corrosion and creep performance at very high temperatures (Soltani et al., 2008; Zhu & Miller, 1997). Thus, from the materials perspective, the corrosion is influenced by several parameters, for example surface and bulk microstructures, thermodynamic stability of the phases, microstructural constituents, electrochemical potentials, protective phases and residual stresses etc. Thereby, it becomes user's responsibility to select an appropriate material system for a given operating condition either to avoid or slow down the deterioration during the service period.

The continued pursuit for increased efficiency in power generation and propulsion systems led to the development of functionally engineered coatings with multiple attributes. For example, an alternative method of combating the effects of coal ash corrosion is to install a material that contains sufficient amount of oxide stabilizing elements such as aluminum or silicon (NiCrAl, NiCrBSi NiCrMoBSi and NiCrBSiFe) to resist the dissolution of the oxide film when the molten ash is deposited. Similarly, functionally gradient materials (FGM) were proposed (Niino & Maeda, 1990) to obtain multifunctional properties with a combination of different metallic and ceramic systems in an engineered fashion. These materials were found to be very promising candidates for high temperature applications because of the reduced thermal stresses between the interfaces, resulting in enhanced thermal fatigue life (Bahr et al., 2003). The high temperature creep strength of metals is also greatly improved by the addition of high temperature stable dispersoid phases, due to grain boundary pinning such as the oxide dispersion strengthened super alloys ($Ni-ThO_2$ and $NiCr-ThO_2$) (Clauer & Wilcox, 1972).

Various approaches have been adopted to disperse the second phase particles into bulk matrix phase, such as mechanical alloying/powder metallurgy (Kang & Chan, 2004), in situ formation of dispersoids via a chemical reaction within the matrix phase (Cui et al., 2000), spray synthesis (Chawla, 1998), casting techniques (Rohatgi et al., 1986) and elecrodeposition (Clark et al., 1997). Processing methods, such as powder metallurgy (Heian et al., 2004; Kawasaki & Watanabe, 1997, 2002) and thermal spraying (Hamatani et al., 2003; Khor et al., 2001, 2003; Polat et al., 2002; Prchlik et al., 2001), cannot easily tailor the composition in a functional manner. Typically, thermal sprayed composite coatings are made using premixed powders with a given ratio of the constituent phases. This limits the production as well as the design flexibility. Further, a spray deposition approach involving direct spraying of nano-sized powders, has a number of limitations (Rao et al., 1997; Skandan et al., 2001). The primary issue is the introduction of nano-sized powders into the high velocity thermal spray jet and their impingement on the substrate. Nano-sized powders tend to agglomerate, resulting in plugged particle feed line, and the extremely small particles do not readily penetrate the jet. Also, impingement on to the substrate is difficult as the small powders follow the gas streamlines. An alternative methodology is to introduce a liquid or gaseous precursor, which reacts in flight to form nanosized particles (Rao et al., 1997; Xie et al., 2004). This approach is very promising, and has worked well for several material systems. Combustion synthesis using liquid precursors has been used to deposit a number of different high temperature oxide coatings, including Al_2O_3, Cr_2O_3, SiO_2, CeO_2, some spinel oxides ($MgAl_2O_4$, $NiAl_2O_4$), and yttria stabilized zirconia (YSZ) (Hampikian & Carter 1999). For example, using a solution of aluminum acetylacetonate in ethanol, alumina was deposited at temperatures of approximately 850, 1050, and 1250°C (Hendrick et al., 1998). Similarly, SiO_2 has been deposited by combustion synthesis of ethanol containing tetrathyloxysilicate precursor.

As for the production of nanoparticle dispersed microcrystalline coating by thermal spray technique, different approaches have been adopted, such as agglomeration of nano-sized particles with a binder used in the Co-WC cermet (Skandan et al., 2001) or premixing of dispersoid phase with the matrix powder (Laha et al., 2004, 2007; Bakshi et al., 2008). However, these approaches also suffer from the same design inflexibility mentioned above. This case study presents an innovative approach to synthesize ultrafine/nano particulate dispersed (Al_2O_3, SiO_2) NiCr alloy coatings. A novel process called "Hybrid Spray Technique" (Kosikowski et al., 2005) has been employed to fabricate these functionally engineered coatings in a single step. The rationale behind the selection of the dispersoid phases, their liquid precursors and the particulate distribution layout is presented. The influence of these dispersoid phases on the functional characteristics of the resulting coatings is discussed.

5.1 Processing and testing of high temperature coatings

The "hybrid spray" process utilized in this study was conceptualized in our laboratory at the University of Michigan (Kosikowski et al., 2005). This process combines the arc and high velocity oxy fuel (HVOF) spray techniques; molten metal at the arcing tip is atomized and rapidly propelled to the substrate by a HVOF jet. This so called "hybrid" concept shown in Fig. 5 offers many advantages.

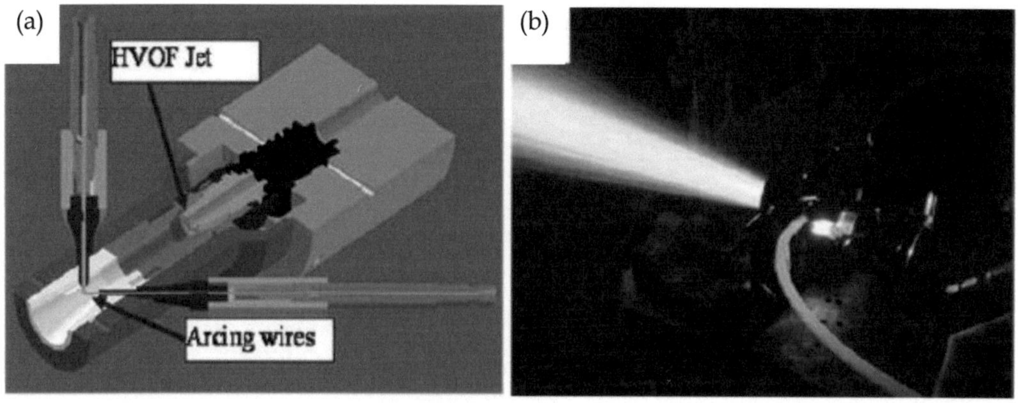

Fig. 5. (a) Schematic of hybrid gun and (b) hybrid gun in operation.

The hybrid process offers all benefits of wire stock and productivity of electric arc spraying combined with noticeably improved coating density of HVOF. In addition to introducing material through arcing mechanism, if desired, powder/liquid/gas precursors can also be fed through the HVOF coaxial feed line (Fig. 5a). This enables us to tailor the composition inflight by introducing particles into the HVOF jet, to cater to specific property requirements of a composite coating. This unique capability completely eliminates the necessity of processing and handling of the ultrafine particulates prior to feeding them into the hybrid gun. Synthesizing and introducing ultrafine and nano dispersoids inflight in a functional manner to produce composite coatings by the hybrid technique is quite unique in terms of simplicity compared to any other processes. A comparative picture of the steps involved in processing of particulate reinforced composites by conventional routes versus our approach is presented in Table 1.

Target Material	**Precursor Materials**	**Percentage**
NiCr-Matrix	NiCr wire	(55/45 wt. %)
Al_2O_3 particulate	Aluminum nitrate -$Al(NO_3)_3 \cdot 9H_2O$	1:1 by weight in isopropyl alcohol (70%)
Cr_2O_3 Stabilizer	Chromium nitrate -$Cr(NO_3)_3 \cdot 9H_2O$	Up to 50% by weight of aluminum nitrate
SiO_2 particulate	Tetraethoxysilane	100 %

Table 1. List of precursors used.

Following the above mentioned approach of inflight synthesis, different oxide ceramic particles were introduced into the NiCr (55/45 wt.%) alloy coating. The following coatings were deposited onto mild steel coupons for characterization: (a) NiCr only, (b) NiCr + Cr_2O_3, (c) NiCr + Al_2O_3, and (d) NiCr + SiO_2. Along with these coatings, NiCr coatings using a twin wire arc spray process (TAFA 3830, Praxair Surface Technologies, Indianapolis, IN) were also deposited for comparison purposes. The arc current and voltage for both the processes were kept at 100 amps and 36 volts, respectively. The HVOF gas pressures were maintained at 50/65/80 psi of propylene/oxygen/air, respectively. The Aluminum nitrate and Tetraethoxysilane precursors were fed from separate reservoirs, however, they were mixed together prior to the injection into the combustion jet. The atomization of the liquid was achieved by a two fluid injector. Liquid precursors up to 100 cc/min were fed to the HVOF jet coaxially.

Table 1 lists the liquid precursors employed for the synthesis of the dispersoid phase particles. The rationale behind the selection of the dispersoids (SiO_2, and Al_2O_3,) and their influence on the properties is as follows.

- The silica particles are expected to provide both creep and crack resistance. It has also been demonstrated that the presence of SiO_2 enhances the high temperature resistance of chromia scale (Carter et al., 1995; Liu et al., 2004).
- The presence of alumina is expected to provide enhanced high temperature corrosion resistance. Also, the introduction of SiO_2 into alumina based coatings has been found to form mullite and reduce the cracking within the coating (Marple et al., 2001). Mullite is known for its excellent creep resistance (Dokko et al., 1977; Lessing et al., 1975).
- It has also been found that the presence of chromia aids in α - alumina formation, as well as limits the phase transformations during heating to temperatures below 1200 ^{0}C (Marple et al., 2001; Chraska et al., 1997). Therefore, chromium nitrate was added to the aluminum nitrate precursor to stabilize the α - alumina phase.

Microstructural analysis of the coatings was done using electron microscopy (SEM/TEM). The oxidation characteristics of the coatings were characterized on a TA instruments SDT Q600 model for thermogravimetric analysis (TGA). For functional property characterization, coatings were tested for hot erosion, wet corrosion and hot corrosion; and compared with 304 stainless steel, as well as alloy 625 overlay cladding. The hot erosion test setup consisted of a grip for holding and rotating (80 rpm) the coated samples while heating with a heat source (HVOF flame), and an alumina grit (250 mesh) delivering system at a fixed angle (45^0) as shown in Fig. 6a. The flow rate of the grit was 60 gm/min and the applied grit carrier air pressure was 15 psi at a rate of 42 SCFM. Testing was done at 750 ^{0}C for 3 minutes on spray-coated cylinders. Wet corrosion tests were done at room temperature in a dilute 0.1% NaCl solution. NiCr coatings sprayed by the hybrid and arc techniques were tested using an electrochemical cell shown in Fig 6b. Electrochemical experiments were performed using a Solartron (Hampshire, England) SI 1287 potentiostat at the open circuit potential for two different time periods (0hrs and 24 hrs). Hot corrosion tests were carried out by applying film of sulfates and chlorides (potassium, sodium and iron) on to the surface of coated samples (304 stainless steel caps) as shown in Fig. 6c. Samples were initially weighed, and then their surfaces were coated with a solution of sulfate/chloride mixed with water in a weight ratio of 1:1. The samples were carefully masked to ensure salt solution only covered the sprayed coating and the area coated with salt solution was also measured. The solution was dried to leave a film of salt on the surface of the sample. The masking material was removed and the sample was weighed again. Samples were then placed in an oven at 900^0 C for 24 hours. This test also included samples of bare 304 stainless steel cap as well as alloy 625 overlay cladding. After the hot corrosion test, weight loss/gain of the samples was measured to evaluate the corrosion resistance.

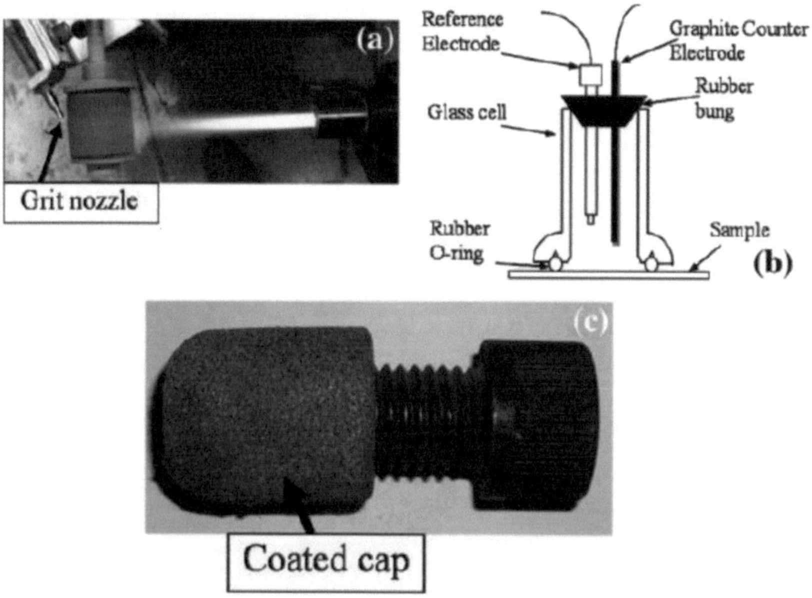

Fig. 6. (a) Hot erosion test set up, (b) corrosion cell, and (c) sample for hot corrosion test.

5.2 Microstructural analysis of high temperature coatings

Fig. 7a, presents the general cross section microstructure of a NiCr coatings with embedded alumina particles produced by the hybrid spray process. The coating is very dense and exhibits the characteristics of an HVOF coating rather than of an arc sprayed coating. The hybrid spray process is unique in the sense that while it yields comparable density to that of the HVOF process, the deposition rate is closer to that of an arc spray process. The observed density is advantageous for high temperature corrosion and erosion performance of the coatings. Details on the corrosion and erosion performance of the coatings are discussed in the forthcoming sections. The dispersion of the alumina particles (dark phase) in the NiCr matrix is shown in Fig. 7b.

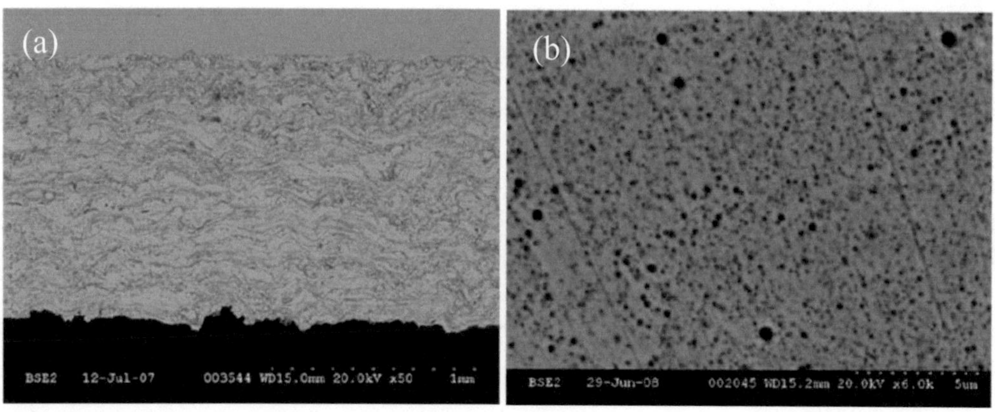

Fig. 7. (a) SEM picture of NiCr coating with dispersed Al_2O_3 particulates and (b) higher magnification SEM picture of NiCr coating with dispersed Al_2O_3 particulates.

The atomization of the liquid precursor (for oxide particles) prior to the injection into the combustion jet plays an important role on the size as well as on the distribution of the particles in the final coating. The requirements for the atomization system include: controlled and uniform flow, ability to operate against a back pressure of 30 psi pressure that exists in HVOF flame at the point of injection and the ability to generate mono-dispersed micron sized droplets. Details on the atomization and optimization of parameters could be found elsewhere (Mohanty et al., 2010). From Fig. 7(b), it is apparent that the distribution of the particles was uniform across the cross section. Similar observations were made in the case of NiCr + Cr_2O_3 and NiCr + SiO_2 systems also. It is to be noted that composites made from premixed powders commonly exhibit large clusters of nanoparticles. Fig. 8 presents the TEM picture of a NiCr coating with embedded silica particles. Many fine particles are observed in the matrix, as well as along the grain boundaries. For enhanced creep resistance resulting from grain boundary pinning, the particles must be small and coherent with the matrix. Especially alloys with very high chrome content can substantially benefit from such ultrafine particle embedment as observed in Fig. 8.

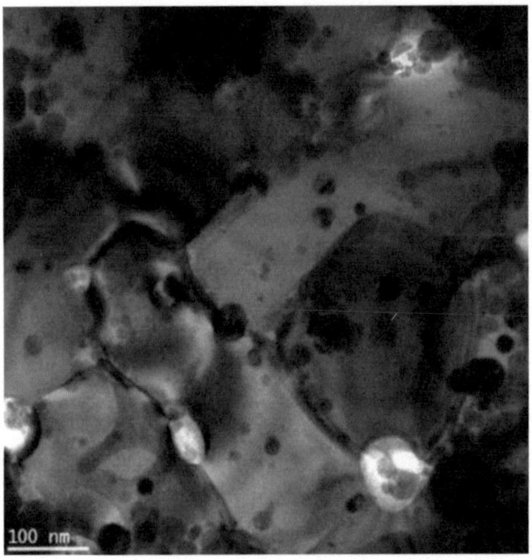

Fig. 8. TEM picture of NiCr coating with dispersed SiO_2 particulates.

5.3 Characterization of high temperature coatings

5.3.1 Oxidation studies

The oxidation characteristics of all the coatings (a), (b), (c) and (d) (refer to page 9) including the arc sprayed NiCr coating, were investigated by TGA studies in air after removing them from the substrate. The TGA curves shown in Fig. 9 indicate an overall weight gain for all the coatings while heating, although there was an initial weight loss for most samples.

The weight gain can be attributed to the oxidation of Cr in the NiCr matrix, as well as the changing oxidation state of the existing oxides. The later phenomenon can also lead to a weight loss in the initial stages because of the changing stoichiometry. Literature (Eschnauer et al., 2008; Hermansson et al., 1986; Richard et al., 1995; Schutz et al., 1991; Vippola et al., 2002) suggests that the oxidation of chromium during thermal spray processes could lead to nonstoichiometric compounds or metastable oxides (CrO_2, CrO and Cr_3O_4) which can undergo changes upon reheating. If CrO_2, which has higher oxygen content compared to Cr_2O_3, forms during the spray process; it can undergo stoichiometric changes to a stable oxide (Cr_2O_3) upon reheating and this could lead to an initial weight loss in the coatings. However, part of the initial weight loss could also be attributed to the evaporation of moisture absorbed by porosity in the coatings. According to Lars Mikkelsen (2003), the specimens may also lose weight due to vaporization of chromium containing species from the chromia scale. Whereas the oxidation of pure Cr to Cr_2O_3 and also the transformation of CrO and Cr_3O_4 to Cr_2O_3 will lead to weight gain because of increasing oxygen content in the coatings.

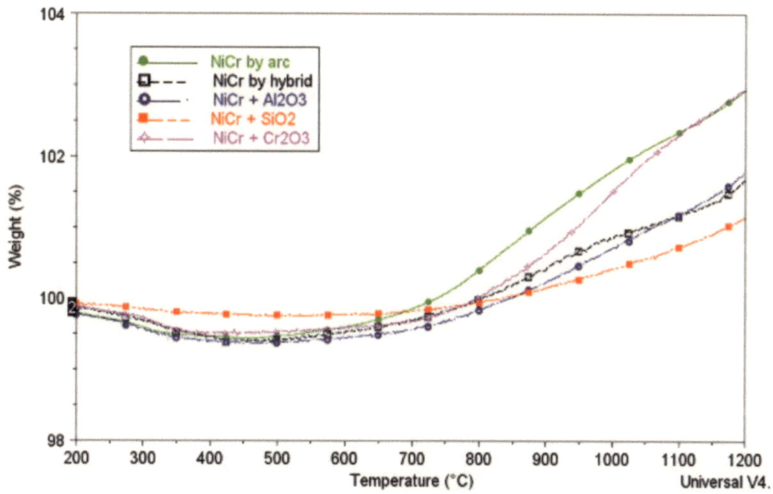

Fig. 9. Weight gain measured using TGA for oxidation studies.

The weight gain for arc sprayed NiCr coating was the highest compared to all other coatings and this could be due to the inherent porosity in the arc spray coatings. The pores in the coatings enhance the oxidation rate. The weight gain in the hybrid NiCr coatings (without any particulate) was much lower than the arc sprayed coating because of their dense splat structure. NiCr + SiO_2 showed the lowest weight gain. The weight gain by NiCr + Al_2O_3 was comparable to that of the plain hybrid NiCr coatings. A large weight gain by the NiCr + Cr_2O_3 could be due to the changes associated in the chromium oxide composition. It is to be noted that there is no need to add Cr_2O_3 particles into NiCr coating using a precursor. The role of chromium nitrate precursor here is to stabilize the α -alumina phase. However, excess addition could lead to undesirable consequence as observed in the case of the NiCr + Cr_2O_3 sample. Determining the appropriate level of chromium nitrate is beyond the scope of this study. From these studies we conclude that addition of SiO_2 has the most remarkable effect on the oxidation behavior of NiCr coatings. It has been demonstrated that the presence of SiO_2 enhances the high temperature resistance of the chromia scale, which helps to improve the oxidation and corrosion resistance of the coatings (Carter et al., 1995).

5.3.2 Hot erosion test

The setup utilized for evaluating the hot erosion behavior was shown in Fig. 6a. The weight of cylinders was measured before and after the hot erosion test. Also, the amount of grit used for each test was measured. The measured weight loss of each sample was based on

200 gm of grit being used. Samples tested included arc sprayed coatings, plain hybrid coatings, and hybrid coatings with alumina, chromia and silica, respectively. The results of the tests, shown in Fig. 10, indicate that the hybrid coatings are up to 30% more resistant to erosion than the arc sprayed coatings at 750⁰C and this is thought to be due to the higher density of the hybrid coatings. However, the weight loss was slightly higher in the case of oxide particulate embedded coatings. This is contrary to the observation of Jiang Xu et al. (Xu et al., 2008), who have reported improved erosion resistance with the addition of nanoparticles in Ni based alloys. Especially, in the case of chromia embedment, the difference was evident. This may be linked to the large bubble shaped features with internal voids that were observed in chromia particles (which are not shown here).

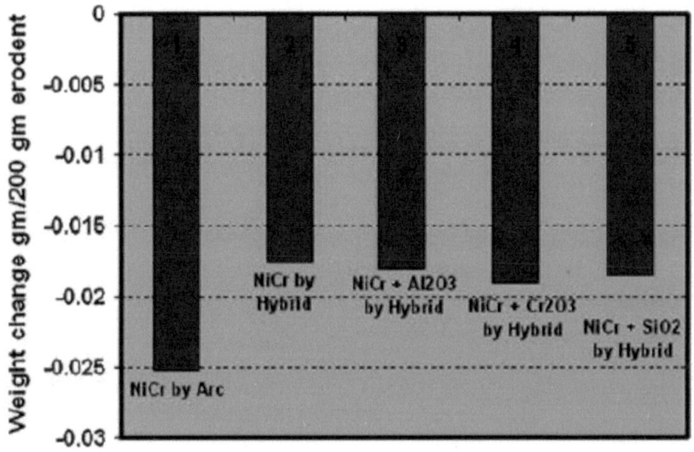

Fig. 10. Weight loss measured in hot erosion test.

5.3.3 Wet corrosion test

The corrosion currents measured from the electrochemical tests are shown in Fig. 11. At zero hours, although the hybrid coating showed less current compared to the arc sprayed coating;

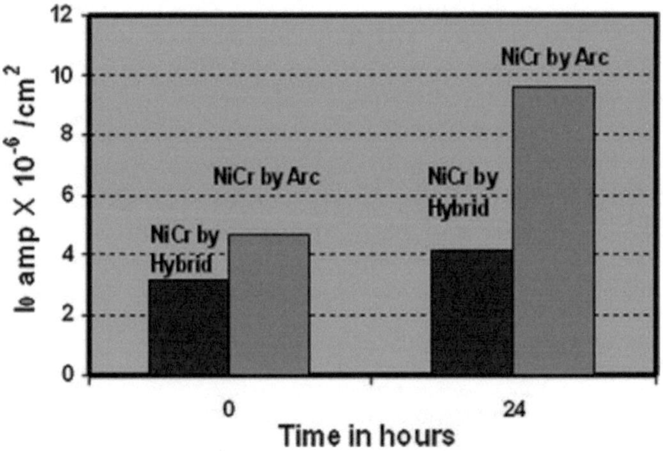

Fig. 11. Wet corrosion of thermally sprayed NiCr coatings in NaCl solution at room temperature.

the difference was not that significant. After 24 hours of immersion, the current values significantly differed between the coatings. The arc spray coating measured two times greater current, Io, after 24 hours. Io is a measure of the corrosion resistance of a material and higher current values indicate lower corrosion resistance. These results confirm that the hybrid coating being denser than the arc spray coating restricts the migration of the corrosive solution/ions to the substrate interface and, therefore, provides more protection to the substrate. Although aqueous corrosion is not an issue for these high temperature coatings, this test has some significance in terms of molten deposit (sulfates) migration through the coating in a coal fired boiler environment.

5.3.4 Hot corrosion test

The hot corrosion test results are shown in Fig. 12. This chart compares the weight loss data obtained on weld overlay coating (with and without salt), 304 stainless steel (304 SS) sample and the coatings - NiCr by arc spray, NiCr and NiCr + SiO_2 by hybrid gun . The chromia stabilized alumina embedded coatings were not included in the test due to their unfavorable oxidation results presented in Fig. 9. NiCr + SiO_2 coatings showed the lowest weight loss compared to all the other samples. Plain NiCr coating by hybrid spray also exhibited lesser weight loss compared to the arc spray coating and this could be attributed to the improved density of the hybrid spray coatings. The superior corrosion resistance of the NiCr + SiO_2 coating is possibly due to the enhanced stability of the chromia scale and the improved oxidation resistance caused by SiO_2. Weld overlay coating showed least weight loss in the absence of the salt; however, when salt was present, it showed poor corrosion resistance compared to the hybrid spray coatings.

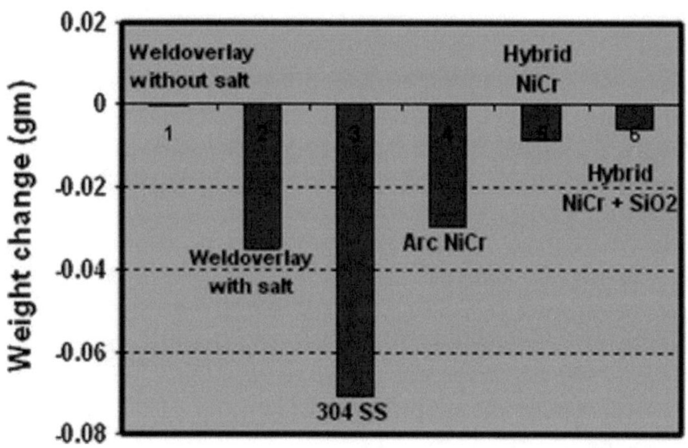

Fig. 12. Weight loss measured in hot corrosion test.

The case study demonstrates that even the base NiCr hybrid spray coatings outperformed the alloy 625 coatings in the presence of corrosive salts. The presence of second phase particles, especially SiO_2, showed improved oxidation and corrosion characteristics. Incorporation of ultrafine and nano sized oxide particles is expected to improve the creep properties by pinning the splat boundaries and reduce the oxidation rate. Chromia addition by itself did not help improve the properties significantly. However, it could act as a stabilizer for alumina and limit the phase transformations up to 1200ºC (Marple et al., 2001; Chraska et al., 1997). Combination of SiO_2 and Al_2O_3 can help improve high temperature creep and corrosion resistance (Carter et al., 1995; Liu et al., 2004; Stollberg et al., 2003).

6. Summary

It is apparent from the brief review and case study that oxide systems could be quite helpful to reduce or slow down the corrosion phenomenon of metallic systems. However, it could be dependent on the oxide system employed. Importantly the stability of oxide phases is very vital as it could undergo several changes during the operation depending on the temperatures of application environment. Dispersion of SiO_2 showed marked enhancement in the oxidation and corrosion resistance of the hybrid coatings at high temperatures.

7. References

Aramaki, K. (2001). The inhibition effects of chromate-free, anion inhibitors on corrosion of zinc in aerated 0.5 M NaCl. *Corrosion Science,* Vol. 43, No. 3, pp. 591-604, ISSN 0010938X

Aruna, S. T.; Bindu, C. N.; Ezhil Selvi, V.; William Grips, V. K. & Rajam, K. S. (2006). Synthesis and properties of electrodeposited Niceria nanocomposite coatings. *Surface & Coatings Technology,* Vol. 200, No. 24, pp. 6871–6880, ISSN 2578972

Aruna, S. T.; William Grips, V. K. & Rajam, K. S. (2009). Ni-based electrodeposited composite coating exhibiting improved microhardness, corrosion and wear resistance properties. *Journal of Alloys and Compounds,* Vol. 468, No. 1-2, pp. 546–552, ISSN 09258388

Aruna, S. T.; Ezhil Selvi, V. ; William Grips, V. K. & Rajam, K. S. (2011). Corrosion- and wear-resistant properties of Ni–Al2O3 composite coatings containing various forms of alumina. *Journal of Applied Electrochemistry,* Vol. 41, pp. 461-468, ISSN 15728838

Bahr, H. A.; Balke, H.; Fett, T.; Hoffinger, I.; Kirchhoff, G.; Munz, D.; Neubrand, A.; Semenov, A. S.; Weiss, H. J. & Yang, Y. Y. (2003). Cracks in Functionally Graded Materials. *Materials Science and Engineering A,* Vol. 362, No. 1-2, pp. 2-16, ISSN 09215093

Bakshi, S. R.; Singh, V.; Balani, K.; Graham McCartney, D.; Seal, S. & Agarwal, A. (2008). Carbon Nanotube Reinforced Aluminum Composite Coating via Cold Spraying, *Surface & Coatings and Technology,* Vol. 202, No. 21, pp. 5162-5169, ISSN 2578972

Bertolini, L.; Brunella, M. F. & Candiani, S. (1999). Corrosion Behavior of a Particulate Metal-Matrix Composite. *Corrosion,* Vol. 55, No. 4, pp. 422-431, ISSN 00109312

Bose, S. (2007). *High Temperature Coatings,* Butterworth-Heinemann Publishers, ISBN 0750682523, Boston, MA, USA

Carter, W. B.; Hampikian, J. M.; Godfrey, S. H. & Polley, T. A. (1995). Thermal Aging of Combustion Chemical Vapor Deposited Oxide Coatings. *Materials Manufacturing Processes,* Vol. 10, No. 5, pp. 1007-1020, ISSN 15322475

Chawla, K. K. (1998). *Composite Materials – Science and Engineering,* Springer-Verlag Inc., ISBN 0387984097, New York, USA

Chraska, P.; Dubsky, J.; Neufuss, K. & Pisacka, J. (1997). Alumina Based Plasma Sprayed Materials, Part I: Phase Stability of Alumina and Alumina-Chromia. *Journal of Thermal Spray Technology,* Vol. 6, No. 3, pp. 320-326, ISSN 15441016

Clark, D.; Wood, D. & Erb, V. (1997). Industrial Applications of Electrodeposited Nanocrystals. *Nanostructured Materials*, Vol. 9, No. 1-8, pp. 755-758, ISSN 09659773

Clauer, A. H. & Wilcox, B. A. (1972). The Role of Grain Size and Shape in Strengthening of Dispersion Hardened Nickel Alloys. *Acta Metallurgica*, Vol. 20, No. 5, pp. 743-757, ISSN 00016160

Cramer, S. D & Covino, B. S. Jr. (2005). *ASM Handbook Vol. 13 B, Corrosion: Materials*, ASM International, ISBN 0871707071, Materials Park, OH, USA

Cui, C., Shen, Y. & Meng, F. (2000). Review of Fabrication Methods of In Situ Metal Matrix Composites. *Journal of Materials Science and Technology*, Vol. 16, No. 6, pp. 619-642, ISSN 10050302

Dokko, P. C.; Pask, J. A. & Mazdiyasni, K. S. (1977). High-Temperature Mechanical Properties of Mullite under Compression. *Journal of the American Ceramic Society*, Vol. 60, No. 3-4, pp. 150-155, ISSN 15512916

Eschnauer, H. (2008). Hard Material Powders and Hard Alloy Powders for Plasma Surface Coating. *Thin Solid Films*, Vol. 73, No. 1, pp. 1-17, ISSN 0040-6090

Erler, F.; Jakob, C.; Romanus, H.; Spiess, L.; Wielage, B.; Lampke, T. & Steinhouser, S. (2003). Interface behaviour in nickel composite coatings with nano-particles of oxidic ceramic. *Electrochimica Acta*, Vol. 48, No. 20-22, pp. 3063-3070, ISSN 00134686

Fang, C.; Huang, C. & Chuang, T. (1999). Synergistic effects of wear and corrosion for Al_2O_3 particulate-reinforced 6061 aluminum matrix composites. *Metallurgical and Materials Transactions A*, Vol. 30, No. 3, pp. 643-651, ISSN 10735623

Goward, G. W. (1986). Protective Coatings Purpose, Role and Design. *Materials Science and Technology*, Vol. 2, No. 3, pp. 194-200, ISSN 17432847

Hamatani, H., Shimoda, N. & Kitaguchi, S. (2003). Effect of the Composition Profile and Density of LPPS Sprayed Functionally Graded Coating on the Thermal Shock Resistance. *Science and Technology of Advanced Materials*, Vol. 4, No. 2, pp. 197-203, ISSN 1878-5514

Hamdy, A. S.; Beccaria, A. M. & Traverso, P. (2005). Effect of surface preparation prior to cerium pre-treatment on the corrosion protection performance of aluminum composites. *Journal of Applied Electrochemistry*, Vol. 35, No. 5, pp. 473-478, ISSN 15728838

Hamdy, A. S. (2006). Advanced nano-particles anti-corrosion ceria based sol gel coatings for aluminum alloys. *Materials Letters*, Vol. 60, No. 21-22, pp. 2633–2637, ISSN 0167577X

Hampikian, J. M. & Carter, W. B. (1999). The Fabrication, Properties and Microstructure of Cu–Ag and Cu–Nb Composite Conductors. *Materials Science and Engineering A*, Vol. 267, No. 1, pp. 7-18, ISSN 09215093

He, Y. & Stott, F. H. (1996). The effects of thin surface-applied oxide coating films on the selective oxidation of alloys. *Corrosion Science*, Vol. 38, No. 11, pp. 1853-1868, ISSN 0010938X

Heian, E. M.; Gibeling, J. C. & Munir, Z. A. (2004). Synthesis and Characterization of Nb5Si3/Nb Functionally Graded Composites. *Materials Science and Engineering A*, Vol. 368, No. 1-2, pp. 168-174, ISSN 09215093

Hendrick, M. R.; Hampikian, J. M. & Carter, W. B. (1998). Combustion CVD-Applied Alumina Coatings and Their Effects on the Oxidation of a Ni-Base Chromia Former. *Journal of the Electrochemical Society*, Vol. 145, No. 11, pp. 3986-3994, ISSN 00134651

Hermansson, L.; Eklund, L.; Askengren, L. & Carlsson, R. (1986). On the Microstructure of Plasma Sprayed Chromium Oxide. *Journal De Physique*, Vol. 47, pp. 165-169, ISSN 1155-4304

Higuera, V., Belzunce, F. J. & Ferna´ndez Rico, E. (1997). Erosion Wear and Mechanical Properties of Plasma-Sprayed Nickel- and Iron-Based Coatings Subjected to Service Conditions in Boilers. *Tribological International*, Vol. 30, No. 9, pp. 641-649, ISSN 0301679X

Hinton, B. R. W.; Arnott, D. R. & Ryan, N. E. (1987). Cationic film-forming inhibitors for the corrosion protection of AA 7075 Aluminum alloy in chloride solutions. *Materials Performance*, Vol. 8, No. 8, pp. 42-47, ISSN 00941492

Hinton, B. R. W.; Arnott, D. R. & Ryan, N. E. (1986). Cerium conversion coatings for the corrosion protection of aluminum. *Materials Forum*, Vol. 9, pp. 162-173, ISSN 0883-2900

Ikeda, Y.; Nii, K. & Yata, M. (1993). Y_2O_3 Dispersion Effect on Al_2O_3 Protective Coating Examined on the Basis of Five Models. *ISIJ International*, Vol. 33, No. 2, pp. 298-306, ISSN 09151559

Jones, Denny A. (1992). *Principles and Prevention of Corrosion*, Prentice Hall Inc., ISBN 0133599930, Upper Saddle River, NJ, USA

Kainer, Karl U. (2006). *Metal Matrix Composites: Custom-made Materials for Automotive and Aerospace Engineering*, John Wiley & Sons, ISBN 9783527608270, Betz-Druck GmbH, Darmstadt, Germany

Kang, Y. C. & Chan, S. L. (2004). Tensile Properties of Nanometric Al_2O_3 Particulate-Reinforced Aluminum Matrix Composites. *Materials Chemistry and Physics*, Vol. 85, No. 2-3, pp. 438-443, ISSN 02540584

Kawasaki, A. & Watanabe, R. (1997). Concept and P/M Fabrication of Functionally Graded Materials. *Ceramics International*, Vol. 23, No. 1, pp. 73-83, ISSN 02728842

Kawasaki, A. & Watanabe, R. (2002). Thermal Fracture Behavior of Metal/Ceramic Functionally Graded Materials. *Engineering Fracture Mechanics*, Vol. 69, No. 14-16, pp. 1713-1728, ISSN 00137944

Khor, K. A.; Gu, Y. W. & Dong, Z. L. (2001). Mechanical Behavior of Plasma Sprayed Functionally Graded YSZ/NiCoCrAlY Composite Coatings. *Surface and Coatings Technology*, Vol. 139, No. 2-3, pp. 200-206, ISSN 2578972

Khor, K. A.; Gu, Y. W.; Quek, C. H. & Cheang, P. (2003). Tensile Deformation Behavior of Plasma-Sprayed Ni–45Cr Coatings. *Surface and Coatings Technology*, Vol. 168, No. 2, pp. 195-201, ISSN 2578972

Kosikowski, D.; Batalov, M. & Mohanty, P. S. (2005). Functionally Graded Coatings by HVOF-Arc Hybrid Spray Gun, *Proceedings of the International Thermal Spray Conference*, ISSN 10599630, Basel, Switzerland, May 2-4, , pp. 444-449

Laha, T.; Agarwal, A.; McKechnie, T. & Seal, S. (2004). Synthesis and Characterization of Plasma Spray Formed Carbon Nanotube Reinforced Aluminum Composite. *Materials Science and Engineering A*, Vol. 381, No. 1-2, pp. 249-258, ISSN 09215093

Laha, T.; Kuchibhatla, S.; Seal, S.; Li, W. & Agarwal, A. (2007). Interfacial Phenomena in Thermally Sprayed Multiwalled Carbon Nanotube Reinforced Aluminum Nanocomposite. *Acta Materialia*, Vol. 55, No. 3, pp. 1059-1066, ISSN 13596454

Lang, Zhou; Ruizeng, Ye; Shouhua, Zhang & Lian, Gao. (1991). The behaviour of a sputtered chromia dispersed Co-Cr coating. *Corrosion Science*, Vol. 32, No. 3, pp. 337-346, ISSN 0010938X

Lessing, P. A.; Gordon, R. S. & Mazdiyasni, K. S. (1975). Creep of Polycrystalline Mullite. *Journal of the American Ceramic Society*, Vol. 58, No. 3-4, pp. 149- 150, ISSN 15512916

Lianga, J.; Gaoa, W.; Lia, Z. & He, Y. (2004). Hot corrosion resistance of electrospark-deposited Al and Ni Cr coatings containing dispersed Y_2O_3 particles. *Materials Letters*, Vol. 58, No. 26, pp. 3280-3284, ISSN 0167577X

Li, J.; Sun, Y.; Sun, X. & Qiao, J. (2005). Mechanical and corrosion-resistance performance of electrodeposited titania-nickel nanocomposite coatings. *Surface & Coatings Technology*, Vol. 192, No. 2-3, pp. 331- 335, ISSN 2578972

Lin, S.; Shih, H. & Mansfeld, F. (1992). Corrosion Protection of Metal Matrix Composites by Polymer Coatings. *Corrosion Science*, Vol. 33, No. 9, pp. 1331-1349, ISSN 0010938X

Liu, Y.; Zha, S. & Liu, M. (2004). Novel Nanostructured Electrodes for Solid Oxide Fuel Cells Fabricated by Combustion Chemical Vapor Deposition. *Advanced Materials*, Vol. 16, No. 3, pp. 256-260, ISSN 15214095

Liu, Y.; Compson, C. & Liu, M. (2004). Nanostructured and Functionally Graded Cathodes for Intermediate Temperature Solid Oxide Fuel Cells. *Journal of Power Sources*, Vol. 138, No. 1-2, pp. 194-198, ISSN 03787753

Lowell, Carl E.; Deadmore, Daniel L. & Whittenberger, Daniel, J. (1982). Long-term high-velocity oxidation and hot corrosion testing of several NiCrAl and FeCrAl base oxide dispersion strengthened alloys. *Oxidation of Metals*, Vol. 17, Nos. 3-4, pp. 205-221, ISSN 15734889

Marple, B. R.; Voyer, J. & Becharde, P. (2001). Sol Infiltration and Heat Treatment of Alumina-Chromia Plasma-Sprayed Coatings. *Journal of the European Ceramic Society*, Vol. 21, pp. 861-868, ISSN 09552219

Martinez-Villafan, A.; Almeyara, M. F.; Gaona, C.; Gonzalez, J. C. & Porcayo, J. (1998). High-Temperature Degradation and Protection of Ferritic and Austenitic Steels in Steam Generators. *Journal of Materials Engineering and Performance*, Vol. 7, No. 1, pp. 108-113, ISSN 15441024

Meadowcroft, D.B. (1987). High Temperature Corrosion of Alloys and Coatings in Oil- and Coal-Fired Boilers. *Materials Science and Engineering*, Vol. 88, pp. 313-320, ISSN 09215093

Michels, H. T. (1976). The effect of dispersed reactive metal oxides on the oxidation resistance of nickel-20 Wt pct chromium alloys. *Metallurgical Transactions*, Vol. 7, No. 3, pp. 379-388, ISSN 03602133

Mikkelsen, L. (2003). High Temperature Oxidation of Iron-Chromium Alloys, (Ph.D. Thesis, Riso National Laboratory Roskilde, Denmark, Date of access: 20th of August 2011, Available at http://130.226.56.153/rispubl/reports/ris-phd-2.pdf

Mohanty, P. S.; Roche, A. D.; Guduru, R. K. & Varadaraajan, V. (2010). Ultrafine Particulate Dispersed High-Temperature Coatings by Hybrid Spray Process. *Journal of Thermal Spray Technology*, Vol. 19, No. 1-2, pp. 484-494, ISSN 15441016

Monticelli, C.; Zucchi, F.; Brunoro, G. & Trabanelli, G. (1997). Stress corrosion cracking behaviour of some aluminium-based metal matrix composites. *Corrosion Science*, Vol. 39, No. 10-11, pp. 1949-1963, ISSN 0010938X

Muhamed Ashraf, P. & Shibli, S. M. A. (2007). Reinforcing aluminium with cerium oxide: A new and effective technique to prevent corrosion in marine environments. *Electrochemistry Communications*, Vol. 9, No. 3, pp. 443–448, ISSN 13882481

Niino, M. & Maeda, S. (1990). Recent Development Status of Functionally Gradient Materials. *ISIJ International*, Vol. 30, No. 9, pp. 699-703, ISSN 09151559

Nunes, P. C. R. & Ramanathan, L. V. (1995). Corrosion behavior of alumina-aluminium and silicon carbide-aluminium metal-matrix composites. *Corrosion*, Vol. 51, No. 8, pp. 610-617, ISSN 00109312

Patnaik, P. C. (1989). Intermetallic Coatings for High Temperature Applications—A Review. *Materials Manufacturing Processes*, Vol. 4, No. 4, pp. 133-152, ISSN 1532-2475

Peng, X.; Ping, D. H.; Li, T. F. & Wu, W. F. (1998). Oxidation Behavior of a Ni-La_2O_3 Codeposited Film on Nickel. *Journal of the Electrochemical Society*, Vol. 145, No. 2, pp. 389-398, ISSN 00134651

Polat, A.; Sarikaya, O. & Celik, E. (2002). Effects of Porosity on Thermal Loadings of Functionally Graded Y_2O_3 ZrO_2/NiCoCrAlY Coatings. *Materials and Design*, Vol. 23, No. 7, pp. 641-644, ISSN 02613069

Prchlik, L.; Sampath, S.; Gutleber, J.; Bancke, G. & Ruff, A. W. (2001). Friction and Wear Properties of WC-Co and Mo-Mo_2C Based Functionally Graded Materials. *Wear*, Vol. 249, No. 12, pp. 1103-1115, ISSN 00431648

Qu, N. S.; Zhu, D. & Chan, K. C. (2006). Fabrication of Ni–CeO_2 nanocomposite by electrodeposition. *Scripta Materialia*, Vol. 54, No. 7, pp. 1421-1425, ISSN 13596462

Quadakkers, W. J.; Halzbrechen, H.; Brief, K. G. & Beske, H. (1989). Differences in growth mechanisms of oxide scales formed on ODS and conventional wrought alloys. *Oxidation of Metals*, Vol. 32, No. 1-2, pp. 67-88, ISSN 15734889

Rao, N. P.; Lee, H. J.; Kelkar, M.; Hansen, D. J.; Heberline, J. V. R.; McMurry, P. H. & Girshick, S. L. (1997). Nanostructured Materials Production by Hypersonic Plasma Particle Deposition. *Nanostructured Materials*, Vol. 9, No. 1-8, pp. 129-132, ISSN 09659773

Richard, C.; Lu, J.; Be´ranger, G. & Decomps, F. (1995). Study of Cr_2O_3 Coatings Part I: Microstructures and Modulus. *Journal of Thermal Spray Technology*, Vol. 4, No. 4, pp. 342-346, ISSN 15441016

Rohatgi, P. K.; Asthana, R. & Das, S. (1986). Solidification, Structure and Properties of Metal-Ceramic Particle Composites. *International Metals Reviews*, Vol. 31, No. 3, pp. 115-139, ISSN 03084590

Schutz, H.; Gossmann, T.; Stover, D.; Buchkremer, H. & Jager, D. (1991). Manufacture and Properties of Plasma Sprayed Cr_2O_3. *Materials and Manufacturing Processes*, Vol. 6, No. 4, pp. 649-669, ISSN 10426914

Skandan, G.; Yao, R.; Kear, B.; Qiao, Y.; Liu, L. & Fischer, T. (2001). Multimodal Powders: A New Class of Feedstock Material for Thermal Spraying of Hard Coatings. *Scripta Materialia*, Vol. 44, No. 8-9, pp. 1699-1702, ISSN 13596462

Soltani, R.; Coyle, T. W. & Mostaghimi, J. (2008). Microstructure and Creep Behavior of Plasma-Sprayed Yttria Stabilized Zirconia Thermal Barrier Coatings. *Journal of Thermal Spray and Technology*, Vol. 17, No. 2, pp. 244-253, ISSN 15441016

Stack, M. M.; Chacon-Nava, J. & Stott, F. H. (1995). Relationship between the Effects of Velocity and Alloy Corrosion Resistance in Erosion-Corrosion Environments at Elevated Temperatures. *Wear*, Vol. 180, No. 1-2, pp. 91-99, ISSN 00431648

Stollberg, D. W.; Hampikian, J. M.; Riester, L. & Carter, W.B. (2003). Nanoindentation Measurements of Combustion CVD Al_2O_3 and YSZ Films. *Materials Science and Engineering A*, Vol. 359, No. 1-2, pp. 112-118, ISSN 09215093

Stringer, J. & Wright, I. G. (1972). The high-temperature oxidation of cobalt-21 wt.% chromium-3 vol.% Y_2O_3 alloys. *Oxidation of Metals*, Vol. 5, No. 1, pp. 59-84, ISSN 15734889

Stringer, J.; Wilcox, B. A. & Jafee, R. A. (1972). The high-temperature oxidation of nickel-20 wt. % chromium alloys containing dispersed oxide phases. *Oxidation of Metals*, Vol. 5, No. 1, pp. 11 - 47, ISSN 15734889

Szczygieł, Bogdan & Kołodziej, Małgorzata. (2005). Composite Ni/Al2O3 coatings and their corrosion resistance. *Electrochimica Acta*, Vol. 50, pp. 4188-4195, ISSN 00134686

Uusitalo, M. A.; J. Vuoristo, P. M. & Mantyla, T. A. (2004). High-Temperature Corrosion of Coatings and Boiler Steels Below Chlorine-Containing Salt Deposits. *Corrosion Science*, Vol. 46, No. 6, pp. 1311-1331, ISSN 0010938X

Vaidya, R. U.; Butt, D. P.; Hersman, L. E. & Zurek, A. K. (1997). Effect of Microbial Corrosion on the Tensile Stress-Strain Response of Aluminum and Al_2O_3-Particle Reinforced Aluminum Composite. *Corrosion*, Vol. 53, No. 2, pp. 136-141, ISSN 00109312

Varma, S. K. & Vasquez, G. (2003) Corrosive wear behavior of 7075 aluminum alloy and its composite containing Al_2O_3 particles. *Journal of Materials Engineering and Performance*, Vol. 12, No. 1, pp. 99-105, ISSN 15441024

Vippola, M. ; Vuorinen, J.; Vuoristo, P.; Lepisto, T. & Mantyla, T. (2002). Thermal Analysis of Aluminum Phosphate Sealed Plasma Sprayed Oxide Coatings. *Journal of the European Ceramic Society*, Vol. 22, No. 12, pp. 1937-1946, ISSN 09552219

Whittle, D. P.; El-Dahshan, M. E. & Stringer, J. (1977). The oxidation behavior of cobalt-base alloys containing dispersed oxides formed by internal oxidation. *Corrosion Science*, Vol. 17, No. 11. pp. 879-891, ISSN 0010938X

Wright, I. G.; Wilcox, B. A. & Jafee, R. A. (1975). The high-temperature oxidation of Ni-20%Cr alloys containing various oxide dispersions. *Oxidation of Metals*, Vol. 9, No. 3, pp. 275-305, ISSN 15734889

Xie, L.; Jordan, E. H.; Padture, N. P. & Gell, M. (2004). Phase and Microstructural Stability of Solution Precursor Plasma Sprayed Thermal Barrier Coatings. *Materials Science and Engineering A*, Vol. 381, No. 1-2, pp. 189-195, ISSN 09215093

Xu, J.; Tao, J.; Jiang, S. & Xu, Z. (2008). Investigation on Corrosion and Wear Behaviors of Nanoparticles Reinforced Ni-Based Composite Alloying Layer. *Applied Surface Science*, Vol. 254, No. 13, pp. 4036-4043, ISSN 0169-4332

Yoshiba, M. (2003). Effect of Hot Corrosion on the Mechanical Performances of Superalloys and Coating Systems. *Corrosion Science*, Vol. 35, No. 5-8, pp. 1115-1121, ISSN 0010938X

Yu, X. Q.; Fan, M. & Sun, Y. S. (2002). The Erosion–Corrosion Behavior of Some Fe_3Al-Based Alloys at High Temperatures. *Wear*, Vol. 253, No. 5-6, pp. 604-609, ISSN 00431648

Zhu, D. & Miller, R. A. (1997). Determination of Creep Behavior of Thermal Barrier Coatings Under Laser Imposed Temperature and Stress Gradients, NASA Technical Memorandum 113169, Report Number ARL-TR-1565, Date of access: 15th of August 2011, Available at http://gltrs.grc.nasa.gov/reports/1997/TM-113169.pdf

13

Improvement of Corrosion Resistance of Steels by Surface Modification

Dimitar Krastev
University of Chemical Technology and Metallurgy
Bulgaria

1. Introduction

The corrosion of metals is a destructive process regarding to the basic modern constructional material with a great importance for the nowadays industry and in many cases represents an enormous economic loss. Therefore, it is not a surprise that the research on the corrosion and corrosion protection of metallic materials is developed on a large scale in different directions and a wide range of engineering decisions. For all that, the improvement of corrosion behaviour of metals and alloys still stays as one of the most important engineering problems in the area of materials application and it is one of the fundamental parts of modern surface engineering.

Special attention is usually focused on the corrosion behaviour of steels as the most commonly used engineering material, because of the limited corrosion resistance for many basic types of these alloys. In more cases they are selected not for their corrosion resistance and important properties are strength, easy fabrication and cost, but there are a lot of exploitation conditions requiring high corrosion resistance. For such a purpose is developed the special group of stainless steels which covers with a high level of certainty these requirements. The stainless steels have an excellent corrosion resistance, but it is not always attended with high strength, hardness and wear resistance. Together with the higher price of the high-alloy steels these are the main restriction for many applications and open up a wide field of opportunities for the surface modification as a method for combination of corrosion resistance along with high strength, hardness and wear resistance.

Surface modification in a wider sense includes all types of surface treatments and coatings that result in change in composition and microstructure of the surface layer. There are different methods for modifying the surfaces of structural alloys, dictated by the performance requirements of the alloy in its service environment. One of the approaches, traditional for the steels, is to modify the surface region of engineering alloys via diffusion of different elements and forming a layer with determinate chemical composition, microstructure and properties. These are the commonly used in practice methods for thermochemical treatment of metals which extended with the methods for physical vapor deposition and chemical vapor deposition form the basic modern techniques for surface engineering regarding to metals. Another approach involves coating of alloy surfaces via plasma spraying, electrospark deposition, modifying the surface by ion implantation or sputter deposition of selected elements and compounds, etc. In recent years a particular attention is directed to the advance methods for surface modification of metals such as laser surface treatment, ion beam surface treatment and electrical discharge machining, which give a modified surface with specific combination of properties in result of nonequilibrium microstructural characteristics.

The obtained by all these methods surface layers can be classified in several ways. Based on the mechanism of the treating process, they can be categorized as:

- Overlay coatings;
- Diffusion coatings;
- Recast layers.

In the overlay coatings, an additional material is placed on the substrate by techniques such as physical vapor deposition (PVD), flame or plasma spraying, etc. The coating in these cases has a mechanical bond with the metallic surface, without much diffusion of the coating constituents into the substrate.

In diffusion coatings a chemical bond is formed with the metallic surface and is obtained a diffusion layer with modifying chemical composition in the depth of the layer. These coatings are formed generally at high temperatures and include such methods as thermochemical treatment and chemical vapor deposition (CVD). Thermochemical treatment is one of the fundamental methods for surface modification of metals and alloys by forming of diffusion coatings. The plain carbon steels and low-alloy steels are mainly treated by these methods to form on the surface layers with high hardness, wear resistance and corrosion resistance, but these methods are also often used to modify the surface of high-alloy steels, cast irons, nonferrous metals and for obtaining of layers with determinate chemical composition, structure and properties.

The recast layers are obtained after attacking the metallic surface with high energy stream such as laser, ion beam or electrical discharge for a very short time and pulse characteristics that involve local melting of the surface and after that rapidly cooling. The recast layer can be with the same chemical composition as the substrate, but with different microstructure and properties in result of nonequilibrium phase transformations during the rapidly cooling, or with a different chemical composition, microstructure and properties in result of attending diffusion process of surface alloying. In recent years of scientific and practical interests is the electrical discharge machining (EDM) for obtaining of recast layers with different characteristics and properties, mainly high hardness, wear resistance and corrosion resistance.

Typical cases of surface modification are diffusion coatings and recast layers, which will be the objectives of this chapter.

2. Diffusion coatings

The diffusion coating process is one of the most effectively and with a great practical application method for improvement of corrosion resistance together with wear resistance, hardness and working live of metals and alloys. This is very important for the carbon steels as the most widely used engineering material accounts more than 80% of the annual world steel production. Despite its relatively limited corrosion resistance, carbon steel has a wide application in whole nowadays industry and the cost of metallic corrosion to the total economy is remarkable high. Because of that the carbon steels and in many cases low-alloy steels are the most used constructional metallic materials for surface engineering on the base of diffusion coatings.

The diffusion coatings are products of thermally activated high temperature processes, that form on the metallic surface chemically bonded layer with determinate chemical composition, structure and properties. For decades a variety of diffusion coatings have been developed and used to improve the properties of metallic surface. There are several kinds of coating methods among which the most commonly used and with the most widely industrial application is thermochemical treatment.

Thermochemical treatment technologies for surface modification of steels have been very well investigated and developed on research and industrial level. These are methods by which nonmetals or metals are penetrated into the metallic surface by thermodiffusion after chemical reaction and adsorption. By thermochemical treatment the surface layers change their chemical composition, structure and properties and in many cases this modified surface can work in conditions which are impossible for the bulk material. Carburizing, nitriding, carbonitriding, nitrocarburizing, boronizing, chromizing, aluminizing and zinc coating are the most popular methods for industrial application. Only carburizing from all these methods could not perform the requirements to form coatings on the surface with high corrosion resistance in the most cases of steel treatment. The other techniques often are used for improvement of the corrosion behaviour of steels by surface modification and more of them increase the wear resistance and hardness of the treated materials.

2.1 Diffusion coatings obtained by nitriding

Nitriding is a thermochemical treatment in which nitrogen in atomic or ionic form is introduced by diffusion process into the metallic surface and in the case of steels is based on the solubility of nitrogen in iron (Davis, 2001, 2002; Pye, 2003). The unique of the nitriding process were recognized by the Germans in the early 1920s. It was used in the applications that required:

- High torque
- High wear resistance
- Abrasive wear resistance
- Corrosion resistance
- High surface compressive strength

Nitrided steels offer improved corrosion and oxidation resistance. The nitrided surface of an alloy steel or tool exhibits increased resistance to saltwater corrosion, moisture and water.

The treatment temperature is usually between 500 and 550 °C for periods of 1 to 100 h depending of the nitriding method, type of steel and the desired depth of the layer. Since nitriding does not involve heating the steel to austenitic temperatures and quenching to martensite is not required, nitriding can be carried out at comparatively low temperatures and thus produce diffusion coating with high quality without deformations of the workpiece.

This technique is of great industrial interest as it forms structures with hard nitride surface layers, so that the global mechanical performance, hardness, wear resistance and corrosion resistance of steels are greatly improved. In recent years new and innovative surface engineering technologies have been developed to meet the rapidly increasing demands from different extreme applications, but gas and plasma (ion) nitriding remain as one of the most widely used techniques for surface engineering.

The case structure of nitrided steel depends on its type, concentration of alloying elements and particular conditions of nitriding treatment. The diffusion zone is the original core microstructure with the addition of nitride precipitates and nitrogen solid solution. The surface compound zone is the region where γ' (Fe_4N) and ε ($Fe_{2-3}N$) intermetallics are formed. The corrosion resistance of steel varies with nitrided layer structure. The surface "white layer" can contain ε nitride, γ' nitride or a two phase mixture ε+γ', below that is the diffusion zone. In acid solutions the iron nitrides corrode more slowly than iron and when the "white zone" is formed on the steel surface the improvement of the corrosion resistance is a fact. In Fig. 1 is shown the typical structure of nitrided plain carbon steel (Minkevich, 1965).

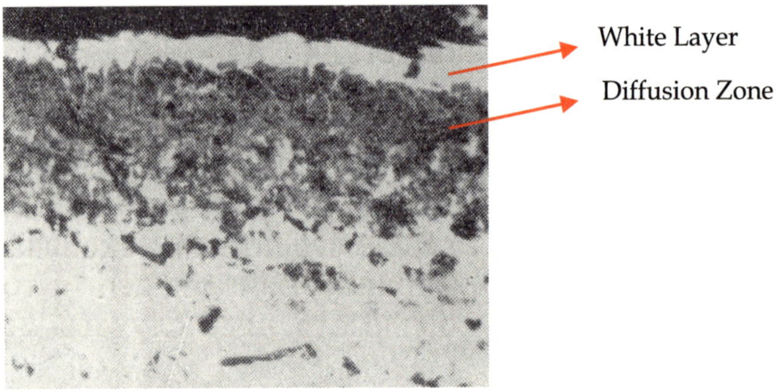

Fig. 1. Microstructure of nitrided GOST 10 steel (x340).

The commonly used steels for nitriding are generally medium-carbon steels that contain strong nitride-forming elements such as aluminium, chromium, vanadium, tungsten and molybdenum. These alloying elements are beneficial in nitriding because they form nitrides that are stable at nitriding temperatures. Other alloying elements such as nickel, silicon and manganese are not so important for the characteristics of the nitrided diffusion coatings. Although these alloy steels are capable to form iron nitrides in the presence of nascent nitrogen, the properties of the nitrided layer are better in those steels that contain one or more of the major nitride-forming alloying elements.

Gas and plasma nitriding are the main methods for obtaining of nitrided diffusion coatings on steels with widely industrial application. The times of gas nitriding can be quite long, that is from 10 to 130 h depending on the application and the depth of the layer is usually less than 0.5 mm. Plasma nitriding allows faster nitriding process and quickly attained surface saturation on the base of the activated nitrogen diffusion. The process provides excellent dimensional control of the white-layer, its composition and properties.

Gas nitriding of steels (Davis, 2001, 2002; Pye, 2003; Smith, 1993) is a thermochemical treatment that takes place in the presence of ammonia gas which dissociates on the steel surface at the operating temperatures. The atomic nitrogen produced is adsorbed at the steel surface, and depending on the temperature and concentration of nitrogen, iron nitrides form at and bellow the steel surface. The patent for gas nitriding was first applied for by Adolph

Machlet and was for nitrogenization of iron and steel in an ammonia gas atmosphere diluted by hydrogen. Either a single-stage or a double-stage process can be used when nitriding with anhydrous ammonia. The temperature of the single-stage process is usually between 495 and 525 °C and it is produced a nitrogen-rich compound zone in a form of white nitride layer on the surface of the nitrided steel. For successful nitriding, it is necessary to control the gas flow so that there is a continuous fresh supply of ammonia at the steel surface. An oversupply of nitrogen may result in the formation of a thick layer of iron nitrides on the steel surface. Independently of some brittleness this nitride layer has a very good corrosion resistance. The principle purpose of double-stage nitriding is to reduce the depth of the white layer on the steel surface, but except for the reduction in the amount of ammonia consuming per hour, there is no advantage in using the double-stage process unless the amount of the white layer produced in the single-stage nitriding cannot be tolerated on the finished parts.

The gas nitriding for improvement of corrosion resistance of plain carbon steels and low-alloy steels can be carried out for shorter times at elevated temperatures (Minkevich, 1965). The purpose is to obtain on the steel surface non-etched nitrided layer without pores and thickness about 0.015 - 0.030 mm. In Table 1 are given the conditions of this process for some plain carbon steels and free-cutting steels.

Type of Steels (GOST)	Temperature, °C	Time, min	Dissociation of ammonia, %
08, 10, 15, 20, 25, 40, 45, 50, A12, A15, A20	600	60 - 120	35 - 55
	600	45 - 90	45 - 65
	700	15 - 30	55 - 75

Table 1. Gas nitriding process conditions for improvement of the corrosion resistance for plain carbon steels and free-cutting steels.

Plasma nitriding (ion nitriding) is a thermochemical treatment process in which nitrogen ions alone or in combination with other gases react at the workpiece surface to produce hardened and corrosion resistance surface on a variety of steels (Buchkov & Toshkov, 1990; Pye, 2003; Smith, 1993). The process is realized on the creation of gaseous plasma under vacuum conditions. The gases can be selected in whatever ratio to provide required surface metallurgy and the layer can consists of single phase, dual phase, or diffusion zone only. The surface metallurgy can be manipulated to suit both the application and the steel. Ion nitriding has many advantages and is appropriate to many applications that are not possible with the conventional nitrided techniques. The nitrided layer on the steel is of the order of 0.1 mm in depth and is harder than nitrided surface layers produced by gas nitriding. The process requires both hydrogen and nitrogen at the workpiece surface. The hydrogen makes certain that the surface of the steel is oxide-free and the chemical reaction takes place between the steel and nitrogen ions. The oxide-free surface enables the nitrogen to diffuse rapidly into the steel and sustains the nitriding actions. A major advantage of the plasma-nitriding process is the enhanced mass transfer of high-energy nitrogen ions to the surface of the steel under the action of an electrical field. The kinetics of the nitrogen ions into the bulk of the steel is controlled by the solid-state diffusion and nitride precipitation. Advantages of plasma nitriding include reduced nitriding cycles, good control of the γ' white iron nitride layer, reduced gas consumption, clean environmental operation, excellent surface quality and reduced distortion of the workpiece. The white layer on the surface of the ion nitrided medium-carbon steel contains mainly from γ' iron nitride with a small amount ε iron nitride. With the control of the process can be obtained layer from single γ' nitride phase. In Table 2 is made a comparison of the white layer structures of gas and ion nitrided GOST 10 steel (Buchkov & Toshkov, 1990).

Type of nitriding	Temperature, °C	Time, min	ε-nitride, %	γ'-nitride, %
Gas	540	30	10	90
	540	300	10	90
	540	720	20	80
Ion	540	30	-	100
	540	300	-	100
	540	720	-	100

Table 2. Comparison of the white layer structures of gas and ion nitrided GOST 10 steel

Our investigations on the microstructure of ion nitrided EN 31CrMoV9 steel show that on the surface is formed nitrided white layer consists from ε-nitride and γ'-nitride (Krastev et al., 2010) - Fig. 2. The thickness of the nitride white layer is 10 - 12 μm and in depth follows 150 - 200 μm diffusion zone with nitride precipitates - fig. 3.

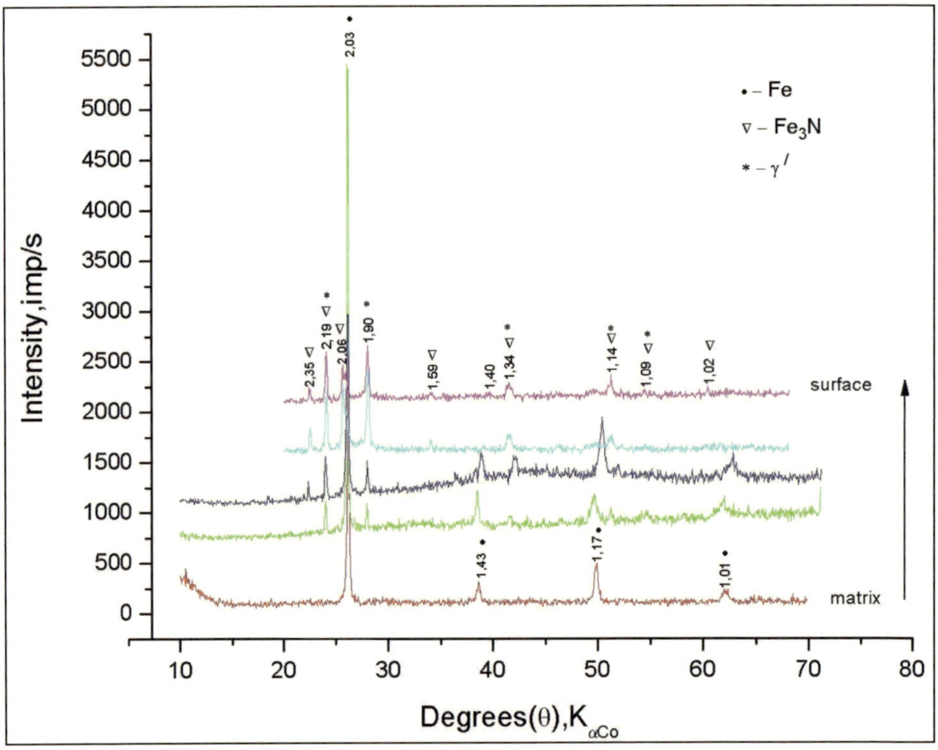

Fig. 2. X-ray diffraction patterns of ion nitrided surface layer on steel 31CrMoV9.

The microhardness of the nitride white layer is about 1050 - 1100 HV which together with the improve corrosions resistance provides high wear resistance.

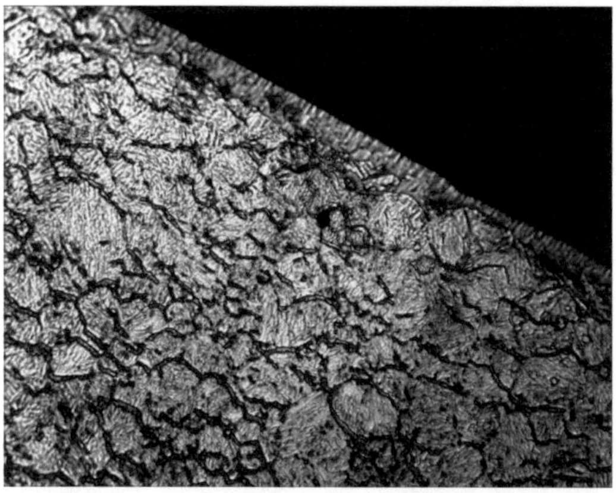

Fig. 3. Microstructure of ion nitrided surface of steel 31CrMoV9 (x600).

By contrast to plain carbon steels, the corrosion resistance of stainless steels can be reduced by nitriding, due to breakdown of the surface chrome oxide barrier to enable nitrogen diffusion into the steel. The stainless steels exhibit generally poor tribological properties, because of that treatments such as nitriding can enhance the surface hardness and improve the wear resistance. Plasma nitriding can be carried out for this purpose at temperatures from 350 to 500 °C (Castaletti et al., 2008). While giving significant improvement in wear resistance, the higher treatment temperatures tend to adversely affect on the corrosion performance of the stainless steels in result of formation of CrN. Improved corrosion resistance of plasma-nitrided layers on stainless steels are observed when the nitriding process is carried out at a lower temperature (400 °C) with presence in the layer of "S - phase", which is supersaturated with nitrogen austenite. The expanded austenite layer in nitrided YB 1Cr18Ni9Ti steel has a good pitting corrosion resistance in 1 % NaCl solution and an equivalent homogeneous corrosion resistance in 1 N H_2SO_4 solution, compared with the original stainless steel (Lei & Zhang, 1997).

2.2 Diffusion coatings obtained by boronizing

Boronizing or boriding is a thermochemical treatment that involves diffusion of boron into the metal surface at high temperatures (Davis, 2002; Liahovich, 1981; Minkevich, 1965; Schatt, 1998). The boriding process is carried out at temperatures between approximately 850 and 1050 °C by using solid, liquid or gaseous boron-rich atmospheres. Boronizing is an effective method for significant increasing of surface hardness, wear and corrosion resistance of metals. The basic advantage of the boronized steels is that iron boride layers have extremely high hardness values between 1600 and 2000 HV. The typical surface hardness of borided carbon steels is much greater than the produced by any other conventional surface hardening treatments. The combination of a high surface hardness and a low surface coefficient of friction of the borided layer provides also for these diffusion coatings a remarkable wear resistance. Boronizing can considerably enhance the corrosion-erosion resistance of ferrous materials in nonoxidizing dilute acids and alkali media and is increasingly used to this advantage in many industrial applications. It is also important that the borided diffusion coatings have a good oxidation resistance up to 850 °C and are quite resistant to attack by molten metals.

On the surface of the boronized steels generally a boron compounds layer is formed. It can be a single-phase or double-phase layer of borides with definite composition. The single phase boride layer consists of Fe_2B, while the double-phase layer consists of an outer phase of FeB and an inner phase of Fe_2B. The FeB phase is brittle and harder, forms a surface under high tensile stress and has a higher coefficient of thermal expansion. The Fe_2B phase is preferred because it is less brittle and forms a surface with a high compressive stress, the preferred stress state for a high-hardness, low-ductility case. Although small amounts of FeB are present in most boride layers, they are not detrimental if they are not continuous. Continuous layers of FeB can be minimized by diffusion annealing after boride formation. In Fig. 4 is shown the typical microstructure of borided layer on the surface of plain carbon steel (Schatt, 1998).

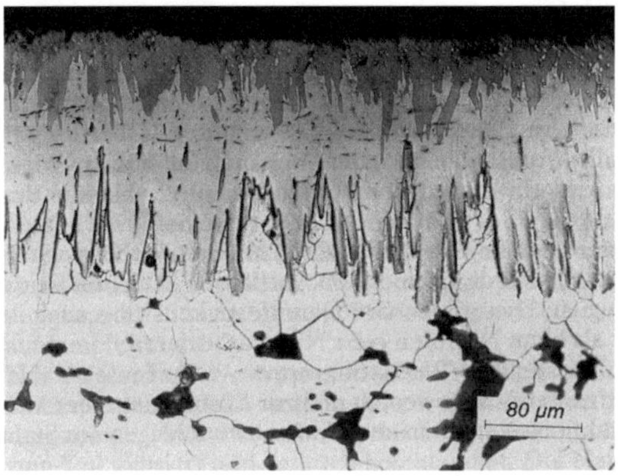

Fig. 4. Microstructure of borided layer on the EN C15 steel consisting of FeB (dark) and Fe_2B (light) phases.

Boriding can be carried out on most ferrous materials such as plain carbon steels, low-alloy steels, tool steels, stainless steels, cast irons and sintered steels. There are a variety of methods for producing of boride diffusion coatings on steel surface. Thermochemical boronizing techniques include:

- Pack boriding
- Paste boriding
- Liquid boriding
- Gas boriding
- Plasma boriding
- Fluidized bed boriding

Only pack and paste boriding from these methods have reached commercial success. Because of environmental problems gas and liquid boriding have a very limited application. Pack boriding is the most common boriding method with a wide development. The process involves packing the steel parts in a boriding powder mixture from ferroboron, amorphous boron or B_4C, fluxes and activators ($NaBF_4$, KBF_4, $Na_2B_4O_7$), and heating in a heat-resistant steel box at 900 to 1050 °C for one to twelve hours depending on the required layer thickness. The commonly produced case depths are 0.05 to 0.25 mm for carbon steels and low-alloy steels and 0.025 to 0.080 mm for high-alloy steels.

Paste boriding is an attractive technique for producing of boride diffusion coatings on steels surface because of lower cost and less difficulty in comparison with pack boriding. It is carried out usually in a paste from B_4C as a boriding agent, Na_3AlF_4 as an activator, fluxes, and binding agent for the paste formation. The temperature of the process is from 800 to 1000 °C and heating is mostly inductively or resistively. A layer in excess of 50 µm thickness may be obtained after inductively or resistively heating to 1000 °C for 20 min. The relationship between the boride layer thickness and time for iron and steel boronized with B_4C-$Na_2B_4O_7$-Na_3AlF_6 based paste at 1000 °C is shown in Fig. 5 (ASTM Handbook, 1991).

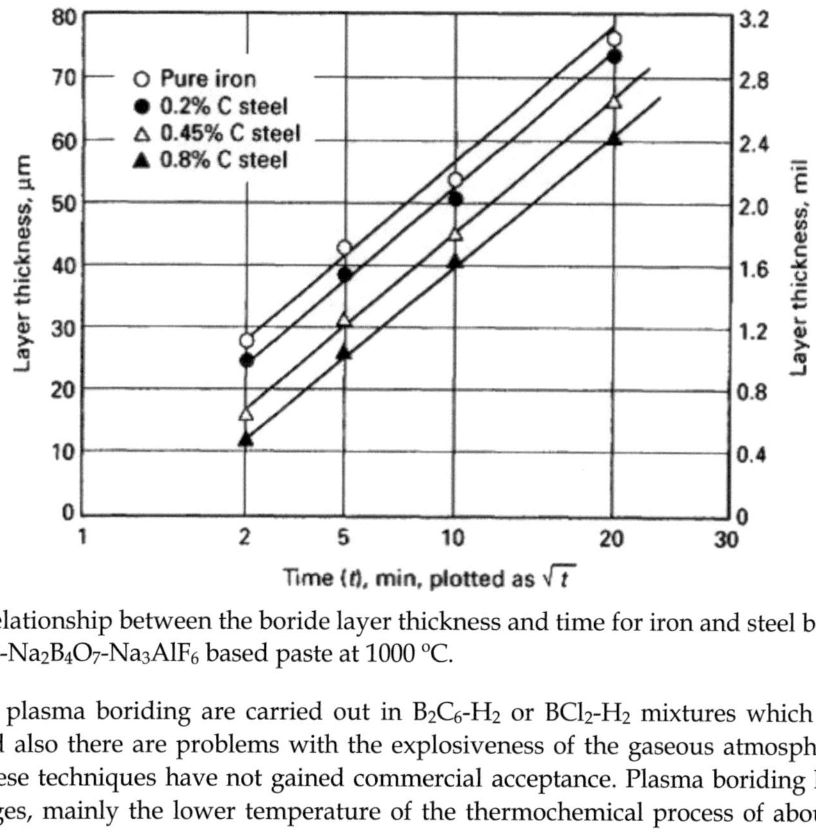

Fig. 5. Relationship between the boride layer thickness and time for iron and steel boronized with B_4C-$Na_2B_4O_7$-Na_3AlF_6 based paste at 1000 °C.

Gas and plasma boriding are carried out in B_2C_6-H_2 or BCl_2-H_2 mixtures which are high toxic and also there are problems with the explosiveness of the gaseous atmosphere. As a result these techniques have not gained commercial acceptance. Plasma boriding has some advantages, mainly the lower temperature of the thermochemical process of about 650 °C and reduction of the duration.

Fluidized bed boriding is the recent innovation on the area of boriding technologies. It is carried out in special retort furnace and involves a bed material of coarse silicon carbide particles, a special boride powder and oxygen-free gas atmosphere from nitrogen-hydrogen gas mixture. The process offers several advantages, can be adaptable to continuous production and has low operating costs due to reduced processing time and energy consumption for mass production of boronized parts.

Our investigations on pack boriding show that it is possible to change the traditional type and amount of activator, and provide a successful diffusion process with high quality of obtained boride layers. The compositions of powder mixtures were from 55 % B_4C; 1 to 3 % $NaBF_4$ or Na_3AlF_6 and diluter Al_2O_3. The thermochemical treatments are carried out with

EN C60 steel at 950 °C and time from two to six hours. The results show that it is possible to exchange the traditional for the process $NaBF_4$ with the inexpensive Na_3AlF_6 as an activator, and amount of 3 % is enough to provide boride layer with the required thickness, structure and hardness. In Fig. 6 are given the structures of borided surface of steel C60 obtained for 2 and 6 h in powder mixture containing 3 % Na_3AlF_6. The XRD analysis shows that the boride layers consist mainly from Fe_2B with a small amount of FeB. The microhardness of the boride diffusion coatings is 1600 – 1800 HV.

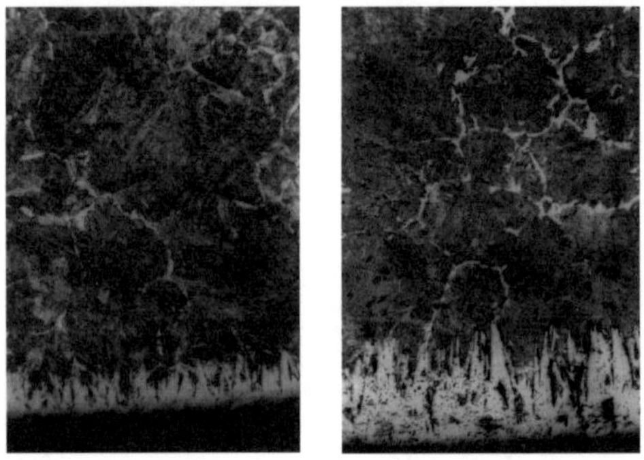

a b

Fig. 6. Microstructure of boride layers on C60 steel obtained for 2 (a) and 6 (b) hours pack boriding at 950 °C in powder mixture containing 3 % Na_3AlF_6 (x150).

The borided steels have a higher corrosion resistance together with the high hardness and wear resistance. In Fig. 7 is given a comparison in the corrosion resistance of 0.45 % C plain carbon steel before and after boronizing (ASTM Handbook, 1991).

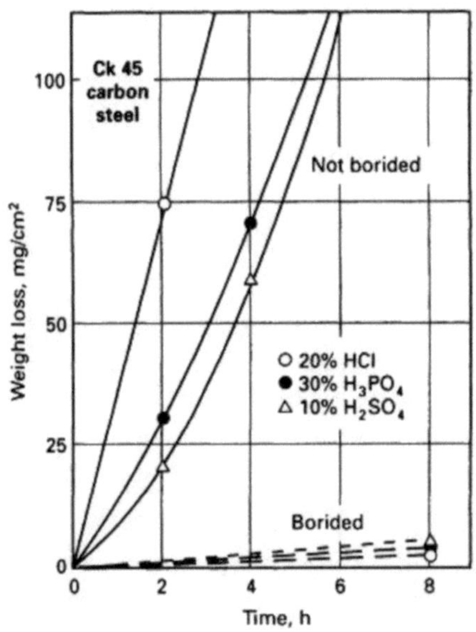

Fig. 7. Corroding effect of mineral acids on boronized and unboronized Ck45 steel.

After boronizing, the corrosion rate of boronized low carbon steel AISI 1018 is about 100 times lower than the corrosion rate of unboronized one based on the electrochemical measurement (Suwattananont, 2005). The boronized high strength alloy steel AISI 4340 and austenitic stainless steel AISI 304 have corrosion rate about several times lower than the corrosion rate of unboronized steels. The comparison of the tafel plots between boronized and unboronized AISI 1018 steel is shown in Fig. 8.

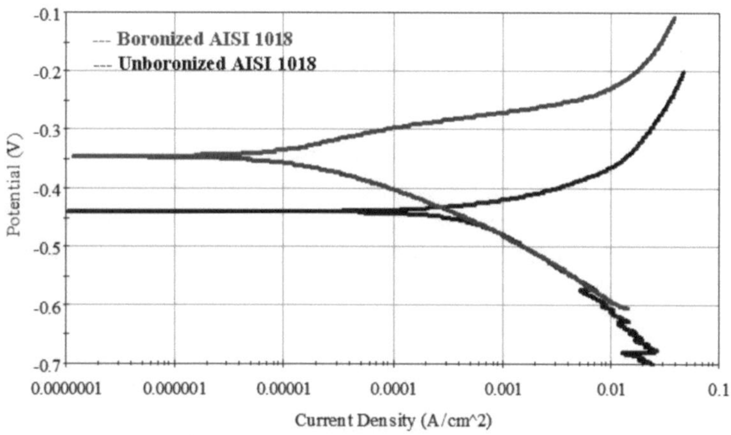

Fig. 8. Tafel plots of boronized and unboronized low carbon steel AISI 1018.

After boronizing the steel becomes nobler and has lower corrosion rate than the unboronized one which proves that the corrosion resistance of the boronized steels is improved.

2.3 Diffusion coatings obtained by carbonitriding and nitrocarburizing

Carbonitriding and nitrocarburizing are those thermochemical treatments which involve diffusional addition of both carbon and nitrogen to the surface of steels for production of diffusion coatings with determinate structure and properties (ASTM Handbook, 1991; Chatterjee-Fischer, 1986; Davis, 2002). Carbonitriding is a modified form of gas carburizing rather than a form of nitriding. The modification consists of introducing ammonia into the gas carburizing atmosphere to add nitrogen to the carbonized case as it is being produced. Nascent nitrogen forms at the steel surface by the dissociation of ammonia in the furnace atmosphere and diffuses simultaneously with carbon. Typically, carbonitriding is carried out at a lower temperatures and shorter times than is gas carburizing, producing a shallower case than is usual in production carburizing. The temperature range for the process is normally 750 - 950 °C and the properties of the obtained diffusion coating are similar with those obtained by carburizing. After the next heat treatment they have high hardness and wear resistance, but the corrosion resistance enhance is unessential. For the improvement of corrosion behaviour of steels the carbonitriding should be carried out at lower temperatures of about 700 °C for obtaining on the surface a carbonitride compound layer. In this case the thermochemical process transforms into nitrocarburizing. Nitrocarburizing, as definition, is thermochemical treatment that is applied to a ferrous object in order to produce surface enrichment in nitrogen and carbon which form a compound layer. This technique has a wide application in the industry and is carried out as gaseous, plasma and liquid process. There is a tendency for limitation of the liquid process because of its toxicity and environmental problems. Based on the temperature range of the thermochemical treatment, nitrocarburizing can be classified as:

- Ferritic nitrocarburizing
- Austenitic nitrocarburizing

Ferritic nitrocarburizing is this thermochemical treatment which is realized at temperatures completely within the ferrite phase field. The primary object of such treatments is usually to improve the anti-scuffing characteristic of ferrous engineering components by producing a compound layer in the surface which has good tribological characteristics. In addition, the fatigue characteristics can be considerably improved, particularly when nitrogen is retained in solid solution in the diffusion zone beneath the compound layer. This is normally achieved by quenching into oil or water from the treatment temperature, usually 570 °C. The obtained at these temperatures compound white layer provided the enhancing of the corrosion resistance of the nitrocarburized surface. The compound layer produced by ferritic nitrocarburizing consists mainly from ε carbonitride because of low carbon solubility in γ' nitride. In Fig. 9 is shown the typical microstructure of nitrocarburized surface of low carbon steel (Chatterjee-Fischer, 1986).

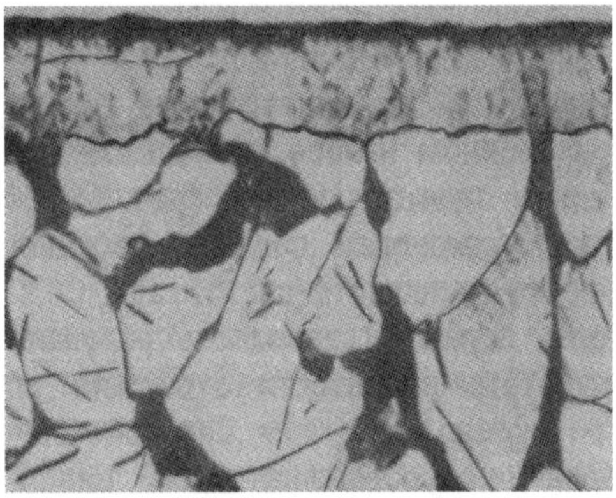

Fig. 9. The microstructure of nitrocarbonized EN C15 steel at 570 °C for 2 hours.

On commercial basis post nitrocarburizing oxidation treatments have been used to enhance the aesthetic properties of gaseous nitrocarburized components. However it is proved that these additional techniques improve the fatigue, wear and corrosion resistance of steel surface and can be successfully combined for this purpose.

When the treatment temperature is such that partial transformation of the matrix to austenite occurs through enrichment with nitrogen, than the treatment is referred to as austenitic nitrocarburizing. With austenitic nitrocarburizing the subsurface is transformed to iron-carbon-nitrogen austenite, which is subsequently transformed to tempered martensite and bainite, with hardness in the range of 750 to 900 HV. The keeping of compound layer from ε carbonitride on the nitrocarburized steel surface together with the transformed subsurface after the nitrocarburizing process at temperatures of about 700 °C provide enhance of the fatigue resistance of treated parts together with high corrosion and wear resistance. Typical transformed austenite case thicknesses are in the range 50 to 200 μm. However, much deeper cases can be achieved by employing a precarburized treatment prior to nitrocarburizing.

2.4 Diffusion coatings with high corrosion resistance on metals basis

The diffusion coatings on metals basis for improvement the corrosion resistance of steels have a wide industrial application. The main aim of the thermochemical treatment in this case is to form on the steel surface a layer from metals with high corrosion resistance, their solid solution in the metal matrix, or their compounds. The metals that are usually used for this thermochemical treatment are chromium, aluminium and zinc.

Chromizing is a surface treatment process of developing a chromized layer on metals and alloys for heat-, corrosion-, and wear resistance (Davis, 2001; Liahovich, 1981; Minkevich, 1965). The technique is applied principally for different types of steels and cast irons, but it is also of interest for surface modification of nickel, molybdenum, tungsten, cobalt and their alloys. Chromized steels offer considerably improved corrosion and oxidation resistance of the surface and can work successfully in complicated conditions combining wear, high temperature, corrosion, erosion and cavitation. If the plain carbon or alloy steel for chromizing contains carbon more than 0.4 %, a corrosion and wear resistant compound layer from $Cr_{23}C_6$ and Cr_7C_3 with thickness 0.01 - 0.03 mm will be formed on the surface. On steels with low carbon content compact chromium carbides layer cannot be formed, but because of high solubility of chromium in iron it will be formed on the steel surface a solid solution with chromium content up to 60 % which provide the high corrosion resistance of the diffusion layer. There are a variety of methods for producing of chromium diffusion coatings on steel surface, such as gaseous, liquid, pack and vacuum chromizing, but only vacuum and pack processes are developed as thermochemical treatment technologies with a wide industrial application.

The pack chromizing is often preferred because of its easily process conditions and low cost. The components to be chromized are packed with fine chromium powder and additives. A typical chromizing mixture consists of 60 percent chromium or ferrochromium powder, up to 2 percent halide salt as an activator and about 38 percent aluminium oxide as inert filler. The process is carried out at 900 – 1050 °C for 6 to 12 hours.

The aluminizing pack-cementation thermochemical treatment has also the most widely industrial application for production of aluminium diffusion coatings. The process is commercially practiced for a wide range of metals and alloys, including plain carbon steels, low-alloy steels and high-alloy steels, cast irons, nickel- and cobalt-base superalloys. Sample aluminide coatings have high corrosion resistance and resist high-temperature oxidation by the formation of an aluminium oxide protective layer and can be used up to about 1000 °C. The powder mixture for pack aluminizing usually consists from about 50 % aluminium or frroaluminium powder, 1 to 2 % NH_4Cl as an activator and about 48 % aluminium oxide as inert filler. As the other processes of pack-cementation, the aluminizing technique consists of packing the steel parts in the powder mixture and heating in a heat-resistant steel box at 800 to 1100 °C for three to fifteen hours, depending on the alloy type and required layer thickness. The aluminized diffusion coatings on plain carbon steels and low-alloy steels are usually about 0.05 - 0.8 mm thick and represent a white layer with a complicated composition from iron aluminides (Fe_3Al, FeAl, $FeAl_2$ and Fe_2Al_5) with high corrosion and oxidation resistance and solid solution of aluminium in α-iron (Davis, 2001; Liahovich, 1981; Minkevich, 1965; Springer et al., 2011).

Our research group has been carried out investigations on high-temperature corrosion and abrasive resistance of chromium and aluminium diffusion coatings on EN C45 steel as parts of sintering machine in an agglomeration process of iron ores. The specimens are produced by pack-cementation process at the optimal characteristics for the both heat treatment techniques. The structure of the obtained diffusion coatings is given in Fig. 10. The comparison shows that better behaviour in these work conditions have the chromium diffusion coatings despite of their smaller thickness.

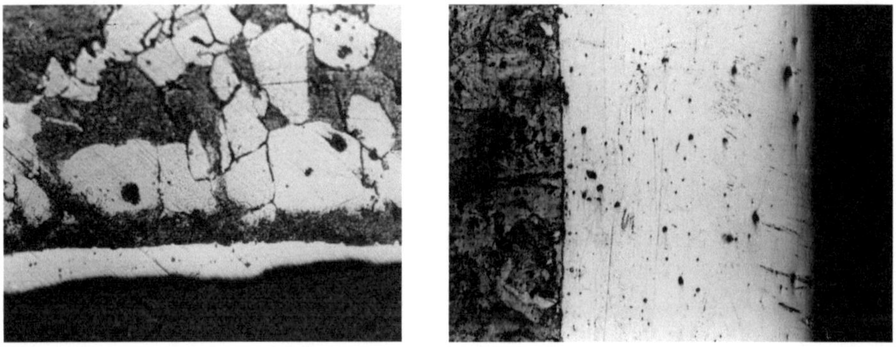

a b

Fig. 10. Diffusion coatings on C45 steel produced by chromizing (a) x150, and aluminizing (b) x250.

Zinc diffusion coatings are traditional method for corrosion protection of steels with a great practical importance and a wide industrial application (Proskurin et al., 1988). Zinc has a number of characteristics that make it a well-suited corrosion protective coating for iron and steel products in most environments. The excellent field performance of zinc coatings results from its ability to form dense, adherent corrosion product films and a rate of corrosion considerably below that of ferrous materials. Many different types of zinc coatings are available and each has unique characteristics, which affect not only on the applicability, but also on the relative economics and expected service live. Hot-dip galvanizing and sherardizing are the main thermochemical treatment techniques for producing of zinc diffusion coatings on steel surface.

The hot-dip galvanizing process, also known as general galvanizing, produces a zinc coating on iron and steel products by immersion of the material in a bath of liquid zinc. Before the coating is applied, the steel surface is cleaned to remove all oils, greases, soils, mill scale, and rust. Galvanized coatings are used on a multitude of materials ranging in size from small parts such as nuts, bolts, and nails to very large structural shapes. The process is usually carried out at 440 to 470 °C for 1 to 10 minutes and in result is obtained zinc diffusion coating which consists of a series of zinc-iron compound layers from $FeZn_{13}$ (ξ-phase), $FeZn_7$ –$FeZn_{10}$ (δ-phase), Fe_5Zn_{21} (Γ_1-phase) with a surface layer of solid solution of iron in zinc (η-phase) or pure zinc.

Sherardizing is a diffusion process in which the steel parts are heated in the presence of zinc dust or powder in inert medium. Aluminium oxide or sand in amount of 20 % is added to the zinc powder as inert filler and 1 to 2 % halide salts are used as activator. The thermochemical treatment can be carried out in retort, rotated drum or as a pack-cementation process at 350 to 500 °C for three to twelve hours. The structure of the obtained layer is the same as the structure on steel surface after hot-dip galvanizing with a thickness about 50 - 400 μm.

3. Recast layers

The recast layers on metals and alloys are created by treating the surface with high energy stream such as laser, ion beam or electrical discharge for a very short time and pulse characteristics. The high energy attack on the surface involves local melting and in many cases vaporizing of metal microvolumes. After the cooling, on the treated metal surface a recast layer with different structure and properties from the substrate is formed. This recast layer can be with the same chemical composition as the substrate or with different one if in the thermal process suitable conditions for surface alloying are created. When the recast process is not controlled there are on the surface microcracks and pores which have negative influence on the surface properties and the recast layer must be removed. In the controlled recast processes it is possible to produce surface layer with determinate chemical composition, thickness, structural characteristics and properties, which are unique for the material with the very high hardness, corrosion- and wear resistance. The basic techniques that give opportunities in this direction are laser surface treatment and electrical discharge machining.

Laser surface treatment is widely used to recast and modify localized areas of metallic components. The heat generated by the adsorption of the laser light provides a local melting and after controlled cooling is obtained a recast layer on the metal surface with high hardness, wear resistance and corrosion resistance. The laser surface melting is based on rapid scanning of the surface with a beam focused to a power density scale of 10^4 W/cm^2 to 10^7 W/cm^2. Quench rates up to 10^8 - 10^{10} K/sec provide the formation of fine structures, the homogenization of microstructures, the extension of solid solubility limits, formation of nonequilibrium phases and amorphous phases or metallic glasses, with corrosion resistance 10–100 time higher compared to crystalline (Bommi et al., 2004). Laser surface melting is a simple technique as no additional materials are introduced, and it is especially effective for processing ferrous alloys with grain refinement and increase of the alloying elements content in solid solution. In fact the process has been employed for improving the cavitation erosion and corrosion resistance of a number of ferrous alloys.

The laser surface melting can be combined with a simultaneous controlled addition of alloying elements. These alloying elements diffuse rapidly into the melt pool, and the desired depth of alloying can be obtained in a short period of time. By this means, a desired alloy chemistry and microstructure can be generated on the sample surface and the degree of microstructural refinement will depend on the solidification rate. The surface of a low-cost alloy, such as low carbon steels, can be selectively alloyed to enhance properties, such as resistance to wear and corrosion (Davis, 2001).

Electrical discharge machining is a thermoelectric process that erodes workpiece material by series of discrete but controlled electrical sparks between the workpiece and electrode immersed in a dielectric fluid (Asif Iqbal & Khan, 2010). It has been proven to be especially valuable in the machining of super-tough, electrically conductive materials, such as tool steels, hard metals and space-age alloys. These materials would have been difficult to machine by conventional methods, but EDM has made it relatively simple to machine intricate shapes that would be impossible to produce with conventional cutting tools. In EDM process, the shapes of mold cavities are directly copied from that of the tool electrode, so time-consuming preparation work must be done on the fabrication of the corresponding tool electrode.

The basis of EDM can be traced as far back as 1770, when English chemist Joseph Priestly discovered the erosive effect of electrical discharges (Ho & Newman, 2003). In 1943 Russian scientists Boris Lazarenko and Natalya Lazarenko (Satel, 1956) applied the destructive effect of electrical sparks for manufacturing and developed a controlled process of machining difficult-to-machine metals by vaporizing material from the surface. At the recent years the research interests and practice are directed to the novel application of electrical discharge machining in the area of surface modification.

3.1 Surface modification by EDM

The electrical discharge machining uses electrical discharges to remove material from the workpiece, with each spark producing temperature of about 8000-20000 °C. This causes melting and vaporizing of small volumes of the metal surface and after cooling in the dielectric fluid the melted zones are transformed in recast layer with specific structure. The EDM modified surface consists from two distinctive zones (Kumar et al., 2009; Ho & Newman, 2003):

- Recast layer
- Heat affected zone

The recast layer is also named white layer and it crystallizes from the liquid metal cooled at high rate in the dielectric fluid. The depth of this top melted zone depends on the pulse energy and duration. Below the top white layer is the heat affected zone with changes in the average chemical composition and possible phase changes. In Fig. 11 is shown the typical microstructure of EDM modified steel surface.

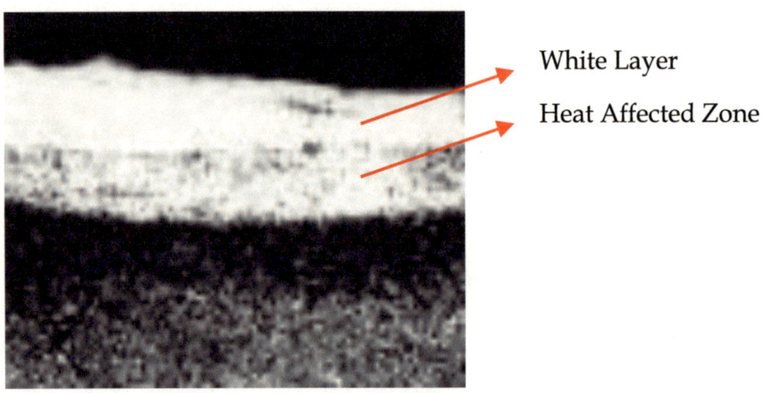

Fig. 11. Microstructure of EDM modified steel surface.

The recast white layer as well as the other discussed white layers can not be etched and has very high hardness, corrosion resistance and wear resistance. The phenomenon of surface modification by EDM has been observed for over four decades. Under the high temperature of the discharge column, the white layer can dissolve carbon from the gases formed in the discharge column from the hydrocarbon dielectric and receives higher carbon content than the base material and hence show increased resistance to abrasion and corrosion. Moreover electrode material has been found in the workpiece surface after machining with conventional electrode. Better surface properties have been obtained by machining with powder metallurgy electrodes containing alloying elements which diffuse in the workpiece surface. Fine powders mixed in the dielectric offer another way for achieving desirable surface modification. All this determines the three main directions for surface modification by electrical discharge machining (Kumar et al., 2009):

- Surface modification by conventional electrode materials
- Surface modification by powder metallurgy electrodes
- Surface modification by powder-mixed dielectric

In the EDM process with conventional electrode has been observed material transfer from the electrode to workpiece surface which is a function of the various electrical parameters of the circuit. The high energy machining results in lower surface deposition, but there is more diffusion in depth. Also it is found that the negative polarity is desirable for increase of material transfer from the tool electrode. The improvement of the surface integrity, wear- and corrosion resistance of the workpiece material can be realized by surface alloying during sparking, using sintered powder metallurgy electrode. With the alloying there is a potential to increase workpiece hardness from two to five times and significant enhance the corrosion resistance that of the bulk material. It is possible remarkable to increase the corrosion resistance of carbon steel by using of composite electrodes containing cooper, aluminium, tungsten carbide and titanium. The material from the electrode is transferred to the workpiece and the characteristics of the surface layer can be changed significantly. The same results can be achieved with the addition of metallic and compound powders in the dielectric. In this case are used Ni, Co, Fe, Al, Cr, Cu, Ti, C (graphite), etc.

3.2 Surface modification by electrical discharge treatment in electrolyte

Such a method as EDM is the electrical discharge treatment in electrolyte, where the modification goes by a high energy thermal process in a very small volume on the metallic surface, involving melting, vaporisation, activation and alloying in electrical discharges and after that cooling of this surface with high rate in an electrolyte. The high energy process put together with the nonequilibrium phase transformations in the metallic system causes considerable modifications of the metallic surface and obtaining of layers with finecrystalline and nanocrystalline structure (Krastev at al., 2009; Krastev & Yordanov, 2010). The metallic surface after electrical discharge treatment in electrolyte has a different structure in comparison with the metal matrix which determines different properties. It is observed remarkable increasing of hardness, strength and corrosion resistance related to the nonequilibrium phase transformations and the obtained finecrystalline microstructure. The investigations show that obtained on tools layers have higher hardness, wear resistance, tribocorrosion resistance and corrosion resistance, which give better performance, considerable increasing of working life and wide opportunities for industrial application.

For the electrical discharge treatment in electrolyte is developed a laboratory device, shown in Fig. 12, giving opportunities for treatment of cylindrical workpieces with diameter up to 20 mm. The electrolyte *3* is in active movement by mixing from a magnetic stirrer *4*. After passing of electric current with determinate characteristics through the electrolyte between the workpiece *1* and electrode *2* starts an active sparking on the workpiece surface. The sparking characteristics depend on different factors such as parameters of the electric current, type and composition of the electrolyte, movement of the workpiece and electrolyte.

The workpieces are made from high speed steel HS 6-5-2 with structure after the typical heat treatment for tools of this steel and hardness about of 950 HV. The choice of high-alloy steel is founded on the opportunity for higher effectiveness of treatment on structure and properties of modified surfaces after the nonequilibrium phase transformations from liquid state.

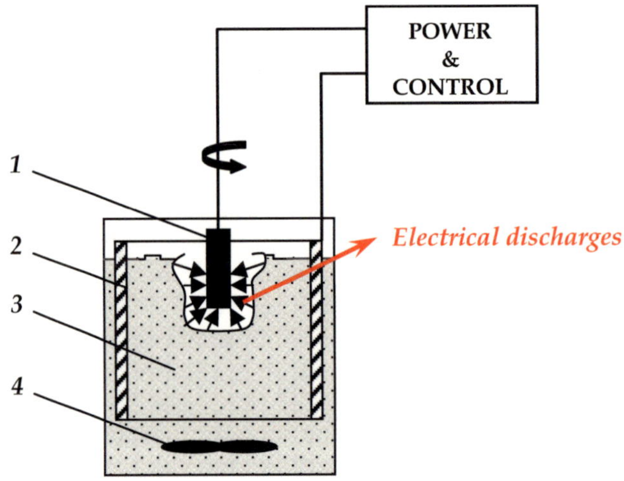

Fig. 12. Installation for electrical discharge treatment in electrolyte: 1 - workpiece, 2 - electrode, 3 - electrolyte, 4 - magnetic stirrer.

The principle changes that occur on the modified steel surface by the high speed quenching from liquid state in the treatment process can be described as:

- Expansion of the solubility in solid state
- Grain refinement with possibilities for obtaining of nanocrystalline structure
- Formation of metastable phases
- High concentration of crystalline imperfections

Some studies at similar conditions show significant increasing of solubility of carbon in steel up to 2 % in the martensite and 3.5 % in austenite which is a precondition for high strength of the treated surface.

By the investigations were obtained layers on the workpiece surface with approximately equal thickness, depending of the electrical current characteristics and time of treatment. The modified surfaces can be observed as a light layer on the workpieces. The melted and resolidified layer during this process can be also referred as a "white layer", since generally no etching takes place in these areas at the metallographic preparation because of its high corrosion resistance which is one of the important characteristics of nanocrystalline structures.

The electric discharges generate an enormous amount of heat, causing local melting on the workpiece surface and thereupon it is rapidly quenched from the liquid state by the electrolyte. This recast area has a specific structure which is composed of several microscopic metallurgical layers, depending of machining conditions. In Fig. 13 is shown an optical micrograph of the modified surface of high speed steel obtained by electrical discharge treatment in electrolyte.

Fig. 13. Microstructure of recast layer obtained by electrical discharge treatment in electrolyte of HS 6-5-2 steel, x800.

The high rate of the recasting process gives opportunities for formation of metastable phases and considerable decreasing of grain size. The electrolyte type is of great importance for the chemical composition, microstructure and properties of the modified layer. By these experiments the electrolyte is on water basis and contains boron or silicon compounds. At short times of treatment it is not observed diffusion of elements from the electrolyte in the modified surface, but it is available diffusion process inside the workpiece between the white layer and the matrix - Table 3. The strong carbide-formed elements such as Mo, W, and V diffuse from the white layer to the matrix and Cr, Co in the opposite side.

Chemical element	Matrix of workpiece	White layer
Si	<0.01	<0.01
Mo	5.58	4.87
V	2.30	1.63
Cr	4.25	4.52
Co	<0.01	0.19
Ni	<0.01	<0.01
W	8.34	5.75

Table 3. EDS analysis of modified workpiece from HS 6-5-2 steel

The thickness and integrity of obtained recast white layer on the steel surface by electrical discharge treatment in electrolyte depend on the electrical current characteristics and time duration of the treatment. At higher voltage are observed thicker white layers by equal durations of the treatment. In Fig. 14 are shown the light microscopy micrographs of steel workpiece surfaces, modified for two minutes at 80 V and 100 V. The optical measured thicknesses of the layers are 0.01 mm and 0.02 – 0.03 mm respectively.

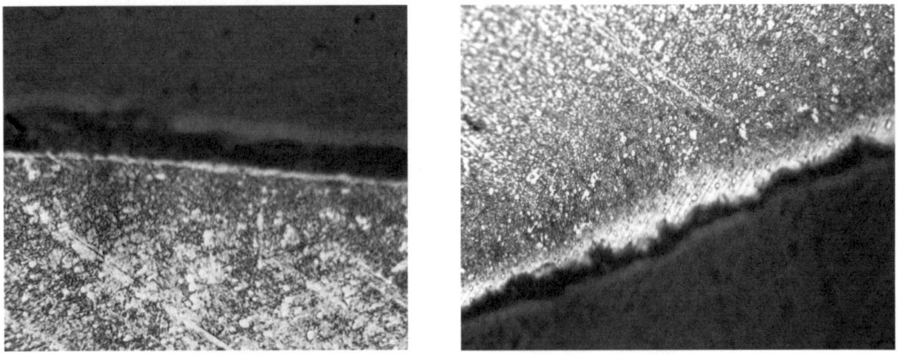

a b

Fig. 14. Microstructure of layers, obtained for 2 min at 80 V, and for 2 min at 100 V (b) x800.

At higher voltage and longer duration of the treatment it is observed increasing of roughness of the white layer surface which is illustrated with SEM micrographs in Fig. 15.

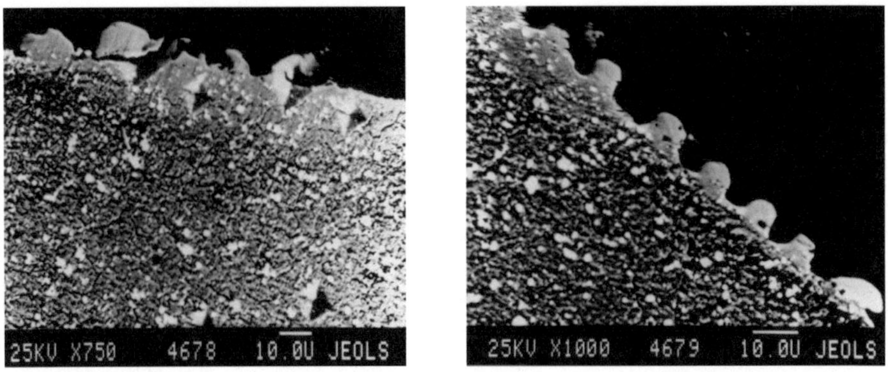

a b

Fig. 15. SEM micrograph of modified steel surface by duration of 2 min (a) and 3 min (b) at 100 V.

The fine structures of modified layers with specific etching are shown after SEM investigation on Fig. 16. By modification on the workpiece surface can be observed two specific zones:

- White zone
- Phase transformations zone

The "Phase transformations zone" has different structures depending on the temperature and cooling rate. When the temperature of the steel surface is above the melting point and cooling rate is lower a zone with dendritic structure is formed – Fig. 16a. In the other case (Fig. 16b), when the temperature is in the austenitic region and the cooling rate is higher than the critical one a martensite is formed.

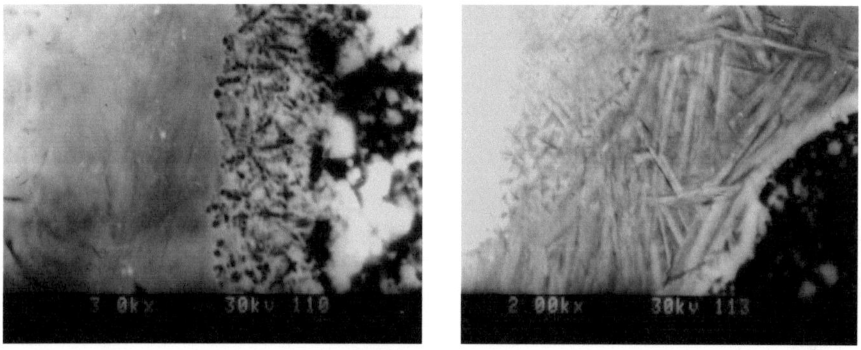

a b

Fig. 16. SEM images of different microstructures of the modified layers on HS 6-5-2 steel surface: a - dendritic microstructure in the phase transformations zone, b – martensitic microstructure in the phase transformations zone.

The hardness of the modified layers can vary considerably and depend of the treatment conditions, electrolyte composition and microstructure, but in principle it is higher then the hardness of the typical microstructure of this steel. The microhardness of the modified layers is measured by Hanneman test and shows values after the different treatments up to 1600 HV which are very higher than the microhardness of HS 6-5-2 steel microstructures after the typical heat treatment. The experiments show that tools with such surface hardness have higher wear resistance and working capacity.

4. References

Asif Iqbal, A. K. M. & Khan A. A. (2010). Influence of Process Parameters on Electrical Discharge Machined Job Surface Integrity. *American Journal of Engineering and Applied Science,* Vol. 2, No. 3, (2010), pp. 396-402, ISSN 1941-7020

ASTM Handbook. (1991). *Heat Treating, Volume 4,* ASTM International, ISBN 0-87170-379-3, Materials Park, Ohio, USA

Bommi, V. C., Mohan, K. M., & Prakash, S. (2004). Surface Modification of Martensitic Stainless Steel Using Metal Working CO_2 Laser, *Proceedings of International Symposium of Research Students on Materials Science and Engineering,* Chennai, India, December 20-22, 2004, 10.07.2011, Available from http://metallurgy.iitm.ac.in/isrs/isrs04/cd/content/Papers/SE/PO-SE-6.pdf

Buchkov, D. & Toshkov, V. (1990). *Ion Nitriding,* Technika, UDC 621.785.5, Sofia, Bulgaria

Castelleti, L. C., Neto, A. L., & Totten G. E. (2008). Plasma Nitriding of Stainless Steels, In: *Industrial Heating,* 05.08. 2011, Available from http://www.industrialheating.com/Articles/Feature_Article/

Chatterjee-Fischer, R. (Ed). (1986). *Wärmebehandlung von Eisenwerkstoffen: Nitrieren und Nitrocarburieren,* Expert Verlag, ISBN 3-8169-0076-3, Sindelfingen, Germany

Davis, J. R. (2001). *Surface Engineering for Corrosion and Wear Resistance,* ASM International, ISBN 0-87170-700-4, Materials Park, Ohio, USA

Davis, J. R. (2002). *Surface Hardening of steels: understanding the basics,* ASM International, ISBN 0-87170-764-0, Materials Park, Ohio, USA

Ho, K. H., Newman, S. T. (2003). State of the Art Electrical Discharge Machining (EDM). *International Journal of Machine Tools & Manufacture,* Vol. 43, (2003), pp. 1287-1300, 2011, 10.08.2011, Available from http://www.sciencedirect.com/science/article/pii/S0890695503001627

Krastev, D., Stefanov, B., Yordanov, B., Angelova, D. (2009). Electrical Discharge Surface Treatment in Electrolyte of High Speed Steel, *Proceedings of VI International Congress on Machines, Technologies and Materials,* Sofia, Bulgaria, 18-20 February 2009

Krastev, D., Yordanov, B. (2010). About the Surface Hardening of Tool steels by Electrical Discharge Treatment in Electrolyte, *Solid State Phenomena,* Vol. 159 (February 2010) pp 137-140, ISSN 1662-9779

Krastev, D., Yordanov, B., & Lazarova, V. (2010). Microstructural Characterization of Diffusion Layer of Nitrided Steel. *Scientific Proceedings of STUME,* Vol. 115, No 5, (June 2010), pp. 389-394, ISSN 1310-3946

Kumar, S., Singh, R., Singh, T. P., Sethi, B. L. (2009). Surface Modification by Electrical Discharge Machining: A Review. *Journal of Materials Processing Technology,* Vol. 209, (21 April 2009), pp. 3675-3687, 10.07.2011, Available from http://www.sciencedirect.com/science/article/pii/S092401360800705X

Lei, M. K., Zang, Z. L. (1997). Microstructure and Corrosion Resistance of Plasma Source Ion Nitrided Austenitic Stainless Steel. *Journal of Vacuum Science & Technology A,* No. 2, (March 1997), pp. 421-427, ISSN 0734-2101

Liahovich, L. S. (Ed.). (1981). *Thermochemical Treatment of Metals and Alloys,* Metallurgia, UDC 621.793.4, Moscow, USSR

Minkevich, A. N. (1965). *Thermochemical Treatment of Metals and Alloys,* Mashinostroenie, UDC 621.78.794, Moscow, USSR

Proskurin, E. V., Popovich, V. A., & Moroz, A. T. (1988). *Galvanizing,* Metallurgia, ISBN 5-229-00112-7, Moscow, USSR

Pye, D. (2003). *Practical Nitriding and Ferritic Nitrocarburizing,* ASM International, ISBN 0-87170-791-8, Materials Park, Ohio, USA

Satel, E. A. (Ed.). (1956). *Handbook of Mechanical Engineer,* Vol. 6, State Research Publisher of Mechanical Engineering Literature, Moscow, USSR

Schatt, W., Simmchen E., & Zuuhar, G. (1998). *Konstruktionswerkstoffe des Maschinen- und Anlagenbaues,* Deutscher Verlag für Grundstoffindustrie, ISBN 3-342-00677-3, Stuttgart, Germany

Smith, F. W. (1993). *Structure and Properties of Engineering Alloys,* McGraw-Hill, New York, USA

Springer, H., Kostika, A., Payton, E. J., Raabe, D., Kaysser-Pyzalla, A., & Eggeler, G. (2011). On the formation and growth of intermetallic phases during interdiffusion between low-carbon steel and aluminum alloys. *Acta Materialia,* Vol. 59, No 4, (February 2011), pp. 1586-1600, ISSN 1359-8454

Suwattamanont, N., Petrova R. S., Zunino III, J. L., & Scmidt, D. P. (2005). Surface treatment with Boron for Corrosion Protection, *Proceedings of 2005 Tri-Service Corrosion Conference,* November 2005, Orlando, USA

14

Low Temperature Thermochemical Treatments of Austenitic Stainless Steel Without Impairing Its Corrosion Resistance

Askar Triwiyanto[1], Patthi Husain[1],
Esa Haruman[2] and Mokhtar Ismail[1]
[1]*Universiti Teknologi PETRONAS,*
[2]*Bakrie University,*
[1]*Malaysia*
[2]*Indonesia*

1. Introduction

Austenitic stainless steel (ASS) is used applied widely owing to its very good corrosion resistance. However, the application of this material as a bearing surface is severely limited by very poor wear and friction behaviour. Consequently, Surface Engineering treatments for austenitic stainless steel are an interesting alternative way to increase the surface hardness and improve the wear resistance. For the purpose of this works, the Surface Engineering design will be classified, very broadly, into three groups : (a) those which coat the substrate: PVD, CVD, etc, (b) those which modify only the structure of the substrate, (c) those which modify the chemical composition and the structure of the substrate: thermochemical, ion implantation, plasma, etc. It is nowadays widely accepted that hard, wear and corrosion resistant surface layers can be produced on ASS by means low temperature nitriding and/or carburizing in a number of different media (salt bath,gas or plasma), each medium having its own strengths and weaknesses. In order to retain the corrosion resistance of austenitic stainless steel, these processes are typically conducted at temperatures below 450°C and 500°C, for nitriding and carburizing respectively. The result is a layer of precipitation free austenite, supersaturated with nitrogen and/or carbon, which is usually referred to as S-phase or expanded austenite.

2. Enlarging application of Austenitic Stainless Steel

Starting from the mid of 1980's, investigations have been performed to improve surface hardness of ASS and thus enlarging their possibility of wider application, but led significant loss of its corrosion resistance. This tendency occur due to the sensitivity effect. Sensitization is a common problem in austenitic steel where precipitation of chromium carbides ($Cr_{23}C_6$) occurs at the grain boundaries at elevated temperatures, typically between 450 to 850°C; diffusional reaction in forming chromium nitride/carbide leads to the depletion of Cr in the austenitic solid solution and consequently unable to produce Cr_2O_3 passive layer to make stainless feature. As a result, it reduces the corrosion resistance property of the stainless steel. This phenomenon causes reduction in ductility, toughness and aqueous corrosion resistance (Clark & Varney, 1962).

The efforts have been made in the past decades to modify the surfaces of these materials to improve their surface hardness, wear resistance as well as corrosion resistance which is shown in Fig. 1.

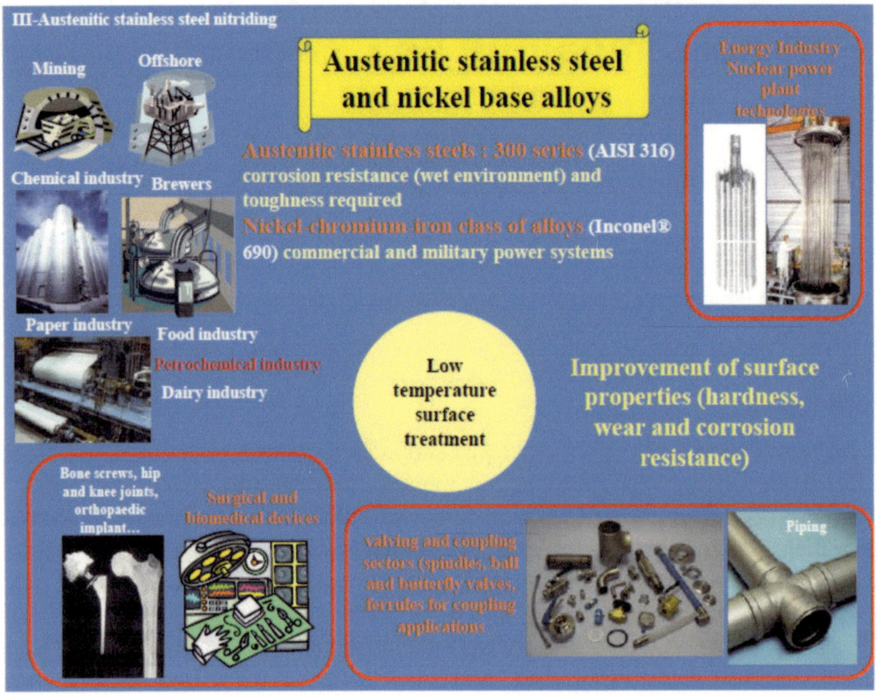

Fig. 1. Enlarging application of Austenitic Stainless Steel (Czerwiec, 2010).

Bell et al. (T. Bell, 2001) suggested that a low temperature nitriding can eliminate the formation of chromium nitrides but at the expense of strengthening effects made by CrN precipitates. Alternatively, the strengthening effect will be replaced by supersaturation of interstitial species in austenite matrix which leads to the hardening of the surface region several tens micro meter thick. This precipitation-free nitride layer not only exhibits high hardness but also possesses good corrosion resistance due to the availability of retaining chromium in solid solution for corrosion protection.

In relation with the functional properties of a part, such as fatigue and static strength, or wear and corrosion resistance, are the basis for specifying the proper process and steel as illustrated in Fig. 2. (T. Bell, 2005). The functional part properties that essentially depend on the compound layer are wear resistance, tribological properties, corrosion resistance and general surface appearance. Both abrasive and adhesive wear resistance increase with hardness and with minimised porosity of the compound layer. Porosity can be positive in lubricated machinery parts as the pores act as lubricant reservoirs. The compound layer depth has to be deep enough not to be worn away. The diffusion layer (depth, hardness and residual stress) determines surface fatigue resistance and resistance to surface contact loads.

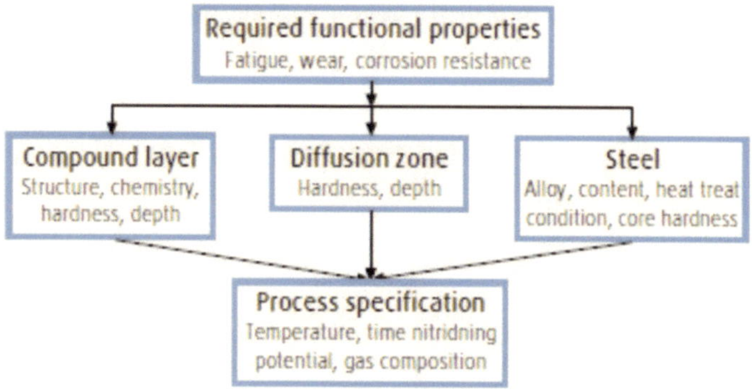

Fig. 2. The steps to process specification starting from required part properties.

2.1 Thermochemical and diffusion surface engineering treatments

Thermochemical treatments, sometimes referred to as case hardening or cementation, are based on the modification of the chemical composition of the substrate material. These treatments can be succeeded by a change in the structure through heat treatment. The formal definition available in BS EN 10052:1994 reads as follows (British standard, 1994):

Thermochemical treatment: Heat treatment carried out in a medium suitably chosen to produce a change in the chemical composition of the base metal by exchange with the medium.

In the case of diffusion treatment, the definition in that same standard is:

Diffusion treatment: Heat treatment or operation intended to cause the diffusion towards the interior of the ferrous product of elements previously introduced into the surface (for example, following carburizing, boriding or nitriding).

The two major low temperature thermochemical processes developed for austenitic stainless steels are nitriding and carburizing (Lewis et al, 1993; Bell. T, 2002). The former is normally carried out at temperatures below 450ºC and the later below 500ºC. The purpose of using low temperatures is to suppress the formation of chromium nitrides and carbides in the alloyed layers, such that chromium is retained in solid solution for corrosion protection (Sun et al, 1999; Thaiwatthana et al, 2002). Hardening of the nitrided layer and the carburised layer is due to the incorporation of nitrogen and carbon respectively in the austenite lattice, forming a structure termed expanded austenite, which is supersaturated with nitrogen and carbon respectively (Lewis et al, 1999; Thaiwatthana et al, 2002). More recently, a hybrid process has also been developed, which combines the nitriding and carburizing actions in a single process cycle by introducing nitrogen and carbon simultaneously into the austenite lattice to form a hardened zone comprising a nitrogen expanded austenite layer on top of a carbon expanded austenite layer (Tsujikawa et al, 2005; Sun et al, 2008; Li et al, 2010). There exist some synergetic effects between nitrogen and carbon: under similar processing conditions, the hybrid treated layer is thicker, harder and possesses better corrosion resistance than the individual nitrided layer and carburised layer.

From these definitions it becomes clear that two main factors will govern the process, namely: the exchange or absorption reaction with the medium, and the diffusion in the metal (ASM, 1977). As it is illustrated in Fig. 3, the medium will determine the way in which the diffusing elements are delivered to the metal surface. A number of different media are available (solid, liquid, gas and plasma), and a detailed account of the media used for carburizing will be given in a following section.

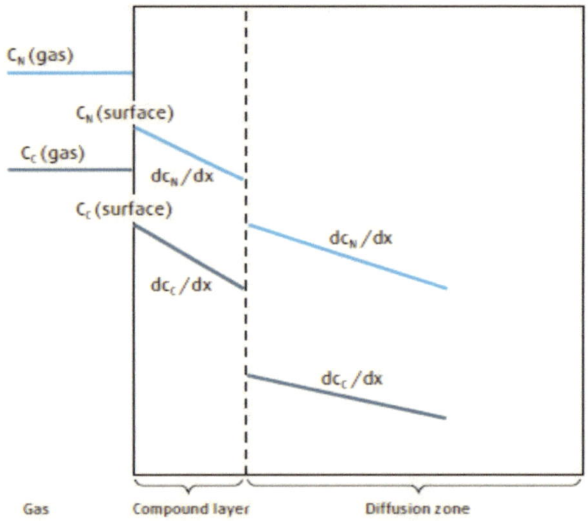

Fig. 3. Concentrations and concentration gradients of nitrogen and carbon. (Christiansen & Somers, 2005)

Once in the metal, the transport of the absorbed substance takes place by diffusion, and follows Fick's laws:

$$J = - D\,[\,dC/\,dx] \tag{1}$$

$$\mathrm{J} = - \mathrm{D}\,[\,\partial \mathrm{C}\,/\,\partial \mathrm{x}] \tag{2}$$

$$\partial \mathrm{C}\,/\,\partial \mathrm{t} = [\,\partial^2 \mathrm{C}\,/\,\partial \mathrm{x}^2\,]\,/\,\partial \mathrm{x}^2 \tag{3}$$

where J is the flux of diffusing substance, D is the diffusion coefficient, and $\partial C\,/\,\partial x$ is the concentration gradient (ASM, 1977). Therefore, the transport of the substance in solution is driven by its concentration gradient and the diffusion coefficient which, at the same time, depends on the temperature, the chemical composition and phase structure of the substrate. For a given alloy, kept at constant temperature in a medium with a consistent concentration of the substance of interest, the case depth will only depend on the time, according to equation (9):

$$x = a\,(Dt)^{½} = Kt^{½} \tag{4}$$

where x is the case depth, a is a constant, D is the element diffusivity, t is the treatment time and K is a factor determined by a and D (ASM, 1977). Higher treatment temperatures yield the same case depth in shorter time, although there are technical limitations related with life of the furnaces, and metallurgical considerations regarding the side effects of keeping the substrate material at high temperatures (Parrish & Harper, 1994). Consequently, diffusion treatments are slower when compared to other surface deposition techniques (Hurricks,

1972), and treatments as long as 72 hours are common practice in industry. On the other hand, thermochemical treatments produce smooth case-core interfaces, which are beneficial for not only the wear and fatigue performance, but also the load bearing capacity (Sun & Bell, 1991).

During nitriding, ammonia (NH_3) in the furnace atmosphere decomposes into hydrogen and nitrogen at the surface, enabling nitrogen atoms to be adsorbed at the steel surface and to diffuse further into the steel as illustrated in Fig. 3. In nitrocarburizing it is additionally necessary to have a carbonaceous gas transferring carbon to the steel surface.

The flux of nitrogen and carbon from the gas to the steel surface is proportional to the concentration differences between the gas and the surface:

$$dm_N/dt_{(surface)} = k_1 \, [\, c_{N\,(gas)} - c_{N\,(surf)} \,] \tag{5}$$

$$dm_C/dt_{(surface)} = k_2 \, [\, c_{C\,(gas)} - c_{C\,(surf)} \,] \tag{6}$$

Here m denotes mass, t time, c concentration per volume unit and k_1 and k_2 are reaction rate coefficients.

The transfer of nitrogen and carbon from the surface further into the steel is controlled by diffusion. Diffusion rates follow Fick's first law, which for the compound layer and diffusion zone are respectively:

$$dm/dt_{(comp\ layer)} = -\, D_{Comp}\, dc/dx \tag{7}$$

$$dm/dt_{(diff\ zone)} = -\, D_{Diff}\, dc/dx \tag{8}$$

Balance of mass requires that all three mass transfer rates are equal:

$$dm/dt_{(surface)} = dm/dt_{(comp\ layer)} = dm/dt_{(diff\ zone)} \tag{9}$$

The slowest of the three stages controls the nitrogen and carbon transfer rates. For a compound layer consisting of alternating ε-γ'-ε layers, the rate will be determined by the phase with the slowest diffusion properties.

2.2 Diffusion in austenitic stainless steel

The mechanisms of nitriding and carburizing involve the transfer of the diffusing species to the surface, the establishment of a diffusing species activity gradient which drives the diffusion process, and the diffusion for itself, may be accompanied by the formation of nitrides or carbides (on the surface or in the core). The diffusion of interstitial species into a metal can only proceed if it exists a chemical potential (or activity) gradient of those species between the surface and the core of the material.

The first step of a thermochemical treatment therefore leads to enrichment of the treated substrate surface with active species. This process makes it necessary to decompose or activate (thermally or in plasma) the gaseous atmosphere and to bring the active species to the surface, so that they can be initially absorbed and afterwards diffuse into the substrate.

The diffusion of the nitrogen and/or carbon elements successively leads to the following steps: (i) the formation of a diffusion layer enriched with the diffusing elements and if the solubility of the latter in the substrate is sufficient then this diffusion layer can be out of equilibrium at low temperatures (ii) at higher temperatures the follow steps occur. The surface formation of nitride, carbide or carbonitride layers of the main element of the substrate and (iii) the subsurface precipitation of nitrides, carbides or carbonitrides of alloying elements in the substrate (e.g. Fe, Ti, Al, Cr, Mo, V). In addition to the law of thermodynamics, the formation of the various phases is also govern by the nitrogen and carbon surface activities, and therefore are related to the temperature of the process used (gaseous or plasma), and to the composition of the gas.

Tables 1. (a) and (b) summarize the possible nitriding and carburizing configuration as described by Hertz, et al. (2008).

Substrate	N solubility	Potential nitrides	Compound layer+ diffusionlayer+ precipitation	Compound layer+ diffusion layer	Diffusion layer + precipitation	Diffusion layer only
Engineering steels	A little in α, more in γ	ε-$Fe_{2\text{-}3}N$ γ'-Fe_4N CrN, with alloys elements	Yes	Yes	Yes	Yes
Stainless steels	A little in α, more in γ	ε-$Fe_{2\text{-}3}N$ γ'-Fe_4N CrN, with alloys elements	Yes: but with reduced corrosion resistance	Yes	Not of industrial interest	Yes

(a)

Substrate	N solubility	Potential nitrides	Compound layer+ diffusion layer+ precipitation	Compound layer+ diffusion layer	Diffusion layer+ precipitation	Diffusion layer only
Engineering steels	A little in α, more in γ	Fe_3C, Cr–C, with alloys elements	Yes	Yes	Yes	Yes
Stainless steel	A little in α, more in γ	Fe_3C, Cr–C, with alloys elements	Yes: but with reduced corrosion resistance	Yes	Not of industrial interest	Yes

(b)

Table 1. Possible configurations of (a) nitriding, (b) carburizing.

To reduce further the potential of distortion and to avoid structural modifications of the substrate, and without repeating the quench and tempering treatments, these carburizing and nitriding treatments have evolved, in the past few years, towards lower temperature processes (350–450°C for austenitic stainless steels). This reduction in the treatment temperatures had to include specific treatments for removing oxide layers, which act as a barrier to the diffusion of nitrogen and carbon.

2.3 Diffusivity of simultaneous nitrogen and carbon in austenitic stainless steel

From diffusion experiments performed by Million et al, (1995), the interesting interactions of nitrogen and carbon are known, indicating that the presence of nitrogen enhances the activity of carbon and thus, its diffusion. It should, therefore, be possible to produce expanded austenite and to enhance the layer growth by simultaneous carbon and nitrogen implantation. Treatment of austenitic stainless steel in either nitrogen or methane plasma at 400 °C results in the formation of expanded austenite (Zhang et al, 1985 & Ueda et al, 2005). The different amounts of nitrogen or carbon in solid solution can be explained by the strength of the interaction between nitrogen or carbon and chromium. Williamson et al, (1994) noted that the strong interaction of nitrogen with chromium results in the trapping of nitrogen at chromium sites. This leads to a much higher supersaturation but reduced diffusivity in comparison to a methane treatment. However, the interaction is not really strong to form CrN. Carbon has a weaker interaction with chromium, so it diffuses inwards faster and a lower supersaturation is attained under similar treatment conditions. In both cases, nitrogen and carbon remain in solid solution, presumably on interstitial sites.

3. Thermochemical surface treatment to produce expanded austenite

As it has been known that the chemical composition of austenitic stainless steel makes them fully austenitic up to room temperature, and thus no phase transformation hardening takes place upon quenching. Consequently, surface treatments are an interesting alternative way to increase the surface hardness and improve the wear resistance. However, surface treatment of this steel has traditionally been considered bad practice (ASM , 1961), as it poses two main problems: the passive oxide film and the precipitation of chromium carbides (Sun et al, 1999). The passive chromium oxide film on austenitic stainless steel is stable under a wide range of conditions and isolates the substrate from the environment. This effect has been of interest for austenitic stainless steel components exposed to carburizing gas mixtures, either in service (Christ, 1998 & Yin, 2005) or for surface engineering purposes (Ueda et al, 2005). In the latter case, the oxide layer impairs diffusion of the hardening elements and, consequently, needs to be removed by applying some sort of surface activation process prior to the surface engineering treatment (Parascandola et al, 2001 & Sommers et al, 2004). Furthermore, traditional surface engineering treatments are conducted at high temperature, around 500-600 °C in the case of nitriding, and 900-1000 °C for carburizing (Zhang et al, 1985 & Ueda et al, 2005). At these temperatures, and with increasing availability of nitrogen and carbon from the hardening medium, profuse precipitation of chromium nitrides and carbides occurs, leading to a marked deterioration of the corrosion resistance of Austenitic stainless steel. However, low temperature thermochemical diffusion treatments with nitrogen and/or carbon have been reported to increase the surface hardness without affecting or even improving the corrosion resistance (Bell. T & Sun, 1998).

The most popular technology used to achieve the aforementioned low temperature thermochemical treatments of stainless steels is plasma technology, namely plasma nitriding (Rie & Broszeit, 1995; Stinville et al, 2010), plasma carburizing (Sun, 2005, Tsujikawa et al, 2007) and plasma hybrid treatments (Sun, 2008; Li et al, 2010). Due to the formation of a native oxide film stainless steel surface when exposed to air or residual oxygen before and during the treatment process, it is rather difficult to facilitate nitrogen and carbon mass transfer from the treatment media to the component surface. However, during plasma processing, due to the sputtering effects of energetic ions, the oxide film can be removed easily and effective mass transfer is obtained. This makes the plasma technology unique for surface treatment of stainless steels. An alternative is using the more conventional gaseous processes like gas nitriding (Gemma et al, 2001) and gas carburizing (Ernst et al, 2007).

These have proven feasible and industrially acceptable for performing low temperature nitriding and carburizing of stainless steels, provided that the component surface is activated before the gaseous process by special chemical treatments and the oxide film formed during the gaseous process is disrupted by introducing certain special gas components (Gemma et al, 2001).

Fluidized bed as one method of thermochemical surface treatments could employed as the expanded austenite (EA) layer formation on source of interest. To obtain the structure, thickness and and quality of the alloyed zone of γ_N and γ_C can be controlled by the processing parameters, such as temperature, time and gas composition in the fluidized bed. The duplex surface layer by combined carburizing and nitriding of 316L steel should be thick and mildly dropping hardness profile. Focusing in the concentration of hybrid process in terms of surface morphology, elemental profiles/structural characteristics, hardness and tribological properties, and corrosion behavior were placed in this presentation.

The use of fluidized bed furnace in heat treating operation has been introduced by Reynoldson which offers several advantages, including faster treatment time, precise control of treatment parameters, despite its economic benefits of low investment and operational cost (Reynoldson, 1995; Haruman & Sun, 2005). The schematic picture of fluidized bed furnace is shown on Fig. 4. Recent work has shown that low temperature nitriding of austenitic stainless steel is possible in a fluidized bed furnace (Haruman & Sun, 2005).

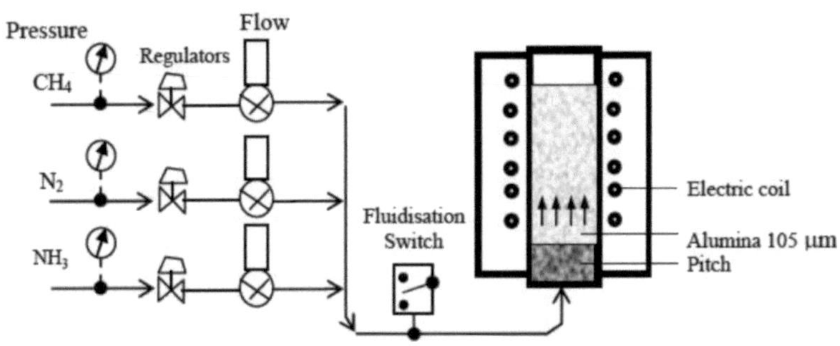

Fig. 4. Schematic picture of Fluidized bed furnace.

It is nowadays widely accepted that hard, wear and corrosion resistant surface layers can be produced on Austenitic stainless steel by means low temperature nitriding and/or carburizing in a number of different media (salt bath, gas or plasma), each medium having its own strengths and weaknesses (Bell, 2002). In order to retain the corrosion resistance of austenitic stainless steel, these processes are typically conducted at temperatures below 450 °C and 500 °C, for nitriding and carburizing respectively. The result is a layer of precipitation free austenite, supersaturated with nitrogen and/or carbon, which is usually referred to as S-phase or expanded austenite (Sun et al, 1999; Li, 2001; Li, et al., 2002; Christiansen, 2006).

Expanded austenite is the microstructural feature which responsible for the highly demanded combination of excellent corrosion and wear performances. Expanded austenite γ_X (X = N, C) hitherto also called S-phase (Ichii et al, 1986; Thaiwatthana et al, 2002;

Christiansen et al., 2004). Expanded austenite without nitrides/carbides is obtained when high amounts of atomic nitrogen and/or carbon are dissolved in stainless steel at temperature below 450 °C for nitrogen and about 550 °C for carbon. The nitrogen/carbon atoms are presumed to reside in the octahedral interstices of the f.c.c. lattice (Christiansen et al, 2004). Long range order among the nitrogen/carbon atoms has so far not been confirmed with X-ray diffraction techniques. Typically, nitrogen contents in expanded austenite range from 20 to 30 at% N; carbon contents range from 5 to 12 at% C (Sun et al, 1999 & Blawert et al, 2001). In terms of N:Cr ratio the homogenity range of nitrogen-expanded austenite spans from approximately 1:1 to 3:1 (Christiansen & Somers, 2005). Expanded austenite is meta-stable and tends to develop chromium nitrides/carbides (Li et al, 1999; Jirásková, et al., 1999; Christiansen & Somers, 2005,). The high interstitial content of C/N is obtained because of the relatively strong affinity of Cr atoms for N and (to a lesser extent) C atoms, leading to anticipated short range ordering of Cr and N/C. Due to the low mobility of Cr atoms as compared to interstitial N/C atoms at lower treatment temperatures, chromium nitrides/carbides do not precipitate until after long exposure times and N/C is kept in solid solution by the Cr "trap sites".

The improvement in wear resistance is perhaps the most outstanding feature of EA. The degree of improvement depends on the sliding conditions, but volume losses between one and two orders of magnitude lower than the untreated ASS are commonly reported for dry sliding (Thaiwatthana et al, 2002). This improvement is attributed to the increased surface hardness, with a typical ratio 4:1 compared to the untreated ASS (Qu et al, 2007). The EA layer prevents the surface from undergoing plastic deformation, and changes the wear mechanism from adhesion and abrasion, to a mild oxidational wear regime (Qu et al, 2007). However, under heavier loads, deformation of the subsurface occurs and leads to catastrophic failure, through propagation of subsurface cracks and spallation of the EA layer (Sun & Bell, 2002). In this way, the carbon EA layers, being thicker and tougher than their nitrogen counterparts, show some advantage.

With regard to corrosion, the results vary significantly depending on the testing conditions. Surprisingly, most researchers found that low temperature nitriding and/or carburizing do not harm the corrosion resistance of ASS, or even improve it. No conclusive explanation has been found for this improved corrosion behaviour, although it is clear that the benefit stands as long as nitrogen and carbon remain in solution and EA is free of precipitates (Li & Dong, 2003). In NaCl solutions, it is generally reported that EA remains passive under similar or wider range of potential compared to the untreated ASS, carbon EA showing a marginal advantage over nitrogen EA (Martin et al, 2002). Similar or slightly higher initial current densities have usually been measured on EA, together with the absence of pitting potential, in contrast to what is usual for ASS (Aoki & Kitano, 2002). Regarding repassivation, the evidence indicates that the passive film heals slower on EA than on ASS (Dong et al, 2006).

3.1 The influence of process variables and composition of expanded austenite

The depth profiles for thermochemically hardened stainless steels typically show a trend of increasing depth with higher temperatures and longer process durations. The very hard layer of nitrogen-expanded austenite exhibits a relatively shallow depth with an abrupt transition to the softer substrate material. The high hardness values associated with nitrided layer formation are consistent with the large compressive stresses in the residual stress profiles which were determined by XRD.

Previous investigation which regards to the influence temperature and time of thermochemical treatments using Fluidized bed shows that nitriding at 400°C for times up to 6 h could not produce a continuous nitrided layer on the substrate surface. When temperature was increased to 450°C, a uniform layer was formed after 6 h nitriding and was not effective for shorter treatments due to only a very thin discontinuous layer formation after 3 h nitriding, whilst after nitriding for 6 h a layer about 13 μm thick was formed, which has a bright appearance and is resistant to the etchant used to reveal the microstructure of the substrate. Increasing the temperature to 500°C resulted in the formation of a relatively thick nitrided layer after 3h and 6 h nitriding. The morphology of the nitrided layers formed at this temperature for longer treatment times is different from that formed at 450°C for 6h. Some dark phases were formed in the layers which is similar to those observed for plasma nitrided product and can be attributed to the decomposition of the S phase and the formation of chromium nitrides, which is believed reduce the corrosion (T. Bell, 1999).

The microhardness measurement on the nitrided subtrate showing a function of processing time for the three different temperatures. It can be concluded that no obvious hardening was achieved after nitriding at 400°C. The hardening effect is also insignificant after nitriding at 450°C for 1 h and 3h, and at 500°C for 1h. This corresponds well with the above metallographic examinations that no effective nitriding was achieved under these conditions.

According to these, both structural analysis and hardness measurement indicate that under the fluidized bed nitriding conditions, there exists an incubation time for the initiation of nitriding reactions. Nitriding must be carried out for a duration longer than the incubation time in order to produce an effective nitrided layer. From the experimental results, it is also evident that the incubation time is temperature dependent: increasing nitriding temperature reduces the incubation time.

This incubation time phenomena, which has not been reported for other nitriding processes, such as plasma nitriding, may be related to the nature of the fluidized bed process, where the disruption of the native oxide film on the specimen surface, which is required to effect the nitriding reactions, has to rely on thermal dissociation. The higher the temperature, the faster is the dissociation of the oxide film and thus the shorter the incubation time.

4. Experimentals method

The substrate material used in this work was AISI 316L type austenitic stainless steel of following chemical compositions (in wt.%): 17.018 Cr, 10.045 Ni, 2.00 Mo, 1.53 Mn, 0.03 C, 0.048 Si, 0.084 P, 0.03 S and balance Fe. This steel was supplied in the form of 2 mm thick hot-rolled plate. Samples of 20 mm x 70 mm size rectangular coupon were cut from the plate. The sample surface was ground on 320, 600, 800, 1000, 1200 grit SiC papers, and then polished using 1 μm Al_2O_3 pastes to the mirror finish. Before treating, these specimens were cleaned with acetone. The treatments were performed at 450°C for a total duration of 8 hours in an electrical resistance heated fluidized bed furnace having 105 μm particulate alumina as fluidized particles which flow inside the chamber due to the flow of nitriding or carburizing gases. The fluidized bed furnace, which was manufactured by Quality Heat Technologies Pty Ltd has a working chamber of 100mm diameter x 250mm deep with maximum worksize of 70mm diameter x 150mm high. Before charging the samples, the chamber was heated to the treatment temperature of 450oC with the flow of nitrogen gas at 1.05 m^3 per hour. Then the samples were charged to the furnace and the treatment gases were introduced and their flow rates were adjusted to meet the required composition, with the total gas flow rate maintained at 0.62 m^3 per hour. Table 2. summarizes the process conditions employed in this work. Four different treatments were conducted, including low temperature nitriding, carburizing, hybrid process, and sequential carburizing-nitriding. The hybrid process involved treating the sample in an atmosphere containing both NH_3 (for nitriding) and CH_4 (for carburizing) for a total duration of 8 h, whilst the sequential process involved treating the sample in the carburizing atmosphere for 4 h and then in the nitriding atmosphere for further 4 h.

Nitriding, carburizing, and hybrid treatments were performed at 450°C in a fluidized bed furnace having particulate alumina as fluidized particles which flow inside the chamber due to the flow of nitriding or carburizing gases.

	Symbol	Temp. (°C)	Gas (%)			Time	Temp.(°C)	Gas (%)			Time	Layer Thickness (μm)
			CH_4	N_2	NH_3			CH_4	N_2	NH_3		
Nitriding	**8N**	450°		85	15	8	No treatment	0	0	0	No treatment	8,35
	5N	450°		85	15	5	No treatment	0	0	0	No treatment	5,10
	2N	450°		85	15	2	No treatment	0	0	0	No treatment	3,26
Carburising	**8C**	450°	5	95		8	No treatment	0	0	0	No treatment	3,92
	5C	450°	5	95		5	No treatment	0	0	0	No treatment	1,63
	2C	450°	5	95		2	No treatment	0	0	0	No treatment	1,20
Nitrocarburising	**8(C+N)**	450°	5	80	15	8	No treatment	0	0	0	No treatment	4,00
	5(C+N)	450°	5	80	15	5	No treatment	0	0	0	No treatment	2,16
	2(C+N)	450°	5	80	15	2	No treatment	0	0	0	No treatment	1,25
			First Step				Second Step					
Hybrid	**4C-4N**	450°	5	95		4	450°		85	15	4	5.2
	2C-3N	450°	5	95		2	450°		85	15	3	1.6
	1C-1N	450°	5	95		1	450°		85	15	1	1.37

Table 2. Treatment conditions and their corresponding layer thicknesses.

The specimens were heated by electrical resistance heating. Prior to treating, the specimens were soaked in concentrated HCl (2 M) solution for 15 minutes duration with the purpose to remove the native oxide film that commonly forms on austenitic stainless steel and protects the metal matrix from corrosion. This oxide layer is believed to act as a barrier for diffusional nitrogen transport (Rie, 1996). After thermochemical treatments, the specimens were quenched in water. The treated specimen cross sections were first characterized by metallographic examination. To reveal the microstructure, the polished surface was etched in Marble's solution (4 g $CuSO_4$ + 20 ml HCl + 20 ml distilled water). The schematic picture of fluidized bed furnace is shown on Fig. 4.

The specimens were further characterized by microhardness indentation, elemental analysis by FESEM and X-ray diffraction (XRD) analysis using Cu-Kα radiation. Tribological properties were evaluated with a Taber® Linear abraser model 5750 dry slide tribo-tester using an 5-mm diameter AISI 316L collet nut as mate material. The stroke length applied was 25.4 mm under a constant load 600 g. After 3600 cycles of sliding (completed in 60 minutes) having maximum velocity of 79.76 mm/sec, the specimen wear loss was measured by balance to evaluate cumulative weight loss. Microstructures of treated layers were investigated by X-Ray diffraction analysis (XRD) using Cu-Ka (40 kV, 150 mA) and Field Emission Scanning Electron Microscopy (FESEM). The electrochemical corrosion behaviour of the as-treated surfaces was evaluated by measuring the anodic and cathodic polarisation curves in aerated 3.0 % NaCl solution at a scan rate of 1 mV/min. The tests were conducted at room temperature by using a three electrode potentiostat with a computer data logging, requisition and analysis system. Potentials were measured with reference to the standard calomel electrode (SCE).

Corrosion tests were performed electrochemically at room temperature in a flat cell with 3.0% NaCl in distilled water. The flat cell, as schematically shown in Fig. 5, was a three-electrode set-up consisting of a saturated calomel reference electrode (SCE), a platinum auxiliary electrode and a working electrode (sample). Sample to be tested was placed against a Teflon ring at one end of the flat cell, leaving a theoretical circle area of 67.5 mm^2 on the sample surface in contact with the testing solution through a round hole in the Teflon ring. Test control, data logging and data processing were achieved by a "Sequencer" computer software. The scanning potential was in the range of -0.5 to + 1.4 V, and the scan rate was 1 mV/s. From the polarization curves, the average values of the corrosion potential (E_{corr}), the corrosion current density (I_{corr}) and the polarisation resistance (LPR) were calculated.

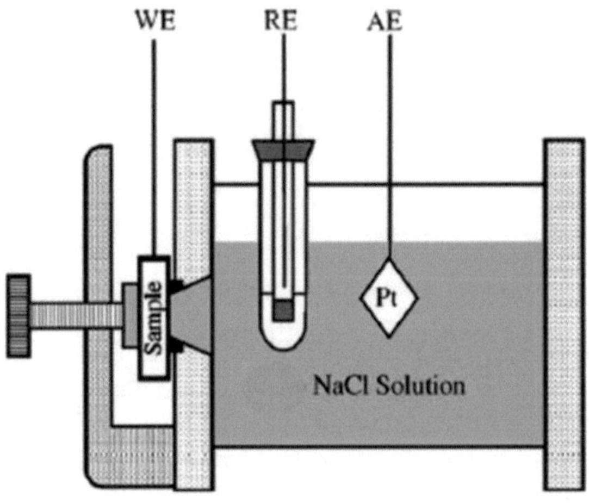

Fig. 5. Schematic diagram of the flat cell used for polarization corrosion test (Li & Bell, 2004).

5. Key results

5.1 Layer morphology and hardness profile

Hardened layers with different morphologies were observed as a result of the various treatment conditions and the thicknesses of the layers produced in different conditions are shown in Table 2. The layer thicknesses are found to be different at different treatments, and their growths against time in Fig. 6. show that layer thickness increases with processing time.

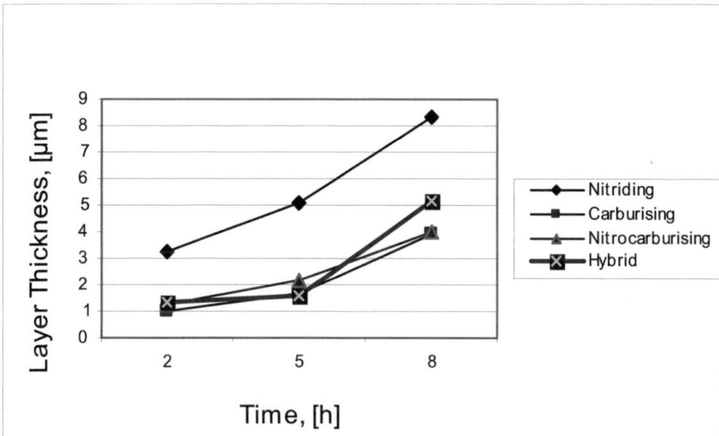

Fig. 6. Thickness of treated layers measured from micrographs.

Micrographs in Fig. 7 show that the morphology of the layer changed with treatment conditions. The two specimens processed under combined treatment conditions, 8(C+N), and 4C-4N produced duplex layers irrespective of whether they were processed simultaneously (Fig. 7b) or sequentially (Fig. 7d). The processed layer thicknesses in Table 2. show that the nitrided specimen, 8N, has a thickness between 3.26 to 8.35 µm and the carburized specimen, 8C, is in between 1.00 to 3.92 µm.

Furthermore, the nitrided-only 8N specimens have deeper layers than combined processed specimens. The depth of the simultaneously carburized and nitrided specimen, 8(C+N), reaches only 50% that of the nitrided specimen, and the thickness of 4C-4N specimen had only about 45% compared to the nitrided-only 8N specimen after being processed for the same duration of 8 h due to the half nitriding duration.

For a similar treatment duration, the Plasma process is reported by Tsujiwaka on 2005 which is produce about 18 µm thick layer which is much higher compared to that of the present conventional nitriding treatment in fluidized bed furnace. In plasma process the native oxide layer is removed mostly by bombardment of the plasma gas which is completely absent in conventional fluidized process. This is one of the reasons why convention fluidized bed treatment produced small layer thickness compared to the corresponding plasma nitriding. Previous investigation revealed that nitriding at 450°C became effective after treatment for 6 h where a continuous nitrided layer was produced (Sun, 2006). This is due to the fact that the incubation time phenomena which may be related to the nature of the fluidized bed process, where the disruption of the native oxide film to cause the nitriding reactions, has to be by thermal dissociation. The higher the temperature, the faster is the dissociation of the oxide film and thus the shorter the incubation time. According to this hypothesis it is understood why a very thin discontinuous layer was formed after a 2 h treatment time ; the layer was about 8 µm thick after 8 h nitriding, which gave a bright appearance and was resistant to the etchant used to reveal the microstructure of the substrate. The treated layers for 4C-4N consist actually of two separate zones with a somewhat diffused interface which was clearly observed under microscope, but not revealed in the micrographs. The outer zone is γ_N and the inner zone is γ_C. Conversely, the nitrocarburised sample shows a distinct separation of the γ_N and γ_C layers, with the γ_C layer closest to the austenite substrate (as indicated in the micrograph in Fig. 7b). The X-ray diffractograms of carburised and nitrided AISI 316L show that two different types of expanded austenite are present (Fig. 6).

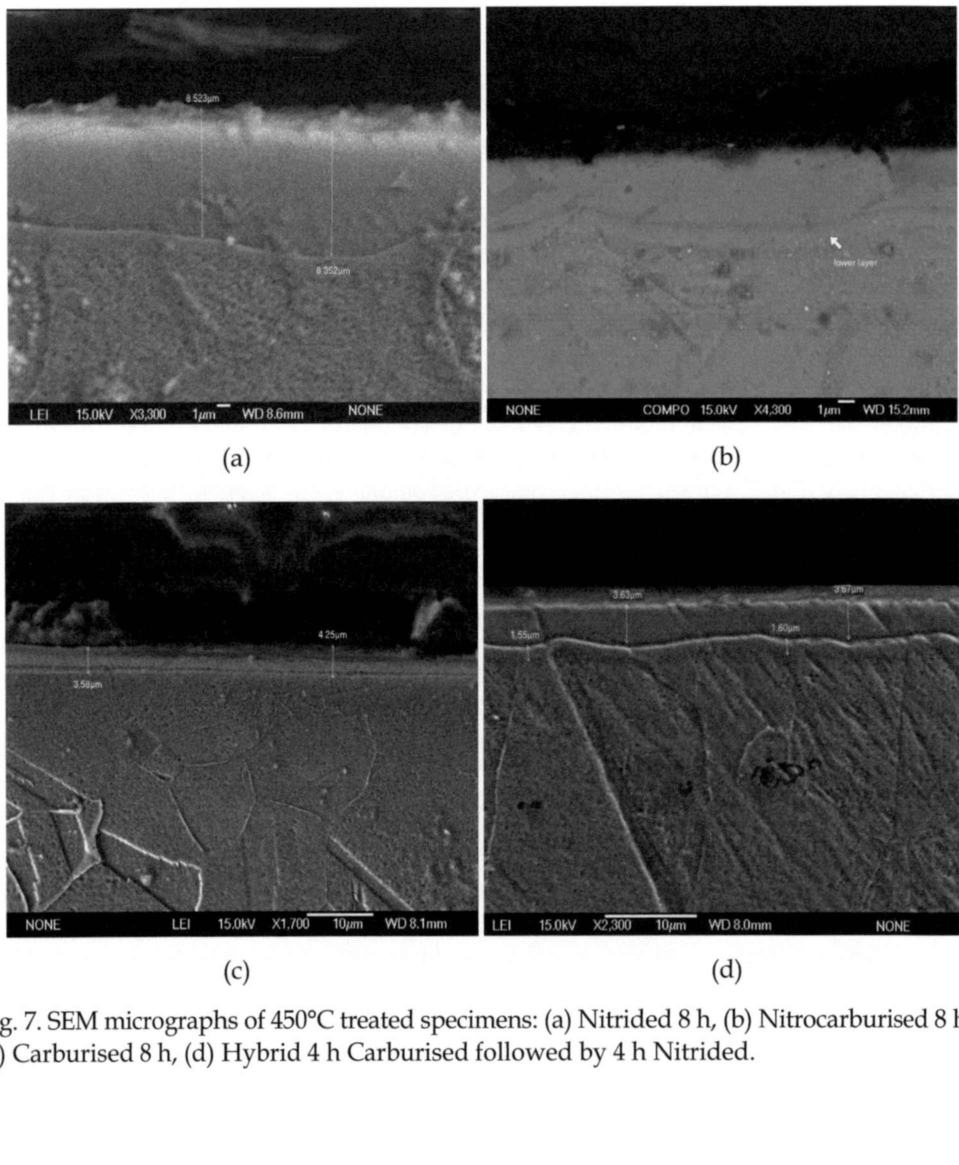

Fig. 7. SEM micrographs of 450°C treated specimens: (a) Nitrided 8 h, (b) Nitrocarburised 8 h, (c) Carburised 8 h, (d) Hybrid 4 h Carburised followed by 4 h Nitrided.

Fig. 8 shows the hardness depth profiles of the treated specimens. The carburized 8C specimen developed a maximum hardness of about 500 Hv, which is much lower than the hardnesses of 1230 to 1588 Hv for other three nitrided and nitrocarburised specimens. The nitrided layer of the 8N specimen produced a hard layer of 1588 Hv with an abrupt layer–core interface, while the 8C carburizing produced a gradually decreased hardness profile. Two combined carburized and nitrided specimens, 8(C + N) and 4C–4N developed a similar tendency to bulge in hardness profiles at inner carburized layer as shown in Fig. 7. The most gradual decrease in hardness from 1230 Hv level to substrate hardness was displayed by the 4C–4N specimen.

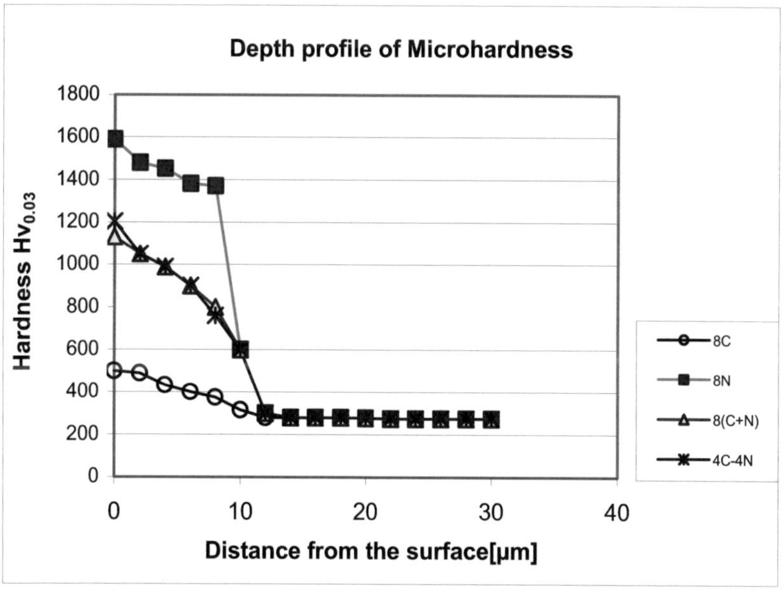

Fig. 8. Depth profiles of Microhardness.

5.2 Elemental profile of carbon and nitrogen

The carbon profile investigated on 4C-4N specimen using EDS FE-SEM is depicted in Fig. 9. It is found that higher carbon at the deeper layer which indicates that carbon pushed-ahead by the incoming nitrogen atom and the dissolved carbon is accumulated at the front of the nitride layer which has also been reported in the literature (Lewis et al, 1993).

Elemental analysis of the Hybrid specimen CN gave more carbon beyond the nitrided layer, but some carbon was also observed at the surface. The variations of chemical concentration in the hybrid nitrocarburized layer were also measured with EDS. These figures shows the typical nitrogen and carbon profiles produced in treated 316L steel. It can be seen from these Figures that there are two features in the nitrogen and the carbon profiles. Firstly, the nitrogen profile on the surface of the treated layer is similar to that of nitrided 316L steel. Secondly, the maximum nitrogen concentration is on the surface and the maximum carbon concentration appears beneath as if carbon was 'pushed' to the middle of the layer by nitrogen.

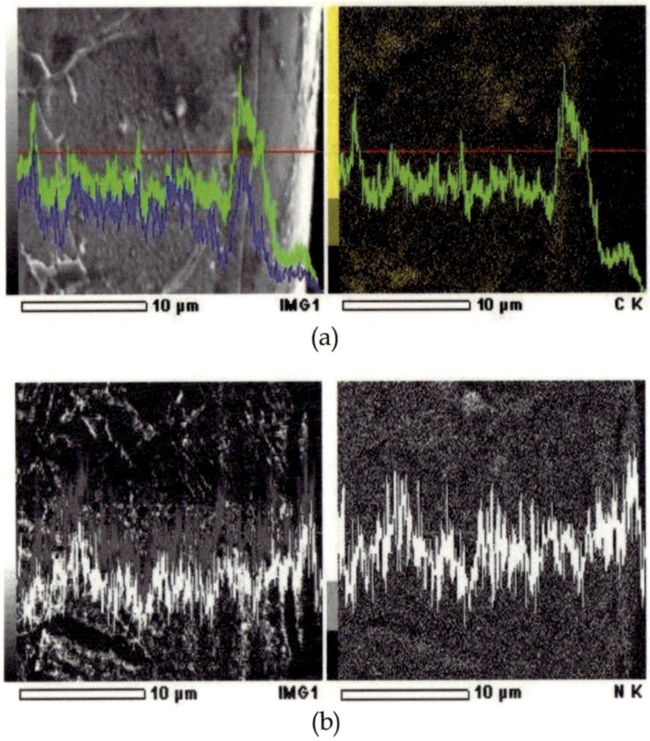

Fig. 9. Carbon profiles (a) and nitrogen profiles (b) along the depth.

Such distributions of nitrogen and carbon in the surface layer are likely to produce some beneficial influences upon the properties of hybrid treated 316L steel. Figure 6b shows results of nitrogen concentration on hybrid dual-stages obtained from energy dispersive X-ray (EDS) analysis. According to these curves, it can be clearly seen that the surface hybrid layer contains very high amount of nitrogen, and nitrogen concentration is gradually reducing from surface to the core with distance increasing due to a low diffusion rate in the case of samples at low temperature. However, some carbon remains in the sub-surface layer.

Fig. 10 summarizes the phase compositions in the treated specimens as determined by XRD from the specimen treated at 450°C for 8h. As confirmed by XRD analysis in Fig. 10, the nitriding treated surface layer comprises mainly the S phase or the expanded austenite. For the hybrid process, consisting of dual layers (Figs. 7d & 7e), revealed another thin interfacial layer. This interfacial layer is believed to be due to the accumulation of carbon as has also been reported in literature (Sun, 2006). One interesting aspect of the diffraction displayed in Fig. 10 regards the variation of the (200) diffraction line width in relation with 2θ angle. This behaviour can be explained by the lattice distortion caused by the greater amount of nitrogen in the interstitial sites and/or only by crystallographic orientation present in this phase. The XRD analysis did not show any peak from nitride or carbide phase.

In accordance with the findings for plasma nitriding (Lewis, 1993; Rie, 1995), the S phase layer produced in this fluidized bed furnace process has minimal chromium nitride/carbide precipitation. Comparing the diffractograms for the nitrided samples with the untreated material, it clearly shows that Bragg reflections (peaks) are shifted to lower 2θ angles. It was caused primarily by the dissolution of nitrogen which causes a dilation of the fcc lattice (hence the name expanded austenite), although residual stress and stacking faults also play a role in this respect (Somers, 2005). The X-ray diffraction pattern of carburised AISI 316L is shown in Fig. 10. γ_C is identified as the only phase present in the surface adjacent region, i.e. within the information depth for the probing X-ray beam. A marked difference is observed as compared to nitrided AISI 316L; a smaller shift of the austenite peaks to lower 2θ, which indicates a substantially lower content of the interstitially dissolved atoms, provided that nitrogen and carbon induce a similar distortion in the fcc lattice. The asymmetrical (200) austenite peak in Fig. 10 indicates a depth-gradient of the carbon content in the near surface zone. The distinct peaks for the carburised sample indicate a smooth concentration gradient and lower defect density in γ_C layers as compared to γ_N layers.

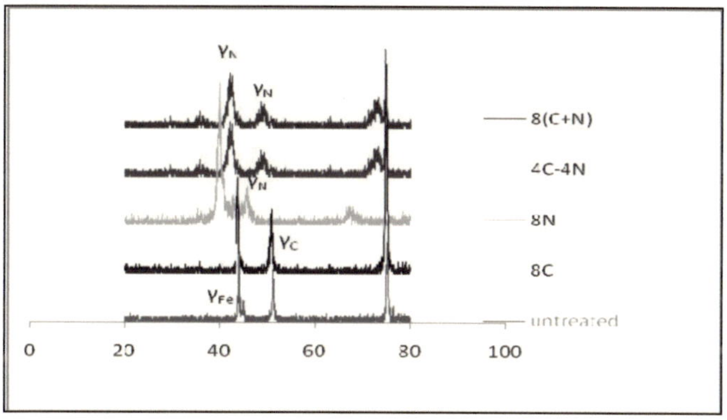

Fig. 10. Comparison of XRD patterns of treated specimens.

5.3 Wear property

The wear properties of the low temperature surface treatment specimens as weight loss under dry sliding friction are presented in Fig. 11 along with an untreated specimen for comparison purpose. The results suggest that the fluidized bed thermochemical-treated specimens have excellent wear resistance. The 8N specimen has the highest wear resistance compared to the values of 4C-4N, 8(C+N), and 8C specimens.

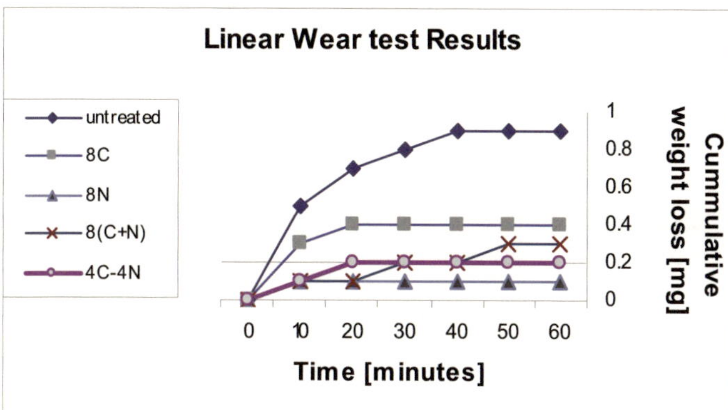

Fig. 11. Wear under dry sliding condition.

The highest hardness for 8N specimen is considered to be responsible for best wear resistance property among the treated specimens used in this investigation. However the findings suggest that nitriding, hybrid, nitrocarburizing and carburizing the austenitic stainless steel at 450^{0}C using a fluidized bed furnace can improve surface hardness and wear resistance of austenitic stainless steel. It is to be noted that at the initial stage of sliding, all the specimens in Fig. 11 gave accelerated weight loss and then leveling off after certain period. It is presumed that at the initial stage of sliding, the 600g load of the sliding mate material was encountered by the asperities of substrate surface, which effectively caused high load sliding and thus more wear loss. The eventual dropping off may be related to smoothening of the asperities at the wear surface, which produced more contact area for the sliding load of 600g and hence reduced or constant wear rate.

The work hardening effect may also cause this tendency together with possibilities of surface oxide or carbide/nitride formation at a certain period of sliding, thus leading to an equilibrium condition of constant wear rate. However, no evidence is available to explain the exact reasons of these wear phenomena.

5.4 Corrosion properties

Corrosion tests using the electrochemical technique demonstrated that the precipitation free carburized and nitrided layers have very good corrosion resistance in the corrosive environments.

The most subtantial improvement in properties of austenitic stainless steels by the hybrid process lies in corrosion resistance as evaluated by electrochemical testing (Li & Bell, 2004). Fig. 12 shows the anodic polarization curves measured for several specimens in 3.0% NaCl solution. As expected, both individual nitriding and carburizing reduce the current density of the steel in the anodic region, indicating improved corrosion resistance. After the hybrid treatment, the anodic polarization curve is shifted towards lower current density by several orders of magnitude as compared to that for the untreated and individually nitrided and carburised steel. This registers an improvement in corrosion resistance by several orders of magnitude and signifies the excellent corrosion resistance of the hybrid treated surface. The much enhanced corrosion resistance observed for the hybrid treated surfaces may be attributed to the extremely large supersaturation of the upper part of the nitrogen-enriched layer with both nitrogen and carbon (see Fig. 9). This would contribute to the observed higher hardness and better corrosion resistance as compared to those achieved by individual nitriding and carburizing.

The treatment conditions are the same as those in Fig. 7. The electrochemical test results for Hybrid-NCT, Nitriding-NT, Carburizing-CT were described in Fig. 12. The NT and CT showing that the current density of treated stainless steel were decreased in the anodic region which indicating positive effect regarding the improvement of corrosion resistance compared to the substrate. After Hybrid-NCT treatment, the anodic polarization curved is shifted towards lower current density which explain that the corrosion rate was decreased and the polarization current measurement gave 0.00003 mA/cm^2 and demonstrate an improvement in corrosion resistance as compared to that untreated and individually nitrided and carburized steel, while passivation current of NCT is the lowest followed by CT, NT and untreated respectively. This trend also similar to the maximum potential passivation behaviour since the dissolution current density increased slowly and gradually with applied potential (Y. Sun & E. Haruman, 2008). Although in the first 250 mV/SCE scan of CT show small increases in current densities where the re-passivation behavior start to occur.

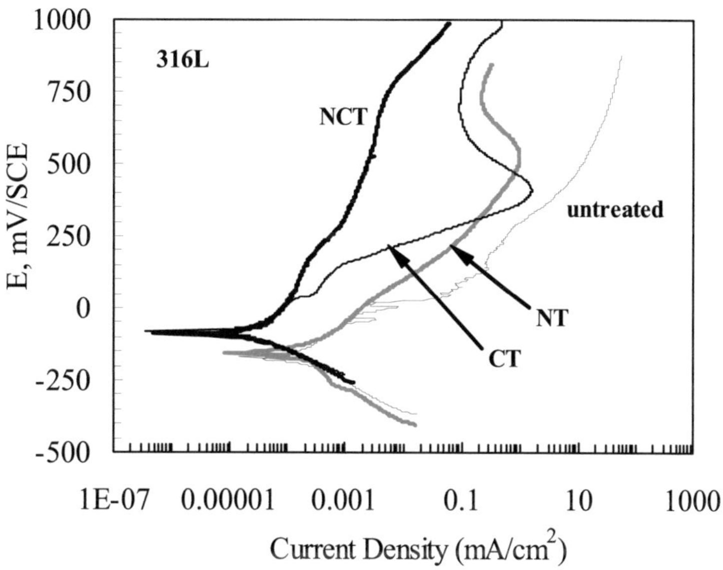

Fig. 12. Anodic polarization curves measured in 3.0%NaCl.

Although, the corrosion behaviour of low temperature nitrided, carburized and hybrid stainless steel thermochemical treatment have been investigated by several investigators (Zhang et al, 1985; Rie et al, 1995; Sun et al, 1999), the reason for improvement of corrosion resistance due to nitrogen and carbon supersaturation in austenite has not been fully understood. A possible mechanism is that supersaturation of nitrogen and carbon promotes the improvement of the passivation ability of austenite, and this effect seems to give beneficial with increasing degree of supersaturation (Munther et al, 2004). Thus, the improvement of corrosion resistance for hybrid-NCT treated material may be attributed to the extremely large supersaturation of the upper part of nitrogen-enriched layer with both nitrogen and carbon. This would be contribute to the observed higher hardness and better corrosion resistance as compared to those achieved by individual nitriding-NT and carburizing-CT.

6. Conclusion

The thermochemical treatments of AISI 316L stainless steel in a fluidized bed process at 450°C demonstrate that it is possible to produce hard layer of an expanded austenite phase without precipitation of chromium carbide/nitride. For nitriding and carburizing treatments the expanded layers consisted of a single layer γN or γC phase while specimens treated by nitrocarburizing or hybrid process gave dual layers consisting of γN at the surface and γC ahead of γN. The layer produced in fluidized bed process is not uniform in thickness under the same treatment conditions. The nitriding treatment produced 8.35 μm thick layers after 8 h duration while carburizing and nitrocarburizing gave much smaller thicknesses for the same processing time. However the layer thickness is found to increase with the treatment time for all the processes. The nitrided treatment developed the highest hardness of 1600 Hv; 1150 Hv and 500 Hv was found for the nitrocarburised and carburized specimens, respectively. All treated specimens gave very good wear resistance compared to the untreated specimen; however, the nitrided specimen produced very high wear resistance which corresponds to highest hardness among the specimens tested. Thermochemical surface treatments of 316L were capable to produce expanded austenite layers which considerably improved the corrosion properties of 316L austenitic stainless steel. The electrochemical test results show that hybrid-NTC can significantly improve the corrosion resistance of austenitic stainless steel which is much better than that measured for the untreated stainless steel and the individually nitrided-NT and carburized-CT layers.

7. References

Akita, M. and Tokaji, K.: *Effect of carburizing on notch fatigue behaviour in AISI 316 austenitic stainless steel*, Surface and Coatings Technology (2006) 200, 20-21, 6073-6078

Aoki, K. and Kitano, K.: *Surface hardening for austenitic stainless steels based on carbon solid solution*, Surface Engineering (2002) 18, 6, 462-464

C.X. Li and T.Bell, Corrosion Science 46, pp. 1527-1547 (2004).

Ceschini, L. and Minak, G.: *Fatigue behaviour of low temperature carburised AISI 316L austenitic stainless steel*, Surface and Coatings Technology (2007) 202, *9*, 1778-1 784

Clark. D.S. and Varney. W.R., Physical Metallurgy for Engineers, Litton educational publishers, (1962).

Committee on gas carburizing, A. S. M.: *Carburizing and carbonitriding*, 1st Ed. (1977) Metals Park, Ohio, American Society for Metals

D. B. Lewis, A. Leyland, P. R. Stevenson, J.Cawley and A. Matthews, Metallurgical study of low-temperature plasma carbon diffusion treatments for stainless steels, *Surf. Coat. Tech.* 60 (1993) 416-423.

D. Munther, H.-J. Species, H. Biermann, Chr. Eckstein, Trans. Mater. Heat Treat. 25 (5) (2004) 311-315.

Dong, H., Qi, P.-Y., Li, X. Y. and Llewellyn, R. J.: *Improving the erosion-corrosion resistance of AISI 316 austenitic stainless steel by low-temperature plasma surface alloying with N and C*, Materials Science and Engineering A: Structural Materials: Properties, Microstructure and Processing (2006) 431, 137-145

E. Haruman and Y. Sun, Proc. 3rd Asian Conf. on Heat Treat. of Mater., Gyeongju, Korea, 10-12 Nov, 2005.

E. Haruman, Y. Sun, H. Malik, A.G.E. Sutjipto, S. Mridha, K. Widi, Low Temperature Fluidized Bed Nitriding of Austenitic Stainless Steel, *Solid State Phenomena.*, Vol. 118, pp. 125-130, 2006.

F. Borgioli, A. Fossati, E. Galvanetto and T. Bacci, "*Glow-discharge nitriding of AISI 316L austenitic stainless steel: influence of treatment temperature*", Surfaces and Coatings Technology, 200, 2474 - 2480, (2005).

F. Ernst, Y. Cao, G.M. Michal, A.H. Heuer, Carbide precipitation in austenitic stainless steel carburized at low temperature, *Acta Mater.* 55 (2007) 1895-1906.

Hertz, et al., (2008) Technologies for low temperature carburizing and nitriding of austenitic stainless steel, *International Heat Treatment and Surface Engineerng*, vol. 2, No. 1.

Hurricks, P. L.: *Some aspects of the metallurgy and wear resistance of surface coatings*, Wear (1972) 22, *3*, 291-319

J. Qu, P. J Blau and Jolly, B. C.: *Tribological properties of stainless steel treated by colossal carbon supersaturation*, Wear (2007) 263, 1-6, 719-726

J.C. Stinville, P. Villechaise, C. Templier, J.P. Riviere, M. Drouet, Plasma nitriding of 316L austenitic stainless steel: Experimental investigation of fatigue life and surface evolution, *Surf. Coat. Tech.* 204 (2010) 1947-1951.

K. Gemma, T. Obtruka, T. Fujiwara, M. Kwakami, Prospects for rapid nitriding in high Cr austenitic alloys, in *Stainless Steel 2000*, p159-166, Ed. Tom Bell and Katruya Akamatsu, Maney Publishing, Leeds, 2001.

K.-T. Rie, E. Broszeit, Plasma diffusion treatment and duplex treatment – recent development and new applications, *Surf. Coat. Tech.* 76–77 (1995) 425-436.

Li, X. Y. and Dong, H.: *Effect of annealing on corrosion behaviour of nitrogen S phase in austenitic stainless steel*, Materials Science and Technology (2003) 19, *10*, 1427-1434.

M. Tsujikawa, D. Yoshida, N. Yamauchi, N. Ueda, T. Sone, S. Tanaka, Surface material design of 316 stainless steel by combination of low temperature carburizing and nitriding, *Surf. Coat. Tech.* 200 (2005) 507-511.

M. Tsujikawa, S. Noguchi, N. Yamauchi, N. Ueda and T. Sone, Effect of molybdenum on hardness of low-temperature plasma carburized austenitic stainless steel, *Surf. Coat. Tech.* 201 (2007) 5102-5107.

Martin, W. C. and Wiese, W. L.: *Atomic, molecular, and optical physics handbook* in 2.1st Ed. (2002) Gaithersburg, NIST,

Parrish, G. and Harper, G. S.: *Production gas carburizing*, 1st Ed. (1985) Oxford, Pergamon

Reynoldson R.W, Advances in surface treatments using Fluidized beds, *Surface and Coatings Technology.*, Vol. 71; 2, pp. 102-107, 1995.

Somers, M.A.J, Christiansen, T., and Møller, P. Case-hardening of stainless steel European Patent 1521861 (2004) EU.

Somers, M.A.J., and Christiansen, T., (2005) Kinetics of microstructure evolution during gaseous thermochemical surface treatment. *J. Phase Equilibria and Diffusion*, No. 5, vol. 26, p. 520-528.

Standard British Standard EN 10052:1994, Vocabulary of heat treatment terms for ferrous products, BSI, London, www.bsi-global.com

Sun, Y. and T. Bell.: *Plasma surface engineering of low alloy steel*, Materials Science and Engineering A: Structural Materials: Properties, Microstructure and Processing (1991) A140, 419-434

T. Bell and Y. Sun, Low temperature plasma nitriding and carburizing of austenitic stainless steels, *Heat Treatment of Met*als 29 (3) (2002) 57-64

T. Bell, *Bodycote-AGA* Seminar, Lidingö, 2005.

T. Czerwiec, Presentation in International Symposium on Surface Hardening Corrosion Resistant Alloys - ASM, Case Reserve Western University, Cleveland, Ohio USA. May, 2010.

Thaiwatthana, S., Li, X. Y., Dong, H. and Bell, T.: *Comparison studies on properties of nitrogen and carbon S phase on low temperature plasma alloyed AISI 316 stainless steel*, Surface Engineering (2002) 18, *6*, 433-437

Thaiwatthana, S., Li, X. Y., Dong, H. and Bell, T.: *Corrosion wear behaviour of low temperature plasma alloyed 316 austenitic stainless steel*, Surface Engineering (2003) 19, *3*, 211-216

X.Y. Li, J. Buhagiar, H. Dong, Characterisation of dual S phase layer on plasma carbonitrided biomedical austenitic stainless steels, *Surf. Eng*. 26 (2010) 67-73.

Y. Sun and E. Haruman, Influence of processing conditions on structural characteristics of hybrid plasma surface alloyed austenitic stainless steel, *Surf. Coat. Tech.* 202 (2008) 4069-4075.

Y. Sun and T. Bell.: *Dry sliding wear resistance of low temperature plasma carburised austenitic stainless steel*, Wear (2002) 253, *5-6*, 689-693

Y. Sun and T. Bell.: *Effect of layer thickness on the rolling-sliding wear behavior of low- temperature plasma-carburized austenitic stainless steel*, Tribology Letters (2002) 13, *1*, 29-34

Y. Sun, Kinetics of low temperature plasma carburizing of austenitic stainless steels, *J. Mater. Proc. Tech.* 168 (2005) 189-194.

Y. Sun, X.Y. Li and T. Bell, Low temperature plasma carburizing of austenitic stainless steels for improved wear and corrosion resistance, *Surf. Eng*. 15 (1999) 49-54.

Y. Sun, X.Y. Li and T. Bell, X-ray diffraction characterisation of low temperature plasma nitrided austenitic stainless steels, *J. Mater. Sci.* 34 (1999) 4793-4802

15

Corrosion Performance and Tribological Properties of Carbonitrided 304 Stainless Steel

A.M. Abd El-Rahman[1,2], F.M. El-Hossary[1], F. Prokert[2], N.Z. Negm[1], M.T. Pham[2] and E. Richter[2]
[1]Physics Department, Faculty of Science, South Valley University, Sohag Branch, Sohag,
[2]Institut für Ionenstrahlphysik und Materialforschung, Helmholtz-Zentrum Dresden-Rossendorf,
[1]Egypt
[2]Germany

1. Introduction

In general, the solid solution austenitic phase (γ) with high chromium content (12 % - 20 %) is responsible about the excellent corrosion performance of austenitic alloys. This advantage allows these alloys to use in biomedical, food and chemical, pulp and paper chemical, petrochemical, heat exchange and nuclear power plant industries [1-4]. However, most of these applications are suffering from their relatively low hardness and poor tribological properties.

Various surface modification technologies such as nitriding, carburizing and nitrocarburizing are used to improve the mechanical and tribological properties of austenitic stainless steels [5-12]. In most cases an increase in surface hardness is accompanied by a decrease in corrosion resistance [13]. The decrease in the corrosion resistance is caused by heavy precipitations of chromium carbide and chromium nitride on the grain boundaries, which are surrounded by chromium-depleted zones [14]. More investigations are succeeded to maintain and sometimes to improve the corrosion resistance of stainless steels after nitriding [15-16]. It is well known that the formation of nitrogen supersaturated solid solution phase without CrN precipitations should maintain the good corrosion resistance of stainless steel [5, 17].

In this paper we present the effect of N_2 to C_2H_2 gas pressure ratio on the corrosion performance and tribological properties of AISI 304 austenitic stainless steel after rf plasma carbonitriding at a relatively low pressure.

2. Experimental work

The samples were treated at a fixed input plasma power of 450 W and for a processing time of 10 min. The gas pressure related to N_2/C_2H_2 ratio was varied from 100% N_2 to 100% C_2H_2. The pressure was increased from an atypical base pressure of $1.3\text{x}10^{-2}$ mbar to a total gas pressure of 8.4x10^{-2} mbar. The sample was heated mainly by the rf field. The sample temperature was measured during the rf plasma process by a Chromel-Alumel thermocouple, attached to the sample holder. As shown in Fig. 1, the substrate temperature was influenced by the effect of gas compositions. It was found that the temperature gradually increases from 475 °C for pure nitriding up to 550 °C for carbonitriding (50% C_2H_2, 50% N_2) and raises up to 600 °C for pure carburizing. Grazing incidence X-ray diffraction (GIXRD) with Cu Kα radiation was used to determine the phases, present in the treated layers. For the chosen incidence angle of 2° the (1/e)-penetration depth of the X-rays was approximately 700 nm. The recorded diffraction pattern shows therefore mainly the structure of the phases formed in this near-surface region. In this paper we concentrate on the study of corrosion resistance, surface morphology before and after corrosion, and tribological properties of the treated samples. The surface roughness was measured by use of the rough machine (Dektak 8000, Veeco Instrument GmbH). Wear and friction measurement were performed at room temperature in laboratory air with low humidity of 16 to 24 % using an oscillating ball-on-disk type tribometer wear tester without lubrication. The 3-mm ball of cobalt tungsten carbide was moved at mean sliding speed of 15 mm/sec with different normal loads of 3, 5 and 8 N. The corrosion properties were evaluated using the electrochemical testing technique. The corrosion tests were performed in a 1 wt. % NaCl solution by application of the potentiodynamic polarization method. A three-electrode electrochemical cell has been used, counter and reference electrode were related to Pt and saturated calomel electrode, respectively.

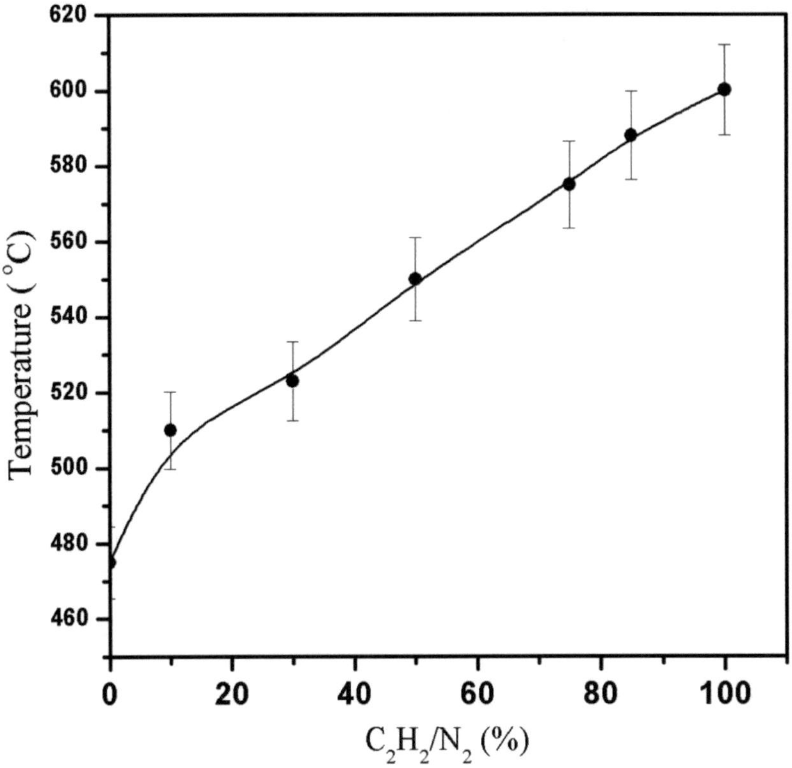

Fig. 1. Temperature variation as a function of the gas composition.

The anodic polarization curves were recorded with potential scan rate of 10 mV /sec. The potentiodynamic polarization curve is plotted using AutoLab PGSTAT 12 + GPES software. The surface morphology before and after corrosion tests and the chemical composition of selected parts of the as-prepared layers were examined by scanning electron microscopy (SEM) and energy dispersive X-ray analysis (EDAX), respectively.

3. Results

3.1 Phase formation

The microstructure of the untreated sample and samples treated at different gas compositions (N_2/C_2H_2) obtained by GIXRD have been studied by us before [11] and it is shown in Fig. 2. It is briefly described here to correlate tribological and corrosion results to the microstructure of the modified surface layers. Only fcc austenitic stainless steel (γ) and bcc ferritic iron (α) were detected in the untreated sample. After treatment at 100 % N_2, the Fe_2N and CrN phases are observed. The formation of CrN phase is typical for such a high treatment temperature (475 °C). Residual signals from fcc γ-austenite and bcc ferritic iron are present. At high percentage of nitrogen (90 %) iron nitride phases of Fe_2N, Fe_3N and chromium nitride phase CrN are detected beside the main phase/phases, cubic Fe_4N and/or nitrogen-expanded austenite (γ_n). Due to an overlapping of the strong reflections, the existence of both phases is possible. The intensities of the CrN are lower in comparison to the case of pure nitriding. This might be due to nitrogen atoms, which are dissipated in favor of the formation of the iron nitrides (Fe_3N, Fe_4N) and γ_n phases. In the sample treated at high carbon content (75 % C_2H_2 and 25 % N_2), most of the peaks are correlated to Fe_3C, carbon-expanded austenite (γ_c) beside the CrN phase. For the sample treated at 100 % C_2H_2, the γ_c and CrC phases are only detected.

3.2 Surface roughness

Fig. 3 shows the relative surface roughness, determined by the ratio of the roughness of treated samples to untreated one, as a function of different C_2H_2/N_2 gas pressure ratios. The value of the surface roughness of the untreated sample was 46.3 nm. Due to pure nitriding the surface roughness is increased only by a factor of 1.33. By addition of C_2H_2, the surface roughness increases abruptly up to a maximum value of 4.12, reached at 30 % C_2H_2. The value is nearly the same up to a gas content of 50 % C_2H_2 and decreases significantly for samples treated at high carbon content (75 % C_2H_2) and at pure carburising.

3.3 Wear test and friction coefficient

The friction coefficient is a mechanical parameter, which depends on the surface material composition and the nature of the surface itself. Fig. 4 presents the relative friction coefficient for the samples treated at different gas composition. It relates the friction coefficient of the treated sample to the value of the untreated stainless steel (0.78). The measurement of the friction coefficient has been done for different number of tracks. For pure nitriding, after the first 2000 tracks, at which the wear depth is lower than 0.6 μm, in all examined treated samples, the friction coefficient is reduced to 59 %. While the C_2H_2/N_2 gas ratio increases, the values of the friction coefficient decrease significantly and reaching approximately 14 % for pure carburizing. As a function of gas composition, the friction coefficient for 20000 numbers of tracks has nearly the same values. This reveals the homogeneity and the mechanical stability of the microstructure of the treated layer within the examined range in the near surface region.

The sliding wear behaviour of the untreated and treated samples was assessed using oscillating ball-on-disk type tribometer. The depth of the wear tracks of examined samples as a function of wear path at a load of 3 N is shown in Fig. 5. Generally, the wear resistance of the untreated samples in comparison to the treated is extremely poor. For all treated samples, examined up to 320 m wear path, maximum one micrometer wear depth has been observed and the wear depth slowly increases with increasing wear path. Otherwise the wear rates have been accounted as total volume loss in mm^3 divided by the total sliding distance in meters. The wear rates for the untreated material were accounted in order to know the improvement in the wear rates for treated one. The wear rate for the untreated 304 austenitic stainless steel was 2.4×10^{-4} mm^3/m at 20000 numbers of tracks (80 m wear path).

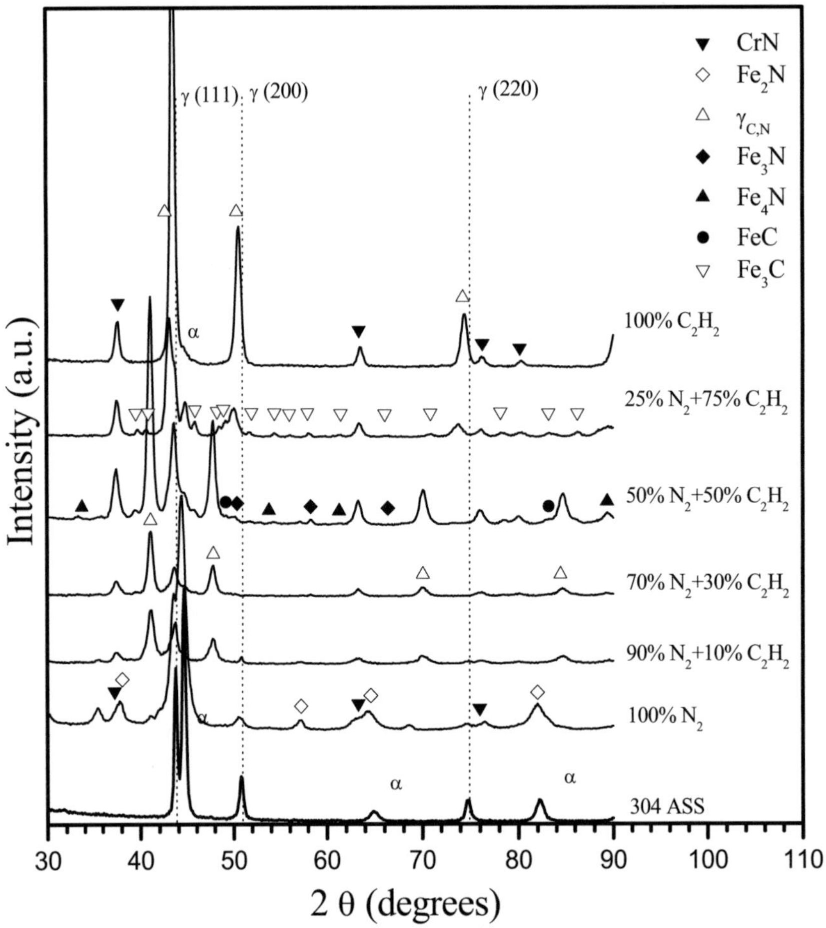

Fig. 2. X-ray diffraction pattern obtained at 2° grazing incidence from 304 stainless steel of untreated samples and samples treated at different C_2H_2/N_2 gas pressure ratios.

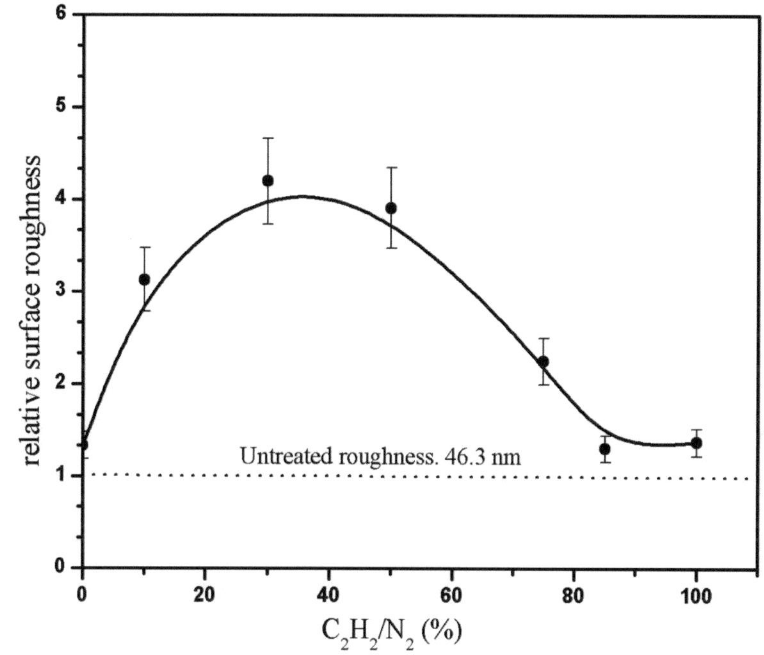

Fig. 3. Relative surface roughness for 304 ASS samples treated at different C_2H_2/N_2 gas ressure ratios.

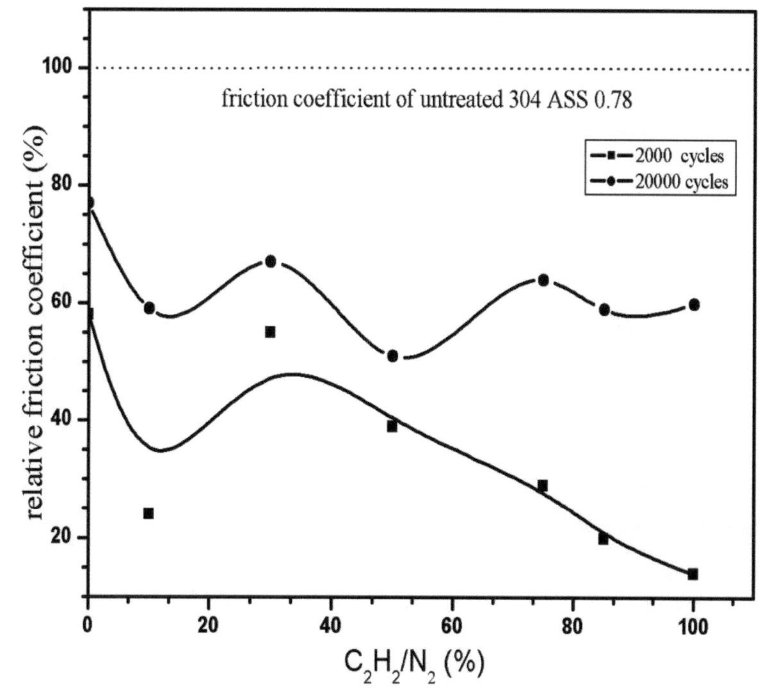

Fig. 4. Relative coefficient of friction as a function of gas composition at different number of cycles.

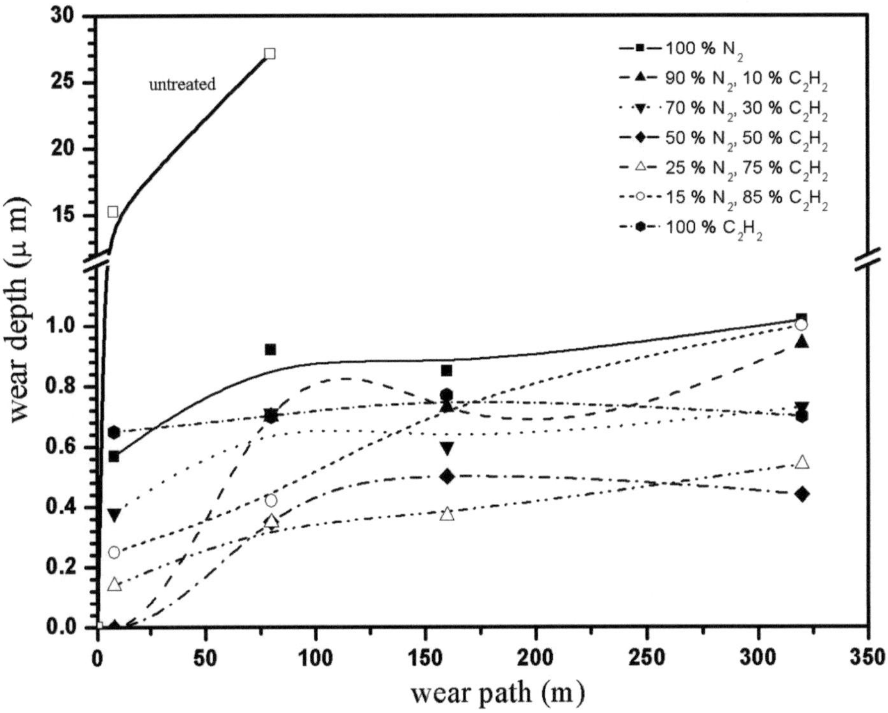

Fig. 5. Wear depth of treated samples at different gas composition compared to an untreated sample.

Fig. 6 presents the variation in the wear rate at different number of tracks as a function of gas composition ratios for a load of 3 N. The wear rate after treatment was reduced by more than two orders in comparison with the untreated material. At 20000 wear tracks, a decrease of the wear rate was observed as much as the C_2H_2 gas ratio increases. However, this improvement was continued up to 75 % C_2H_2 and 50 % C_2H_2 at 40000 and 80000 numbers of tracks, respectively. The decrease in the wear rate by increasing the C_2H_2 content is related to some improvement in the friction coefficient of treated layers due to fine carbon precipitations which work as solid lubricant on the first few hundred nanometers. However, for samples treated at high carbon content, a small increase in the wear rate can be seen with increasing the sliding distance.

Fig. 7 shows the resistance of treated samples toward physical wearing by accounting the wear rate at different loads of 3 N, 5 N and 8 N. The wear rate of untreated sample is increased by one order from 2.7 x 10^{-4} mm^3/m to 2.7 x 10^{-3} mm^3/m at 5 N and 8 N, respectively. Generally, for all treated samples the wear rate increases in the same order with the load. The wear rate decreases significantly with the increase of the C_2H_2/N_2 gas ratio and for relatively high load (8 N) it reaches a minimum at 50 %.

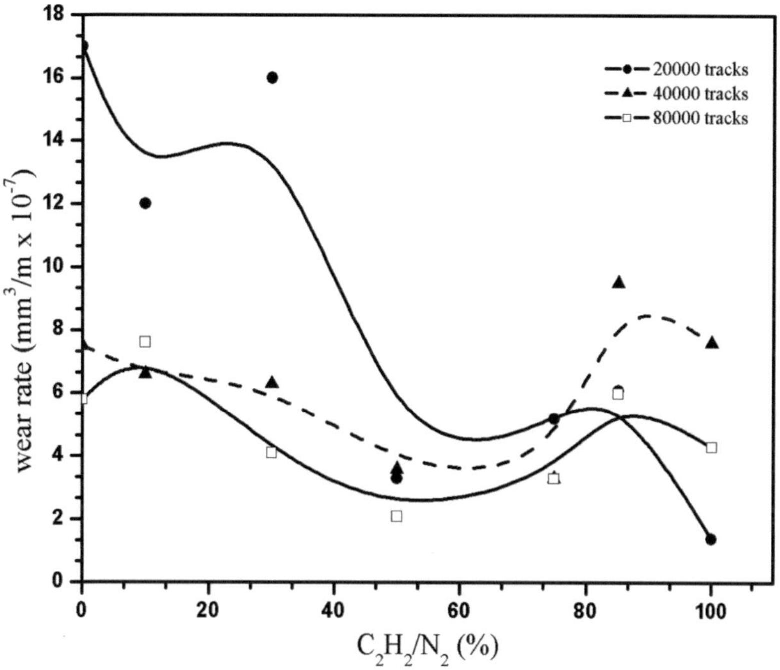

Fig. 6. Wear rate of treated samples as a function of gas composition for a load of 3 N.

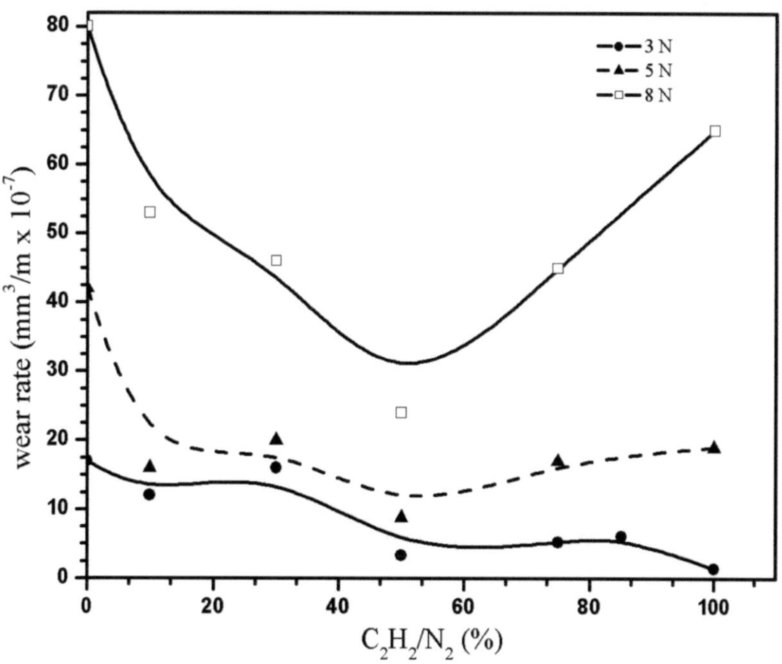

Fig. 7. Wear rate of treated samples as a function of gas composition at different loads and fixed numbers of tracks (200000).

3.4 Corrosion test and surface morphology

Fig. 8 shows the anodic polarization curves obtained from treated and untreated 304-AISI samples. These results was published elsewhere and represented here to make a correlation to the study of the surface morphology before and after corrosion effect [12]. The increase in the corrosion current and decrease in corrosion potential indicate a degradation of the corrosion resistance for the treated samples. The highest degradation in comparison to the untreated sample is observed for the sample treated in pure nitrogen and carbon plasma. The lowest degradation in the corrosion resistance is observed for the sample treated at the gas composition of 70 % N_2 and 30 % C_2H_2.

The SEM pictures, given in Fig. 9, show the surface morphology of the untreated in comparison to treated samples. Moreover the treated surfaces have been scanned after corrosion test. The original material (304-AISI) may be characterized by a non-uniform shapes, thin grain boundaries and very weak links between the grains. In general the treated samples have wider grain boundaries and smaller grain size. Nitrocarburized samples show also a tendency of grain coalescence. At preparation with 100 % C_2H_2, the grain boundaries are not clearly visible, which is caused by the higher carbon precipitations at the surface. However, it seems that the grain size is larger than that obtained in the nitrocarburized samples.

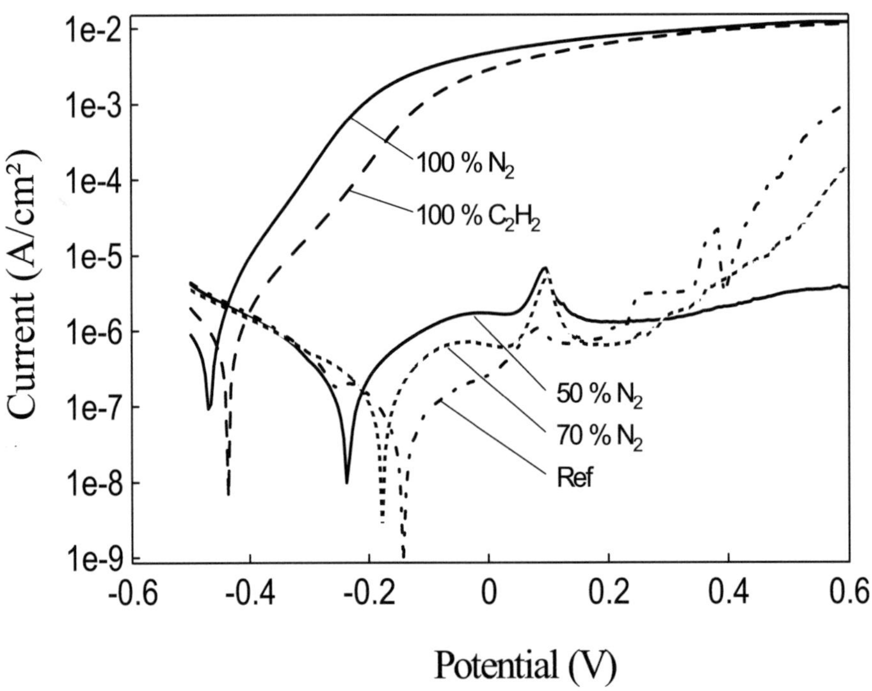

Fig. 8. Anodic polarization curves for untreated and treated samples at different C_2H_2/N_2 gas pressure ratios obtained in 1 wt. % NaCl solution.

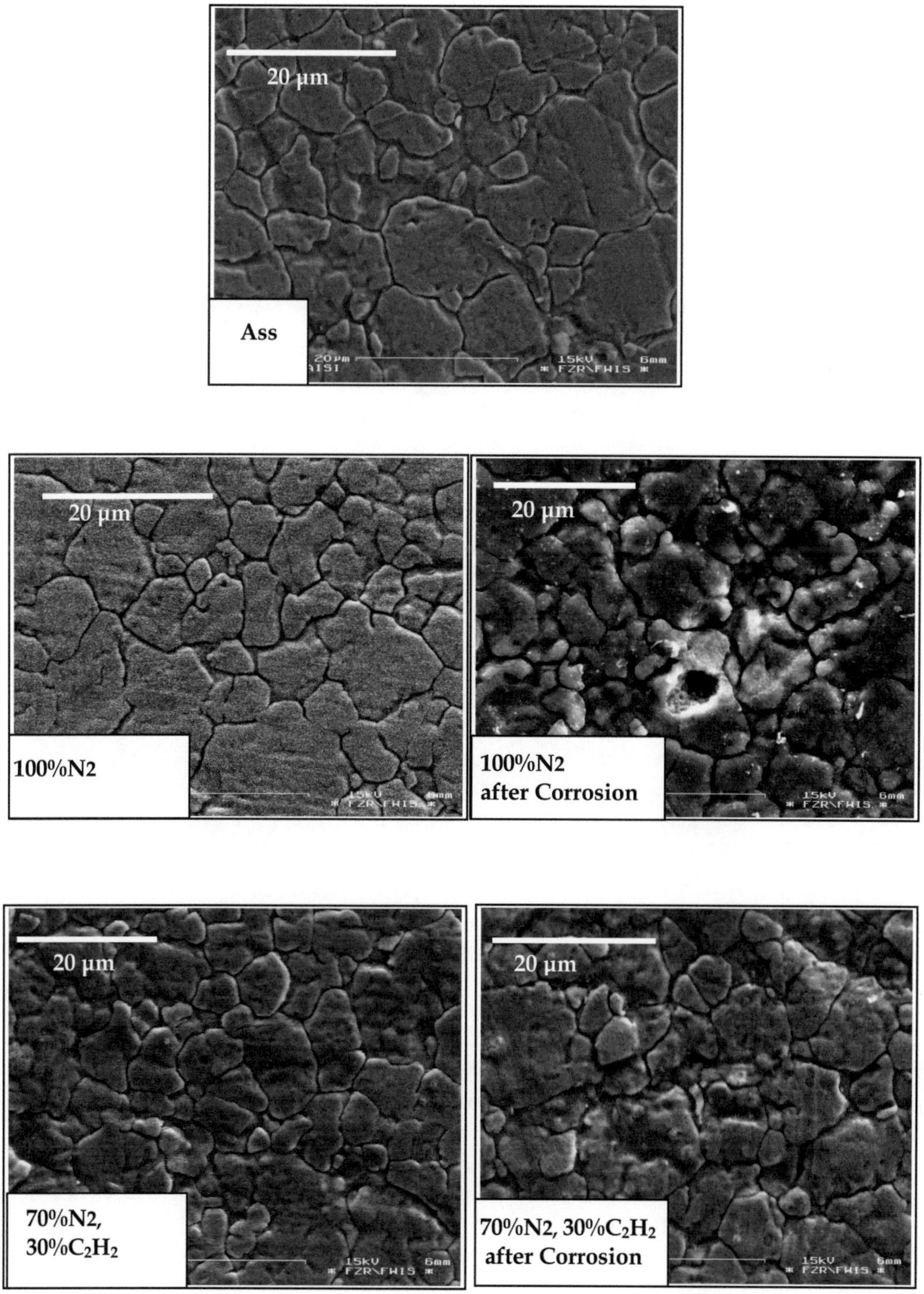
20 µm
Ass
20 µm
100%N2
20 µm
100%N2
after Corrosion
20 µm
70%N2,
30%C_2H_2
20 µm
70%N2, 30%C_2H_2
after Corrosion

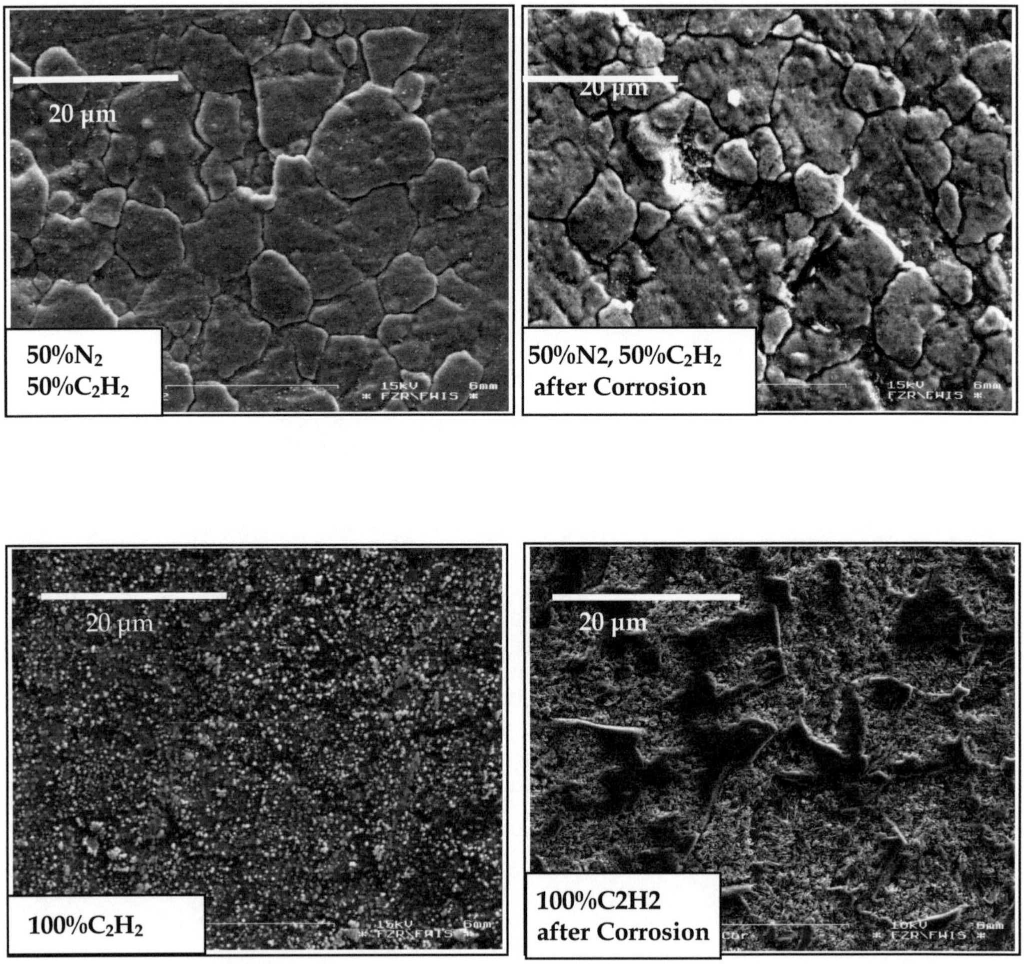

Fig. 9. SEM pictures of the surface of untreated 304 AISI and treated samples, respectively, in comparison with pictures of treated samples after corrosion test. The last image is for treated sample at 100 % C_2H_2 scanned after removing few tens nanometers from the surface.

Two distinguishable regions are observed on the surface of the samples, beside the white points and the black base. For 100% N_2, one observes a very fine white uniform deposition all over the surface. On the surface of sample treated with 100% C_2H_2 the white deposition is larger. For 50% N_2, surface fractures and little particle deposition can be seen. EDAX analysis shows that the white precipitations contain more carbon than the black ones. The elemental composition of the untreated, treated samples and selected parts was analysed by EDAX and the results are listed in Table 1. For nitrocarburizing with 50 % N_2 and 50 % C_2H_2, the nitrogen concentration is higher than that of carbon.

	N (at.%)	C (at.%)	Fe (at.%)	Cr (at.%)	Ni (at.%)
304 ASS	0	0	72.3	19.1	8
100 % N2	15.2	0	58.5	17.6	8
50 % N2	18.8	12.1	46.2	15.5	6.9
100 % C2H2	0	46	37.9	9.3	5.8

Table 1. Surface elemental composition of ASS and selected treated samples determined by EDAX.

The SEM pictures of the treated layer after corrosion test show that the corrosion occurs predominantly by bitting, intergranular or by general attack, depending on the difference of the corrosion rate of the grain boundary zones or the grain faces. This difference in rate is determined not only by the metallurgical structure and the composition of the boundary, but also by the characteristics of the corroding solution. The corrosion attack decreases with increasing the carbon content up to 30 % C_2H_2. After that the attack increases again. At 100 % N_2 the surface undergoes significant corrosion leading to visible intergranular stress corrosion cracking. The surface obtained by treatment with 100 % C_2H_2 exhibits dealloying by selective material dissolution over large surface regions. The substrate is anodic to carbon-bound region and corrodes, leaving behind a mass of carbon compound related areas. In both cases, with pure nitriding or pure carburizing, the loss of corrosion resistance is associated with the depletion of Cr in regions near the grain boundaries. That component, however, is necessary for regeneration of corrosion protective film.

These results are in accordance with those from potentiodynamic polarization curves, as shown in Fig. 8. Samples treated at 100 % N_2 or 100 % C_2H_2 corrode significantly as evidenced by their more negative corrosion potentials and high corrosion currents. Moderate nitrogen content appears to degrade the corrosion resistance insignificantly, especially for the gas ratio 70 % N_2/30 % C_2H_2.

4. Discussion

The plasma efficiency may be increased due to creation of more plasma species such as H, CH, NH, HCN or CN by adding C_2H_2 to N_2 gas during rf plasma carbonitriding. The microstructure of the modified layers depends on the pressure ratio between nitrogen and acetylene plasma gas. The nitride phases and their intensities increased by adding C_2H_2. The effect of adding acetylene has been also observed in [18] where 0.7 % of C_2H_2 was used in addition to the N_2 gas. Even though Blawert et. al. [8, 19] has observed the nitrogen and carbon expanded austenite phases nearly at the same peak positions as in our case. However, we can not ignore that at high N-content by XRD the γ_n phase can not be easily distinguished from the Fe_4N phase. Both nitride and carbide phases contribute to the improvement of the mechanical properties of the treated samples. Nitride phases (γ_n) are harder than the carbide phases (γ_c) [8]. The interplay between the temperature and gas composition might be caused by the effect of hydrogen from the acetylene gas. Compared to carbon and nitrogen the mass difference between plasma species and the ionization potential of atoms can play an important role in the resulting plasma temperature (electron and ion temperature) which has an influence on the temperature of the substrate. In this case, the ionization potential of hydrogen is 6.4 % lower than of that nitrogen and the light hydrogen ions are easily to accelerate by the rf plasma field. These ions itself contribute to the plasma heating due to secondary electrons generation by elastic collisions with the plasma species.

Most probably, the high increase in the surface roughness, especially for the samples treated at high nitrogen content, is correlated with the increase in the sample temperature beside some physical reactions between the heavy plasma species such as CN and HCN and the surface. In a comparable study concerned with nitriding of stainless steel by plasma immersion ion implantation, the formation of expanded austenitic phase in the matrix was accompanied by a high improvement of the microhardness. However, the surface roughness increased by a factor of 4.6 and 7 at 450 °C and 520 °C, respectively [20]. Blawert et. al. [21] has attributed to the increasing in the surface roughness of treated 304-AISI samples to the sputtering surfaces caused by ion bombardment during the treatment. Even though, an increase in the substrate temperature has been observed for samples prepared at high and at pure carbon content, the surface roughness is sharply decreased. It might be due to the decrease in the physical reactions and the high amount of fine precipitations from carbon on the surface.

The decrease in the friction coefficient for samples treated at high C_2H_2 content can be attributed to the fine precipitation from carbon on the surface, which works as solid lubricant between sliding surfaces in wear experiment. After that, as the wear path increased, the effect of fine precipitation of carbon nearly disappeared. Therefore, the friction coefficient increases with the number of wear tracks. This suggestion is supported experimentally by imaging the carburized surface after removing a few hundred nanometres by a fine mechanical polishing (Fig. 9). The high number of carbon precipitations is nearly removed after polishing. Obviously, the wear behavior is influenced by the microstructure of the first sublayer which is created in dependence of the gas composition by the process of nitriding, carburizing or nitrocarburizing. Blawert [8] has reported that nitrogen expanded austenite layers are harder than those of the carbon expanded and therefore this results in smaller wear depth at low load (5 N). The role of the oxide layer is completely different in the wear behavior of untreated and treated samples. In untreated samples the oxidized wear particles lead to severe wear resulting in a high friction coefficient for the sliding of surfaces. However, for all treated samples, the oxide layer works in the opposite direction. The oxide layer of samples treated at high nitrogen content or high carbon content contains low or high amount of fine precipitations from carbon on the surface, respectively. This oxide layer acts as a lubricating layer, which prevents metal-to-metal contact, decreases the friction coefficient, reduces adhesive wear and therefore generally reduces the wear [17].

The high concentration of nitrogen detected by EDAX on the surface of the nitrocarburized sample is supported by the XRD results, which show that more nitride phases are created in the compound layer.

The solid solution phase γ_N should maintain the good corrosion resistance of stainless steel [5, 21]. But in our case, microstructure reveals that high parts of chromium nitride or chromium carbide are detected on the surface precipitated at the grain boundaries, which are responsible for the breakdown in the corrosion resistance [20]. The degradation in corrosion resistance of the modified layers is mostly related to the concentration of CrN and CrC. It is well known that the nature of the corrosion reactions depends on the microstructure of the treated surface which controlled by gas composition and treatment temperature. Using XPS, Borges et al. [18] have recently observed a decrease in the chromium concentration on the nitride surface by adding small amount of acetylene. The authors interpreted this decrease in chromium concentration on the surface by the chemical formation of $Cr\text{-}H_x$ (x = 1, 2) which is partly removed by the vacuum system. Former experiments involving the reaction of chromium atoms with H_2 and matrix isolation of the products have provided the spectroscopic evidence for the molecules CrH_2 and CrH [22]. In our case the intensity of CrN decreases with the increase of C_2H_2 up to 30 %, onward it increases again. This is in a good agreement with the sharp increase in the corrosion resistance of the same sample in comparison with the pure nitrided or pure carburized layer. Maybe the formation rate of the CrH_2 and CrH on the surface is higher than the precipitation rate of CrN in the grain boundaries on the surface of the sample prepared at 30 % C_2H_2. But by the increase of the substrate temperature higher than 525 °C, the balance in the two rates has been changed.

5. Conclusions

The gas composition has a significant influence on the microstructure of the modified layers. At nitrocarburizing, most of phases related to nitride phases (such as γ_n, Fe_2N, Fe_3N, Fe_4N and CrN) and carbide phases (such as γ_c, Fe_3C and CrN or CrC) for samples treated at high nitrogen content and high carbon content, respectively. In dependence on the gas composition ratio (N_2/C_2H_2), the sample temperatures varied from 475 °C to 600 °C. The surface roughness was found to increase as the C_2H_2 content increases up to 50 % but it decreases for higher ratios. The amount of fine precipitations of carbon on the surface is responsible for the gradually decrease in the surface roughness and friction coefficient for samples prepared at high and pure carbon content. In comparison to the untreated samples, the wear rate is reduced by more than two orders. The carbonitrided layer exhibits higher corrosion resistance in comparison to the layers obtained after pure nitriding or pure carburizing treatment. The lower content of the CrN phase leads to a good corrosion resistance. Pure nitriding samples are exposed to a strong biting corrosion surrounded by intergranular corrosion where the carburized layer is exposed to general corrosion.

6. References

[1] N. Yasumaru, Mater. Trans. 39 (1998) 1046.
[2] Y.F. Liu, J.S. Mu, X.Y. Xu, S.Z. Yang, Mater. Sci. Eng. A 458 (2007) 366.
[3] Ajit K. Roy, Vinay Virupaksha, Mater. Sci. Eng. A 452–453 (2007) 665.
[4] Jan Macák, Petr Sajdl, Pavel Kučera, Radel Novotný, Jan Vošta, Electrochim. Acta 51(2006) 3566.
[5] Z. L. Zhang, T. Bell, Surf. Eng. 1 (1985) 131.
[6] E. Menthe, K.-T. Rie, Surf. Coat. Technol. 116-119 (1999) 199.
[7] E. Richter, R. Günzel, S. Parascandola, T. Telbizova, O. Kruse, W. Möller, Surf. Coat. Technol. 128-129 (2000) 21.

[8] C. Blawert, H. Kalvelage, B. L. Mordike, G. A. Collins, K. T. Short, Y. Jirásková, O. Schneeweiss, Surf. Coat. Technol. 136 (2001) 181.
[9] F. M. El-Hossary, N. Z. Negm, S. M. Khalil, A. M. Abd El-Rahman, Thin Solid Films 405 (2002) 179.
[10] A. M. Abd El-Rahman, Surf. Coat. Technol., 205 (2010) 674-681.
[11] F. M. El-Hossary, N. Z. Negm, S.M. Khalil, A. M. Abd El-Rahman, M. Raaif, S. Maendl, Journal of Applied Physics A 99 (2010) 489-495.
[12] A. M. Abd El-Rahman, F. M. El-Hossary, T. Fitz, N. Z. Negm, F. Prokert, M. T. Pham, E. Richter and W. Moeller, Surf. Coat. Technol., 183 (2004) 268- 274.
[13] J. Takada, Y. Ohizumi, H. Miyamura, H. Kuwahara, S. Kikuchi, I. Tamura, J. Mater. Sci. 21 (1986) 2493.
[14] A. Tekin, J. Martin, B. Senpior, Journal of Material Science 26 (1991) 2458.
[15] M. Samandi, B. A. Shedden, D. I. Smith, G. A. Collins, R. Hutchings, J. Tendays, Surf. Coat. Technol. 74-75 (1995) 417.
[16] Wang Liang, Xu Bin, Yu Zhiwei, Shi Yagin, Surf. Coat. Technol. 130 (2000) 304.
[17] P. A. Dearnley, A. Namvar, G. G. A. Hibberd, T. Bell, Surface Engineering: Proceedings of International Conference PSE, DGE (1989) 219.
[18] C. F. M. Borges, S. Hennecke, E. Pfender Surf. Coat. Technol. 123 (2000) 112.
[19] C. Blawert, B. L. Mordike, G. A. Collins, K.T. Short, Y. Jiraskova, O. Schneeweiss, V. Perina, Surf. Coat. Technol. 128-129 (2000) 219.
[20] M. Samandi, B. A. Shedden, D. I. Smith, G. A. Collins, R. Hutchings, J. Tendays, Surf. Coat. Technol. 59 (1993) 261.
[21] C. Blawert, A. Weisheit, B. L. Mordike, F. M. Knoop, Surf. Coat. Technol. 85 (1996) 15.
[22] Z. L. Xiao, R. H. Hauge, J. L. Margrave, J. Phys. Chem. 96 (1992) 636.

16

Corrosion Resistance of High-Mn Austenitic Steels for the Automotive Industry

Adam Grajcar
Silesian University of Technology
Poland

1. Introduction

Significant progress in a field of development of new groups of steel sheets for the automotive industry has been made in the period of the last twenty years. From the aspect of materials, this development has been accelerated by strong competition with non-ferrous aluminium and magnesium alloys as well as with composite polymers, which meaning has been successively increasing. From the aspect of ecology, an essential factor is to limit the amount of exhaust gas emitted into the environment. It is strictly connected to fuel consumption, mainly dependent on a car weight. Application of sheets with lower thickness preserving proper stiffness requires the application of sheets with higher mechanical properties, keeping adequate formability. Figure 1 presents conventional high-strength steels (HSS) and the new generations of advanced high-strength steels (AHSS) used in the automotive industry. Steels of IF (Interstitial Free) and BH (Bake Hardening) type with moderate strength and high susceptibility to deep drawing were elaborated for elements of body panelling. However, the increasing application belongs to new multiphase steels consisting of ferritic matrix containing martensitic islands (DP - Dual Phase) or bainitic-austenitic regions (TRIP - Transformation Induced Plasticity). These steels together with CP (Complex Phase) and MART steels with the highest strength level are the first generation of advanced high-strength steels (AHSS) used for different reinforcing elements (International Iron & Steel Institute, 2006).

Nowadays, apart from limiting fuel consumption, special pressure is placed on increasing the safety of car users. The role of structural elements such as frontal frame members, bumpers and other parts is to take over the energy of an impact. Therefore, steels that are used for these parts should be characterized by high product of UTS and UEl, proving the ability of energy absorption. It is difficult to achieve for conventional HSS and the first generation AHSS because the ductility decreases with increasing strength (Fig. 1).

The requirements of the automotive industry can be met by the second generation of advanced high-strength steels combining exceptional strength and ductile properties as well as cold formability (Fig. 1). These TWIP (Twinning Induced Plasticity) and L-IP (Light - Induced Plasticity) steels belong to a group of high-manganese austenitic alloys but are much cheaper comparing to Cr-Ni stainless steels (AUST SS). Their mean advantage over first generation steels with a matrix based on A2 lattice structure is the great susceptibility of austenite on plastic deformation, during which dislocation glide, mechanical twinning and/or strain-induced martensitic transformation can occur. The group of high-manganese steels includes alloys with 15-30% Mn content. Two mean chemical composition strategies had been worked out so far. The first includes alloys with different Mn concentration and 0.5 to 0.8% C (Ghayad et al., 2006; Jimenez & Frommeyer, 2010). The function of carbon is stabilization of γ phase and hardening of solid solution. In the second group, the concentration of carbon is decreased below 0.1%, whereby there is an addition up to 4% Al and/or 4% Si (Frommeyer et al., 2003; Graessel et al., 2000). The solid solution strengthening caused by Al and Si compensates smaller C content. Sometimes, the steels contain chromium (Hamada, 2007; Mujica Roncery et al., 2010) or microadditions of Nb, Ti and B (Bleck & Phiu-on, 2005; Grajcar et al., 2009; Huang et al., 2006).

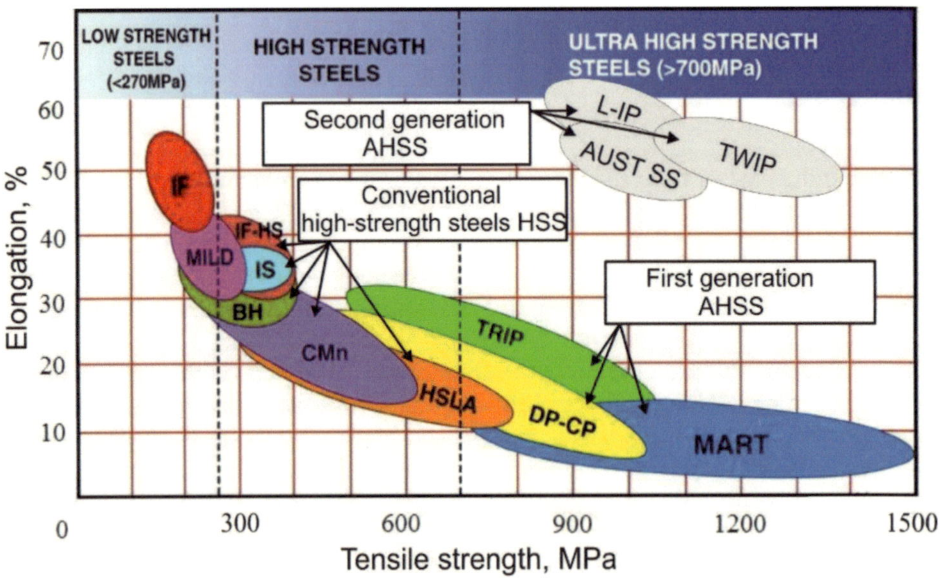

Fig. 1. Conventional high-strength steels (HSS) and the new generations of advanced high-strength steels (AHSS) used in the automotive industry (International Iron & Steel Institute, 2006).

Mechanical properties of high-manganese steels are dependent on structural processes occurring during cold deformation, which are highly dependent on SFE (stacking fault energy) of austenite (De Cooman et al., 2011; Dumay et al., 2007; Vercammen et al., 2002). In turn, the SFE is dependent on the temperature and chemical composition. Figure 2 shows that the stacking fault energy increases with increasing temperature and Al, Cu content whereas Cr and Si decrease it (Dumay et al., 2007; Hamada, 2007). If the SFE is from 12 to 20 mJm^{-2}, a partial transformation of austenite into martensite occurs as a main deformation mechanism, taking advantage of TRIP effect.

Values of SFE from 20 to 60 mJm^{-2} determine intensive mechanical twinning related to TWIP effect. At SFE values higher than about 60 mJm^{-2}, the partition of dislocations into Shockley partial dislocations is difficult, and therefore the glide of perfect dislocations is the dominant deformation mechanism (Hamada, 2007). In TRIPLEX steels with a structure of austenite, ferrite and κ-carbides ($(Fe,Mn)_3AlC$) and for SFE > 100 mJm^{-2}, the SIP (Shear Band Induced

Plasticity) effect is considered as the major deformation mechanism (Frommeyer & Bruex, 2006). High impact on the dominating deformation mechanism have also the temperature, strain rate and grain size (Dini et al., 2010; Frommeyer et al., 2003; Graessel et al., 2000). The key to obtain the mechanical properties regime in Fig. 1 is the high work hardening rate characterizing the plastic deformation of high-Mn alloys. The high level of ductility is a result of delaying necking during straining. In case of the local presence of necking, strain-induced martensitic transformation occurs in such places (in TRIP steels) or deformation twins are preferably generated in locally deformed areas (in TWIP steels). It leads to intensive local strain hardening of the steel and further plastic strain proceeds in less strain-hardened adjacent zones. The situation is repeated in many regions of the sample what finally leads to delaying necking in a macro scale and high uniform and total elongation. The shear band formation accompanied by dislocation glide occurs in deformed areas of TRIPLEX steels and the SIP effect is sustained by the uniform arrangement of nano size κ-carbides coherent to the austenitic matrix (Frommeyer & Bruex, 2006).

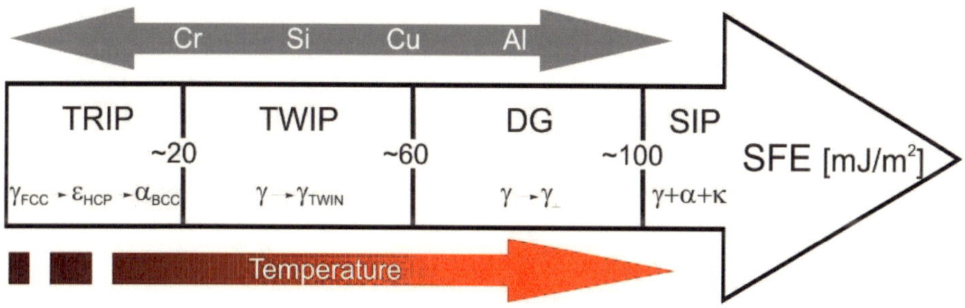

Fig. 2. Schematic drawing of the effects of temperature and chemical composition on the stacking fault energy (SFE) of austenite and the correlation of SFE with a main deformation mechanism in high-Mn alloys.

2. Corrosion behaviour

2.1 General and pitting corrosion

The mean area of studies on high-manganese steels concern their high-temperature deformation resistance (Bleck et al., 2007; Cabanas et al., 2006; Dobrzański et al., 2008; Grajcar & Borek, 2008; Grajcar et al., 2009) and the cold-working behaviour (Dini et al., 2010; Frommeyer & Bruex, 2006; Frommeyer et al., 2003; Graessel et al., 2000; Huang et al., 2006). Much less attention has been paid on their corrosion resistance (Ghayad et al., 2006; Grajcar et al., 2010a, 2010b; Hamada, 2007; Kannan et al., 2008; Mujica et al., 2010; Mujica Roncery et al., 2010; Opiela et al., 2009). The research on Fe-C-Mn-Al alloys (Altstetter et al., 1986) for cryogenic applications that were supposed to substitute expensive Cr-Ni steels was carried out in the eighties of the last century. The role of manganese boils to Ni replacement and obtaining austenitic microstructure, whereas aluminium has a similar impact as chromium. Improvement of corrosion resistance by Al consists in formation of thin, stable layer of oxides. As the result of conducted research it was found that Fe-C-Mn-Al alloys show inferior corrosion resistance than Cr-Ni steels and they can be used as a substitute only in some applications (Altstetter et al., 1986). The addition of 25% Mn to mild steels was found to be very detrimental to the corrosion resistance in aqueous solutions (Zhang & Zhu, 1999).

The Fe-25Mn alloy was difficult to passivate, even in such neutral aqueous electrolytes as 1M Na_2SO_4 solution. With increasing Al content up to 5% of the Fe-25Mn-Al steel, the anodic polarization curves exhibit a stable passivation region in Na_2SO_4 solution, but it shows no passivation in 3.5wt% NaCl solution.

Recently, corrosion resistance of Fe-0.05C-29Mn-3.1Al-1.4Si steel in acidic (0.1M H_2SO_4) and chloride-containing (3.5wt% NaCl) environments was investigated (Kannan et al., 2008). Moreover, the corrosion behaviour of the tested high-Mn steel with that of IF-type was compared. Performing immersion and polarization tests it was found that Fe-Mn-Al-Si steel has lower corrosion resistance than IF steel, both in acidic and in chloride media. The corrosion resistance of the high-manganese steel in chloride solutions is higher compared to that observed in acidic medium. The behaviour of Fe-0.2C-25Mn-(1-8)Al steels with increased concentration of Al up to 8% wt. in 3.5wt% NaCl was also investigated (Hamada, 2007). Hamada reported that the corrosion resistance of tested steels in chloride environments is pretty low. The predominating corrosion type is the general corrosion, but locally corrosion pits were observed. In steels including up to 6% Al with homogeneous austenite structure, places where the pits occur are casually, whereas in case of two-phase structure, including ferrite and austenite (Fe-0.2C-25Mn-8Al), they preferentially occur in α phase. The corrosion resistance of examined steels can be increased trough anodic passivation in nitric acid, which provides modification of chemical composition and constitution of the surface layer (Hamada et al., 2005). This was done by reducing the surface concentration of Mn and enriching the surface layer in elements that improve the corrosion resistance (e.g. Al, Cr).

A better effect was reached by chemical composition modification. It was found that addition of Al and Cr to Fe-0.26C-30Mn-4Al-4Cr and Fe-0.25C-30Mn-8Al-6Cr alloys increases considerably the general corrosion resistance, especially after anodic passivation ageing of surface layers in an oxidizing electrolytic solution (Hamada, 2007). Cr-bearing steels passivated by nucleation and growth of the passive oxide films on the steel surface, where the enrichment of Al and Cr and depletion of Fe and Mn have occurred. The positive role of Cr in obtaining passivation layers in 0.5M H_2SO_4 acidic solution was recently confirmed in Fe-25Mn-12Cr-0.3C-0.4N alloy (Mujica et al., 2010; Mujica Roncery et al., 2010). The steel containing increased Cr, C and N content shows passivity at the current density being five orders of magnitude lower compared to the Fe-22Mn-0.6C steel.

2.2 Effect of deformation

Results of corrosion tests described above concern steels in the annealed or supersaturated state. The influence of cold plastic deformation on corrosion behaviour in 3.5wt% NaCl was studied in Fe-0.5C-29Mn-3.5Al-0.5Si steel (Ghayad et al., 2006). It was found on the basis of potentiodynamic tests, that the steel shows no tendency to passivation, independently on the steel structure after heat treatment (supersaturated, aged or strain-aged). Higher corrosion rate of deformed specimens than that of specimens in supersaturated state, was a result of faster steel dissolution caused by annealing twins, which show a different potential than the matrix. The highest corrosion rate was observed in strain-aged samples, as a result of ferrite formation, which creates a corrosive galvanic cell with the austenitic matrix. The enhanced corrosion attack at the boundaries of deformation twins was also observed in Fe-22Mn-0.5C steels (Mazancova et al., 2010).

2.3 Hydrogen embrittlement and delayed fracture

Generally, increasing the strength of steels, their hydrogen embrittlement susceptibility increases. This is one of the main problem to use AHSS. If hydrogen content reaches the critical value, it can induce a reduction of strength and ductile properties. A critical concentration of hydrogen is various for different steels (Lovicu et al., 2010; Sojka et al., 2010). Hydrogen embrittlement is usually investigated by performing slow strain rate tensile tests on hydrogenerated samples. Austenitic alloys are considered to be immune to this type of corrosion damage. However, the stress- or strain-induced martensitic transformation of austenite taking place in TRIP-aided austenitic alloys can be a reason of their embrittlement. This can happen due to the high difference in solubility and diffusion rate of hydrogen in the BCC and FCC lattice. Austenite is characterized by high solubility and low diffusivity of H in the A1 lattice and thus acts as a sink for hydrogen lowering its mobility and increasing the hydrogen concentration. Due to the slow diffusion rate of hydrogen in austenite, it is hardly to enrich it homogeneously to a hydrogen content causing embrittlement. However, it was shown (Lovicu et al., 2010) that the hydrogen concentration in surface regions of the high-Mn steel is much higher than in the centre zone. It can lead to the intragranular fracture in these regions because of strain-induced or hydrogen-induced martensitic transformation and finally to reduction of strength and ductility.

When the formed automotive element is exposed to the air the delayed fracture can occur. The technological formability is usually investigated in cup forming tests (Otto et al., 2010; Shin et al., 2010). It was observed (Shin et al., 2010) that the 0.6C-22Mn steel cup specimen underwent the delayed fracture when exposed to the air for seven days, even though the specimen was not cracked during forming. This is because the strain-induced martensitic transformation occurred during the cupping test in places of stress concentration. When the addition of 1.2% Al was added the steel cup forms with the high share of mechanical twinning instead the $\gamma \rightarrow \alpha'$ transformation. It leads to lower stress concentration and finally to improvement in cup formability (Shin et al., 2010).

3. Experimental procedure

3.1 Material

The chapter addresses the corrosion behaviour of two high-Mn steels of different initial structures in chloride and acidic media. Their chemical composition is given in Table 1.

Steel grade	C	Mn	Si	Al	Nb	Ti	S	P	N	O	Structure
26Mn-3Si-3Al-Nb	0.065	26.0	3.08	2.87	0.034	0.009	0.013	0.004	0.0028	0.0006	γ
25Mn-3Si-1.5Al-Nb-Ti	0.054	24.4	3.49	1.64	0.029	0.075	0.016	0.004	0.0039	0.0006	$\gamma+\varepsilon$

Table 1. Chemical composition of the investigated steels, wt. %

The vacuum melted steels have similar C, Mn and Si concentration. Significant impact on the SFE of austenite has the difference in Al and Ti content. The lower SFE of 25Mn-3Si-1.5Al-Nb-Ti steel compared to 26Mn-3Si-3Al-Nb steel is a result of the lower Al content. Moreover, several times higher Ti content in a first steel provides a decrease of γ phase stability, as a result of fixing the total nitrogen and some carbon (Grajcar et al., 2009).

The steels were delivered in a form of sheet segments of 340x225x3.2 mm, obtained after the thermo-mechanical rolling. The thermo-mechanical processing consisted of:

- heating the charge up to the temperature of 1100°C and austenitizing for 15 minutes,
- rolling in a range from 1050°C to 850°C in 3 passes (relative reduction: 20, 15 and 15%),
- holding of the rolled sheet segments at the temperature of finishing rolling for 15s,
- solution heat treatment of the flat specimens in water.

The microstructures of the steels after the thermo-mechanical treatment are shown in Figs. 3 and 4. The 26Mn-3Si-3Al-Nb steel exhibits a homogeneous austenite structure with grains elongated in the rolling direction (Fig. 3a). The susceptibility to twinning confirms the presence of a great number of annealing twins. The single-phase structure of the steel is confirmed by X-ray diffraction pattern in Fig. 3b. The lower SFE of the 25Mn-3Si-1.5Al-Nb-Ti steel results in the presence of the second phase with a lamellar shape, distributed in the austenite matrix (Fig. 4a). The number of annealing twins is much lower. The X-ray diffraction analysis confirms the presence of ε martensite (Fig. 4b).

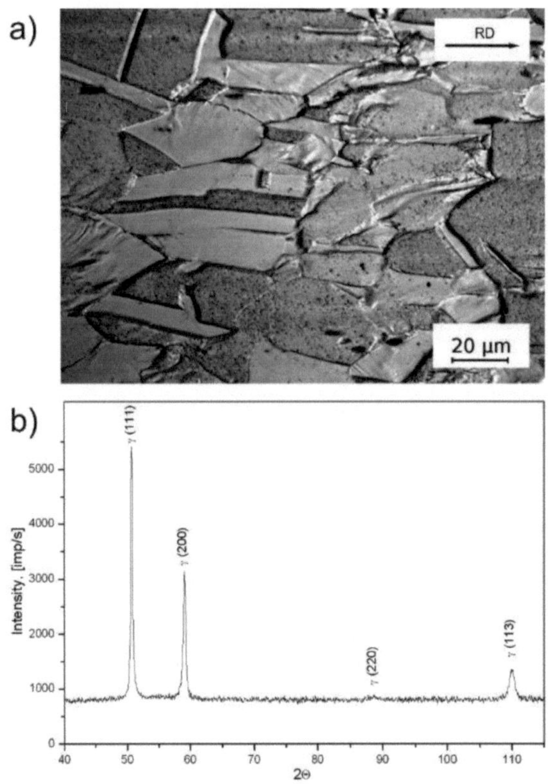

Fig. 3. Austenitic structure with annealing twins of 26Mn-3Si-3Al-Nb steel after the thermo-mechanical rolling and immersion in 1N H_2SO_4 (a) and X-ray diffraction pattern (b).

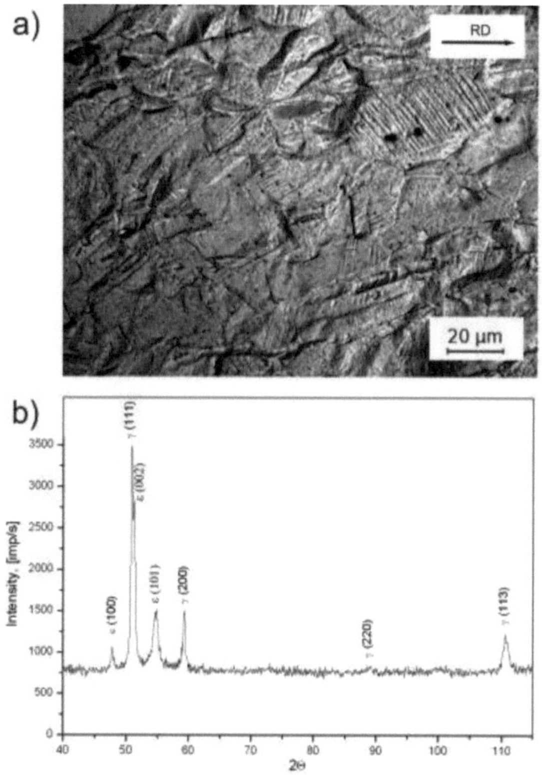

Fig. 4. Austenitic matrix with ε martensite of 25Mn-3Si-1.5Al-Nb-Ti steel after the thermo-mechanical rolling and immersion in 1N H_2SO_4 (a) and X-ray diffraction pattern (b).

3.2 Immersion tests

The immersion tests were used to assess the corrosion resistance of the steels at the initial state (after the thermo-mechanical rolling) and after cold deformation. The corrosion resistance was investigated in two solutions: 1N H_2SO_4 and 3.5wt% NaCl. Prior to the corrosion tests, three samples of the each steel with the size 3.2x10x15 mm were ground to the 1000-grit finish and then they were washed in distilled water, ultrasonically cleaned in acetone and finally rinsed with ethanol and dried. The specimens were weighed with the accuracy of 0.001g and put into the solution for 100 hours at the temperature of 23±1°C. After the test the specimens were weighed and analysed using optical microscopy and SEM. Corrosion loss was calculated in a simple way as the difference between final and initial mass of the samples. Percentage mass decrement was also calculated. Cold deformation was applied by bending at room temperature. Samples with a size of 10x15 mm and a thickness of 3.2 mm were bent to an angle of 90°, with a bending radius of 3 mm.

Metallographic observations of non-metallic inclusions and corrosion pits were carried out on polished sections, whereas the microstructure observations on specimens etched in nital. The investigations were performed using LEICA MEF 4A light microscope, with magnifications from 100 to 1000x. Fractographic investigations were carried using scanning electron microscope SUPRA 25 (Zeiss) at the accelerating voltage of 20kV. In order to remove corrosion products, the specimens were ultrasonically cleaned before the analysis.

3.3 Potentiodynamic polarization tests

Investigation of the electrochemical corrosion behaviour was done in a PGP 201 potentiostat using a conventional three-electrode cell consisting of a saturated calomel reference electrode (SCE), a platinum counter electrode and the studied specimen as the working electrode. To simulate the corrosion media, 0.5N H_2SO_4 and 0.5N NaCl solutions were used. The solution temperature was 23°C±1°C. The corrosion behaviour was studied first by measuring the open circuit potential (OCP) for 30 min. Subsequently, anodic polarization curves were registered. The curve started at a potential of ~100 mV below the corrosion potential. The potential has been changed in the anodic direction at the rate of 1 mV/s. After the anodic current density being equal 1mA/cm^2 was achieved, the direction of polarization has been changed. Thus, the return curve was registered. The corrosion current densities and the polarization resistance were obtained on the basis of the Tafel analysis.

4. Results and discussion

4.1 Corrosion behaviour in the initial state

4.1.1 Results of immersion tests

The results of the immersion tests in two media are given in Table 2. After 100 hours immersion in 1N H_2SO_4 both steels showed a significant percentage mass decrement, among 38 and 41%. Mass loss of samples dipped in 3.5wt% NaCl is about 100 times lower. The difference is due to different corrosion mechanisms. When the solution is acidic, the corrosion process is running according to hydrogen depolarization, whereas in chloride media the specimens are corroding with oxygen depolarization.

	Corrosion medium	
Steel grade	1N H_2SO_4	3.5% NaCl
26Mn-3Si-3Al-Nb	38.4 ± 5.2	0.40 ± 0.03
25Mn-3Si-1.5Al-Nb-Ti	41.3 ± 9.6	0.48 ± 0.03

Table 2. Mean percentage mass loss of samples after the immersion tests, %.

In 26Mn-3Si-3Al-Nb steel immersed in 1N H_2SO_4 many deep corrosion pits along the whole specimen surface were observed (Fig. 5). Moreover, in places with higher density of non-metallic inclusions, microcracks locally occur. Similar pits are present in the steel with lower aluminum content. Slightly smaller corrosion pits are formed in specimens after the immersion test in 3.5wt% NaCl, regardless of a steel type. Places privileged to creation of corrosion pits are pointwise aggregations and chains of non-metallic inclusions (Fig. 6).

Characteristically for the structure of 25Mn-3Si-1.5Al-Nb-Ti steel dipped in 3.5wt% NaCl solution are small microcracks located along ε martensite lamellas (Fig. 7). They are propagated from significantly elongated in a rolling direction, sulphuric non-metallic inclusions. In specimens with the single-phase austenitic structure, microcracks were not observed.

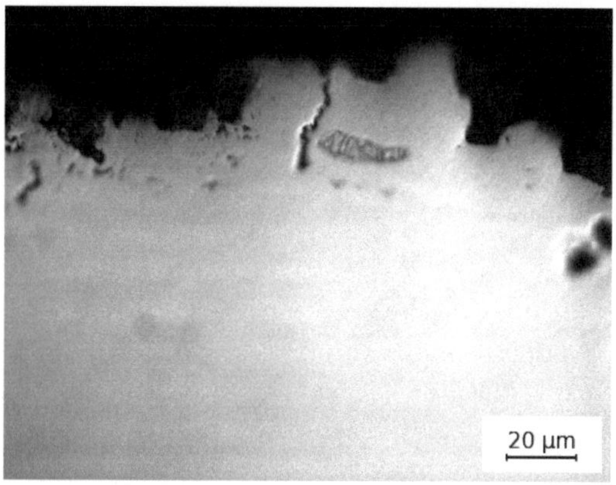

Fig. 5. Corrosion pits and microcracks in 26Mn-3Si-3Al-Nb steel after the immersion test in 1N H_2SO_4.

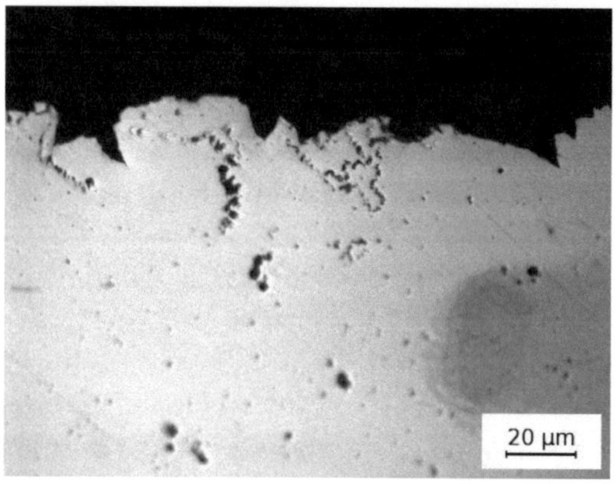

Fig. 6. Corrosion pits in 25Mn-3Si-1.5Al-Nb-Ti steel after the immersion test in 3.5wt% NaCl.

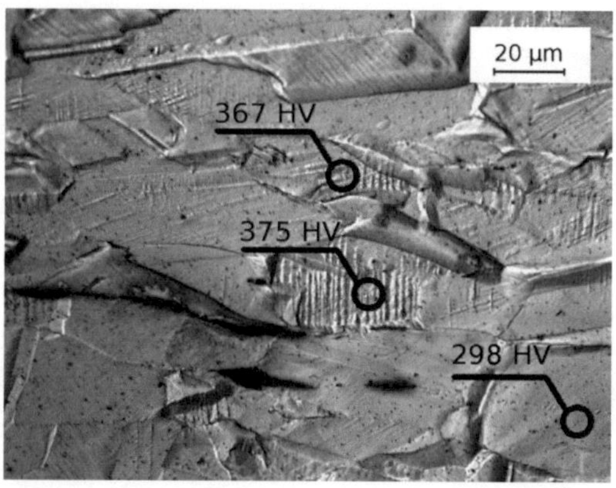

Fig. 7. Austenitic matrix containing plates of ε martensite and microhardness test results of 25Mn-3Si-1.5Al-Nb-Ti steel.

A precise definition of the character of corrosion damages was possible on the basis of SEM observations. On the surface of specimens of both steels dipped in NaCl solution, a layer of corrosion products is occurring, protecting the metal against further corrosive medium penetration. Created layer includes many cracks, especially in surroundings of non-metallic inclusions (Fig. 8). In case of the specimens dipped in H_2SO_4 solution, the number of created surface cracks is much higher (Fig. 9). Apart from corrosion products residues, many craters formed as a result of corrosion pitting. They are occurring both in the steel with ε martensite lamellas (Fig. 9), as well as in the steel with a single-phase austenitic structure (Fig. 10).

The results confirmed the low corrosion resistance of high-manganese steels in acidic and chloride media. Especially low corrosion resistance the investigated steels show in 1N H_2SO_4, where the mass decrement is about 40%, what is about 100 times higher than for specimens dipped in 3.5wt% NaCl (Table 2). The similar order of magnitude of corrosion progress was observed for the Fe-0.05C-30Mn-3Al-1.4Si steel (Kannan et al., 2008).

The high difference in corrosion resistance is because of different corrosion mechanisms in both environments. The big mass loss in the H_2SO_4 solution is due to the hydrogen depolarization mechanism, which is typical for corrosion in acidic media. Hydrogen depolarization is a process of reducing hydrogen ions (from the electrolyte) in cathodic areas by electrons from the metal, to gaseous hydrogen, resulting in continuous flow of electrons outer the metal and consequently the corrosion progress. Due to this process, numerous corrosion pits occur in examined steels (Figs. 5, 6). Corrosion pits are occurring most intensively in the places containing non-metallic inclusions. They are less precious than the rest of material, fostering potential differences and galvanic cell creation. This causes the absorption of hydrogen ions, which, due to increasing pressure and temperature can recombine to a gaseous form and get out of the metal accompanying formation of corrosion pits (Fig. 10). This process is accompanied by local cracking of corrosion products layer (Figs. 9, 10), uncovering the metal surface and causing further penetration of the corrosive medium and the intensive corrosion progress.

In chloride solution, the corrosion process is running according to the oxygen depolarization. In this mechanism, oxygen included in the electrolyte is being reduced by electrons from the metal to hydroxide ions. On the surface of the alloy appears a layer of corrosion products (Fig. 8), protecting the material before further penetration of the corrosion medium. This is why the mass loss in chloride solution is much lower compared to acidic medium. At less corrosion-resistant places (e.g. with non-metallic inclusions) potential differences are occurring. This enables the absorption of chloride ions, which are forming chlorine oxides of increased solubility. This results in local destructions of corrosion products layer (Fig. 8) and the initiation of corrosion pits. Further pit expansion is running autocatalytic.

As a consequence of small steel softening during static recrystallization (Grajcar & Borek, 2008; Grajcar et al., 2009) after finishing rolling, the state of internal stresses in examined steels can be increased. In specimen areas with internal stresses, crevices are occurring. Due to limited oxygen access and a lack of possibility of corrosion products layer forming, they become susceptible to corrosion. As a result of chloride ions adsorption on the crevice bottom, a concentrated electrolyte solution is forming, fostering the corrosion progress (Cottis & Newman, 1995). As a result, stress corrosion cracking can take place. Microcracks were observed along ε martensite lamellas in 25Mn-3Si-1.5Al-Nb-Ti steel. The microcracks initiation proceeds in places with elongated non-metallic inclusions (Fig. 7), while their propagation runs along plates of the second phase.

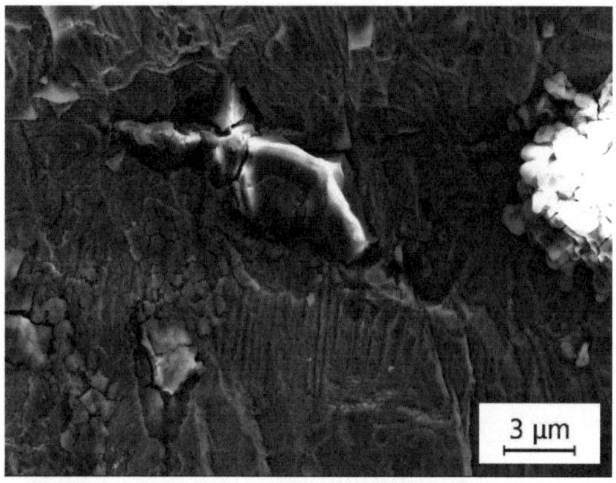

Fig. 8. Bursted corrosion products layer on 26Mn-3Si-3Al-Nb steel surface after the immersion test in 3.5wt% NaCl.

Fig. 9. Craters created as a result of corrosion pitting and bursted corrosion products layer in 25Mn-3Si-1.5Al-Nb-Ti steel after the immersion test in 1N H_2SO_4.

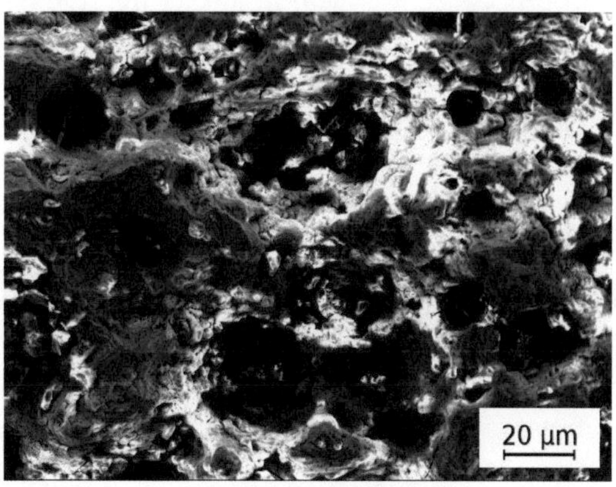

Fig. 10. Craters created as a result of corrosion pitting and corrosion products residues in 26Mn-3Si-3Al-Nb steel after the immersion test in 1N H_2SO_4.

The mass decrement in two steels is comparable (Table 2) both in acidic and chloride media. It indicates that a small ε martensite fraction does not have meaningful impact on the corrosion progress. The observed corrosion products are related rather to the chemical composition than to the phase structure of investigated steels. That confirms a slightly higher mass decrement in the steel with lower Al and somewhat higher Si content. In general, the low corrosion resistance of high-manganese steels is from the fact, that Mn in steels forms unstable manganese oxide due to low passivity coefficient and hence reduces their electrochemical corrosion resistance (Kannan et al., 2008). It leads consequently to the high dissolution rate of manganese and iron atoms both in H_2SO_4 and NaCl solutions (Ghayad et al., 2006; Hamada, 2007; Kannan et al., 2008; Zhang & Zhu, 1999).

The high mass decrement of steels examined in H_2SO_4 solution is a result of fast general corrosion progress and corrosion pits formation (Figs. 5, 6). Much lower mass loss in steels examined in NaCl solution is connected mainly with corrosion pits forming (Fig. 6). The presence of corrosion pits in chloride medium in steels of the type Fe-25Mn-5Al and Fe-0.2C-25Mn-(1-8)Al was also confirmed by other authors (Hamada, 2007; Zhang & Zhu, 1999). However, significant participation of pitting corrosion was not observed in the studies on Fe-0.05C-29Mn-3.1Al-1.4Si steel (Kannan et al., 2008) and on Fe-0.5C-29Mn-3.5Al-0.5Si steel (Ghayad et al., 2006). Localized corrosion attack in the presently investigated steels is enhanced by the lower aluminium and higher silicon concentrations. It means that a character of corrosion damages in high-manganese steels in chloride medium is a complex reaction of the chemical composition and structural state related with a phase composition and the degree of strain hardening.

4.1.2 Results of potentiodynamic polarization tests

Electrochemical corrosion resistance in potentiodynamic tests was carried out on the steel characterized by two-phase structure after the thermo-mechanical rolling. The change of current density as a function of potential for the sample investigated in 0.5N NaCl solution is presented in Fig. 11. The value of corrosion potential was equal -796 mV and the corrosion current density was 8.4 μA/cm^2. Determination of pitting potential was impossible due to the fast course of corrosion processes. It is clear in Fig. 11 that the passivation did not occur. The factors which precluded repassivation inside pits being formed on the surface of the sample were probably the increase of chloride ions concentration as a consequence of their relocation along the corrosion current, what made the contribution to the formation of a corrosion cell inside the pit as well as difficult supply of oxygen into the interior of the pit because of its low solubility in the electrolyte. The change of polarization of samples did not cause any decrease of anodic current.

The corrosion potential of the 25Mn-3Si-1.5Al-Nb-Ti steel investigated in 0.5N H_2SO_4 is equal to -574 mV (Fig. 12). It is shifted towards the more noble direction, as compared to chloride solution. However, the corrosion current density is equal to about 3400 μA/cm^2, what is over two orders of magnitude higher compared to chloride solution.

The similar values of the corrosion potential and corrosion current density both in chloride and acidic media are reported by other authors (Ghayad et al., 2006; Kannan et al., 2008). The sample gains no passivation and the pitting potential was about 57 mV (Fig. 12). It is interesting to note the fast increase of corrosion current after the initiation of pitting corrosion. The escalation of corrosion current usually is more mild.

Fractographic analyses of sample surface after the corrosion tests allowed to evaluate the type and the degree of corrosion damages. On the surface of samples investigated in 0.5N NaCl numerous relatively small corrosion pits and micropores were revealed (Figs. 13, 14). Damaging of a superficial layer occurred around the pits. Cracked passive layer was also observed, what could be a result of rapid penetration of corrosive medium into interior of investigated specimens (Fig. 14). Similar corrosion effects, i.e. pitting, cracked interfacial layer and scaled surface were identified in the specimens after electrochemical tests in 0.5N H_2SO_4 solution (Figs. 15, 16). The results correspond well with those obtained after immersion tests.

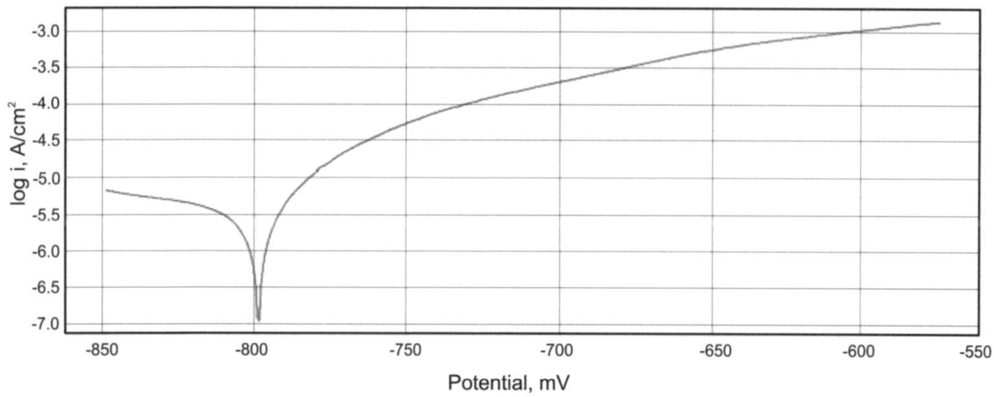

Fig. 11. Anodic polarization curve registered for the sample of 25Mn-3Si-1.5Al-Nb-Ti steel in 0.5N NaCl.

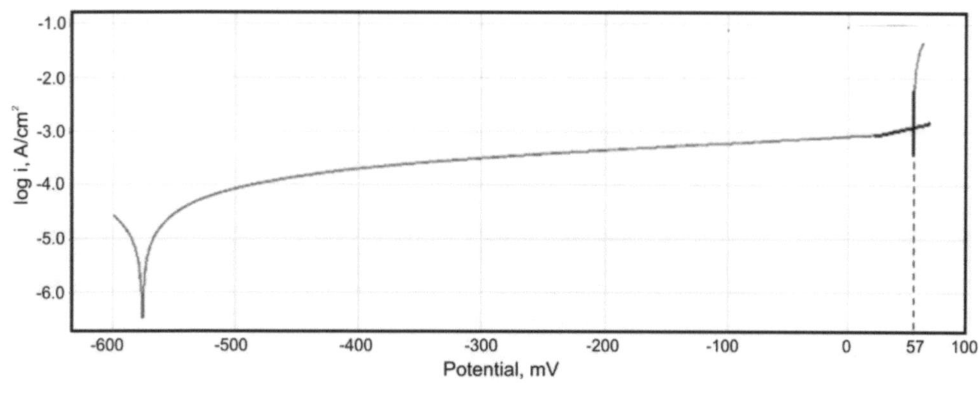

Fig. 12. Anodic polarization curve registered for the sample of 25Mn-3Si-1.5Al-Nb-Ti steel in 0.5N H_2SO_4.

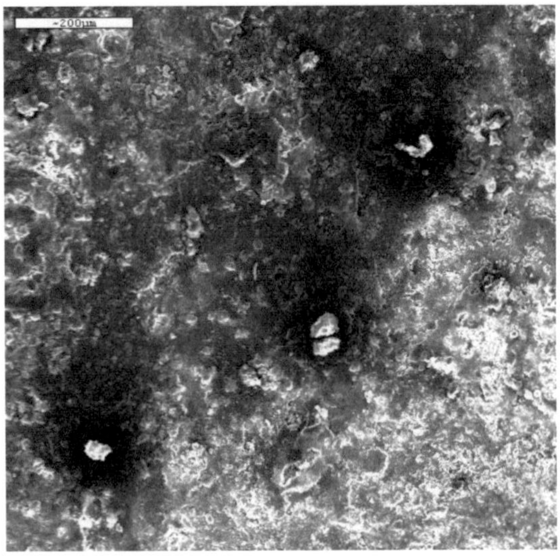

Fig. 13. Numerous corrosion pits on the surface of the specimen after electrochemical tests in 0.5N NaCl.

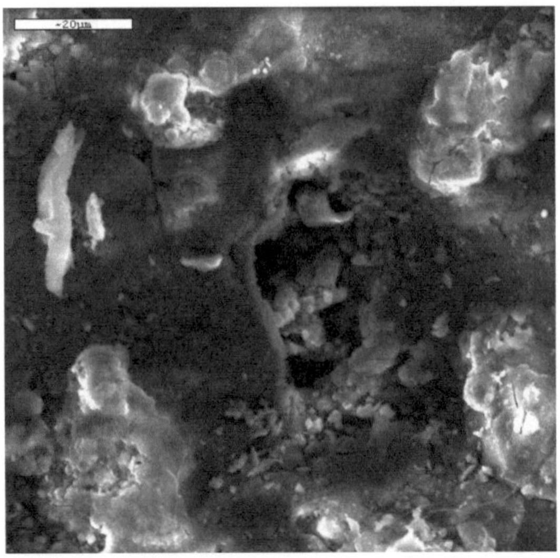

Fig. 14. Corrosion pitting on the surface of the specimen after electrochemical tests in 0.5N NaCl.

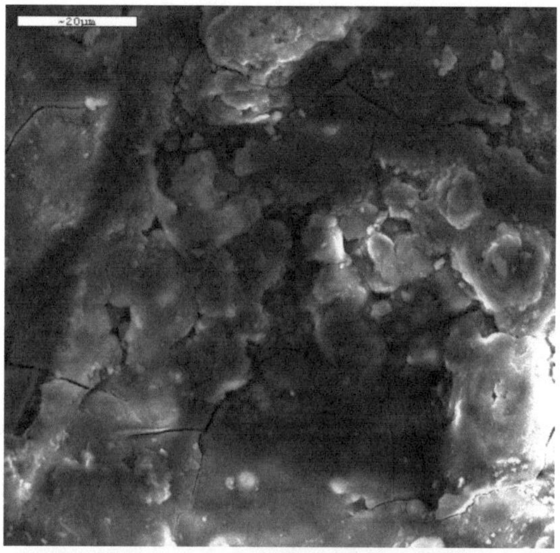

Fig. 15. Scaled and partially cracked surface of the specimen after electrochemical tests in the 0.5N H_2SO_4.

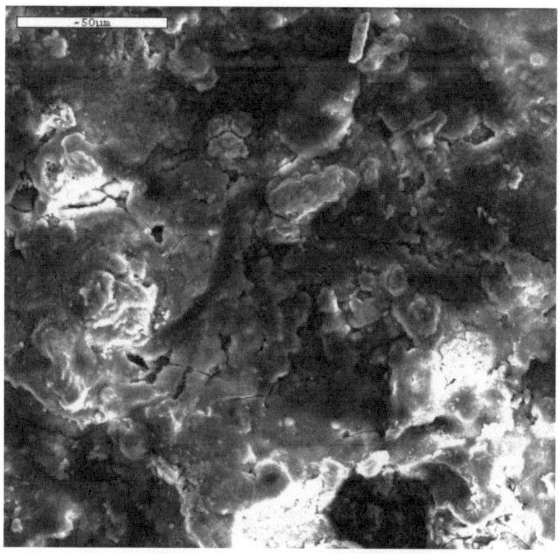

Fig. 16. Scaled surface and corrosion pits of the specimen after corrosion tests in the 0.5N H_2SO_4..

4.2 Corrosion behaviour in the cold-worked state

In both steels, immersed after deformation in 1N H_2SO_4 many corrosion pits of various size were observed (Fig. 17). The amount and the size of pits are very high and they are formed along the entire surface of specimens. Privileged places to pits forming are surface concentrations of non-metallic inclusions, which are also probable place of hydrogen penetration. Hydrogen also penetrates deeper into the steel - probably by ε martesite plates - accumulating in a surroundings of elongated non-metallic sulfide inclusions (Fig. 18).

Hydrogen failures were usually observed to the depth of about 0.3 mm. Places of hydrogen accumulation are also visible on samples revealing the steel structure after cold deformation. Usually, these places are elongated non-metallic inclusions, grain boundary areas and/or twin boundaries (Fig. 19). The 26Mn-3Si-3Al-Nb steel keeps after plastic deformation the austenitic structure, whereas a fraction of martensite in 25Mn-3Si-1.5Al-Nb-Ti steel increases (Fig. 20).

Besides non-metallic inclusions, especially privileged to hydrogen accumulation are lamellar areas of ε martensite. Absorbed atomic hydrogen penetrating the steel, accumulates in places with non-metallic inclusions, lamellar precipitations of the second phase, microcracks and other structural defects, where convenient conditions for recombining of atomic hydrogen to molecular H_2 exist. The recombination of atomic hydrogen to molecular state is a very exothermic reaction, which provides a pressure increase in formed H_2 bubbles as well as nucleation and growth of microcracks in a surface region of the sample (Fig. 20).

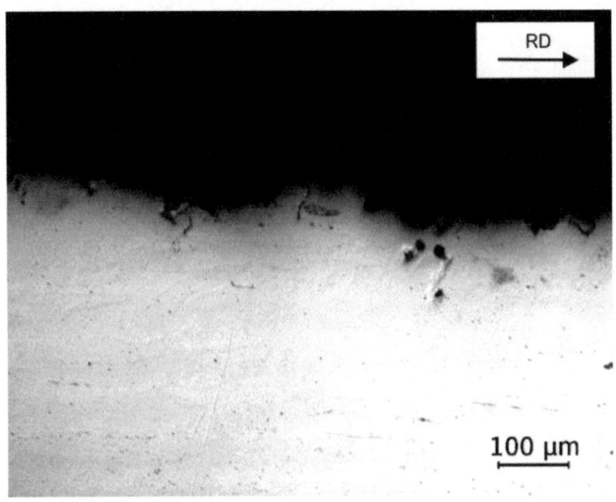

Fig. 17. Wide corrosion pits on the surface of 26Mn-3Si-3Al-Nb steel immersed in 1N H_2SO_4 and probable places of hydrogen penetration (transverse section).

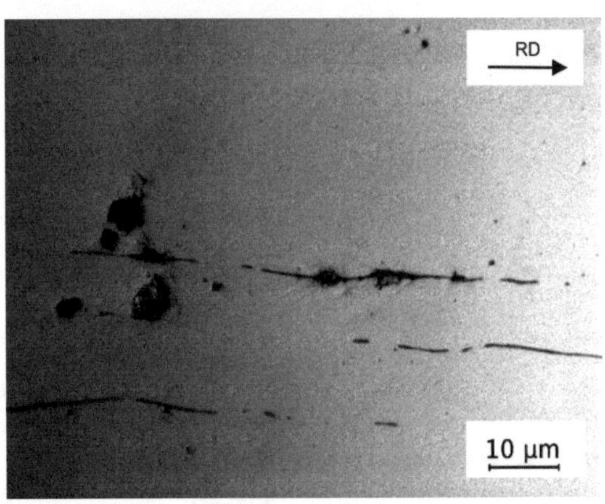

Fig. 18. Regions of hydrogen accumulation around elongated sulfide-type non-metallic inclusions in 25Mn-3Si-1.5Al-Nb-Ti steel (transverse section).

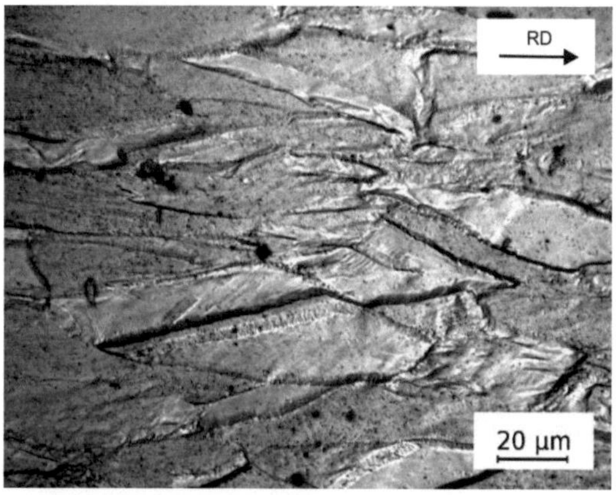

Fig. 19. Elongated austenite grains of 26Mn-3Si-3Al-Nb steel after cold deformation and some places of hydrogen accumulation (transverse section).

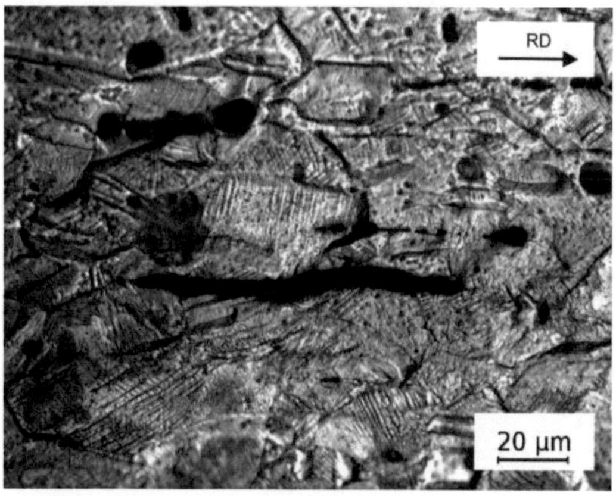

Fig. 20. Regions of hydrogen failures and hydrogen microcracks in 25Mn-3Si-1.5Al-Nb-Ti steel, plastically deformed and immersed in 1N H_2SO_4 (transverse section).

After bending and 100 hours immersion in 1N H_2SO_4 both steels show a meaningful percentage mass decrement, among 47 and 49% and 2 orders of magnitude lower in chloride solution (Table 3). Cold deformation rises slightly the mass decrement in acidic medium, in comparison with the specimens investigated in undeformed state (Table 2). The opposite is true for chloride solution. However, the differences are not significant. Comparable mass loss of two steels both in non-deformed and plastically deformed states indicates that the initial structure does not have meaningful impact on the corrosion progress.

	Corrosion medium	
Steel grade	1N H_2SO_4	3.5% NaCl
26Mn-3Si-3Al-Nb	47.5 ± 1.6	0.33 ± 0.01
25Mn-3Si-1.5Al-Nb-Ti	49.5 ± 2.4	0.37 ± 0.12

Table 3. Mean percentage mass loss of cold-deformed samples after the immersion tests, %

Figures 21-23 present the SEM microstructures of plastically deformed samples immersed in acidic solution. It is characteristic that corrosion cracks were not observed, whereas deep corrosion damages and band arranged corrosion products can be perceived (Fig. 21). The corrosion products layer is not continuous and has many cracks (Fig. 22). Besides remaining corrosion products, a numerous number of craters, created due to intensive corrosion pitting and probably as a result of hydrogen impact, is characteristic. Craters forming is accompanied by local cracking of corrosion products layer (Figs. 21, 22), uncovering the metal surface and causing further penetration of the corrosive medium and finally the intensive progress of general and pitting corrosion.

Hydrogen Induced Cracking (HIC) is a problem in carbon steels and especially in high-strength low-alloy steels. Typical examples are hydrogen failures of gas pipelines containing hydrogen sulfide (Cottis & Newman, 1995; Ćwiek, 2009). Conventional Cr-Ni austenitic steels are not usually liable to such damages. One of the reasons is relatively low diffusion coefficient of hydrogen in austenite as distinguished from steels with ferritic or martensitic structures (Kumar & Balasubramaniam, 1997; Xu et al., 1994). However, enhanced permeation of hydrogen was observed in cold worked austenitic steels what was attributed to strain-induced martensitic transformation leading to promote hydrogen diffusion as the diffusivity is much higher in the bcc martensite lattice (Kumar & Balasubramaniam, 1997). The hydrogen induced surface cracking at the high hydrogen concentration places, i.e. grain and twin boundaries, ε/γ interface was also observed in Cr-Ni steels during hydrogen effusion from the supersaturated sites (Yang & Luo, 2000). Additionally, hydrogen mobility is enhanced by the presence of high-dislocation density due to cold working (Ćwiek, 2009). It is important to note that hydrogen impact occurs both in diphase (Figs. 18, 20) and single phase (Figs. 10, 23) structures of the steels.

In the investigated high-Mn austenitic steels the high corrosion progress and uncovering of metal surface by formed successively corrosion pits (Figs. 17, 23) should be taken into account. Uncovered active metal inside of expanding pits reacts with the acidic solution with hydrogen emission. In this regard hydrogen impact can influence the corrosion behaviour of the investigated steels. Indirect confirmation of this fact are numerous craters formed due to corrosion pitting and probably hydrogen impact (Figs. 10, 23). The effect of hydrogen can be further enhanced by the presence of increased sulphur concentration (for example present as sulfide non-metallic inclusions).

On the surface of specimens dipped in NaCl solution, a layer of corrosion products, which protects the metal against continuous penetration of corrosive media, is forming. Created scaled layer strongly adheres to the base, though numerous surface cracks (Fig. 24). There are many corrosion cracks running from the specimen surface with a maximum value of inner stresses in the steel with martensite plates (Fig. 25). Rectilinear course of cracks, shows the transcrystalline cracking character, to which austenitic steels in media with chloride ions are sensitized. Corrosion cracks were not present in the steel with single-phase austenitic matrix. A few microcracks along martensite lamellas were also revealed (Fig. 26). The microcracks were usually nucleated on elongated non-metallic inclusions and were spread along martensitic plates with a hardness higher compared to the austenitic matrix (Grajcar et al., 2010a).

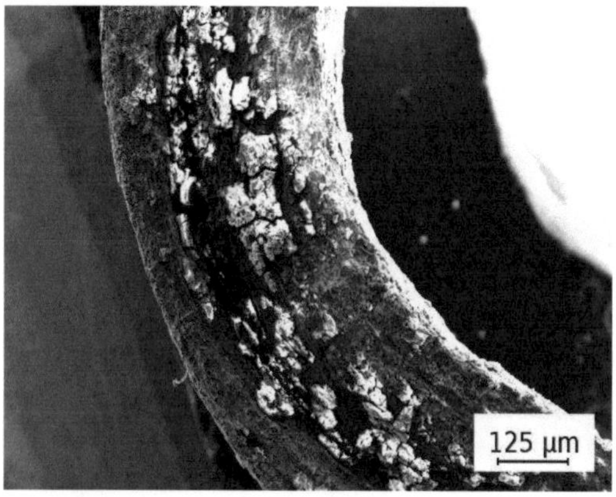

Fig. 21. Deep corrosion decrements and banding-like arrangement of corrosion products in 26Mn-3Si-3Al-Nb steel, plastically deformed and immersed in 1N H_2SO_4.

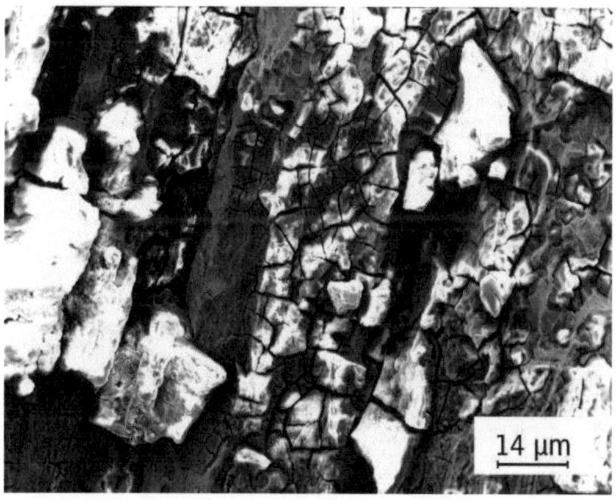

Fig. 22. Cracked layer of corrosion products with banding-like arrangement in 25Mn-3Si-1.5Al-Nb-Ti steel, plastically deformed and immersed in 1N H_2SO_4.

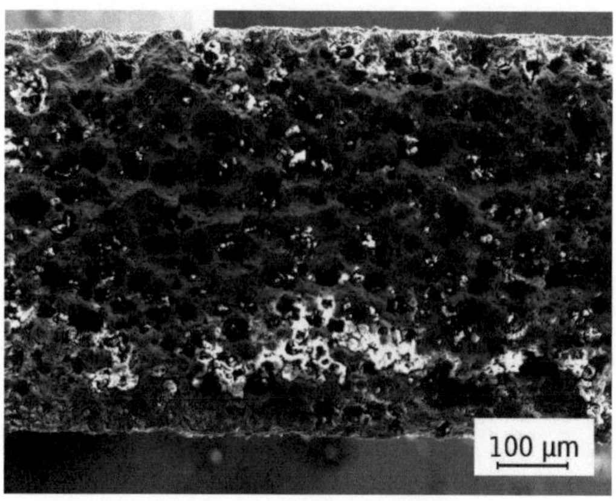

Fig. 23. Numerous craters formed due to corrosion pitting and probable hydrogen penetration in 26Mn-3Si-3Al-Nb steel, plastically deformed and immersed in 1N H_2SO_4.

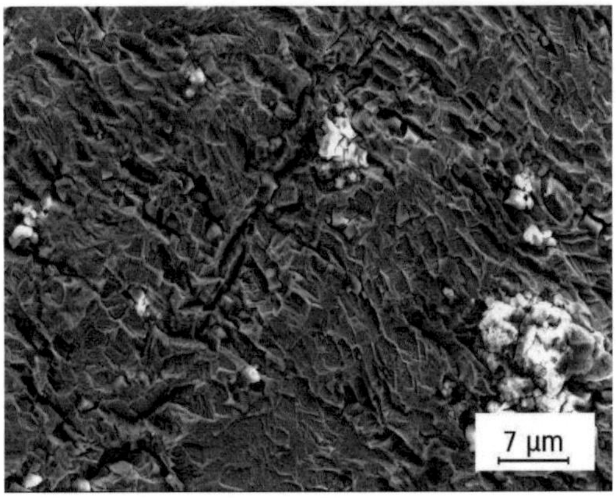

Fig. 24. Scaled and cracked layer of corrosion products in 25Mn-3Si-1.5Al-Nb-Ti steel, plastically deformed and immersed in 3.5wt% NaCl.

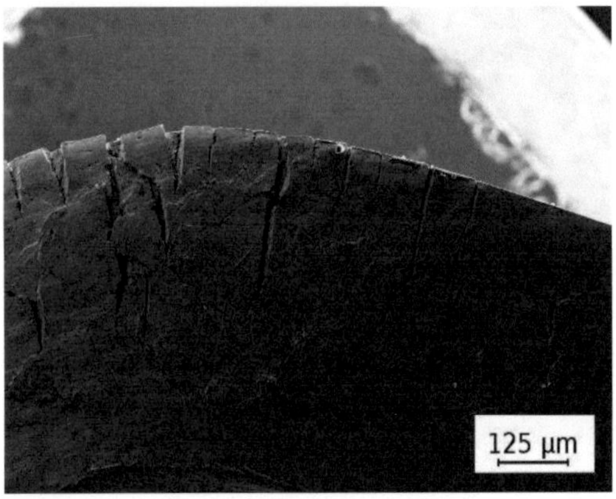

Fig. 25. Deep corrosion cracks in 25Mn-3Si-1.5Al-Nb-Ti steel after bending and the immersion in 3.5wt% NaCl.

Fig. 26. Microcracks running from non-metallic inclusions along hard martensitic plates in 25Mn-3Si-1.5Al-Nb-Ti steel, plastically deformed and immersed in 3.5wt% NaCl.

5. Summary

The automotive industry still requires steel sheets with higher strength, ductility and technological formability. Recently, special pressure is put to the need of increasing the passive safety of passengers what can be met by using specially designed controlled crash zones absorbing the energy during crash events. High-manganese austenitic alloys satisfy these requirements. However, the main disadvantage is their relatively poor corrosion resistance.

The results presented in this study focused on the evaluation of corrosion resistance of two high-Mn steels of the different initial structure in acidic and chloride media. The investigations were carried out on the specimens after the thermo-mechanical rolling and after cold deformation. The results of immersion and potentiodynamic tests as well as structural analysis prove that both examined steels, independent of initial structure, have very low corrosion resistance in acidic medium and low corrosion resistance in chloride solution. In particular it was found that:

- the mass decrement of specimens immersed in 1N H_2SO_4 for 100 hours is equal about 40% and is about 100 higher compared to the specimens immersed in 3.5wt% NaCl; The percentage mass loss of plastically deformed specimens is slightly higher compared to non-deformed specimens in acidic solution and slightly lower in chloride medium;
- the percentage mass decrement in the steel with single-phase austenitic structure is slightly lower than in the steel with ε martensite lamellas. However, any significant impact of the second phase on corrosion process acceleration was not observed;
- the decisive impact on the corrosion resistance of examined steels has their chemical composition, which determines the high rate of manganese and iron dissolution in acidic solution. The oxygen depolarization process results in formation of corrosion products layer on the surface of the steel examined in chloride medium. Therefore, the mass decrement of steels in 3.5wt% NaCl is much lower than in 1N H_2SO_4;
- both steels are liable to general and pitting corrosion, especially intensively in the sulfuric acid solution. A very adverse influence on corrosion pitting initiation has a relative large fraction of non-metallic inclusions, especially of these forming local aggregations. In chloride solution it also results in occurring local microcracks, nucleating at elongated non-metallic inclusions and growing along hard martensite plates;
- the surface layer of band-arranged corrosion products located accordingly to the deformation direction has many cracks, especially in surroundings of corrosion pits and non-metallic inclusions. In case of acidic solution, the cracks are also formed round craters formed due to corrosion pitting;
- the craters identified in both steels examined in acidic medium are combined effect of various corrosion damages in high-Mn austenitic steels. Hydrogen impact is the additional effect accompanying general corrosion and corrosion pits forming. Its influence is enhanced by numerous corrosion pits, sulfide non-metallic inclusions and martensite plates of high hardness;
- mechanical twins formed during cold working in the single-phase austenitic steel accelerate the corrosion progress. However, the special care should concern the steels containing ε martensite plates (in the initial structure, strain-induced or hydrogen-induced), which are especially susceptible to surface cracking in hydrogen containing solution and to corrosion cracking in chloride solution.

The corrosion resistance in chloride solutions in high-Mn alloys containing aluminum can be improved by the anodic passivation in 30% HNO_3 aqueous solution (Hamada et al., 2005). It leads to the modification of the chemical composition of the surface layer, connected with reducing the manganese concentration at the surface layer and the enrichment this region in Al, improving the corrosion resistance. Another way of improving the corrosion resistance of high-manganese steels is to use zinc coatings. Due to alloying problems, the use of the electro-galvanizing process is promising (Hamada, 2007). The best solution seems to be the incorporation of Cr, which promotes the formation of a passivation layer and improves the corrosion resistance (Hamada, 2007; Mujica et al., 2010). However, the changes in SFE of austenite by Cr and the resulting main deformation mechanism should be taken into account. Some chemical composition strategies include alloys with increased nitrogen concentration, which improves the resistance to pitting corrosion (Mujica Roncery et al., 2010).

6. Acknowledgement

The author would like to thank to Dr. Wojciech Krukiewicz, Dr. Marek Opiela, Mr. Sławomir Kołodziej for carrying out some corrosion experiments and to Dr. Witold Walke for his fruitful discussion.

7. References

Altstetter, C.J.; Bentley, A.P.; Fourine, J.W. & Kirkbridge, A.N. (1986). Processing and properties of Fe-Mn-Al alloys. *Materials Science and Engineering A,* Vol.82 (1986), pp. 13-25

Bleck, W. & Phiu-on, K. (2005). Microalloying of cold-formable multi phase steel grades. *Materials Science Forum,* Vol.500-501, (2005), pp. 97-112

Bleck, W.; Phiu-on, K.; Herring, C., & Hirt, G. (2007). Hot workability of as-cast high manganese high-carbon steels. *Steel Research International,* Vol.78, No.7, (2007), pp. 536-545

Cabanas, N.; Akdut, N.; Penning, J. & De Cooman, B.C. (2006). High-temperature deformation properties of austenitic Fe-Mn alloys. *Metallurgical and Materials Transactions A,* Vol.37A, (2006), pp. 3305-3315

Cottis, R.A. & Newman, R.C. (1995). *Stress corrosion cracking resistance of duplex stainless steels,* HSE Books, ISBN 0-7176-0915-4, London, UK

Ćwiek, J. (2009). Hydrogen degradation of high-strength steels. *Journal of Achievements in Materials and Manufacturing Engineering,* Vol.37, (2009), pp. 193-212

De Cooman, B.C.; Chin, K. & Kim, J. (2011). High Mn TWIP steels for automotive applications, In: *New trends and developments in automotive system engineering,* M. Chiaberge, (Ed.), pp. 101-128, InTech, ISBN 978-953-307-517-4, Rijeka, Croatia

Dini, G.; Najafizadeh, A.; Ueji, R. & Monir-Vaghefi S.M. (2010). Tensile deformation behavior of high manganese austenitic steel: The role of grain size. *Materials and Design,* Vol.31, (2010), pp. 3395-3402

Dobrzański, L.A.; Grajcar, A. & Borek, W. (2008). Microstructure evolution and phase composition of high-manganese austenitic steels. *Journal of Achievements in Materials and Manufacturing Engineering,* Vol.31, (2008), pp. 218-225

Dumay, A.; Chateau, J.P.; Allain, S.; Migot, S. & Bouaziz, O. (2008). Influence of addition elements on the stacking-fault energy and mechanical properties of an austenitic Fe-Mn-C steel. *Materials Science and Engineering,* Vol.483-484, (2008), pp. 184-187

Frommeyer, G. & Bruex, U. (2006). Microstructures and mechanical properties of high-strength Fe-Mn-Al-C light-weight TRIPLEX steels. *Steel Research International,* Vol.77, No.9-10, (2006), pp. 627-633

Frommeyer, G.; Bruex, U. & Neumann, P. (2003). Supra-ductile and high-strength manganese-TRIP/TWIP steels for high energy absorption purposes. *ISIJ International,* Vol.43, No.3, (2003), pp. 438-446

Ghayad, I.M.; Hamada, A.S.; Girgis, N.N. & Ghanem, W.A. (2006). Effect of cold working on the aging and corrosion behaviour of Fe-Mn-Al stainless steel. *Steel Grips,* Vol.4, No.2, (2006), pp. 133-137

Graessel, O.; Krueger, L.; Frommeyer, G. & Meyer, L.W. (2000). High strength Fe-Mn-(Al, Si) TRIP/TWIP steels development - properties - application. *International Journal of Plasticity,* Vol.16, (2000), pp. 1391-1409

Grajcar, A. & Borek, W. (2008). The thermo-mechanical processing of high-manganese austenitic TWIP-type steels. *Archives of Civil and Mechanical Engineering,* Vol.8, No.4, (2008), pp. 29-38

Grajcar, A.; Kołodziej, S. & Krukiewicz, W. (2010a). Corrosion resistance of high-manganese austenitic steels. *Archives of Materials Science and Engineering,* Vol.41, No.2, (2010), pp. 77-84

Grajcar, A.; Krukiewicz, W. & Kołodziej, S. (2010b). Corrosion behaviour of plastically deformed high-Mn austenitic steels. *Journal of Achievements in Materials and Manufacturing Engineering,* Vol.43, No.1, (2010), pp. 228-235

Grajcar, A.; Opiela, M. & Fojt-Dymara, G. (2009). The influence of hot-working conditions on a structure of high-manganese steel. *Archives of Civil and Mechanical Engineering,* Vol.9, No.3, (2009), pp. 49-58

Hamada, A.S. (2007). *Manufacturing, mechanical properties and corrosion behaviour of high-Mn TWIP steels,* Acta Universitatis Ouluensis C281, ISBN 978-951-42-8583-7, Oulu, Finland

Hamada, A.S.; Karjalainen, L.P. & El-Zeky, M.A. (2005). Effect of anodic passivation on the corrosion behaviour of Fe-Mn-Al steels in 3.5%NaCl, *Proceedings of the 9th International Symposium on the Passivation of Metals and Semiconductors and the Properties of Thin Oxide Layers,* pp. 77-82, Paris, France, June 27-30, 2005

Huang, B.X.; Wang, X.D.; Rong, Y.H.; Wang, L. & Jin, L. (2006). Mechanical behavior and martensitic transformation of an Fe-Mn-Si-Al-Nb alloy. *Materials Science and Engineering A,* Vol.438-440, (2006), pp. 306-313

International Iron & Steel Institute (September 2006). Advanced High Strength Steel (AHSS) Application Guidelines - version 3, Available from http://worldautosteel.org

Jimenez, J.A. & Frommeyer, G. (2010). Microstructure and texture evolution in a high manganese austenitic steel during tensile test. *Materials Science Forum,* Vol.638-642, (2010), pp. 3272-3277

Kannan, M.B.; Raman, R.K.S., & Khoddam, S. (2008). Comparative studies on the corrosion properties of a Fe-Mn-Al-Si steel and an interstitial-free steel. *Corrosion Science,* Vol.50, (2008), pp. 2879-2884

Kumar, P. & Balasubramaniam, R. (1997). Determination of hydrogen diffusivity in austenitic stainless steels by subscale microhardness profiling. *Journal of Alloys and Compounds,* Vol.255, (1997), pp. 130-134

Lovicu, G.; Barloscio, M.; Botaazzi, M.; D'Aiuto, F.; De Sanctis, M.; Dimatteo, A.; Federici, C.; Maggi, S.; Santus, C. & Valentini, R. (2010). Hydrogen embrittlement of advanced high strength steels for automotive use. *Proceedings of International Conference on Super-High Strength Steels,* pp. 1-13, Peschiera del Garda, Italy, October 17-20, 2010

Mazancova, E.; Kozelsky, P. & Schindler, I. (2010). The TWIP alloys resistance in some corrosion reagents. *Proceedings of International Conference METAL,* pp. 1-6, Roznov pod Radhostem, Czech Republic, May 18-20, 2010

Mujica, L.; Weber, S. & Theisen, W. (2010). Development of high-strength corrosion-resistant austenitic TWIP steel. *Proceedings of International Conference on Super-High Strength Steels,* pp. 1-9, Peschiera del Garda, Italy, October 17-20, 2010

Mujica Roncery, L.; Weber, S. & Theisen, W. (2010). Development of Mn-Cr-(C-N) corrosion resistant twinning induced plasticity steels: thermodynamic and diffusion calculations, production and characterization. *Metallurgical and Materials Transactions A,* Vol.41A, No.10, (2010), pp. 2471-2479

Opiela, M.; Grajcar, A. & Krukiewicz, W. (2009). Corrosion behaviour of Fe-Mn-Si-Al austenitic steel in chloride solution. *Journal of Achievements in Materials and Manufacturing Engineering,* Vol.33, No.2, (2009), pp. 159-165

Otto, M.; John, D.; Schmidt-Juergensen R.; Springub, B.; Cornelissen, M.; Berkhout, B.; Bracke, L. & Patel, J. (2010). HSD-steels, optimized high strength and high ductility austenitic steel. *Proceedings of International Conference on Super-High Strength Steels,* pp. 1-12, Peschiera del Garda, Italy, October 17-20, 2010

Shin, S.Y.; Hong, S.; Kim, H.S.; Lee, S. & Kim, N.J. (2010). Tensile properties and cup formability of high Mn and Al-added TWIP steels. *Proceedings of International Conference on Super-High Strength Steels,* pp. 1-9, Peschiera del Garda, Italy, October 17-20, 2010

Sojka, J.; Mazancova, E.; Schindler, I.; Kander, L.; Kozelsky, P.; Vanova, P. & Wenglorzova, A. (2010). Resistance against hydrogen embrittlement of advanced materials for automotive industry. *Proceedings of International Conference METAL,* pp. 1-6, Roznov pod Radhostem, Czech Republic, May 18-20, 2010

Vercammen, S.; De Cooman, B.C.; Akdut, N.; Blanpain, B. & Wollants, P. (2002). Microstructural evolution and crystallographic texture formation of cold rolled austenitic Fe-30Mn-3Al-3Si TWIP-steel. *Proceedings of International Conference on TRIP-aided High-Strength Ferrous Alloys,* pp. 55-60, Ghent, Belgium, June 19-21, 2002

Xu, J.; Sun, X.; Yuan, X. & Wei, B. (1994). Hydrogen permeation and diffusion in low-carbon steels and 16Mn steel. *Journal of Materials Science Technology,* Vol.10, (1994), pp. 92-96

Yang, Q. & Luo, J.L. (2000). Martensite transformation and surface cracking of hydrogen charged and outgassed type 304 stainless steel. *Materials Science and Engineering A,* Vol.288, (2000), pp. 75-83

Zhang, Y.S. & Zhu, X.M. (1999). Electrochemical polarization and passive film analysis of austenitic Fe-Mn-Al steels in aqueous solutions. *Corrosion Science,* Vol.41, (1999), pp. 1817-1833

17

Improvement of Corrosion Resistance of Aluminium Alloy by Natural Products

R. Rosliza
TATI University College, Jalan Panchor, Teluk Kalong, Kemaman, Terengganu, Malaysia

1. Introduction

Protection of metals from ever progressing corrosion presents one of the topical issues of this century. The increasing industrialization of our life is accompanied with the ever-growing number of metals that corrode and become devalued. Corrosion is a chemical or electrochemical reaction process against certain material, usually metal and its environment which produce the deterioration of the material and its properties. The corrosion reaction produces a less desirable material from the original metal and resulted in the reduced function of a component or system, a significant problem encountered everyday.

Corrosion is a problem that impacts every industry. The serious consequences of the corrosion process have become a problem of worldwide significance. It is estimated that annual loss and damage due to corrosion in the United Kingdom costs about £5000 million; and approximately one tone of steel is lost through corrosion every 90 seconds. Further, it is estimated that 25% of this loss could be avoided by correct design, correct material selection and proper preventive processes (Barbara and Robert, 2006).

Even with the proper application of available countermeasures, the estimated cost by National Association of Corrosion Engineers (NACE) for replacing corroded piping systems in the United States alone stands well in excess of $70 billion annually, which was 4.2% of the gross national product (GNP)-making corrosion one of the most potentially damaging losses to any commercial, private, or industrial property (Barbara and Robert, 2006).

Even though the term of corrosion is usually applied to metals; all materials including ceramics, plastics, rubber and wood deteriorate at the surface to some extent as a result of being exposed to certain combinations of liquids and/or gases. Few practical examples are the rusting of tools and automobiles over many years of use; the failure of pipelines delivering volatile components such as natural gases and environmentally harmful chemicals such as crude oil and hydrochloric acid; bridge failure, ship failure (due to pumps, fuel tanks, boiler and sensors) and aircraft crashes; for example, Aloha Airlines flight 737 jet landing gear failure in 1988 (Radia, 2004). Therefore, the importance of understanding corrosion is clear, especially in the analysis and the design systems that incorporate metal as a major component material which exposed to corrosive environments.

Although much progress has been made in understanding the thermodynamics and kinetics of the corrosion process, the mechanisms of localized corrosion are not well understood, nor are those for imparting resistance or protection against aqueous or gases corrosion. With knowledge of the types of and a better understanding of the mechanism and causes of corrosion, it is possible to make measures to prevent them from occurring. For examples, we may change the nature of the environment by selecting a material that is relatively non reactive and/or protect the material from corrosion. Controlling corrosion in the infrastructure can prevent premature failure and lengthen useful service life, both of which save money and natural resources, promote public safety and protect the environment.

Due to the various industrial applications and economic importance of aluminium and its alloys, its protection against corrosion has attracted much attention (Aballe *et al.*, 2001; Cheng *et al.*, 2004; Hintze and Calle, 2006). Most aluminium alloys have good corrosion resistance towards natural atmospheres and other environments, because aluminium alloy surfaces are covered with a natural oxide film of thickness about 5 nm (Klickic *et al.*, 2000). However, in the presence of aggressive ions, like chloride, the protective layer can be locally destroyed and corrosive attack takes place (Kliskic *et al.*, 2000). Yet, if correctly protected, applications of aluminium alloy may be more reliable and have long service life.

One of the methods to protect metals or alloys against corrosion is addition of species to the solution in contact with the surface in order to inhibit the corrosion reaction and reduce the corrosion rate (Trabenelli *et al.*, 2005) known as corrosion inhibitor. A number of corrosion inhibitors for aluminium alloys have been developed for this purpose such as lanthanide chloride, tolytriazole, bitter leaf, Schiff base compounds and polyacrylic acid (Benthencourt *et al.*, 1997; Onal and Aksut, 2000; Avwiri and Igho, 2003; Yurt *et al.*, 2006; Amin *et al.*, 2009).

Owing to the growing interest and attention of the world towards environmental problems and towards the protection of environment and the hazardous effects of the use of chemicals on ecological balance, the traditional approach on the choice of corrosion inhibitors has gradually changed. Researches are mainly focusing on non-toxic "green" corrosion inhibitors. Therefore, there is a great task to search for suitable natural source to be used as corrosion inhibitor as an alternative for the existing inhibitors.

2. Literature review on corrosion resistance

Corrosion can be controlled by suitable modifications of the environment which in turn stifle, retard or completely stop the anodic or cathodic reactions or both. This can be achieved by the use of inhibitors (Blustein *et al.*, 2005; Emregul *et al.*, 2005; Goa *et al.*, 2008). Corrosion inhibitors are substances which when added in small concentrations to corrosive media decrease or prevent the reaction of the metal with the media. Inhibitors are added to many systems, *e.g.* cooling systems, refinery units, acids, pipelines, chemicals, oil and gas production units, boiler and process waters etc. (Raja and Sethuraman, 2009).

A number of corrosion inhibitors have been developed to mitigate aluminium corrosion for the last two decades. A variety of inhibitors have been tested such as chromates, dichromates, molybdates, nitrate, nitrite and sulfate. Their high efficiency/cost ratio has made them standard corrosion inhibitors for a wide range of metals and alloys (Benthencourt *et al.*, 1997).

Although chromates, dichromates, molybdates, nitrate, nitrite and sulfate were found to be the effective inhibitor for the corrosion processes taking place at the electrode/electrolyte interface of aluminium and some of its alloys in acidic and basic solutions (El-Sobki *et al.*, 1981; Kassab *et al.*, 1987; Badawy *et al.*, 1999), unluckily a major disadvantage is their toxicity and such as their use has come under severe criticism (Bethencourt *et al.*, 1997; Song-mei *et al.*, 2007).

In recent days many alternative eco-friendly corrosion inhibitors have been developed, the range from rare earth elements (Neil and Garrard, 1994; Mishra and Balasubramaniam, 2007), and inorganic (Salem *et al.*, 1978) to organic compounds (Onal and Aksut, 2000; Branzoi *et al.*, 2002; Maayta and Al-Rawashdeh, 2004). Owing to the growing interest and attention of the world towards environmental problems and towards the protection of environment and the hazardous effects of the use of chemicals on ecological balance, the traditional approach on the choice of corrosion inhibitors has gradually changed. Researches are mainly focusing on non-toxic "green" corrosion.

El-Etre and Abdallah (2000) study the natural honey as corrosion inhibitor for carbon steel in high saline water. It was found that natural honey exhibited a very good performance as inhibitor for steel corrosion in high saline water. The effect of fungi on the inhibition efficiency of natural honey is markedly decreased in high saline water. This is due to the high concentration of NaCl that retard the growth of fungi. This finding attracts the author to carry out further investigation on the effect of natural honey in seawater which contains NaCl.

One of the aromatic groups that showed good inhibitive effect is vanillin. The inhibition effects of vanillin on the corrosion of steel in HCl and H_2SO_4 solutions were investigated by Emregul and Hayvali (2002) and Li *et al.* (2008), meanwhile El-Etre (2001) studied the effect of vanillin against acid corrosion of aluminium. They were explored that an aromatic aldehyde containing carbonyl, methoxy and hydroxyl groups arranged around the aromatic ring in vanillin contributed to the inhibition mechanism process. Lack of research on the effect of this inhibitor on the corrosion of aluminium alloy in seawater has motivated the author to explore this research area as contribution to the current interest on environmental-friendly and green corrosion inhibitors.

3. Research methodology

3.1 Materials

The material employed was Al-Mg-Si alloy (AA6061). The chemical composition (weight %) of Al-Mg-Si is listed in Table 1 and the validity of composition was determined by EDS. Extruded shape of Al-Mg-Si alloy was selected in this study because of its well-proven medium strength structural alloy that satisfies the requirements of a number of specifications and most applicable alloy used in marine applications.

The samples were cut into 25 x 25 x 3 mm coupons and mechanically polished using #400, 500 and 600 silicone carbide emery papers (ASTM G 1) and lubricated using distilled water. The polished samples were cleaned with acetone (Merck, 99.8% purity) washed using distilled water, dried in air and stored in moisture-free desiccators prior to use.

Alloys		% Weight
Silicon	Si	0.40–0.80
Iron	Fe	0.7
Copper	Cu	0.15–0.40
Manganese	Mn	0.15
Magnesium	Mg	0.80–1.20
Chromium	Cr	0.04–0.35
Zinc	Zn	0.25
Titanium	Ti	0.15
Others (each)		0.05
Others (total)		0.15
Aluminium	Al	Remainder

Table 1. The chemical-composition of Al-Mg-Si alloy

3.2 Test solution

The study test solution was tropical seawater collected approximately 100 m from the shoreline of Pantai Teluk Kalong, Kemaman, Terengganu. Pantai Teluk Kalong is located near Teluk Kalong Industrial Estate (oil and gas industrial area) and Kemaman Supply Base port which is about 6.78 km from Chukai, Terengganu, Malaysia.

Pantai Teluk Kalong was selected in this research due to widely application of Al-Mg-Si alloy in the shipping, marine, oil and gas industrials surrounding the area. The values of physicochemical properties of seawater such as salinity, dissolved oxygen, pH and temperature were monitored during the immersion test. Average selected physicochemical data of the seawater used are reported in Table 2.

pH	Temperature (°C)	Salinity (g/L)	Dissolved oxygen (mg/L)
7.63	28.3	34.8	7.66

Table 2. Physicochemical properties of seawater

3.3 Corrosion inhibitors

The corrosion inhibitors tested in this study were natural honey, vanillin and tapioca starch. The choice of natural honey (NH), vanillin (VL) and tapioca starch (TS) as corrosion inhibitors for the study were based on the following:

i. the selected natural products contained the possible adsorption functional groups
ii. these natural products are commercially available at low cost
iii. good solubility in water and non-toxic

The Nicolet 380 Fourier transform infrared (FTIR) spectrometer was used to determine the function group for each inhibitor.

3.4 Electrochemical measurements

All electrochemical measurements (PP, LPR and EIS) were accomplished with Autolab frequency response analyzer (FRA) and general purpose electrochemical system (GPES) for

Windows-version 4.9.005 coupled to an Autolab potentiostat connected to a computer. The cell used consists of conventional three electrodes with a platinum wire counter electrode (CE), a working electrode (WE) and a saturated calomel electrode (SCE) as reference to which all potentials are referred.

The WE was in the form of a square cut so that the flat surface was the only surface in the electrode. The exposed area to the test solution was 3.75 cm^2. The WE was first immersed in the test solution and after establishing a steady state open circuit potential, the electrochemical measurements were performed. The cell was exposed to air and the measurement was conducted at room temperature (25.0 ± 0.1 °C). Triplicate experiments were performed in each case of the same conditions to test the validity and reproducibility of the measurements. All procedures for electrochemical measurements were performed in accordance with the Standard Practice for Calculation of Corrosion Rates and Related Information from Electrochemical measurements (ASTM G 102).

4. Results and discussion

Many corrosion phenomena can be explained in terms of electrochemical reactions. Therefore, electrochemical techniques can be used to study these phenomena. Measurements of current-potential relations under carefully controlled conditions can yield information on corrosion rates, coatings and films, pitting tendencies and other important data.

4.1 Electrochemical measurements

The corresponding corrosion potential (E_{corr}), corrosion current density (i_{corr}), anodic Tafel slope (b_a), cathodic Tafel slope (b_c) and *CR* for uninhibited and inhibited systems from PP measurement are listed in Table 3. The data demonstrates that the E_{corr} values shift to more positive values as the concentration of added studied inhibitors are increased. On the other hand, the corrosion current densities are markedly declined upon addition of the studied corrosion inhibitors. The extent of its decline increases with increasing of the corrosion inhibitor concentration. Moreover, the numerical values of both anodic and cathodic Tafel slopes decreased as the concentration of inhibitors were increased. This means that the three natural products have significant effects on retarding the anodic dissolution of aluminium alloy and inhibiting the cathodic hydrogen evolution reaction.

Anodic and cathodic processes of aluminium corrosion in seawater are dissolution of aluminium and reduction of dissolved oxygen, respectively, as

$$4Al \rightarrow 4Al^{3+} + 12e^- \quad (1)$$

$$3O_2 + 6H_2O + 12e^- \rightarrow 12OH^- \quad (2)$$

Hence, Al^{3+} reacts with OH^- to form aluminum hydroxide near the aluminium surface as below

$$4Al + 3O_2 + 6H_2O \rightarrow 4Al(OH)_3 \quad (3)$$

Inhibitor	c_{inh} (ppm)	PP					LPR
		E_{corr} (mV)	i_{corr} (μA cm^{-2})	b_a (mV dec^{-1})	b_c (mV dec^{-1})	CR (10^{-2}mmyr^{-1})	R_p (kΩ cm^2)
Blank		-796	1.622	101	274	1.078	11.71
	200	-554	0.593	65	93	0.381	34.78
	400	-546	0.551	69	85	0.354	37.15
NH	600	-553	0.434	75	89	0.289	55.06
	800	-550	0.258	67	92	0.166	79.23
	1000	-556	0.137	48	59	0.088	112.72
	200	-693	0.538	56	66	0.349	35.28
	400	-642	0.346	83	61	0.230	52.48
VL	600	-590	0.330	32	67	0.215	57.13
	800	-545	0.198	80	76	0.129	90.98
	1000	-530	0.122	40	43	0.079	145.05
	200	-578	0.483	65	133	0.315	41.29
	400	-565	0.237	82	86	0.155	78.43
TS	600	-577	0.199	65	128	0.130	81.67
	800	-567	0.173	44	57	0.113	102.89
	1000	-514	0.103	43	49	0.064	177.00

Table 3. The electrochemical parameters of Al-Mg-Si alloy in absence and presence of different concentrations of NH, VL and TS

and the hydroxide precipitates on the surface due to its low solubility product. Aluminium hydroxide changes gradually to aluminium oxide, resulting in the formation of passive film (Aramaki, 2001):

$$2Al(OH)_3 \rightarrow (Al_2O_3) + 3H_2O \tag{4}$$

However, this nature oxide film does not offer sufficient protection against aggressive anions and dissolution of aluminium substrate occurs when exposed to corrosive solution.

Seawater predominantly consists of about 3.5% of sodium chloride (NaCl) and many other ions. Chloride ions are very strong and could easily penetrate the passive film. Thus, dissolution of the aluminium substrate occurs and results in corrosion. The adsorption of the corrosion inhibitor competes with anions such as chloride. By assuming that the corrosion inhibitor molecules preferentially react with Al^{3+} to form a precipitate of salt or complex on the surface of the aluminum substrate, the anodic and cathodic processes subsequently suppressed by inhibitor molecules. Thus, this result suggests that the protective film that was formed comprise aluminium hydroxide, oxide and salts or complexes of the corrosion inhibitor anions.

Polarization resistance, R_p values for Al-Mg-Si alloy in seawater in the presence and absence of corrosion inhibitor were determined using linear polarization method. The values of R_p are tabulated in Table 3. Generally, the value of R_p increased with increasing inhibitor concentration for all studied inhibitor. The highest R_p values for Al-Mg-Si alloy obtained at

1000 ppm of NH, VL and TS i.e. 112.72, 145.05 and 177.00 kΩ cm², respectively. A higher of R_p indicates the lower of the corrosion rate. The corrosion resistance obtained by LPR measurement was compared between the studied inhibitors. The inhibitor concentration was plotted against the values of R_p for each studied inhibitor (Figure 1). The results show that the values of R_p after addition of inhibitor increase with the following sequence: NH < VL < TS.

The values of R_{ct} and double layer capacitance, C_{dl} for Al-Mg-Si alloy at various concentrations of NH, VL and TS are presented in Table 4. The results show that the R_{ct} values increase with the addition of corrosion inhibitors when compared with those without corrosion inhibitor. Furthermore, the values of R_{ct} are observed to increase with the increasing corrosion inhibitor concentration, which can be attributed to the formation of a protective over-layer at the metal surface. It becomes a barrier for the charge transfers.

The values of R_{ct} for the alloy in inhibited solution with NH were enhanced up to 10 times higher as compared to that of the value of R_{ct} in uninhibited solution. Meanwhile, VL and TS shown better performance in improving the value of R_{ct} for Al-Mg-Si alloy in studied aggressive solution, where they were increased the values up to 13 and 14 times higher, respectively.

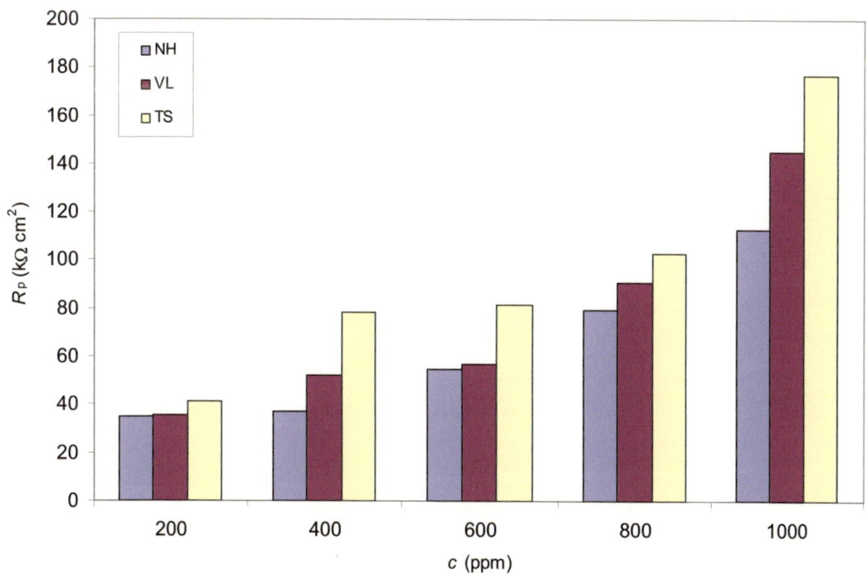

Fig. 1. R_p versus concentration of the studied inhibitors

It should be noted that while R_{ct} values increase with the addition of corrosion inhibitor, the capacitance, C_{dl} values decrease indicating the formation of a surface film. Thus, effective corrosion resistance is associated with high R_{ct} and low C_{dl} values (Yagan *et al.*, 2006). Increase in R_{ct} values and decrease in C_{dl} values by NH, VL and TS indicated that the studied inhibitors inhibit the corrosion of Al-Mg-Si alloy in seawater by adsorption mechanism (Noor, 2009) and the thickness of the adsorbed layer increases with the increase of inhibitor concentration.

The equivalent circuit fitting for these experimental data is a Randles circuit. The Randles equivalent circuit is one of the simplest and most common circuit models of electrochemical impedance. It includes a solution resistance, R_s in series to a parallel combination of resistor, R_{ct}, representing the charge transfer (corrosion) resistance and a double layer capacitor, C_{dl}, representing the electrode capacitance (Badawy *et al.*, 1999). In this case, the value of R_s can be neglected because the value is too small as compared to that of the value of R_{ct}. The equivalent circuit for the Randles cell is shown in Figure 2.

Inhibitor	*c* (ppm)	R_{ct} (kΩ cm^2)	C_{dl} (μF cm^{-2})
Blank		11.76	23.98
NH	200	33.10	8.11
	400	36.06	7.79
	600	57.12	5.16
	800	76.39	3.81
	1000	119.84	2.31
VL	200	39.04	7.06
	400	51.63	5.43
	600	60.27	4.39
	800	98.05	3.16
	1000	155.84	1.95
TS	200	40.73	7.30
	400	69.72	3.86
	600	79.23	3.37
	800	107.45	2.79
	1000	166.09	0.99

Table 4. R_{ct} and C_{dl} of Al-Mg-Si alloy in seawater obtained using impedance method

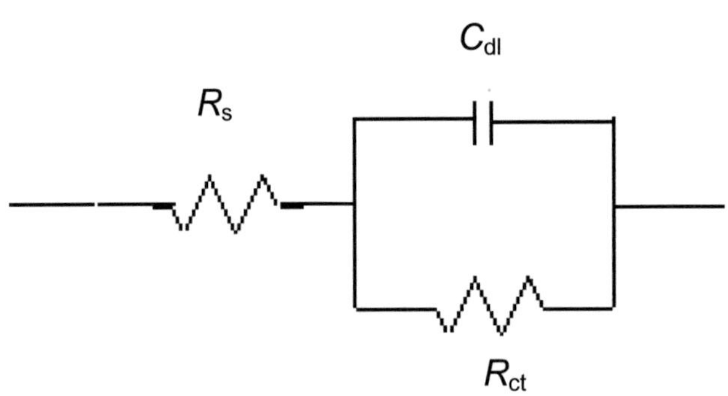

Fig. 2. The equivalent circuit for the Randles cell

4.2 Inhibition Efficiency (IE)

Owing to the adsorption of the corrosion inhibitor molecules onto the surface of Al-Mg-Si alloy, a thin film is formed on the aluminium alloy to retard the corrosion. Thus, in this case, all studied inhibitors worked as the filming corrosion inhibitor to control the corrosion rate. Instead of reacting with or removing an active corrosive species, the filming corrosion inhibitor function by strong adsorption and decrease the attack by creating a barrier between the metal and their environment (Al-Juhni and Newby, 2006).

The values of IE (%) from the PP, LPR and EIS measurements obtained by using following equation:

$$\mathrm{IE_{PP}}\ (\%) = 100\left(\frac{i_{corr} - i'_{corr}}{i_{corr}}\right) \tag{5}$$

$$\mathrm{IE_{PP}}\ (\%) = 100\left(\frac{CR - CR'}{CR}\right) \tag{6}$$

$$\mathrm{E_{RP}}\ (\%) = 100\left(\frac{R'_p - R_p}{R_p}\right) \tag{7}$$

$$\mathrm{IE_{EIS}}\ (\%) = 100\left(\frac{R'_{ct} - R_{ct}}{R_{ct}}\right) \tag{8}$$

The values of the result are presented in Table 5. All these parameters showed a similar trend. In all cases, increasing the inhibitors concentration is accompanied by an increase in the IE (%) and maximum for 1000 ppm. The IE (%) for all the measurements obtained from three different methods; PP, LPR and EIS are in good agreement. The inhibitive properties of the studied natural products can be given by the following order: NH < VL < TS.

4.3 Inhibition mechanism

The protection action of inhibitor substances during metal corrosion is based on the adsorption ability of their molecule where the resulting adsorption film isolates the metal surface from the corrosive medium. Consequently, in inhibited solutions, the corrosion rate is indicative of the number of the free corrosion sites remaining after some sites have been blocked by inhibitor adsorption.

The adsorption of NH, VL and TS compounds on the aluminium alloy surface reduces the surface area available for corrosion. Increases in inhibitor concentration results in amplify the degree of metal protection due to higher degree of surface coverage, θ (θ = IE%/100). This is resulting from enhanced inhibitor adsorption. The higher θ were acquired at 1000 ppm of NH, VL and TS i.e. 0. 9036, 0.9185 and 0.9587, respectively. Further investigation using surface analytical technique i.e. FTIR and SEM-EDS enable to characterize the active materials in the adsorbed layer and identify the most active molecule of the studied inhibitors.

Inhibitor	c (ppm)	IE (%)			
		PP		LPR	EIS
		CR	i_{corr}	R_p	R_{ct}
Honey	200	64.66	63.43	66.34	64.48
	400	67.16	66.04	68.48	67.39
	600	73.21	73.24	78.73	79.42
	800	84.63	84.12	85.22	84.61
	1000	91.85	91.58	89.61	90.19
Vanillin	200	66.84	67.62	66.81	69.88
	400	78.70	78.68	77.69	77.23
	600	79.66	80.02	79.50	80.49
	800	87.81	88.01	87.13	88.01
	1000	92.50	92.67	91.93	92.46
Tapioca	200	70.27	71.84	71.64	71.13
	400	85.42	86.19	85.07	83.14
	600	87.75	88.40	85.66	85.16
	800	89.34	89.90	88.62	89.06
	1000	93.98	93.98	93.38	92.92

Table 5. Values of IE (%) for Al-Mg-Si alloy at various concentrations of NH, VL and TS

4.3.1 Inhibition mechanism of NH

The inhibition performances of honey could be explained as follows: Fourier transform infrared (FTIR) spectrum in Figure 3 demonstrates that honey is a mixture of various compounds containing carbon (C), oxygen (polyphenols), nitrogen and sulphur (glucosinolates) which all can be adsorbed on the corroded metal (Radojcic *et al.*, 2008). The bands at about 1055.3 and 1418.1 cm^{-1} are consists of C, O, H and N atoms, meanwhile the peak at 1255.6 cm^{-1} is due to sulphur (S). A band appearing near 2935.7 cm^{-1} proves the existence of C, O and H atoms in NH. A band located at 3355.1 cm^{-1} corresponds to O, N and H atoms.

The adsorption of NH onto the surface of Al-Mg-Si alloy may take place through all these functional groups. The simultaneous adsorption of the four functional groups forces the natural honey molecule to be horizontally oriented on the surface of Al-Mg-Si alloy (Gao *et al.*, 2008). As the corrosion resistant concentration increases, the area of the metal surface covered by the corrosion resistant molecule also increases, leading to an increase in the IE.

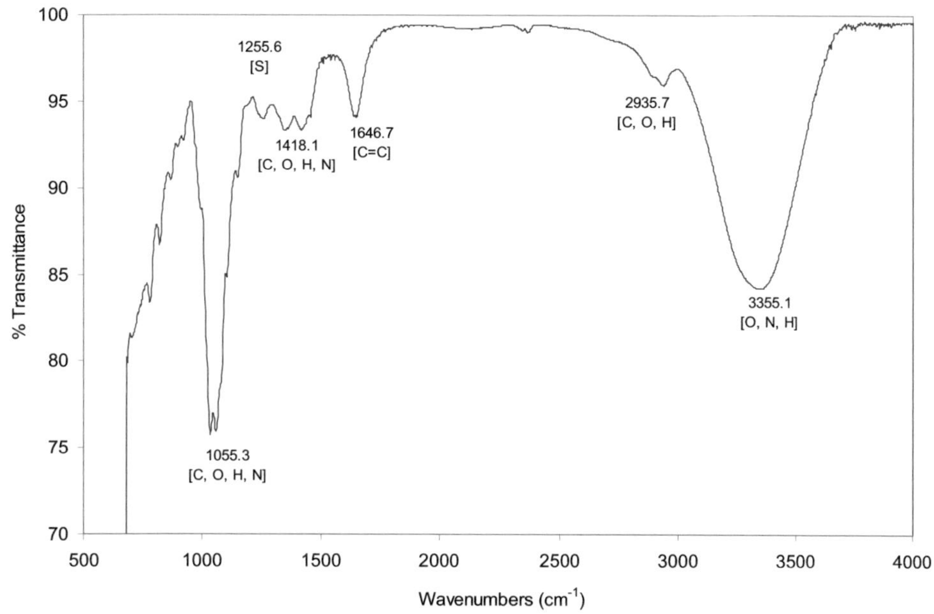

Fig. 3. FTIR spectrum of NH.

4.3.2 Inhibition mechanism of VL

The inhibition process of vanillin could be explained as follows: FTIR spectrum illustrate that vanillin is an aromatic aldehyde containing carbonyl, methoxy, and hydroxyl groups arranged around the aromatic ring (Figure 4). The bands at about 1153.7 to 1199.9 cm^{-1} and 2362.3 cm^{-1} in the spectrum are assigned to carbonyl group, meanwhile the bands located between 1429.8 to 1664.9 cm^{-1} are refers to hydroxyl group and aromatic compound (benzene ring).

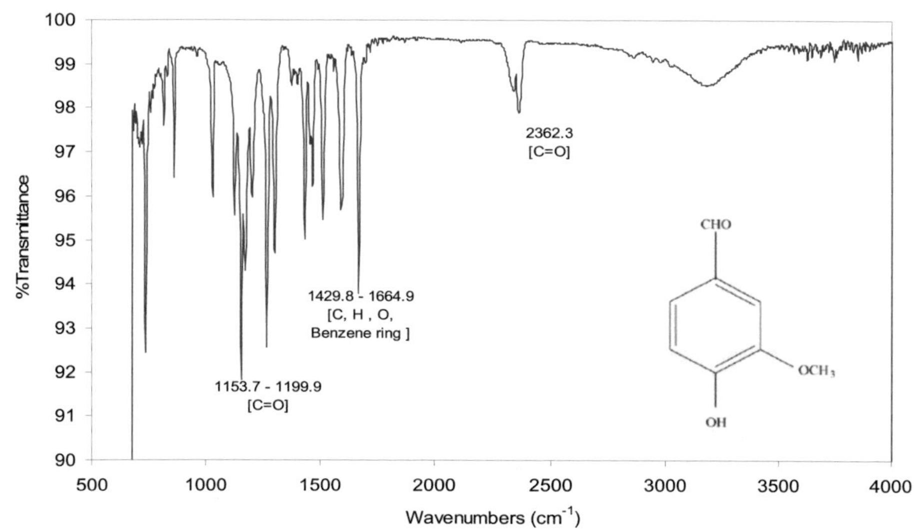

Fig. 4. FTIR spectrum of VL.

The adsorption of vanillin onto the surface of the aluminium alloy may take place through all these functional groups. The simultaneous adsorption of the three functional groups forces the vanillin molecule to be horizontally oriented at the surface of the aluminium alloy (Li *et al.*, 2008). As the inhibitor concentration increases, the area of the metal surface covered by the inhibitor molecule also increases, leading to an increase in the IE.

Similar to the findings reported previously (El-Etre, 2001; Emregul and Hayvali, 2002; Li *et al.*, 2008) the adsorption of vanillin mechanism is related to the presence of carbonyl, methoxy, and hydroxyl groups arranged around the aromatic ring in their molecular structures.

4.3.3 Inhibition mechanism of TS

The inhibition process of tapioca starch can be explained as follows: FTIR spectrum shows that TS is composed of mixture of two molecular entities (polysaccharides), a linear fraction, amylase and highly branched fraction, amylopectin (Figure 5). Both of them are polymers of glucose.

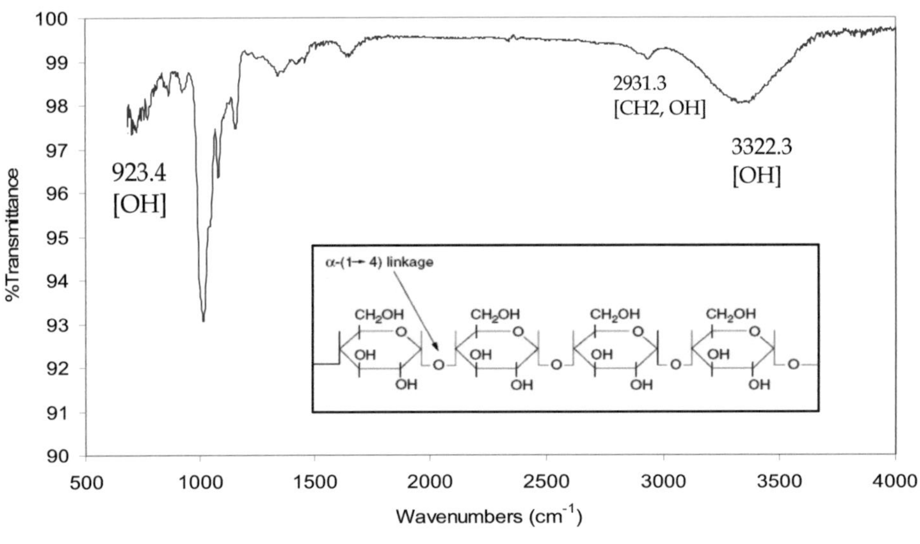

Fig. 5. FTIR spectrum of TS.

Amylose is constituted by glucose monomer units joined to one another head to tail forming alpha-1, 4 linkages; these are linked with the ring oxygen atoms all on the same side. Amylopectin differs from amylase in that branching occurs, with an alpha-1, 6 linkages every 24-30 glucose monomer units. Amylopectin has phosphate groups attached to some hydroxyl group (Wu *et al.*, 2009).

The peak at 923.4, 2931.3 and 3322.3 cm^{-1} in the FTIR spectrum are characteristic of hydroxyl group (OH). The adsorption of TS on Al-Mg-Si alloy surface would take place through all these functional groups. As the concentration of corrosion inhibitor increases, the part of the metal surface covered by the corrosion inhibitor molecule also increases, leading to an increase in the IE.

4.4 Surface morphology studies

Uninhibited and inhibited samples were analyzed by SEM and EDS in order to identify the morphology and composition of the corrosion products before and after immersion in seawater at 25°C.

4.4.1 Unexposed specimen

The SEM micrograph of the unexposed Al-Mg-Si alloy is shown in Figure 6. It shows that the surface of the metal is absolutely free from any pits and cracks. Polishing scratches are also visible.

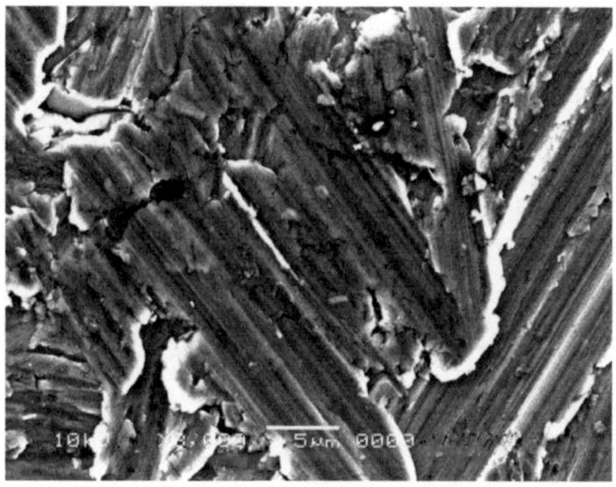

Fig. 6. SEM analysis of the unexposed Al-Mg-Si alloy surface.

4.4.2 Unexposed specimen

Figure 7 corresponds to the SEM of the specimen surface after 60 days of immersion in seawater. Flakes showing corrosion products like metal hydroxides and its oxides can be observed, however no pits or cracks were noticed (Gao *et al.*, 2008). The figures also show the presence of micro organisms (plankton) on the surface of the specimen which contributes to the corrosion process. The corrosion process in deep seawaters occurs under very specific conditions and is characterized mainly by high chloride contents, the presence of CO_2 and H_2S and micro organisms (Yagan *et al.*, 2006).

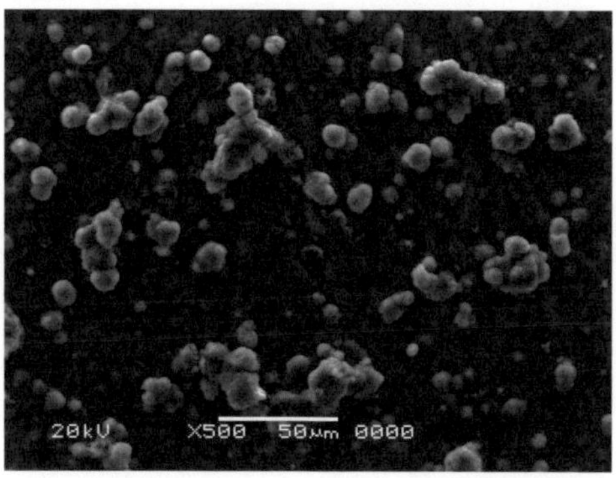

Fig. 7. SEM of Al-Mg-Si alloy surface after 60 days of immersion

The EDS spectrum in Figure 8 presents the elements exist in Al-Mg-Si alloy after immersion test. It shows the presence of sodium (Na) and chlorine (Cl) in the specimen surface. Seawater predominantly consists about 3.5% of sodium chloride (NaCl) and many other ions. Chloride ions are very strong, and could easily penetrate the passive film, and dissolution of the aluminium substrate occurs and results in corrosion.

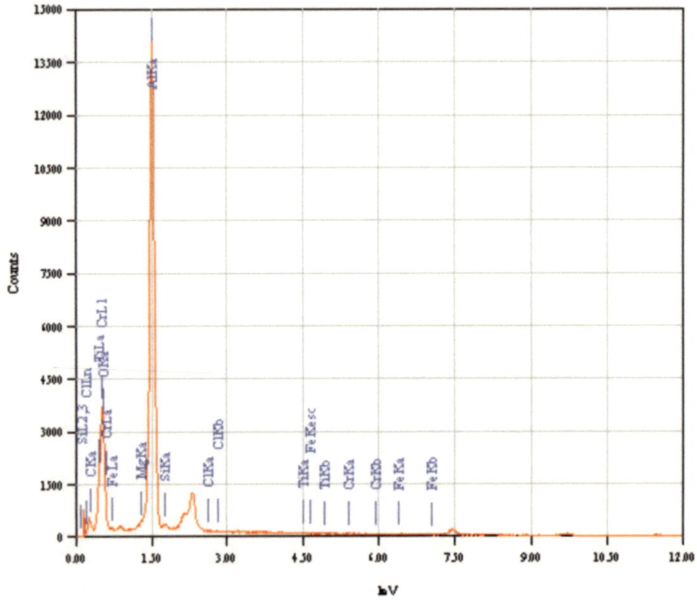

Fig. 8. EDS analysis of Al-Mg-Si alloy surface after 60 days of immersion.

4.4.3 Inhibited specimen with NH

Figure 9 depicts the SEM of the specimen surface after 60 days of immersion in seawater with the addition of 1000 ppm of NH. It can be seen that the flakes in the surface of the specimens decreased as compared to that of the micrograph in Figure 8. The specimen is covered with the inhibitor molecules giving a protection against corrosion, where a thin layer developed on the specimen surfaces.

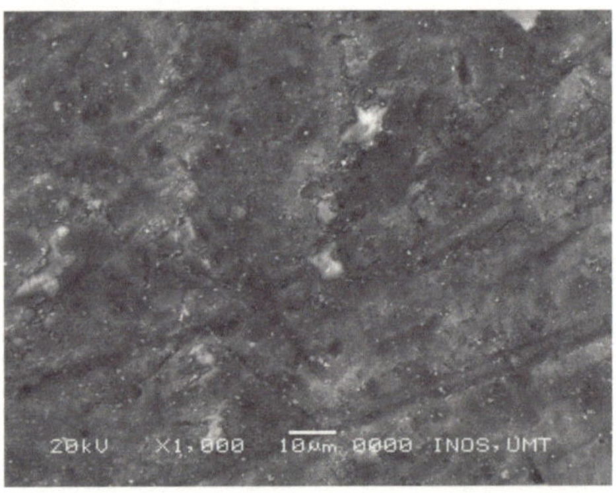

Fig. 9. SEM of specimen surface after immersion in seawater containing 1000 ppm of NH.

The EDS spectrum in Figure 10 shows the existence of carbon (C), oxygen (O) and sulphur (S); due to the carbon, oxygen and sulphur atoms of the NH. These data show that carbonaceous material containing O and S atoms has covered the specimen surface. This layer is absolutely due to the inhibitor, because the carbon signal and the high contribution of the oxygen signal are absent on the specimen surface exposed to uninhibited seawater. It can be seen obviously from the spectra, the Al peaks are dramatically suppressed relative to the samples in uninhibited seawater (Figure 8). This is due to the overlying inhibitor film on the specimen surface area (Amin *et al.*, 2009).

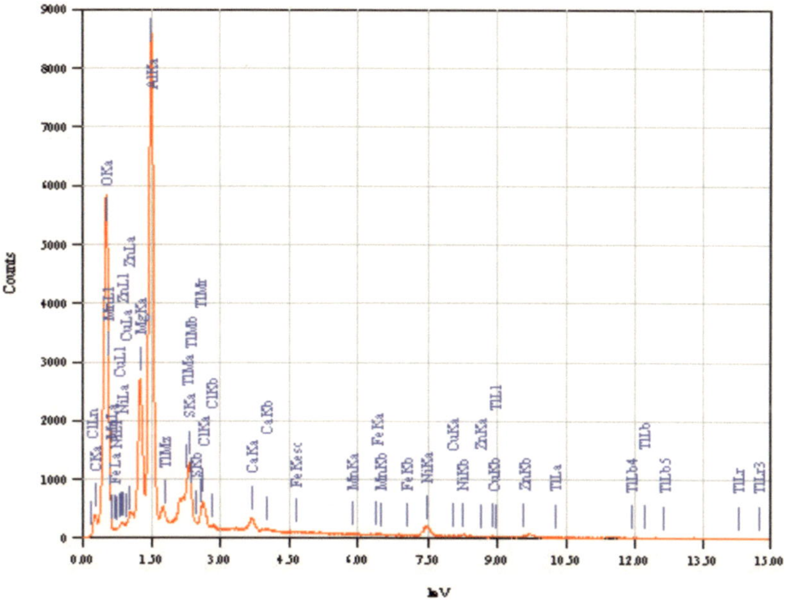

Fig. 10. EDS analysis of Al-Mg-Si alloy immersed in seawater containing 1000 ppm of NH.

4.4.4 Inhibited specimen with VL

Figure 11 portrays the SEM micrograph of Al-Mg-Si alloy immersed in seawater with the presence of 1000 ppm of VL. It was observed that compounds of VL were precipitated on the alloy surface.

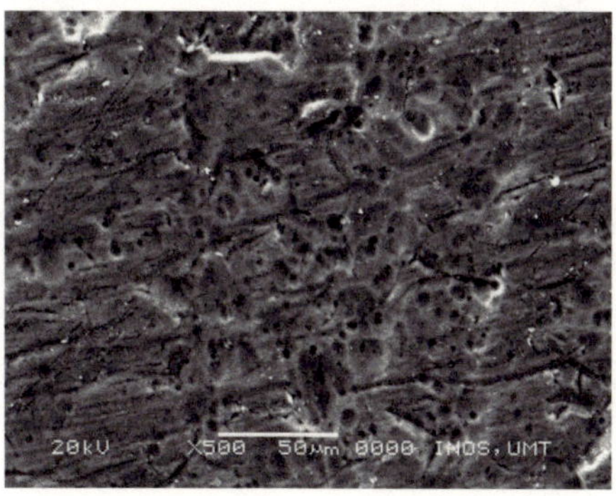

Fig. 11. SEM of specimen surface after immersion in seawater with presence of 1000 ppm VL

The observations mentioned above were confirmed by EDS analysis (Figure 12). The EDS spectra showed an additional line characteristic for the existence of C (due to the carbon atoms of the VL). These data show that carbonaceous material has covered the specimen surface. This layer is absolutely due to the VL compound, because the carbon signal is absent on the specimen surface exposed to uninhibited seawater.

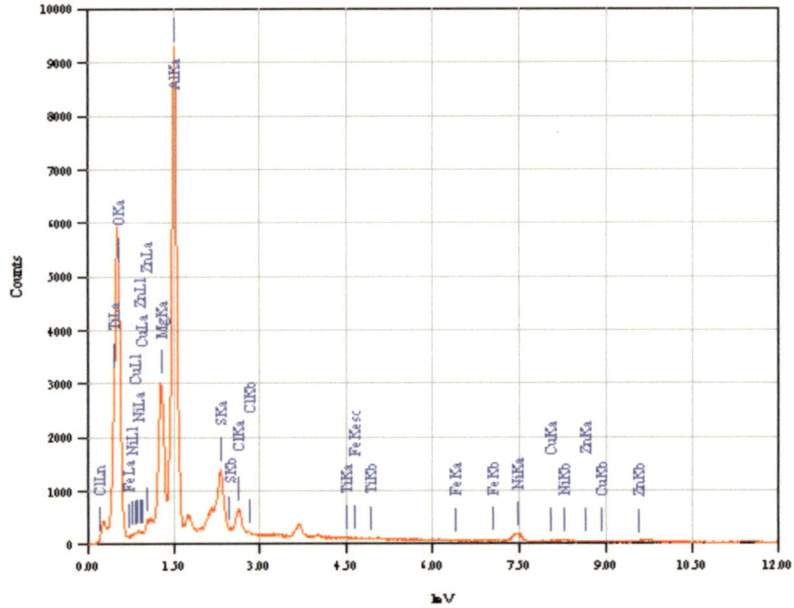

Fig. 12. EDS analysis of Al-Mg-Si alloy immersed in seawater containing 1000 ppm of VL.

4.4.5 Inhibited specimen with TS

Figure 13 depicts the SEM of specimen surface after 60 days of immersion in seawater with the addition of 600 ppm of tapioca starch. TS had shown similar characterization as NH and VL. From the micrograph, it can be seen that the flakes in the surface of the specimen are lessen when compared with that of the micrograph of uninhibited specimen.

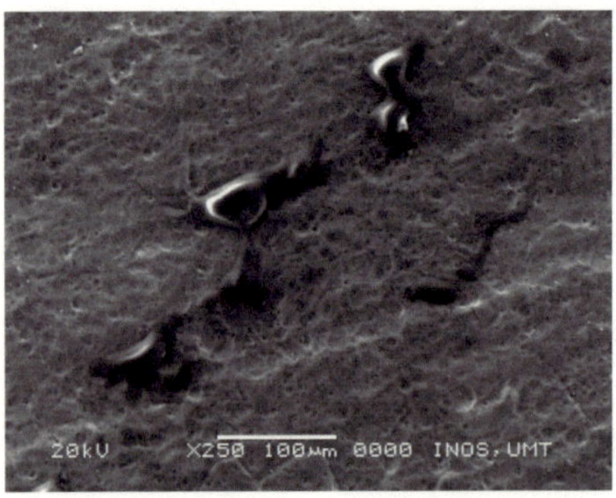

Fig. 13. SEM of specimen surface after immersion in seawater with presence of 1000 ppm TS

The EDS spectrum proved the existence of carbon (due to the carbon atoms of the tapioca starch) and these molecules covered the surface of specimen. These data show that the carbonaceous material has covered the specimen surface (Figure 14).

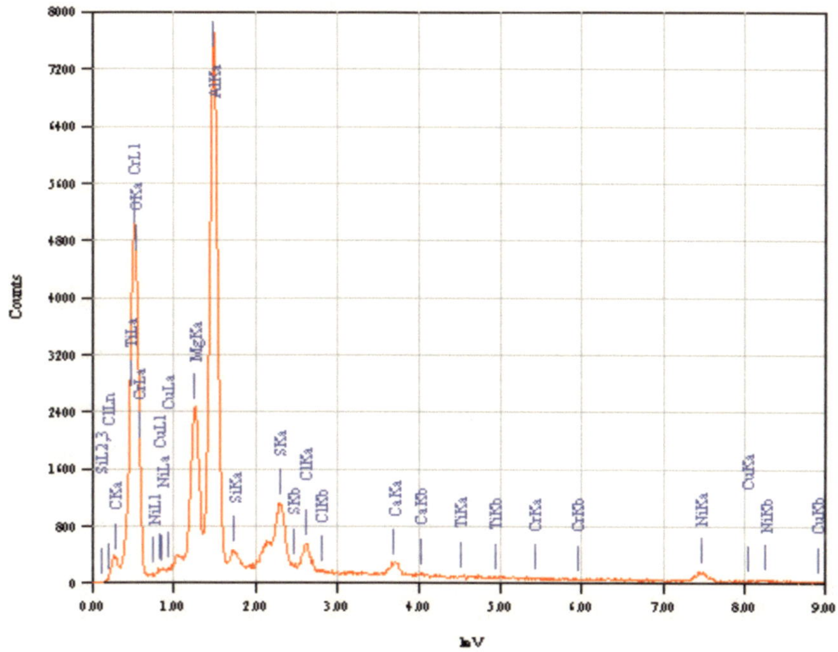

Fig. 14. EDS analysis of Al-Mg-Si alloy immersed in seawater containing 1000 ppm of TS

5. Conclusions

The corrosion inhibition studies of the aluminium alloy were carried out at room temperature using seawater and the results indicate that NH, VL and TS are an effective corrosion inhibitor of aluminium alloy in that particular solution.

The electrochemical studies of the corrosion inhibition process of Al-Mg-Si alloy in seawater using three selected natural products as corrosion inhibitors show that the corrosion rate of the alloy significantly reduced upon the addition of studied inhibitors. PP measurement reveals that the studied inhibitors can be classified as mixed-type inhibitors without changing the anodic and cathodic reaction mechanisms. The inhibitors inhibit both anodic metal dissolution and also cathodic hydrogen evolution reactions.

EIS measurements clarifying that the corrosion process is mainly charge-transfer controlled and no change in the corrosion mechanism occurred due to the inhibitor addition to seawater. It also indicates that the R_{ct} values increase with addition of inhibitor whilst, the capacitance values decrease indicating the formation of a surface film. The EIS measurement also confirms the similar corrosion process and mechanism occurs in PP measurements. According to LPR data, the values of R_p of Al-Mg-Si after addition of the studied inhibitors increase with the following order: NH < VL < TS.

It can be concluded that the corrosion parameters result obtained from PP, LPR and EIS measurements show the inhibitive effect on the corrosion behaviour of aluminium alloy in seawater by the studied inhibitors. The performance of natural products as corrosion inhibitors was evaluated by inhibition mechanism. The results reveal that inhibition efficiency increases with the increase in concentration of the studied inhibitors. Similar to the findings reported previously (Yurt *et al.,* 2006; Bhrara and Singh, 2006; El-Etre and Abdallah, 2000) the adsorption corrosion inhibitor mechanism is related to the presence of heteroatom such as nitrogen, oxygen, phosphorous, sulphur and long carbon chain length, as well as triple bond or aromatic ring in their molecular structures.

The studied natural products give above 90% of inhibition efficiency in their tested solutions. The comparison of present results with the results of reviewed paper by the other researchers on the same studied inhibitors proves that NH, VL and TS are comparable to the other natural products as corrosion inhibitor for aluminum alloy.

The use of SEM-EDS techniques provide good insight into the surface corrosion products grown on Al-Mg-Si alloy during the immersion test in seawater with and without the natural products as corrosion inhibitors. The SEM results indicate that the natural products (NH, VL and TS) absolutely minimized the corrosion products on the specimen surfaces. They also protect the passive film from dissolution in aggressive solution like seawater.

The EDS spectrums reveal that the presence of C, O, and S for NH as elements which take place in the inhibition mechanism. The carbonyl, methoxy and hydroxyl groups arranged around the aromatic ring are determined as functional groups of VL in inhibition process. The C atoms in TS are recognized by the EDS analysis, where these atoms involve in the adsorption process in alloy surface. The formation of precipitates of oxides/hydroxides of these inhibitors results in improved corrosion resistance.

Based on the results from SEM and EDS studies, it can be concluded that the TS gave the best protection of Al-Mg-Si alloy from the corrosion attack in seawater, following by VL and NH. The protection of passive film is increased with the increasing in inhibitor concentrations.

It is explored and proven in this research that NH, VL and TS carry tremendous potential for industrial usage. Unlike the pure synthetic product that requires enormous investment scale; NH, VL and TS can be produced at any type of industrial scale, which is potentially capable of eradicating the disparity among the communities, especially in the third world. Furthermore, the potential usages of these natural products discussed in this research are in line with the recent trend of the environment-friendly concept. However, resolution of the problem of whether the origin of these effects is associated with an application of aluminium alloys must await the results of further experimental studies.

6. References

Aballe, A., Bethencourt, M., Botana, F.J. & Marcos, M. (2001). $CeCl_3$ and $LaCl_3$ binary solutions as environment-friendly corrosion inhibitors of AA5083 Al-Mg alloy in NaCl solutions. *Journal of Alloys and Compounds* 323:855–858.

Al-Juhni, A.A. & Newby, B.Z. (2006). Incorporation of benzoic acid and sodium benzoate into silicone coatings and subsequent leaching of the compound from the incorporated coatings. *Progress in Organic Coatings* 56:135–145.

Amin, M.A., Abd El-Rehim, S.S., El-Sherbini, E.E.F., Hazzazi, O.A. & Abbas, M.N. (2009). Polyacrylic acid as a corrosion inhibitor for aluminium in weakly alkaline solutions. Part I: Weight loss, polarization, impedance EFM and EDX studies. *Corrosion Science* 51:658–667.

Avwiri, G.O. & Igho, F.O. (2003). Inhibitive action of *Vernonia amygdalina* on the corrosion of aluminium alloys in acidic media. *Materials Letters* 57:3705-3711.

Badawy, W.A., Al-Kharafi, F.M. & El-Azab, A.S. (1999). Electrochemical behaviour and corrosion inhibition of Al, Al-6061 and Al-Cu in neutral aqueous solutions. *Corrosion Science* 41:709–727.

Barbara, S. & Robert, K. (2006). What is corrosion? *The electrochemical Society Interface* 6:24–26.

Bethencourt, M., Botana, F.J., Cauqui, M.A., Marcos, M., Rodriguez, M.A. & Rodriguez-Izquierdo, J.M. (1997). Protection against corrosion in marine environments of AA5083 Al-Mg alloy by lanthanide chlorides. *Alloys and Compounds* 250:455–460.

Bhrara, K. & Singh, G. (2006). The inhibition of corrosion of mild steel in 0.5 M sulfuric acid solution in the presence of benzyl triphenyl phosphonium bromide. *Applied Surface Science* 253:846–853.

Blustein, G., Rodriguez, J., Romanogli R. & Zinola, C.F. (2005). Inhibition of steel corrosion by calcium benzoate adsorption in nitrate solutions. *Corrosion Science* 47:369–383.

Branzoi, V., Branzoi, F. & Golgovici, F. (2002). A comparative electrodissolution and localized corrosion study of pure aluminium in acidic solutions with different aggressive anions and organic inhibitors. *Revue Roumaine de Chimie* 47:131–137.

Cheng, Y.L., Zhang, Z., Cao, F. H., Li, J.F., Zhang, J.Q., Wang, J.M. & Cao, C.N. (2004). A study of the corrosion of aluminum alloy 2024-T3 under thin electrolyte layers. *Corrosion Science* 46:1649–1667.

El-Etre, A.Y. (2001). Inhibition of acid corrosion of aluminum using vanillin. *Corrosion Science* 43:1031–1039.

El-Etre, A.Y. & Abdallah, M. (2000). Natural honey as corrosion inhibitor for metals and alloys. II. C-steel in high saline water. *Corrosion Science* 42: 731–738.

El-Sobki, K.M., Ismail, A.A., Ashour, S., Khedr, A.A. & Shalaby, L.A. (1981). Corrosion behaviour of aluminium in neutral and alkaline chloride solution containing some anions. *Corrosion Prevention and Control* 28:7–12.

Emregul, K.C., Akay, A.A. & Atakol, O. (2005). The corrosion inhibition of steel with Schiff base compounds in 2 M HCl. *Materials Chemistry and Physics* 93:325–329.

Emregul, K.C. & Hayvali, M. (2002). Studies on the effect of vanillin and photocatechualdehyde on the corrosion of steel in hydrochloric acid. *Materilas Chemisty Physics* 83:209–216.

Gao, B., Zhang, X. & Sheng, Y. (2008). Studies on preparing and corrosion inhibition behaviour of quaternized polyethyleneimine for low carbon steel in sulfuric acid. *Materials Chemistry and Physics* 108:375–381.

Hintze, P.E. & Calle, L.M. (2006). Electrochemical properties and corrosion protection of organosilane self-assembled monolayers on aluminum 2024-T3. *Electrochimica Acta* 51:1761–1766.

Kassab, A., Kamel, K.M., Abdel Hamid, E. (1987). Effect of molybdate ion on the corrosion behaviour of aluminium in NaOH solutions. *J. Electrochemical Society of India* 36:27–30.

Kliskic, M., Radosevic, J., Gudic, S., & Katalinic, V. (2000). Aqueous extract of Rosmarinus officinalis L. as inhibitor of Al-Mg alloy corrosion in chloride solution. *Journal of Applied Electrochemistry* 30:823–830.

Li, X., Deng, S., Fu, H. & Mu, G. (2008). Synergism between rare earth cerium (IV) ion and vanillin on the corrosion of steel in 1.0 M HCl solution. *Corrosion Science* 50:3599–3609.

Maayta, A.K. & Al-Rawashdeh, N.A.F. (2004). Inhibition of acidic corrosion of pure aluminum by some organic compounds. *Corrosion Science* 46:1129–1140.

Mishra, A.K. & Balasubramaniam, R. (2007). Corrosion inhibition of aluminium by rare earth chlorides. *Materials. Chemistry and Physics* 103:385–393.

Neil, W. & Garrard, C. (1994). The corrosion behaviour of aluminium-silicon carbide composites in aerated 3.5% sodium chloride. *Corrosion Science* 36:837–851.

Noor, E.A. (2009). Evaluation of inhibitive action of some quaternary N-heterocyclic compounds on the corrosion of Al-Cu alloy in hydrochloric acid. *Materials Chemistry Physics* 114:533–541.

Onal, A.N. & Aksut, A.A. (2000). Corrosion inhibition of aluminium alloys by tolyltriazole in chloride solutions. *Anti-Corrosion Methods and Materials* 47:339–348.

Radia, A. (2004). Examination of the corrosion mechanism and corrosion control of metals in saltwater environments. Department of Chemical Engineering, University of Toronto.

Radojcic, I., Berkovic, K., Kovac, S. & Vorkapic-Furac, J. (2008). Natural honey and black radish juice as tin corrosion inhibitors. *Corrosion Science* 50:1498–1504.

Salem, T.M. Horvath, J. & Sidky, P.S. (1978). The use of soluble corrosion inhibitors for aluminium alloys. *Corrosion Science* 18:363–369.

Song-mei, L., Hong-rui, Z. & Jian-hua, L. (2007). Corrosion behavior of aluminum alloy 2024-T3 by 8-hydroxy-quinoline and its derivative in 3.5% chloride solution. *Transactions of Nonferrous Metals Society of China* 17:318–325.

Raja, P.B. & Sethuraman, M.G. (2008). Natural products as corrosion inhibitor for metals in corrosive media – A review. *Materials Letters* 62:113–116.

Reis, F.M., de Melo, H.G. & Costa, I. (2006). EIS investigation on Al 5052 alloy surface preparation for self-assembling monolayer. *Electrochimica Acta* 51:1780–1788.

Wu, Y., Geng, F., Chang, P.R., Yu, J. & Ma, X. (2009). Effect of agar on the microstructure and performance of potato starch film, *Carbohyd. Polymers* 76:299–304.

Yagan, A., Pekmez, N.O. & Yildiz, A. (2006). Corrosion inhibition by poly (*N*-ethylaniline) coatings of mild steel in aqueous acidic solutions. *Progress in Organic Coatings* 57:314–318.

Yurt, A., Ulutas, S. & Dal, H. (2006). Electrochemical and theoretical investigation on the corrosion of aluminum in acidic solution containing some Schiff bases. *Applied Surface Science* 253:919–925.

18

Studies of Resistance to Corrosion of Selected Metallic Materials Using Electrochemical Methods

Maria Trzaska
Warsaw University of Technology
Poland

1. Introduction

One of the main causes of degradation of metallic products during their operation time is their corrosion. The destruction by corrosion arises from spontaneous adverse chemical reactions in metallic materials with the surrounding environment. Irreversible corrosive processes damage any metallic products both, during their operation, and their storage. Economic losses due to destructive corrosive actions are very important and still growing due to increasing environmental pollution. The annual cost of corrosion and corrosion protection in the world is estimated to be in excess of hundred billion dollars. Reducing the continuing degradation of metallic materials by corrosion is one of the fundamental objectives of modern technological solutions and is still the subject of intensive research in many research centers in the world (Yang, 2008).

Metallic products in operational conditions are primarily exposed to electrochemical corrosion. Corrosion processes, which include oxidation and reduction reactions, mainly occur at the interface between the metal and the environment. Both, the structure and the properties of the metal as well as the characteristics of the environment affect the corrosive processes. The rate of corrosion processes depends on the electrochemical susceptibility of a given metal, its chemical composition, homogeneity and surface topography, on the type and chemical composition of the environment, the concentration of aggressive agents, temperature, as well as the type of corrosion products themselves. Electrochemical corrosion processes are accompanied by mass transport and flow of electric charge through the metal - corrosive environment boundary. To characterize the susceptibility of metallic materials to electrochemical degradation modern research techniques increasingly use the relationship between voltage and current intensity occurring in corrosive systems. Such studies rely on computerized measuring system, in which suitable electrical stimulation is generated numerically while the system analyzes simultaneously the response. The results are presented in the form of graphs showing the current - voltage relationships (Trzaska, 2010).

In this chapter we present the results of investigations of corrosion properties of metallic materials with different chemical susceptibilities and different crystalline structures. The main focuses are materials playing important roles in current technologies.

Two electrochemical methods were used to characterize the corrosion properties of the materials under investigation: potentiodynamic polarization and impedance spectroscopy.

2. Electrochemical methods for testing the susceptibility to corrosion of metallic materials

Both processes of oxidation and reduction simultaneously occur at the metal-corrosive environment interface during an electrochemical corrosion. The basic processes occurring during the electrochemical corrosion of metallic materials are:

- oxidation of the metal atoms: $M - ne \rightarrow M^{n+}$
- reduction of ions present in the corrosive environment: $X^{n-} - ne \rightarrow X$.

In the reduction processes in natural environments hydrogen ions H^+ and oxygen O_2 are most often involved, and the related reduction processes can be written as following:

- reduction of hydrogen ions: $2H^+ + 2e \rightarrow H_2$
- reduction of oxygen, depending on the pH of the corrosive environment:
  - in alkaline and neutral environment: $O_2 + 2H_2O + 4e \rightarrow 4OH^-$
  - in an acidic environment: $O_2 + 4H^+ + 4e \rightarrow 2H_2O$.

Oxidation and reduction processes are accompanied by the flow of electric charge through the interface metal-corrosive environment. In metals the charge carriers are electrons while in the corrosive environment charge flow is due to ions. Thus an active assessment of electrochemical corrosion processes can be achieved by assessing the electrical charge transfer process. In the reactions of corrosion that are controlled by the rate of charge transfer, the current - potential relationship can be described by the Butler-Volmer equation:

$$j = j_0\left[\exp\left(\frac{\alpha_A nF}{RT}\eta\right) - \exp\left(-\frac{\alpha_K nF}{RT}\eta\right)\right] \tag{1}$$

where: j - current density, j_0 - exchange current density, $\eta = E\text{-}E_0$ - overpotential (voltage), n - number of electrons, α_A and α_K - transfer coefficients, respectively, at the anode and cathode, F - Faraday constant, R - gas constant and T – absolute temperature (Marcus, 2011).

The disruption of the steady state corrosion by the electrical signal and the measurement of its response to the stimulation allow determining the set of electrical quantities providing valuable information about electrochemical processes occurring in the system under study. The most common approaches for the electrochemical characterization of corrosion processes are non-stationary methods, which are easy to automate and computer-control (Trzaska & Trzaska, 2007).

The present study of corrosion processes of metallic materials uses variable current technology, namely the electrochemical polarization potentiodynamic and electrochemical impedance spectroscopy (EIS) techniques. The basis of polarization potentiodynamic electrochemical technique is the stimulation of the corrosion system by a potential, whose value varies linearly in time and the recording of the instantaneous value of current flowing in the system. The electrochemical impedance spectroscopy consists of a perturbation of the steady state corrosion by applying the sinusoidal alternating potential signal of small amplitude, but in a wide range of frequencies and the automatic recording of current intensity responses of the system. Investigations of corrosion by those methods, based on a change in the relationship between potential and current were implemented in the three-electrode system (Fig. 1).

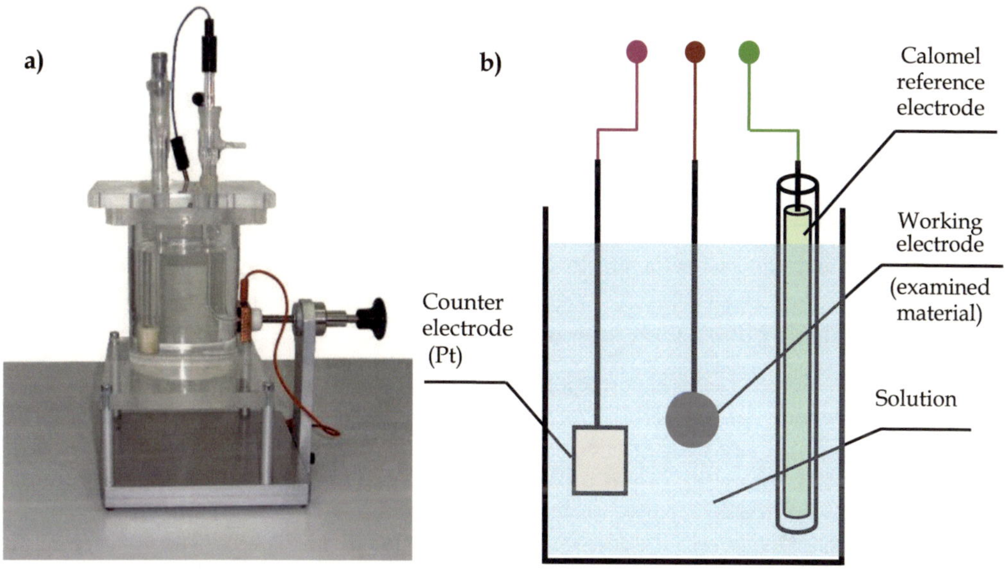

Fig. 1. Three-electrode system for corrosion studies: a) measuring system, b) circuit diagram.

Research of metallic materials corrosion was carried out by means of computerized measuring systems, which generated in a digital form an electrical signal of a certain shape to stimulate the system and simultaneously analyze the response of the corrosion test (Figs. 2 and 3).

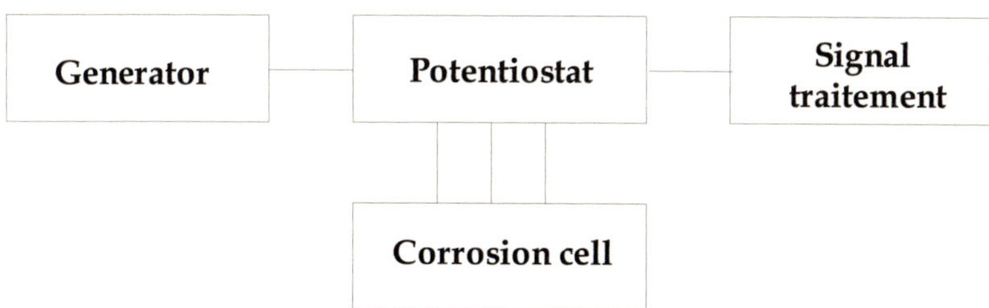

Fig. 2. Block diagram of the electrochemical corrosion tests.

In the three-electrode system, used in the current study, the examined metal takes the role of the active electrode. A calomel electrode, $Hg/Hg_2Cl_2/KCl$ characterized by the potential +244 mV, was used as the reference electrode. Auxiliary electrode was made of platinum (Pt). The tests were carried out in the corrosive environment of 0.5M NaCl solution at pH = 7 and a temperature of 293K (Trzaska &Trzaska, 2010).

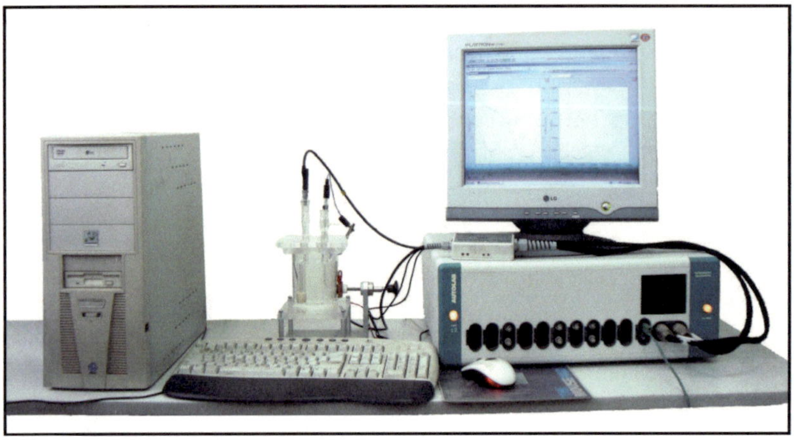

Fig. 3. The set of appliances for the electrochemical corrosion testing by EIS technique.

2.1 Potentiodynamic polarization method

In any process of electrochemical corrosion oxidation processes, i.e. the anodic processes, as well as the reduction i.e. cathodic processes occur on the surface of the metal. In the measurement system these processes occur simultaneously during the polarization for each of the applied potential values but at different speeds. The resultant speed of the processes constituting a sub-inflicted response to the potential changing at prescribed rate is recorded by the measurement system in the form of instantaneous current values. The characteristics $j = f(E)$ of the current intensity as function of the potential obtained in this way are called potentiodynamic polarization curves.

Potentiodynamic polarization method was applied to a wide range of potential changes to characterize the current-potential relationship $j = f(E)$ in corrosion systems under investigation. The corrosion current density j_{cor} and potential E_{cor} of tested metallic materials were further determined based on extrapolation of tangents to the curves of the cathodic and anodic polarization zones.

2.2 Impedance spectroscopy method

The perturbation of the equilibrium of the corrosive systems composed by metal - 0.5M NaCl solution, was obtained through time-varying sinusoidal signal described by the following relationship $E(t) = E_0\cos(\omega t)$, where $E(t)$ is the instantaneous potential value [V], E_0 - potential magnitude [V], t - time [s]. The response of the corrosion system to such an interfering signal was the current intensity signal, which is the effect of transferring electrical charge between the corrosive metal - an electron conductor, and the electrolyte - ionic conductor. This response is described by the time-varying current signal $I(t) = I_0\cos(\omega t + \varphi)$, where $I(t)$ – instantaneous current value [A], $\omega = 2\pi f$ - pulsation [rad/s], f - frequency [Hz], φ - phase shift [rad.]. The measuring system digitally generates the excitation having the above sinusoidal form and measures the current system response as a function of frequency. Then the plots of $|Z(\omega)|$ and $\varphi(\omega)$ were generated , i.e., amplitude and phase spectra of impedance, called Bode plots, and curves $X(\omega) = F(R(\omega))$, called amplitude-phase characteristics or Nyquist plots (Orazem & Tribollet, 2008), (Sword et al., 2007).

The experimental results expressing the dependence of impedance spectra on the applied signal frequency are shown by

- the Bode diagrams in the form of two plots as function of ω: i $| Z | = f_1(\log(\omega))$ and $\varphi = f_2(\log(\omega))$, where $Z(j\omega) = |Z| e^{j\varphi}$, $|Z|$ - impedance magnitude,
- the Nyquist diagrams representing relationship $Z'' = f(Z')$, where Z' is a real component and Z''- imaginary component of input driving impedance $Z(j\omega)=Z'+jZ''$ of the corrosive systems investigated.

Impedance is an essential characterization of the current intensity response of the corrosion system to the sinusoidal perturbation of the potential applied to the metal. The results of impedance measurements made in a suitably wide range of frequencies provide valuable information about the system and electrochemical corrosion occurring therein. The majority of electrochemical as well as physical processes can be interpreted within the impedance spectroscopy method as elements of electrical circuits with appropriate time constants. Thus, to interpret the results of electrochemical impedance measurements surrogate models of electrical circuits, known as Randles models, can be used.

Experimentally determined frequency characteristics were used to map the corrosion processes using models based on suitable equivalent circuits. Each element of such a circuit models the specific process or phenomenon occurring in the corrosion system under investigation.

3. Electrochemical characteristics of corrosion resistance of metallic materials

Metals are very commonly used materials in various technologies and applications. Properties of metallic materials are shaped by their composition and structure. Moreover, most natural metals are found in chemical combination with other elements. In the current study, resistance to electrochemical corrosion tests were applied to metallic materials with different properties and structures: aluminum (Al), aluminum with a surface layer of oxide aluminum (Al_2O_3), iron (Fe), S235JR steel, nickel (Ni), microcrystalline nickel (Ni_m), nanocrystalline nickel (Ni_n), and amorphous alloy of phosphorus-nickel (NiP). The choice of these materials was due to the universality of their applications in technology.

3.1 Identification of the resistance to corrosion of aluminum

Aluminum and its alloys are materials of great technical importance. Attractive physical properties of aluminum such as low density, high ductility, good thermal and electrical conductivities, relatively low production costs and its high abundance in nature make it an indispensable metal in many industries and in numerous areas of daily life, both as a pure metal and in various alloys. Aluminum, as an element of high chemical activity, shows a significant tendency to passivity, leading to high resistance of aluminum and its alloys to corrosion in many environments with low aggressiveness (Vargel, 2004).

However, the processes of alloying and heat treatments are not always sufficient to ensure the qualities of aluminum required in the modern technical applications. One way of modifying the performance of aluminum and its alloys in order to adapt them to the operating conditions is the production on their surfaces of a thin layer of Al_2O_3 by anodic oxidation process (Huang, et al., 2008).

Corrosion test has been applied to technical aluminum (99.9%) and to aluminum with Al_2O_3 surface layer produced by hard anodic oxidation and sealed in boiling-hot deionized water. Images of morphology and topography of the surface layer of aluminum and Al_2O_3 before corrosion tests using scanning electron microscope (SEM) are shown in Fig. 4.

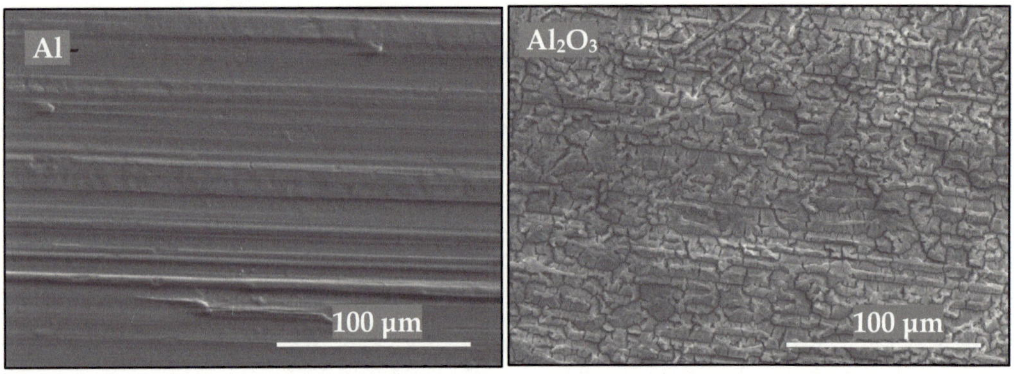

Fig. 4. Morphology and topography of the surface of Al and Al_2O_3 layer before corrosion tests.

Potentiodynamic polarization distortion of the steady state technique at the interface of both Al as well as Al_2O_3 with 0.5M NaCl solution, was applied with the change in the potential ranging from -780mV to -450mV. The rate of the potential change during the test was 0.2mV/s. Current characteristics j=f(E) of test materials in the form of potentiodynamic polarization curves are shown in Fig. 5.

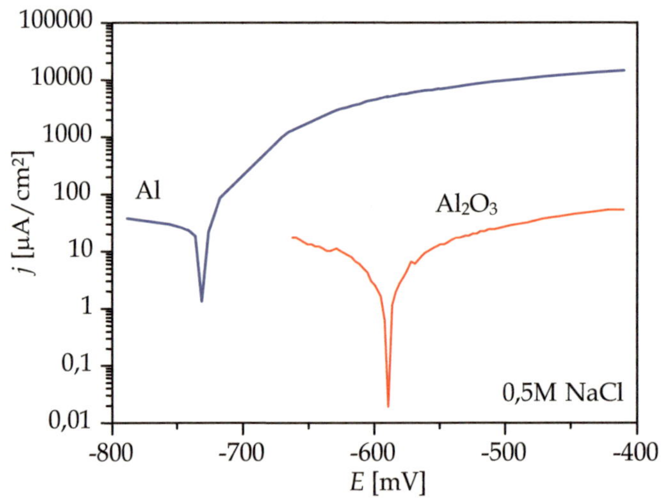

Fig. 5. Potentiodynamic polarization curves of materials of bulk raw Al and Al_2O_3 layer in the corrosive environment of 0.5M NaCl solution.

To determine the corrosion current density j_{cor} and the corrosion potential E_{cor} the tangential extrapolation method was used for the polarization curves j = f(E) from the cathode and anode zones. The values of corrosion current densities and corrosion potentials for tested materials are summarized in Table 1.

Material	E_{cor} [mV]	j_{cor} [μAcm^{-2}]
Al	-719	14.9
Al_2O_3	-583	0.23

Table 1. Corrosion parameters of bulk raw Al and Al_2O_3 layer in 0.5M NaCl.

It is worth mentioning that Al_2O_3 layer has a much higher corrosion resistance compared to aluminum. Corrosion protection of aluminum using Al_2O_3 layer is guaranteed by efficient isolation of the substrate material from the corrosive environment. The effectiveness of the corrosion protection depends on the thickness and tightness of Al_2O_3 layer.

For further characterization of electrochemical corrosion processes at the interface between the environment of 0.5M NaCl solution and Al and Al_2O_3, the electrochemical impedance spectroscopy method was applied. As stated above this method allows considering the corrosion process as a combination of equivalent electric circuits. In the case of Al the study was carried out with the amplitude of the forcing sinusoidal signal of 10mV. However, in the case of Al_2O_3 layer the amplitude of perturbing signal in the corrosion balance was fixed at 20mV. The study was conducted in the frequency range 23kHz ÷ 16mHz. Measured impedance spectra of Al and Al_2O_3 layer in the corrosive environment of 0.5M NaCl solution are presented in the form of Nyquist and Bode diagrams.

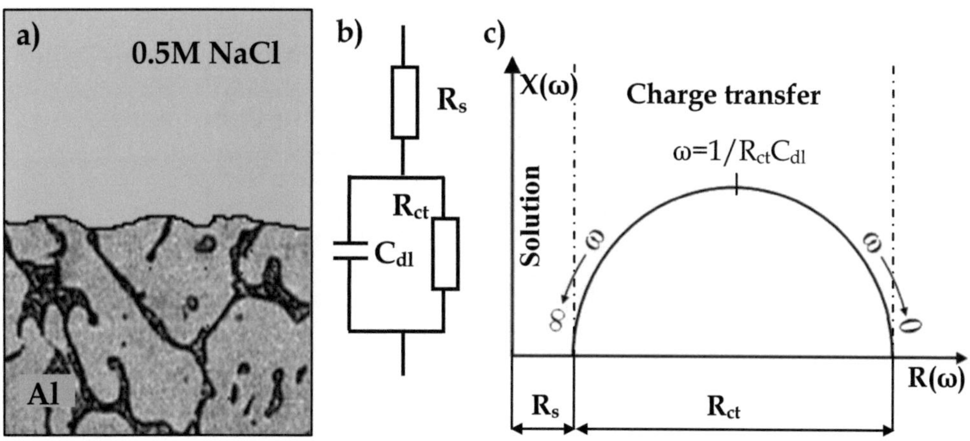

Fig. 6. Equivalent electric circuit for corrosion of bulk Al: a) corrosion system Al - 0.5M NaCl solution, b) an equivalent circuit scheme, c) Nyquist frequency characteristics

Equivalent electrical circuits obtained by minimizing the mean square error were further used for the analysis of experimentally identified frequency characteristics and a description of corrosive processes in the systems under investigation. A simple electric circuit consisting of three elements of type R and C with a single time constant was adopted as system model for Al. Figs. 6 and 7 show the circuits modeling respectively the corrosion systems of bulk Al and Al_2O_3 surface layer deposited on aluminum in a 0.5 M NaCl corrosive environment.

Fig. 6 presents the system of Al corrosion in 0.5 M NaCl solution, its frequency impedance characteristic in the form of Nyquist plot and the equivalent electrical circuit. Individual parts of the electric circuit reflect the electrochemical and electrical characteristics of the corrosion systems. In this arrangement, the spectral characteristic of the impedance in the Nyquist plot has the shape of a semicircle, whose intersection with the real axis in the high-frequency range determines the electrolyte solution resistance R_s. Conversely, the intersection of the real axis in the low-frequency range corresponds to the sum of $R_s + R_{ct}$, where R_{ct} indicates the charge transfer resistance of the boundary metal/electrolyte, and characterizes the rate of corrosion. On the other hand, C_{dl} component of the circuit represents capacity of the double layer at the interface metal/electrolyte.

Layout of the corrosive system for aluminum with a surface layer of Al_2O_3 in 0.5M NaCl corrosive environment, and the designated equivalent circuit are shown in Fig.7.

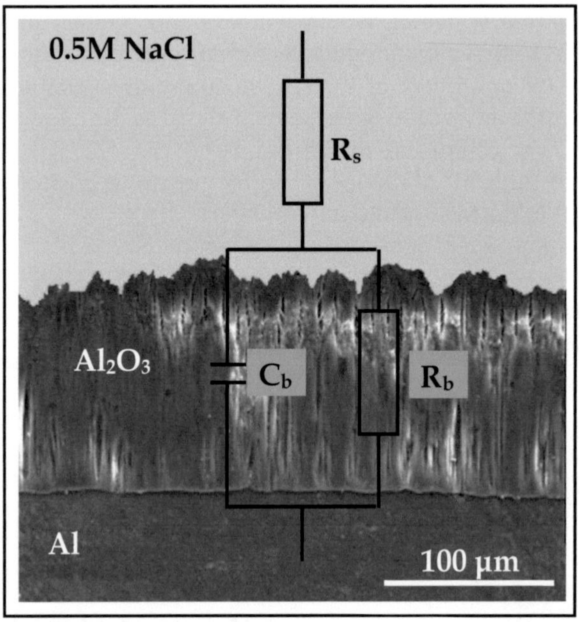

Fig. 7. Layout of the corrosive system for aluminum with a protective layer of Al_2O_3 in 0.5M NaCl environment and its equivalent circuit.

Similarly to the previous case the element R_s of the equivalent electric circuit for this corrosion system represents the resistance of the 0.5M solution of NaCl electrolyte used as the corrosive environment. Elements in parallel in the equivalent electric circuit characterize the protective properties of Al_2O_3 layer deposited on the bulk Al. The element C_b specifies the Al_2O_3 layer capacitance, which depends on the thickness of this layer and on the dielectric properties of the material. The resistor R_b in such a system represents the resistance of the protective layer, and depends on properties of the material forming the layer, and varying with the thickness of the layer and its material composition. The value of the resistance R_b in the test case of Al_2O_3 protective layer is large and amounts to R_b = 100 kΩcm².

The resulting impedance of the equivalent circuit (Figs. 6 and 7) adopted to describe the corrosion processes occurring in systems with bulk Al and Al_2O_3 layer deposited on an aluminum substrate in the corrosive environment of 0.5M NaCl solution is determined by the expression

$$Z = R_s + \frac{1}{\frac{1}{R} + j\omega C} \quad (2)$$

The equivalent electrical circuit approach adopted maps the processes occurring in the corrosion systems and enables the determination of parameters relevant to these processes. The parameter values of individual elements of equivalent electrical circuit representing investigated corrosion systems are summarized in Table 2.

Materiał	R_s [Ωcm²]	C µF/cm²]	R [kΩcm²]
Al	13.3	C_{dl} = 10.0	R_{ct} = 0.8
Al_2O_3	12.5	C_b = 1.4	R_b = 100.0

Table 2. Parameters of equivalent electrical circuit for corrosion processes of bulk Al and Al_2O_3 layer in 0.5M NaCl solution.

The frequency characteristics of corrosion systems of bulk Al and Al_2O_3 layer in 0.5M NaCl solution in the form of Nyquist and Bode plots obtained by the measurements and calculations based on adopted equivalent electrical circuits are shown in Figs. 8 and 9, respectively.

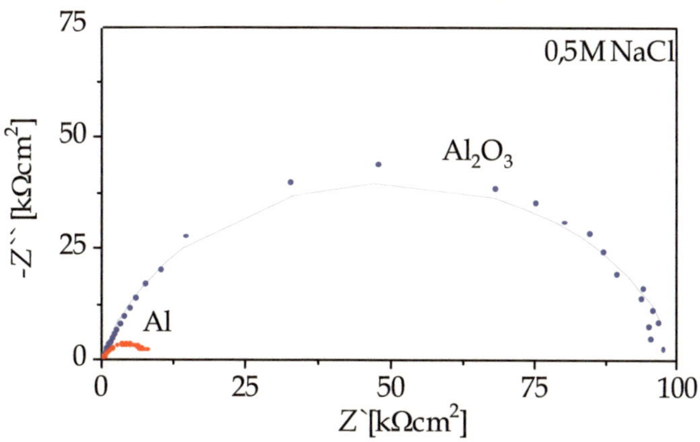

Fig. 8. Nyquist diagrams of impedance spectra of corrosion systems for bulk Al and Al_2O_3 layer in 0.5M NaCl solution determined experimentally (point line) and as a result of calculations (solid lines).

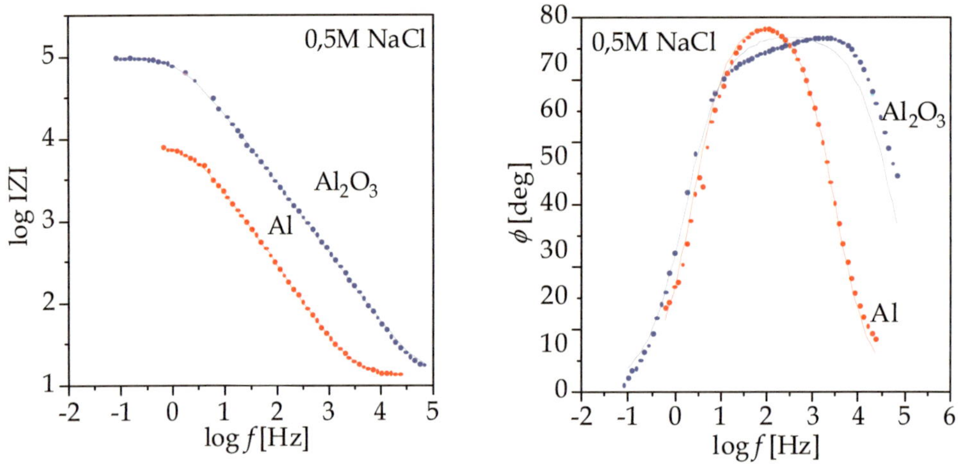

Fig. 9. Bode amplitude and phase spectra of circuit impedances for corrosion of bulk Al and Al_2O_3 layer in 0.5M NaCl solution determined experimentally (point line) and as a result of calculations (solid lines).

Fig. 10. SEM images of the surface of Al and Al_2O_3 materials after corrosion tests in an environment of 0.5M NaCl solution.

The comparison of plots of the frequency characteristics of corrosion systems obtained experimentally and from the calculations (Figs. 8 and 9) attests that the adopted scheme for the equivalent circuits reproduces well the impedance measurements across the whole frequency range of the forcing signal.

In addition, the circuit impedance characteristics for corrosion of bulk Al and Al_2O_3 layer in 0.5M NaCl solution confirm the results obtained from potentiodynamic polarization studies of these systems.

SEM images of material surfaces of bulk Al and surface layers of Al_2O_3 in the study of influences of corrosive environment of 0.5M NaCl solution are shown in Fig. 10. Images of the surfaces after corrosion tests show that in an environment of 0.5M NaCl solution Al substrate material undergoes pitting corrosion. However, Al_2O_3 surface layer provides a good corrosion protection for aluminum substrate.

3.2 Identification of the resistance to corrosion of iron

Iron is a metal with the greatest technical importance. Its useful physical properties include relatively high hardness, ductility and large malleability, relatively low production costs and high prevalence in nature. However, chemically pure iron has practically no direct use, while iron alloys with carbon, silicon and other metals have an enormous technical and practical importance.

Iron is a relatively reactive metal - its standard electrochemical potential is -760mV. It reacts with all diluted acids resulting in the salts of iron (II). Chemically pure iron is relatively less prone to corrosion compared to its commonly used alloys. Steels containing various alloying elements have different chemical compositions in material micro-zones. Such micro-zones in contact with the electrolyte solution lead to different electrochemical potentials and are able to create micro-cells, in which iron is most often an anode. As a result of these electrode processes the iron oxidation occurs and the formation of various corrosion products takes place, in which iron occurs primarily at two and three degrees of oxidation.

Electrochemical corrosion characteristics of iron were determined by potentiodynamic and impedance spectroscopy techniques. Tests were applied to chemically pure iron Fe made by electrocrystalization method and to carbon steel S235JR with the chemical composition shown in Table 3.

Component	C	S	P	Si	Mn	Cr	Ni	Cu
[%] weight	0.22	0.05	0.05	0.30	1.10	0.30	0.30	0.30

Table 3. The content of alloying elements in S235JR steel.

Images of topography and surface morphology of Fe iron and S235JR steel prior to corrosion tests are shown in Fig. 11.

In studies using potentiodynamic polarization perturbation of the steady state of the metal-solution the potential varied between -550mV to -10mV in the case of iron produced electrochemically, and in the range -723mV to -20mV in the case of S235JR steel. The rate of potential changes during the test was 0.3mV/s.

The characteristics j = f(E) for iron Fe and S235JR steel in 0.5M NaCl solution obtained in the above potential ranges are shown in Fig. 12.

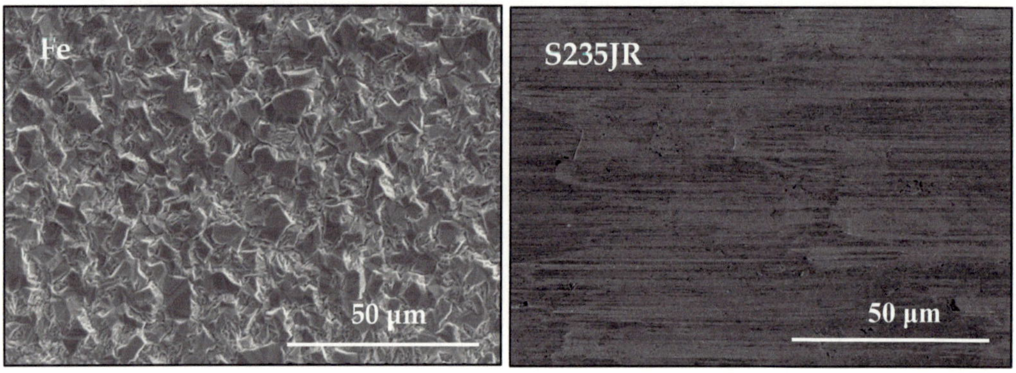

Fig. 11. Topography and surface morphology of Fe and S235JR steel prior corrosion tests.

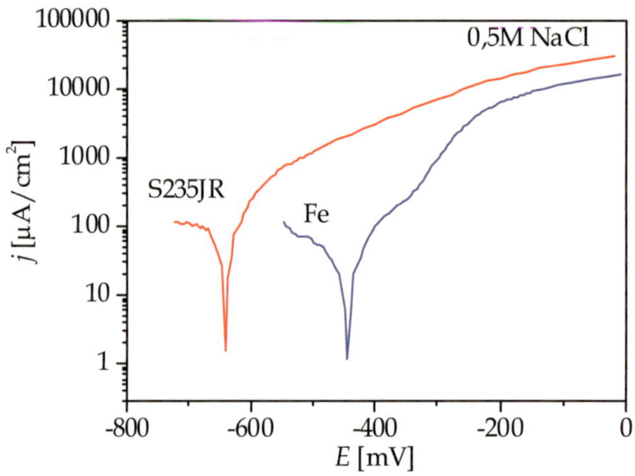

Fig. 12. Potentiodynamic polarization curves of Fe and S235JR steel in 0.5M NaCl solution.

Each of the potentiodynamic polarization curves for iron Fe and S235JR steel in the range of potentials tested consists of two parts: the cathodic and anodic segments. Part of the reduction process corresponds to the cathodic corrosive components of H^+ and O_2 occurring on the metal surface. However, part of anodic potentiodynamic polarization curve is characterized by the oxidation process of metal atoms or the process of corrosion - in the case of iron the reaction is $Fe - 2e \rightarrow Fe^{+2}$.

The corrosion test parameters of iron and S235JR steel in 0.5M NaCl solution are summarized in Table 4.

Material	E_{cor} [mV]	j_{cor} [µA/cm²]
Fe	-445	24
S235JR	-641	93

Table 4. Parameters of Fe and S235JR steel in corrosion environment of 0.5 M NaCl solution.

The results show that pure iron produced electrochemically has much higher corrosion resistance compared to S235JR steel. Thus, alloying elements and the heterogeneity of the material in the case of carbon steel activate electrochemical processes of the material.

For further characterization of electrochemical corrosion processes at the interface of iron and S235JR steel with the 0.5M NaCl solution environment electrochemical impedance spectroscopy was used. The study consisted of perturbing the equilibrium of the corrosion system with a sinusoidal potential signal of small amplitude (15mV) across a wide frequency range (10kHz ÷ 33mHz) and recording the changes in time of system's current intensity response.

The measured frequency characteristics of electrochemically produced iron and S235JR steel in corrosive environment of 0.5M NaCl solution are presented in the form of Nyquist diagrams (Fig. 14) and Bode plots of impedance spectra (Fig. 15).

Effective modeling of complex electrochemical processes of corrosion in the systems based on iron required the use of a more complex equivalent electrical circuit, i.e., circuit containing CPE - constant phase elements. Constant phase element (CPE) is characterized by a constant angle of phase shift. Impedance of the CPE is described by the following expression: $Z_{CPE} = 1/Y_0(j\omega)^n$, where Y_0 and n are parameters related to the phase angle. The more heterogeneous the corrosion processes occurring on the metal surface the smaller value of the parameter n.

Best matching of all designated impedance spectra for experimentally studied systems of corrosion of iron and S235JR steel in the solution of 0.5 M NaCl was obtained by using an equivalent electric circuit with two time constants, whose structure is shown in Fig. 13.

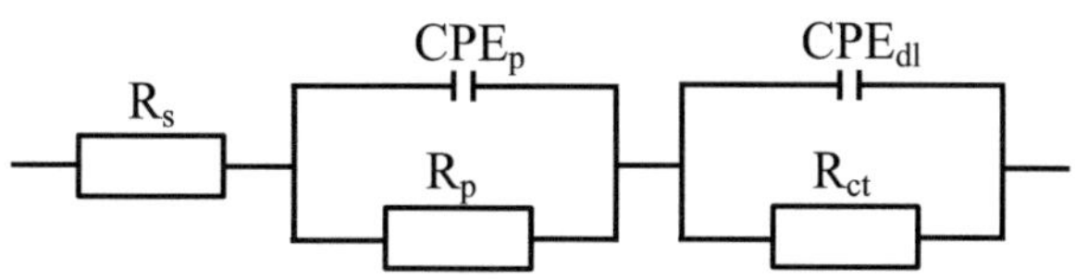

Fig. 13. The equivalent circuit for corrosion of Fe and S235JR steel in the solution of 0.5M NaCl.

This equivalent electric circuit can be described by the following relationship defining the resulting impedance, namely

$$Z = R_s + \frac{1}{\frac{1}{R_p} + Y_p(j\omega)^{n_p}} + \frac{1}{\frac{1}{R_{ct}} + Y_{dl}(j\omega)^{n_{dl}}} \tag{3}$$

Each element of this circuit appropriately models the specific process or phenomenon occurring in the system investigated. In the circuit shown in Fig. 13 the resistance element R_s represents corrosive environment, i.e., 0.5 M NaCl solution. The resistance representing the charge transfer through the interface associated with the process of oxidation of iron, i.e., the corrosion element, is described by R_{ct}, and the electrical double layer at the interface iron - 0.5M NaCl solution is characterized by a constant phase element CPE_{dl}. The use of two constant-phase elements in an equivalent electric circuit improves the quality of model fit to the observations, as shown by Figs. 14 and 15. However, this introduces two additional circuit elements whose physical meaning can be expressed as follows: CPE_p - the capacity of the surface area of materials with high degree of surface development, R_p - resistance of electrolyte contained in the pores of the corroded material zone.

Analysis of impedance spectra with a fitted equivalent circuit allows the assessment of the variability of individual circuit elements with the change of potential and current intensity flowing in the corrosion system.

The values of the parameters of equivalent circuit elements that characterize the processes occurring in the corrosion of iron and S235JR steel in 0.5M NaCl solution are summarized in Table 5.

Material	R_s [Ω cm²]	R_p [Ω cm²]	CPE_p [μFs^{n-1}/cm²]		R_{ct} [Ω cm²]	CPE_{dl} [μFs^{n-1}/cm²]	
			Y_p	n_p		Y_{dl}	n_{dl}
Fe	16.7	5.7	212.6	0.68	1513	136.5	0.76
S235JR	15.4	114	182.4	0.68	386	48.3	0.98

Table 5. Electrical circuit parameters of the corrosion systems of Fe and S235JR steel in 0.5M NaCl solution.

Compatibility of the actual processes in the system under study with a description of the corrosion with an equivalent circuit through which current flows with the same amplitude and same phase angle as in the corrosion system at a given excitation is illustrated in Figs. 14 and 15, respectively.

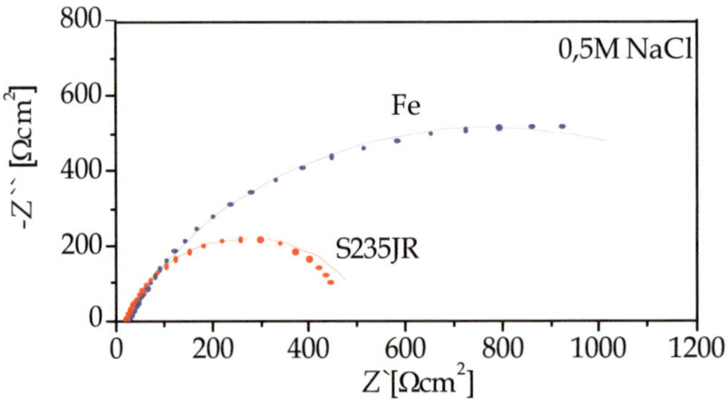

Fig. 14. Nyquist diagrams of impedance spectra of corrosive systems of Fe and S235JR steel in 0.5M NaCl solution determined experimentally (point line) and as a result of calculations (solid lines).

Nyquist diagrams (Fig. 14) in the shape of the characteristic semi-circles indicate the activation process control during corrosive material tests. Much larger diameter of the semi-circle in the case of electrochemically generated iron shows high electrical resistance at the interface metal-solution, which is the result of oxidation of iron and Fe^{+2} ions passing into the solution. This indicates a greater corrosion resistance of iron compared to steel S235JR in the test environment, which also confirms the results obtained with the potentiodynamic polarization method.

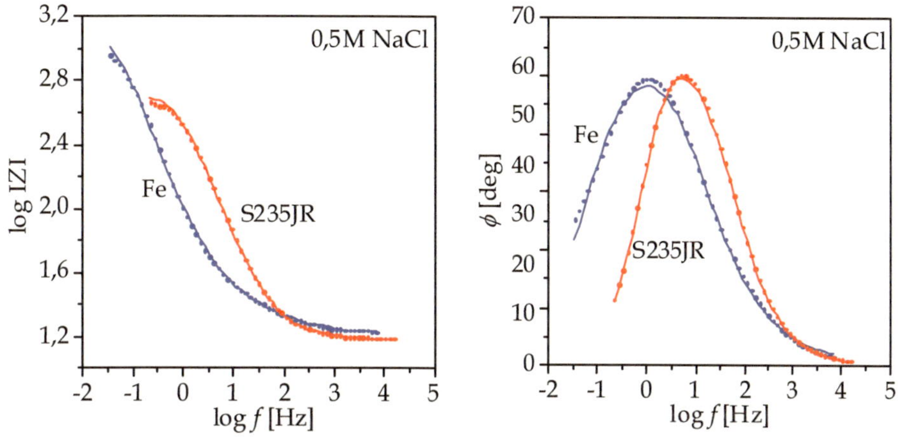

Fig. 15. Bode amplitude and phase spectra of circuit impedance and corrosion systems of Fe and S235JR steel in 0.5M NaCl solution determined experimentally (point line) and as a result of calculations (solid lines).

Images of destruction of corrosion test samples of Fe and S235JR steel after corrosion tests are shown in Fig. 16.

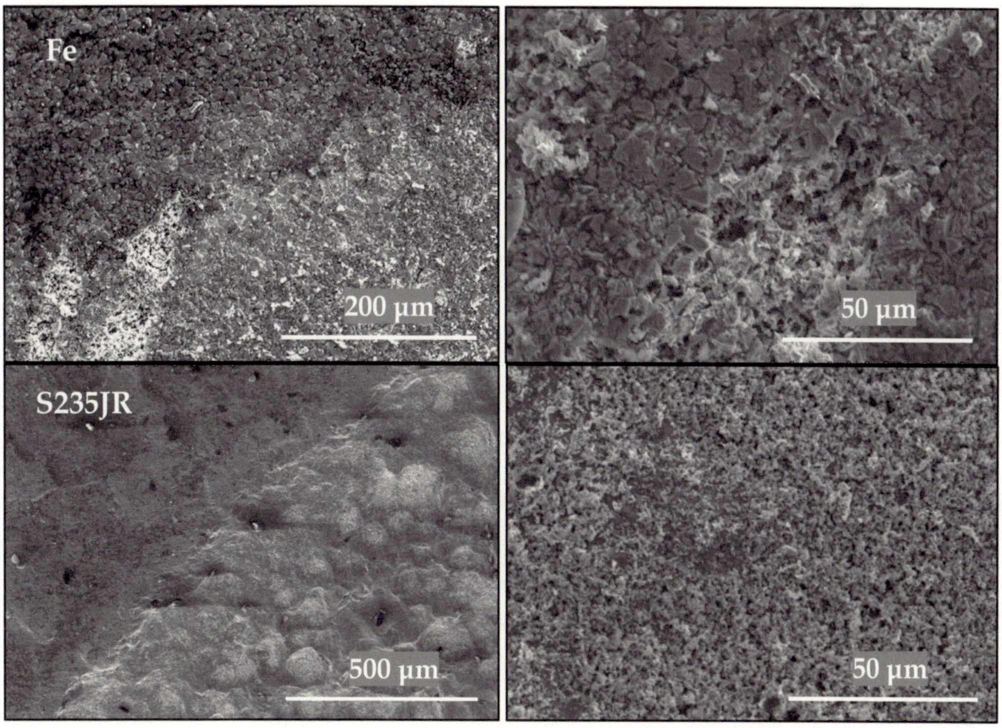

Fig. 16. Images of the surfaces of Fe and S235JR steel after corrosion tests.

Samples of iron Fe and S235JR steel subjected to corrosion test not only differed in structure and chemical composition, but also the morphology and surface topography (Fig. 11), which had also influenced the course of corrosion processes. Both the iron Fe and S235JR steel were submitted to uneven general corrosion (Fig. 16). Material pickling on the grains boundaries is clearly visible in the case of electrochemically produced iron.

3.3 Identification of the resistance to corrosion of nickel

Nickel is a metal characterized by soft, ductile, smelting and converting properties. Its standard electrochemical potential is -0.24V. The chemical compounds of nickel are mainly found in 2nd oxidation states, rather than the 3rd and 4th ones. It dissolves in mineral acids, but insensitive to bases (alkalis). In the atmospheric environment and many aqueous solutions nickel has the ability to passivity in a fairly wide pH range. Thanks to its passivity it has high resistance to corrosion in many environments (Trzaska &Moszczynski, 2008).

Nickel in pure state is used for manufacturing protective coatings of products made of other metals - mainly steel - and in its fine particle form is used as a catalyst for many chemical reactions. It is also one of the major components of many alloys, which are used in a variety of current technologies. However, restrictions in uses of nickel in various products are constantly growing due to its rarity in nature.

Electrochemical corrosion characteristics of nickel were carried out by potentiodynamic polarization and impedance spectroscopy methods. Corrosion tests of nickel produced by electrocrystallization were applied to its micrometric (Ni_m) and nanometric (Ni_n) crystalline structures and for NiP amorphous alloy of nickel with phosphorus at content of 10.7% by weight (Eftekhari, 2008), (Kowalewska & Trzaska, 2006).

Images of surface topography and morphology of nickel with microcrystalline and nanocrystalline structures and of NiP alloy before corrosion tests are shown in Fig. 17.

Potentiodynamic polarization curves of all tested nickel materials were determined in the same conditions for all the above materials: during measurements the polarization potential was increased in a wide range from -750mV to +700 mV with a 0.4mV/s rate.

The potentiodynamic polarization curves j = f(E) of nickel with different crystalline structures and of amorphous NiP alloy determined from measurements are shown in Fig. 18.

Analysis of these curves indicates a noticeable influence of the structure of nickel and other ingredients contained in the material, on the process of corrosion in the test environment. The corrosion parameters of the tested materials obtained from the experiment are summarized in Table 6.

Analysis of these parameters shows that the greatest potential for corrosion and the smallest corrosion current density characterize nickel with the nanocrystalline structure. This highlights its highest resistance to corrosion in the test environment. Increased corrosion resistance of electrochemically produced nickel with the nanocrystalline structure in comparison to microcrystalline nickel may indicate a greater tendency to passivity of the nanocrystalline nickel. A passive layer that forms on the surface of nanocrystalline nickel inhibits the processes of corrosion of nickel in a certain range of the potential. However, differences in the corrosion resistance of pure nickel and its alloy result from both the additive contained in the material alloy, as well as different material structures.

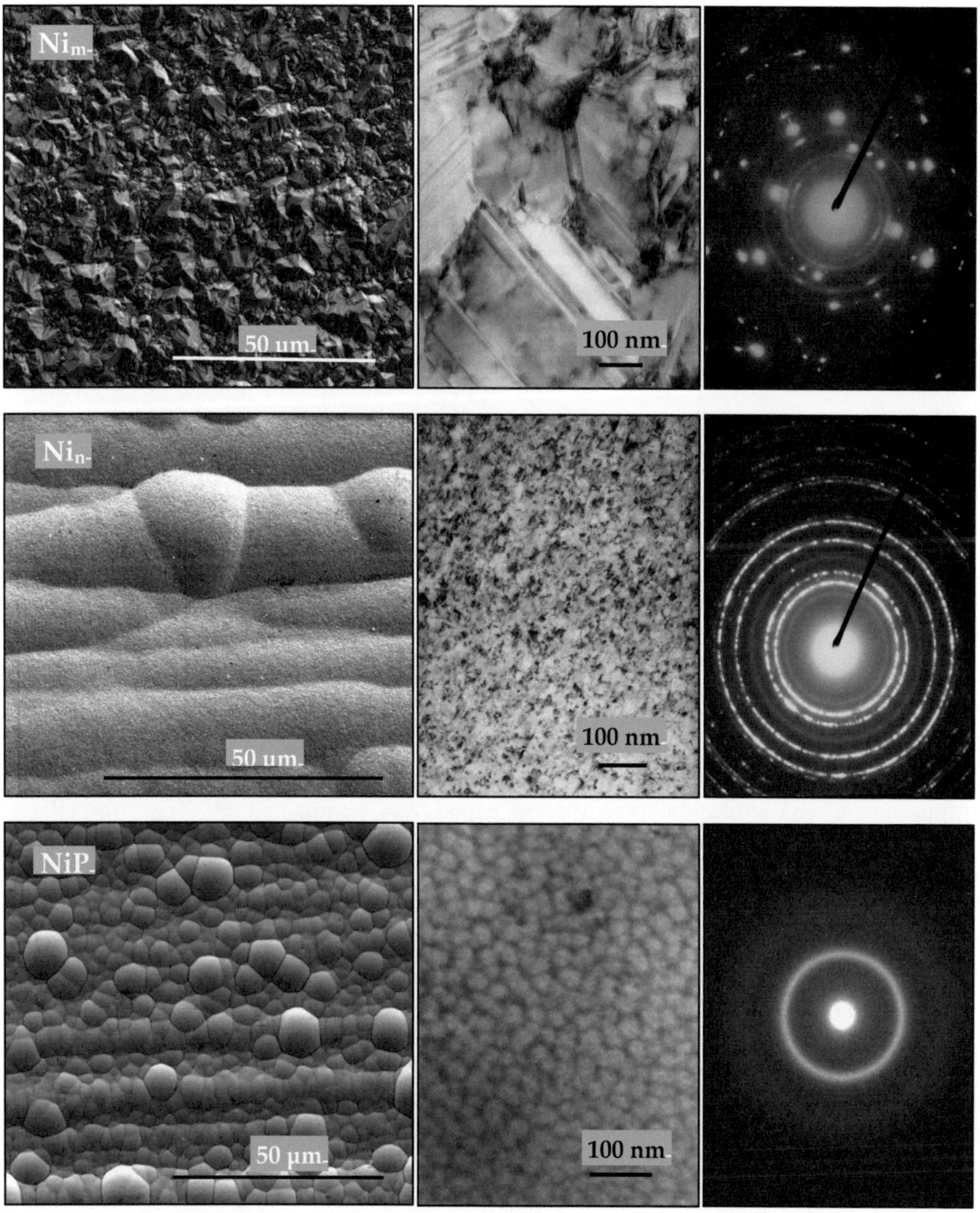

Fig. 17. Images of surface morphology (SEM), structure (TEM) and electron diffraction (SAED) of microcrystalline nickel Ni$_{m}$, nanocrystalline nickel Ni$_{n}$ and NiP amorphous alloy prior corrosion tests.

Material	E_{cor} [mV]	j_{cor} [μA/cm²]
Ni_m	-568	24
Ni_n	-340	1.5
NiP	-390	4.5

Table 6. Corrosion parameters of nickel and alloy NiP in 0.5M NaCl solution.

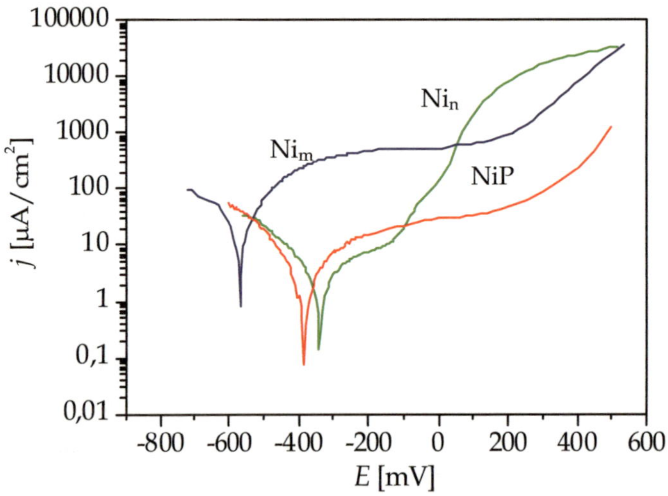

Fig. 18. Potentiodynamic polarization curves of nickel with the microcrystalline structure (Ni_m), and nanocrystalline structure (Ni_n) and of NiP alloy.

Investigations of the processes at the interface nickel-0.5M NaCl solution by impedance spectroscopy were performed with frequency changes in the range of 10kHz ÷ 2mHz. Amplitude of the sinusoidal perturbation signal was maintained at 15mV. Impedance spectra recorded for test materials are shown as Nyquist diagrams (Fig. 22) and Bode diagrams (Fig.23), in the form of two relationships: the impedance magnitude and phase angle versus frequency.

Nyquist diagrams (Fig. 22) indicate significant differences in the course of corrosion processes of the different forms of nickel and its alloy structures in the environment of 0.5M NaCl solution. In the case of nanocrystalline nickel structure and alloy NiP, the impedance spectra obtained are expressed in the form of an arc forming part of the semi-circle of very large radius. This chart indicates good corrosion resistance of nanocrystalline nickel and alloy NiP in the test environment. Impedance spectrum plot of microcrystalline nickel in the shape of semicircle of small radius ends with a fragment of straight line in the low frequency part of the forcing signal. This straight line fragment of the relationship between the imaginary component (Z'') and the real part (Z') of the impedance points to the diffusion control of corrosion processes at low frequency of the forcing signal. The smallest diameter of the semi-circle in the case of nickel with microcrystalline structure corresponds to a small value of resistance electric current flowing through the phase boundaries as a result of oxidation of nickel, which corresponds to a high rate of corrosion processes.

All obtained results of impedance measurements confirm the significant impact of the material structure and the additions of nickel alloy on the resistance to corrosion and are consistent with the results obtained by potentiodynamic polarization method.

The two methods showed that nanocrystalline nickel has the highest corrosion resistance in an environment of 0.5M NaCl solution..

Further assessment of the characteristics of the impedance at the system boundaries for nickel and its alloy in 0.5M NaCl solution was obtained by approximation of experimental data using equivalent electrical circuits. The equivalent electrical circuits most suitable to represent the measured impedance characteristics of studied systems of nickel with different structures and its alloy in corrosive environment of 0.5M NaCl solution are shown in Figs. 19, 20 and 21. The corresponding resulting impedances are described in the expressions (4) ÷ (6). For the analysis of the corrosion of microcrystalline structure nickel with the equivalent electrical circuit a simple layout shown in Fig. 19 was used.

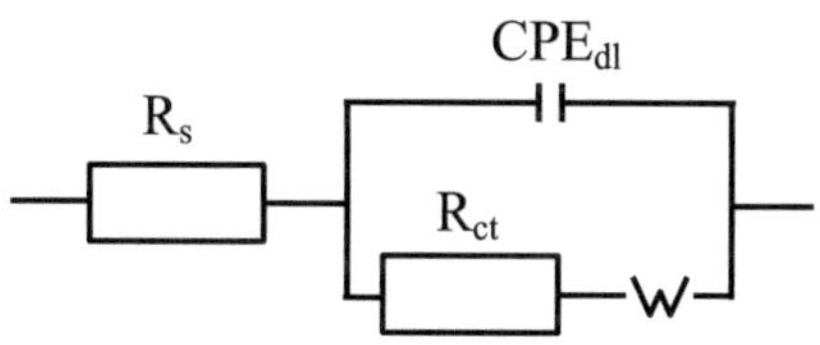

Fig. 19. The equivalent electrical circuit for corrosion of microcrystalline structure nickel in 0.5 M NaCl solution.

This system includes four elements: R_s - resistance of 0.5M NaCl solution, R_{ct} - electric charge transfer resistance for phase boundary of nickel - solution, CPE_{dl} – constant phase element characterizing the electrical properties of the double layer at the interface, and the element W - Warburg impedance, which characterizes the control of corrosion processes by diffusion of mass in the area of the electrolyte at the metal surface.

The equivalent electrical circuit is described by the resulting impedance

$$Z = R_s + \frac{1}{\frac{1}{R_{ct} + W} + Y_{dl}(j\omega)^{n_{dl}}} \tag{4}$$

Experimentally determined impedance spectra of nanocrystalline nickel corrosion (fig. 20) are well mapped by equivalent electrical circuit with two time constants described by equation (5)

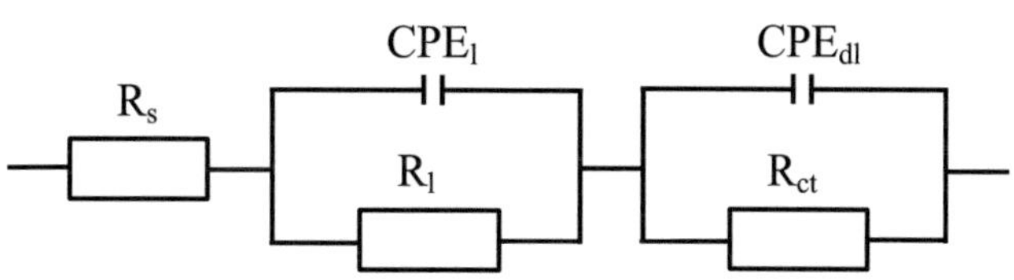

Fig. 20. Equivalent electrical circuit of corrosion of the nanocrystalline structure nickel in 0.5M NaCl solution.

$$Z = R_s + \frac{1}{\frac{1}{R_l} + Y_l(j\omega)^{n_l}} + \frac{1}{\frac{1}{R_{ct}} + Y_{dl}(j\omega)^{n_{dl}}} \qquad (5)$$

This circuit, besides elements such as R_s, R_{ct}, CPE_{dl} which are needed in the equivalent electrical circuit to describe the corrosion of microcrystalline nickel contains two additional elements: CPE_l - modeling capacity of the passive layer on the material surface, and R_l - describing the resistance of the passive layer.

To describe the corrosion processes occurring in the system NiP- 0.5M NaCl solution the equivalent electrical circuit shown in Fig. 21 was designed with the resulting impedance expressed by (6).

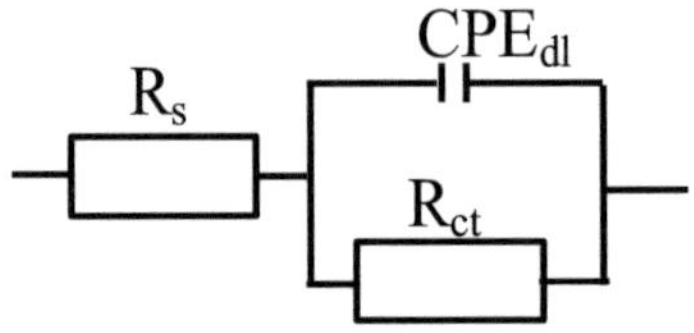

Fig. 21. Equivalent electrical circuit for corrosion of the nanocrystalline structure nickel in 0.5M NaCl solution.

$$Z = R_s + \frac{1}{\frac{1}{R_{ct}} + Y_{dl}(j\omega)^{dl}} \qquad (6)$$

The parameters of the equivalent electrical circuits of corrosive systems of nickel materials tested in this study are summarized in Table 7.

Material	R_s [Ωcm^2]	R_l [Ωcm^2]	CPE_l [$\Omega Fs^{n-1}/cm^2$]		R_{ct} [Ωcm^2]	CPE_{dl} [$\mu Fs^{n-1}/cm^2$]		W [Ωcm^2]
			Y_l	n_l		Y_{dl}	n_{dl}	
Ni_m	14.7	-	-	-	1420	69	0.8	705
Ni_n	14.7	4708	17.6	0.9	15317	30	0.8	-
NiP	12.6	-	-	-	20430	17	0.9	-

Table 7. The parameters of equivalent electrical circuits for corrosive systems of nickel - 0.5M NaCl solution.

The agreement between characteristics predicted by the equivalent circuit methods and those obtained from measurements are illustrated in Figs. 22 and 23.

Images of the damage on the surface of nickel samples with different structure and its alloy after corrosion tests are shown in Fig. 24.

In the case of the microcrystalline structure nickel and NiP alloy in corrosive environment of 0.5M NaCl solution, a pickling of their internal structures occurred over the entire surface exposed and even its internal structures was revealed. On the other hand, corrosion of the nanocrystalline nickel in this environment takes the form of uneven local corrosion.

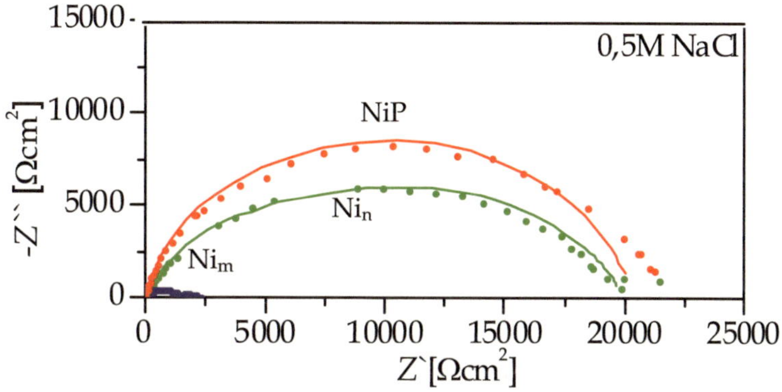

Fig. 22. Nyquist diagrams of impedance spectra of investigated corrosion systems of nickel and its alloy in the environment of 0.5M NaCl solution determined experimentally (point line) and as a result of calculations (solid lines).

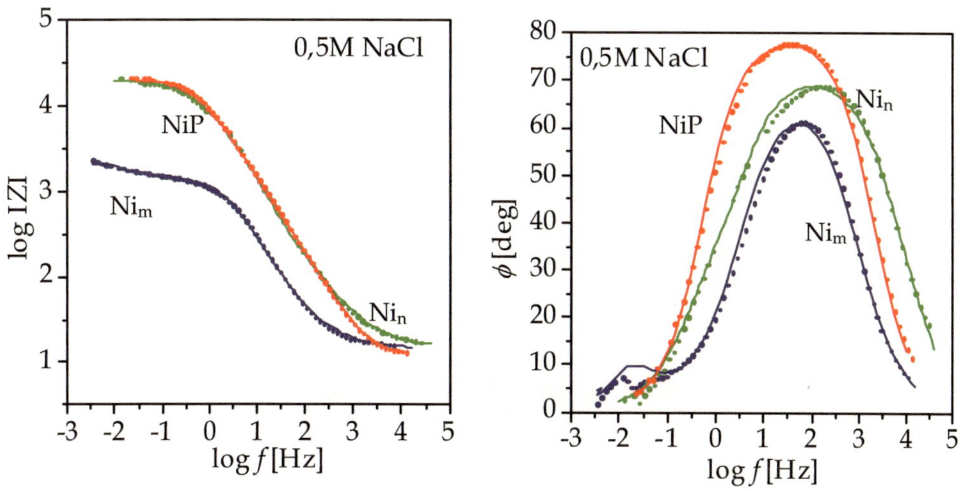

Fig. 23. Bode diagrams of impedance spectra of investigated corrosion systems of nickel and its alloy in the environment of 0.5M NaCl solution determined experimentally (point line) and as a result of calculations (solid lines).

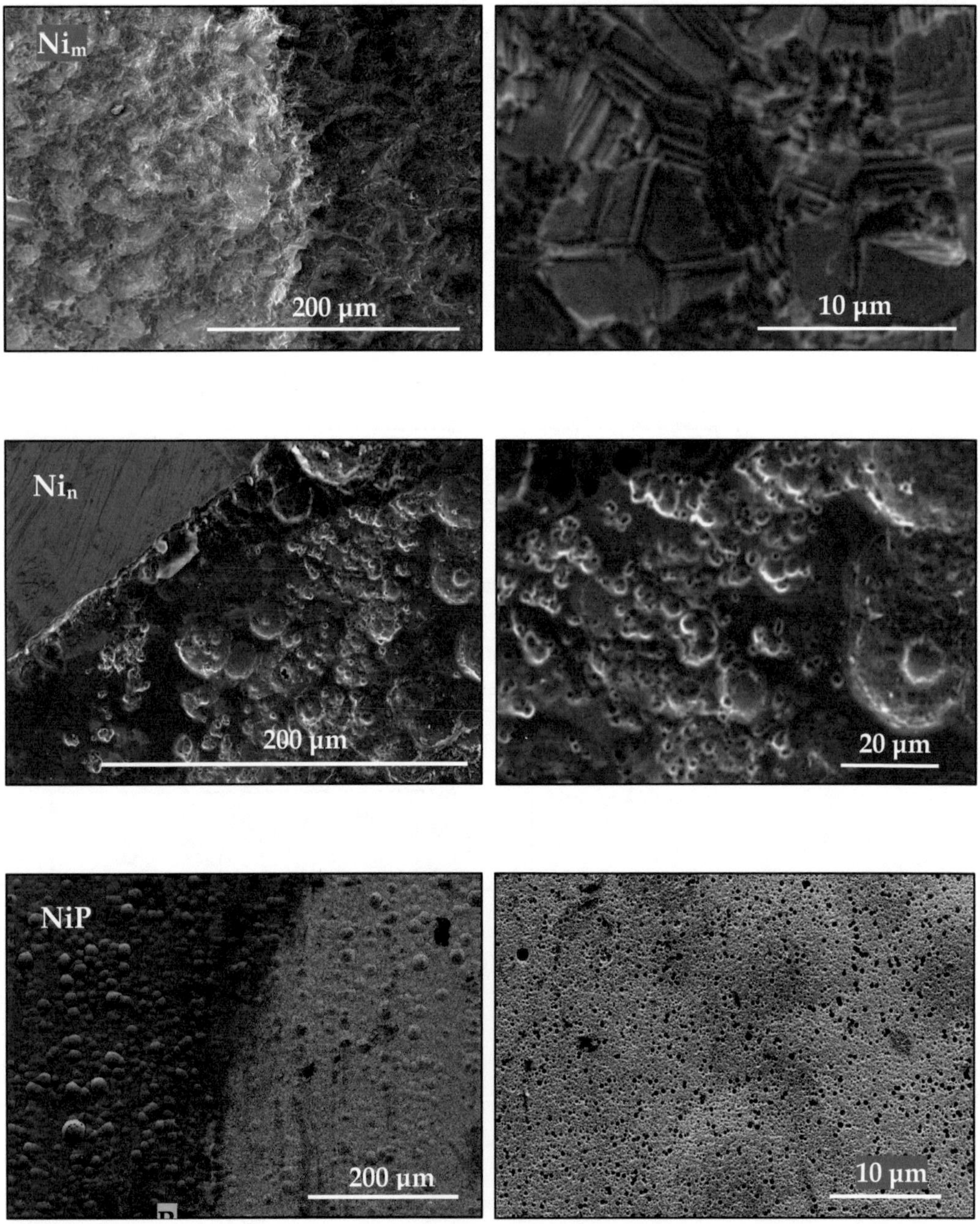

Fig. 24. SEM images of the surface of Ni$_m$, Ni$_n$, and NiP alloy after corrosion tests in an environment of 0.5M NaCl solution.

4. Summary

The rate of corrosion processes of metallic materials in a corrosive environment depends on the chemical activity of the metal and the additional components, the structure of the material as well as the degree of development of their surfaces.

Electrochemical methods for the study of corrosion processes are based on the relationships between electrical, chemical and physical properties, which are used to identify phenomena and processes at the interface metal-corrosive environment. Electrochemical potentiodynamic polarization method allows determining the corrosion potential and corrosion current density of the metallic material in a corrosive environment. Additionally, the precise measurement method using electrochemical impedance spectroscopy (EIS) generates frequency characteristics of corrosion systems and forms the solid basis to design models based on equivalent electrical circuits, which maps the processes occurring in the corrosion system under investigation. Such an equivalent electrical circuit that meets the criteria of a mathematical (and metrological) model can also be considered as a physical model describing the phenomena and the processes occurring in a given system undergoing electrochemical corrosion.

The current research of corrosion phenomena appearing at the interface metal-natural environment showed that chemical re-combination of the metals to form ore-like compounds is a natural process, because the energy content of the metals and alloys is higher than that of their ores. It has to be emphasized that there are number of means of controlling corrosion. The choice of a means of corrosion control depends on economics, safety requirements, and a number of technical considerations. However, it is necessary to learn and recognize the forms of corrosion and the parameters that must be controlled to avoid or mitigate corrosion.

Through the understanding of the electrochemical processes and how they can act to cause the various forms of corrosion, the natural tendency of metals to suffer corrosion can be overcome and equipment that is resistant to failure by corrosion can be designed. In this study we have shown that the measuring methods based on the electrochemical impedance spectroscopy are able to detect the potential corrosion spots in very early stages.

5. References

Eftekhari, A. (ed.). (2008). *Nanostructured Materials in Elektrochemistry*, ISBN 978-3-527-31876-6, Ohio, USA

Huang, Y., Shih, H., Huang, H., Daugherty, J., Wu, S., Ramanathan, S., Chang, Ch. & Mansfeld, F. (2008). Evaluation of the corrosion resistance of anodized aluminum 6061 using electrochemical impedance spectroscopy (EIS). *Corrosion Science*, Vol.50, No.12, (December 2008), pp.3569-3575, ISSN 0010-938X

Kowalewska, M., Trzaska, M., Influence of Si_3N_4 disperse ceramic phase on the corrosion resistance of micro- and nano-crystalline nickel layers. Physico Chemical Mechanics of Materials, Vol. 2, No. 5, (May 2006), pp. 615-619, ISSN 0430-6252

Marcus, Ph., (2011), *Corrosion Mechanism in Theory and Practice.* (2nd ed.), Taylor and Francis, ISBN 1420094629, London, GB

Trzaska, M., Computer modeling of corrosion processes by electrochemical impedance spectroscopy. Electrical Review, Vol. 86, No. 1, (January 2010), pp. 133-135, PL ISSN 0033-2097

Trzaska, M., Trzaska, Z. (2010). *Electrochemical impedance spectroscopy in materials science.* Publ. Office of the Warsaw Univ. Technol., ISBN 978-83-7207-873-5, Warsaw, Poland

Trzaska, M., Trzaska, Z., *Straightforward energetic approach to studies of the corrosion performance of nano-copper thin-layers coatings.* Journal of Applied Electrochemistry, Vol. 37, No. 9 (September 2007), pp. 1009 - 1014, ISDN 0021-891X

Trzaska, M. Moszczynski P. On Influences Of Ionic Liquid Additives To Watts Bath On The Corrosion Resistance of Electrodeposited Nickel Surface Layers. *Journal of Corrosion Measurements (JCM)*, Vol. 6, 2008

Sword, J., Pashley, D. H., Foulger, S., Tay, F. R. , Rodgers R., *Use of electrochemical impedance spectroscopy to evaluate resin-dentin bonds.* Journal of Biomedical Materials Research, Vol. 84B, No.2, (February 2007), pp.468 - 477, ISSN1552-4981

Orazem, M. E., Tribollet B., (2008), *Electrochemical Impedance Spectroscopy.* J. Wiley, ISBN 9780470041406, New York, USA

Vargel, Ch. (2004).*Corrosion of Aluminium,* Elseevier, ISBN 0 08 044495 4, New York, USA

Yang, L. (ed.). (2008).*Techniques for corrosion monitoring,* Press ISBN 978-1-4200-7089-7, Cambridge, England

19

Households' Preferences for Plumbing Materials

Ewa J. Kleczyk[1] and Darrell J. Bosch[2]
[1]*ImpactRx, Inc., Horsham, Pa.,*
[2]*Agricultural and Applied Economics Dept., Virginia Tech, Blacksburg, Va., USA*

1. Introduction

Consumers' decisions on plumbing material selection are dictated by various factors, including state and federal regulations, service providers, and individual household preferences. The regulations and standards of the federal, state, and local governments have major impacts on the plumbing material chosen for installation in a private house. For example, the use of plastic plumbing material, such as PEX, has been approved in all U.S. states except for California and Massachusetts, where the material installation requires local jurisdiction acceptance. Similarly, in some parts of Florida, PEX is preferred due to the seriousness of pinhole leak[1] problems (NSF, 2008). These regulations influence services provided by plumbers, material producers (e.g. pipe manufacturers, interior coating providers), and water utility companies. For example, general contractors are the primary decision-makers of plumbing material installation in new houses, while utility companies respond to corrosion threats by adding corrosion inhibitors to drinking water treatment. Consequently, all service providers influence consumer decisions, regarding the best plumbing material for private properties.

Homeowners have an important stake in finding plumbing system appropriate for their households, and they should rely not only on expert advice, but also acquire information on plumbing material attributes such as price, health impact, longevity, and corrosion resistance, in order to make informed investment decisions about plumbing systems for their homes. For example, health effects, water taste and odor have been found to be the most important factors in consumers' evaluations of plumbing material for home use (Lee et al., 2009). Additionally, households are willing to pay up to $4,000 when guaranteed a leak-free plumbing system for 50 years (Kleczyk et al., 2006). Information on consumer preferences for drinking water plumbing attributes can be useful not only to individual households, but also to policymakers, program managers, water utilities, and firms with interests in drinking water infrastructure.

---

[1] Pinhole Leaks are a small holes that commonly are caused by pitting corrosion, a type of corrosion concentrated on a very small area of an inner pype. In most cases, pinhole leaks are hard to detect, if they are visible, they appear as green, wet area on pipe and porcelain fixtures (Kleczyk & Bosch, 2008).

The public perceptions of corrosion risk and cost of prevention play a fundamental role in consumers' drinking water decisions. Homeowners' perceptions of risk and cost of prevention may affect households' decisions on plumbing material repairs and replacement, as well as the type of material used. When informed about the attributes of each plumbing material alternative, consumers can decide on the most preferred plumbing system. The decision of choosing an appropriate plumbing material is based on various plumbing material attributes, such as cost (material cost plus labor and installation cost), health effects, corrosion susceptibility, strength, property real estate values, and behavior in the case of a fire (Champ et al., 2003).

As it is important to learn household perceptions and preferences for drinking water infrastructure, the chapter objective is to investigate homeowners' preferences for plumbing materials (i.e. copper, plastic, an epoxy coating), as well as preventive techniques against corrosion based on households' experiences with plumbing material failures. In 2007, a survey of a Southeastern Community in the United States was conducted in order to meet these goals, and obtain information on the prevalence of plumbing material failures, householders' experiences with plumbing material failures, the cost of repairs and property damages due to the material failures, and household preferences for plumbing systems.

The objective of the study is fulfilled by analyzing in-depth the information of the prevalence of home plumbing corrosion, preventive measures taken against corrosion, as well as the financial, health, and time costs associated with repairing faulty plumbing systems. In addition, analyses are performed to elicit household preferences for plumbing materials, and to identify the attributes important to choosing home plumbing systems. Summary statistics as well as regression methods, such the Ordered Logit model, are employed to support the study, and provide insight into the scale of corrosion in the community, the financial burden accrued from repairing the problem, and finally recommendation for the best plumbing materials for household use.

The knowledge gained from this chapter can be helpful in the design of public policy aimed at corrosion prevention. The research provides information to federal and state officials, plumbers, plumbing material manufacturers, and utility company managers on the financial burden individual households are willing to take on to avoid corrosion. In addition, the study should help in bridging the gap between the perceptions of the public and drinking water infrastructure experts, regarding the problem of pinhole leaks and other corrosion related issues.

2. Literature review

As mentioned above, the household decision-making process with regards to choosing a plumbing material for a private residency is complicated, and involves several factors, such as federal, state, and local standards and regulations, corrosion risk perceptions of drinking water as viewed by infrastructure service providers, insurance companies, households, as well as the financial impact of corrosion prevention. The regulations and standards of the federal, state, and local governments have major impacts on the plumbing material chosen for installation in a private house. These regulations influence the services provided by plumbers, home builders, material producers, and water utility companies (Lee et al., 2009).

To make an informed decision about the optimal plumbing material for their home, homeowners need information on the various risks involved in choosing plumbing systems. When informed about the plumbing material characteristics, the consumers are able to decide on an alternative most preferable to them based on the preference trade-offs among plumbing materials' attributes. Households make decisions on a plumbing alternative when either replacing an existing system or installing a plumbing system in a new house. Each alternative has advantages and disadvantages that impact health and the overall cost of installation and maintenance. The problem becomes more complex as consumers think in terms of cost (material plus labor charges), taste and odor of the water, corrosion problem, longevity of the pipe system, fire retardance, convenience of installation or replacement, plumbers' and general contractors' opinions or expertise, and proven record in the market. Householders weigh each of these attributes in order to choose the most preferred option for their houses (Lee et al., 2009).

For example, Lee et al. (2005), utilizing the AHP method, studied the preferences for plumbing materials of Virginia Tech potable water experts. Participants ranked the health effects, reliability, taste and odor, and longevity as the most important attributes when choosing a plumbing material. Property value and fire resistance were listed at the bottom of the ranking. These results showed that health, water taste and odor dominate preferences for plumbing materials. Lack of reliability resulting in the need to repair the damage associated with pipe corrosion relates to stress and a worry about future leaks (Lee et al., 2005).

There are several plumbing material types for a householder to choose from when deciding on a plumbing material to be installed in a house: copper, plastic (CPVC and PEX), and stainless steel. According to Marshutz' survey (2000), copper is used in nearly 90% of homes in the U.S. followed by PEX (cross linked polyethylene) with a 7% installation rate, and CPVC (chlorinated polyvinyl chloride) with a 2% installation rate. Telephone surveys of plumbers conducted in 2005 show an increased use of plastic pipes, due to easier handling in installation and lower material cost (Scardina et al., 2007).

Copper is the most widely used material in residential plumbing and has several advantages, including affordability, fire resistance, few health hazards, and durability. Woodson (1999) studied the performance of different plumbing material alternatives: copper, CPVC, and PEX. He found copper pipes generally perform well, except for cases involving major leak problems (Woodson, 1999). Due to increased pinhole leak incidents reported in hotspot areas of the U.S. (eg. Washington, D.C. suburbs and Sarasota, Florida), many consumers replaced copper with other options. Concerns with copper pipes include a metallic taste, especially with long stagnation periods and increased absorption of residual disinfectant by the pipe walls. High levels of copper can cause nausea, vomiting, and diarrhea (ATSDR, 2004). Elevated copper levels in drinking water may increase lead levels when lead solder joints, lead service lines, or brass fixtures are present in plumbing material. It is advised to test for lead when testing for copper levels in drinking water as lead and copper enter drinking water under similar conditions (Lee, 2008).

PEX (polyethylene cross linked) is another type of plumbing material often used in residential plumbing. This material is used to make flexible plastic pipes. A different plumbing design characterized by individual pipe lengths is required for every fixture. The main advantage of PEX is the lack of joints requiring soldering, which decreases the probability of pipe failures. On the other hand, PEX plumbing has raised some concerns regarding possible leaching of MTBE (methyl tertiary butyl ether), ETBE (Ethyl tert-butyl ether), and benzene into drinking water. Other concerns are the negative health impacts associated with PEX's reaction with chlorine, increased water odors (Durand & Dietrich, 2007), the material's ability to withstand fire, and its final disposal (PRNews Wire, 2004). In addition, PEX may become stiff in cold weather, which makes faulty pipe repairs more difficult. PEX use has been approved in all U.S. states (Toolbase News, 2008), and has met all health standards set by NSF/ANSI-61 for potable water supply (NSF, 2008).

CPVC plumbing material is also employed in residential plumbing, but presents many concerns. For example, it can become brittle when exposed to sunlight for an extended period of time, and presents possible negative health effects from microbial growth in the inner pipe. Other possible concerns are cracking in the event of an earthquake, plastic water taste, and melting in the event of fire. The solvents used to join fittings and pipe lengths may contain volatile organic compounds (VOCs), requiring proper ventilation during installation, and causing unpleasant odor problems. However, CPVC by itself has a low odor potential (Heim & Dietrich, 2007).

The last plumbing material type is stainless steel, which is often used in industrial applications. Stainless steel provides excellent resistance to corrosion, due to the presence of 18% chromium and 8% nickel (Roberge, 2000). The stainless steel material is, however, expensive. Due to the cost, its use is limited to specialized industries for conveying chemicals or other similar applications (Lee, 2008). A concern with stainless steel pipes is the possibility of leaching chromium into drinking water; however, all U.S. states have approved stainless steel use (NSF, 2008; Roberge, 2000).

The economically sustainable optimal replacement time for home plumbing systems is about 22 years after installation (Loganathan & Lee, 2005). The estimate, however, is dependent on the source and type of the employed data (Loganathan and & Lee, 2005). When it is time to replace the plumbing system, the homeowners have to decide on a plumbing system to be installed in their homes. For example, several homeowners in a Southeastern Community in the U.S. replaced their copper pipes with PEX. According to them, PEX is less labor intensive in case of installation, resistant against corrosion, and less expensive compared to copper (Plumbing and Mechanical Magazine, 2007).

However, plumbing material replacement or repairs can be rather expensive. Farooqi and Lee (2005) conducted a survey of plumbers in the U.S. and found plumbers to charge their work on an hourly basis. The cost per hour varied from $45 to $75, and the total cost of plumbing material replacement ranged from $3,654 for PEX to $5,680 for copper pipes (Farooqi & Lee, 2005). Furthermore, fixing dry wall, floor tiles, or ceilings affected by plumbing material replacement is not part of the services provided by the plumber, and homeowners have to hire a general contractor to fix the water related damage. Kleczyk and Bosch (2008) have reported the additional costs associated with damage from pipe failures reaching as much as $25,000, and forcing household members to reside in temporary housing during the repair period.

On the other hand, Scardina et al. (2007) (also discussed in Kleczyk et al. 2006) investigated the willingness-to-pay for a leak-free plumbing material in households located in different parts of the U.S., such as Florida and California. They found 47% of all respondents willing to pay a positive amount to ensure that material would remain leak-free, 27% unwilling to pay any amount to ensure that material would remain leak-free, and 25% unsure about how much they would be willing to pay. About 6% of respondents were willing to pay at least $4,000 to ensure that material would remain leak free for 50 years. This amount is 10 times the suggested base material cost for re-plumbing a 2,000 square foot house. The mean willingness-to-pay estimate was higher for respondents with leaks compared to respondents who had no leaks, constituting $1,130 and $1,007 respectively. Finally, 45% of respondents with leaks and 41% of respondents without leaks were not willing to pay for leak-free plumbing materials (Dietrich et al., 2006; Kleczyk et al., 2006; Scardina et al. 2007).

3. Survey design and distribution

The Southeastern Community located in the United States of America was established in 1980s, and spans over 4,700 acres. There are about 3,300 homes, including condos and apartments, with 6,600 residents in total. Most of the resident population is retired, so the community is rather a homogenous group. The first incidents of pinhole leaks were reported in 2001.

In August 2007, a questionnaire was sent to 1600 households in the Southeastern Community. The community's Property Owners' Association provided a list of the residents' names and addresses, and the sample was randomly selected from this list. Members of the Assocaition's Board reviewed the survey questions. The Association encouraged participation of community residents in the study. The survey was distributed following the Dillman technique of mail surveying, which included mailing a questionnaire with postage-paid return envelope, sending a reminder card, and mailing a second copy of the survey to nonresponders (Dillman, 1978).

In 2007, two surveys were conducted by the Virginia Tech researchers to learn about the home plumbing issues and the preventive measures taken against future corrosion incidences. The first survey acquired information on the incidents of pinhole leaks in the residential area, the adoption rate of preventive measures against corrosion, the homeowners' preferences for corrosion risk, and the costs associated with a leak free environment. The second survey elicited preferences for three hypothetical plumbing materials with different attribute levels. The sample of respondents was based on the first Southeastern Community survey respondents, who were willing to participate in the follow-up questionnaire.

A follow-up survey was administered in October 2007 to learn household preferences for home plumbing materials. The follow-up survey was sent 363 Southeastern Community householders who responded to the first survey, and who agreed to participate in future surveys. The respondents were exposed to attributes of three hypothetical to them plumbing system materials, which were left unnamed to avoid a survey exposure bias[2]. The materials represented in the questionnaire were copper, plastic, and epoxy coating. Materials were left unnamed, because most homeowners were familiar with at least one material type (copper,

[2] Survey Exposure Bias represents the ability to skew respondents' responses, based on the information either presented during the study or known prior to the study (Champ et al., 2003).

plastic, or epoxy coating), and positive or negative experiences with these materials could have influenced their responses. The questions included two stimuli, which are compared simultaneously. Each respondent rated each of the two alternatives on a scale from 1 to 9. The scale value of 1 indicates the plumbing material is not preferred, while 9 indicates an extremely preferred plumbing system. The material attributes are listed in Table 1.

Attributes	Material A (Epoxy Coating[a])	Material B (Plastic[a])	Material C (Copper[a])
Corrosion Resistance	Corrosion proof	Same as material A	Some risk of corrosion
Taste / Odor	Compounds released from this material in drinking water plumbing may give a chemical or solvent taste or odor to the water.	Same as material A	Compounds released from this material in drinking water plumbing may give a bitter or metallic taste or odor to the water.
Health Effects	Material meets EPA Standards. There is a very small chance that compounds from this plumbing material that are released into drinking water may lead to microbial growth in water. Microbial growth may cause severe illness.	Same as material A	Material meets EPA Standards. There is a very small chance that compounds from this plumbing material that are released into drinking water may cause vomiting, diarrhea, stomach cramps, and nausea.
Convenience of Installation	No need to tear into the wall and/or floor. Installation takes around 4 days.	Need to tear into some sections of wall for installation. Installation takes 5-6 days.	Need to tear into the wall and/or floor to replace the existing system. 7-9 days required for installation.
Proven performance in market	Less than 10 years in the market	Less than 20 years in the market	More than 50 years in the market
Cost (labor + material)	\$9,000 ~ 14,000 depending on the size of house	\$6,500 ~ 13,000 depending on the size of house	\$9,000 ~ 16,000 depending on the size of house
Warranty	Warranty is 15 years for the material.	Warranty is 10 years for the material.	A 50 year manufacturer's warranty applies. Some exceptions apply (e.g. warranty reduces to one year if compounds in water corrode pipes).

[a]Names of the plumbing materials were not revealed to the study participants

Table 1. Description of plumbing materials.

4. Empirical analysis

The empirical analysis of the Southeastern Community home plumbing data includes several econometric and statistical techniques. The first survey data analysis uses simple descriptive statistics, such as mean (average values), percentages (percent distribution across all responses), and total sums, in order to provide a summary view of the home plumbing issues faced by the Southeastern Community. These issues include the frequency of pipe failure, the location of the failure in the plumbing system, the costs and time associated with fixing pipe failures, and the preventive measure taken to avoid incidences in the future. The analysis preferred plumbing materials concentrates on estimating the household preferences for plumbing types based on the follow-up survey of the Southeastern Community. The data estimation process employs the Ordered Logit regressions, based on which the household preferences for plumbing materials are derived. The paragraphs presented below describe the econometric models in more detail.

4.1 Ordered logit model description

The second Southeastern Community survey data analysis employs the Conjoint Analysis (CA) methodology to analyze the preferences for plumbing materials. This type of analysis includes eliciting the preferred good / service choices based on the presented information / stimuli. Utility Maximization Theory is usually employed to guide the process, design, and analysis of the CA studies, and involves making a choice that yields the greatest satisfaction to the respondents, otherwise known as utility, based on their available financial resources. As a result, the preference maximization problem is defined mathematically, as maximization of a utility function based on a specified financial resource constraint (Varian, 1992):

$$\text{Maximize utility function: } u(x) \tag{1}$$

$$\text{Subject to: } px \leq m, \text{ where } x \text{ is in } X, \tag{2}$$

where u(x) represents the utility function, and $px \leq m$ represents the financial resource constraint, with m being the fixed amount of money available to households (Champ et al., 2003).

In this chapter, a household faces a choice among three plumbing material alternatives. The utility (satisfaction) obtained from choosing a plumbing material, i, by the nth household is U_{ni}. The decision maker chooses the option yielding the highest level of utility, which implies the following behavioral model: $U_{ni} > U_{nj}$, where $i \neq j$. The level of utility is not observed by the researcher, but the attributes of the plumbing alternatives (x_{ni}) in the choice set are observed, as well as the socioeconomic characteristics of the decision maker (z_n). Based on the known variables, a representative utility function can be specified as: $V_{ni} = V(x_{ni}, z_n)$ for all alternatives (Train, 2003).

For this exercise, each respondent pair-wise rated the preferred plumbing material option. The rating scale ranges from 1 to 9, with 1 indicating a not preferred plumbing material option, and 9 indicating the most preferred option. The plumbing material rating exercise is based on the utility-maximizing behavior, as higher plumbing material rating results in an increased level of utility, and therefore, a higher preference level for a given alternative. The rating scale questions require individuals to make judgements about the magnitude of utility associated with plumbing material profiles. These plumbing material evaluations directly transform utility levels into a rating scale. As a result, an employment of rating models in which the rating value for each profile is regressed on a vector of attribute levels is justified (Champ et al., 2003).

To analyze the CA data, an Ordered Logit regression is employed. The Ordered Logit is based upon the idea of the cumulative logit, which relies on the cumulative probability. The cumulative probability CP_{nl} is the probability that the nth individual is in the lth or higher plumbing material valuation category:

$$CP_{nl} = \text{probability } (R_l \leq l) = \sum_{(l=1 \text{ to } L)} \text{probability}(R_l = L). \quad (3)$$

The cumulative probability is transformed into the cumulative logit:

$$\text{logit } CP_{nl} = \log(CP_{nl}(1 - CP_{nl})). \quad (4)$$

The ordered logit simply models the cumulative logit as a linear function of independent variables:

$$\text{logit } CP_{nl} = \alpha_l - \boldsymbol{\beta} \mathbf{x_n}. \quad (5)$$

There is a different intercept for each level of the cumulative logit, but $\boldsymbol{\beta}$ remains constant across rating categories. In addition, the product of $\boldsymbol{\beta}$ and the independent variable, $\mathbf{x_n}$, is subtracted rather than added in the model. As a result, each α_l indicates the logit of the odds of being equal to or less than category l for the baseline group (when all independent variables are zero). The $\boldsymbol{\beta}$ represents the increase in the log-odds of being higher than category l as the independent variable increases by one-unit (Edner, 2005).

The empirical Ordered Logit model is represented by the following regression:

$$R = \alpha_l - \sum_{(n=1 \ldots N)} \sum_{(i \in R)} [\sum_{(k=1 \text{ to } K)} \beta_k x_{jkn} + \beta_p p_{jkn}] + e \quad (6)$$

where R represents the ordered rating scale (1-9), where β_k is the preference parameter associated with the plumbing material attributes , x_{jkn} are the plumbing material attributes in profile j for individual n, β_p is the parameter on profile cost, p_{jkn} is the cost attribute for profile j and e is the error term (Champ et al., 2003).

Although attributes of the plumbing materials vary over alternatives; the characteristics of each household do not differ over the alternatives. As a result, the socioeconomic variables need to enter the model estimation to leverage and explain the differences in utility levels between corrosion preventive options. These characteristics can enter the model through interaction with the plumbing material attributes (Train, 2003).

5. Home plumbing corrosion issues

5.1 Pinhole leak awareness and Incidents

A total of 1,047 survey responses were received, a 65% response rate. Seventy-six percent of respondents reported being very aware of pinhole leaks, 21% said they were somewhat aware of the problem, and 2% said they were unaware of the problem. Nineteen percent reported learning about pinhole problems through their own experience, 65% heard through a neighbor or friend, 48% heard about pinhole leaks through the media, and 42% reported hearing of the problem through the property management.

Two hundred twelve respondents (20%) reported incidents of pinhole leaks in drinking water pipes in their current homes; 780 respondents (74%) reported no incidents of pinhole leaks; and 32 respondents (3%) were not sure of any incidents. One hundred twenty eight respondents (60% percent of the respondents with leaks) had 1 or 2 leaks, 47 respondents (22%) had 3 or 4 leaks, 17 (8%) had 5 or 6 leaks, and 15 (7%) had 7 or more incidents. Over 90% of the leaks had occurred since the year 2000. Of 212 respondents with pinhole leaks, 151 (71%) stated that their first pinhole leak occurred since 2004, and 44 (21%) stated that their first leak occurred between 2000 and 2003.

Respondents with pinhole leaks had somewhat older homes compared to respondents without leaks (Table 2). Fifty-three percent of respondents without leaks lived in homes built since 2000 compared to 4% of respondents with leaks. Five percent of respondents without leaks lived in homes built before 1990 compared to 23% of respondents with leaks.

	Respondents without leaks		Respondents with leaks	
Year house was built	Number	Percent[a]	Number	Percent[b]
Since 2000	441	53	9	4
1995 to 1999	240	29	76	36
1990 to 1994	102	12	75	36
Before 1990	39	5	49	23
Do not know	1	0	2	1
Missing/not reported	12	1	1	0
Total	835	100	212	100

[a]Percent = number divided by total number of respondents without leaks (835).
[b]Percent = number divided by total number of respondents with leaks (212).

Table 2. Year house was built.

Most respondents with leaks had leaks in horizontal pipes, while fewer had leaks in vertical pipes or pipe bends (Figure 1). Most leaks were in the finished or unfinished basement followed by the crawl space and first floor, respectively.

Pinhole leaks occurred in cold water pipes in 138 cases, in both cold and hot water pipes in 14 of the cases, and in hot water pipes only in 33 cases (Table 3). Twenty respondents were not aware of the type of water pipes where leaks occurred.

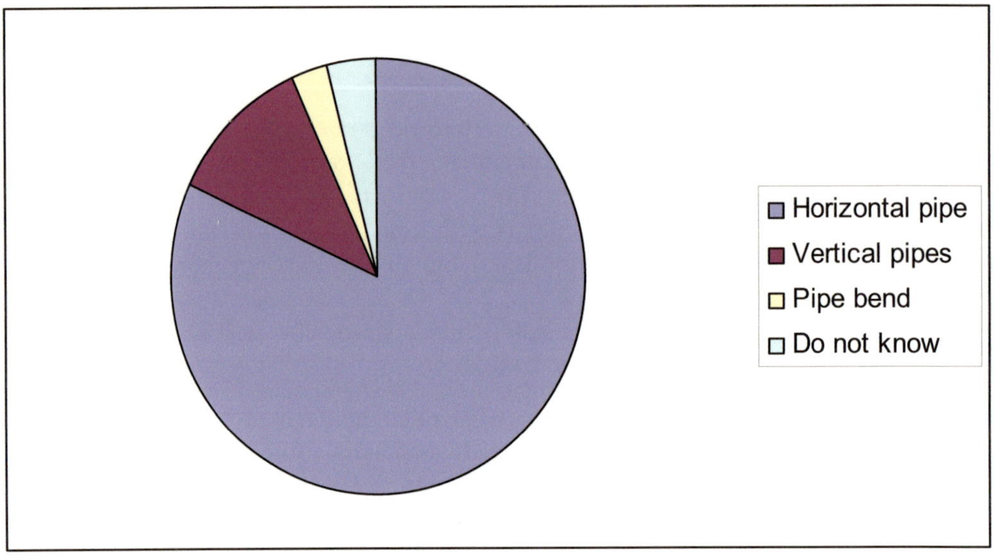

[a]Multiple choices per respondent were accepted. Percent = number reported divided by the total number of respondents with leaks (212).

Fig. 1. Pinhole leaks by type of pipe.

Type of pipe	Number of observations	Percent
Cold water pipes	138	65
Hot water pipes	33	16
Both	14	7
Do not know	20	9
Missing/not reported	7	3
Total	212	100

Table 3. Pinhole leaks occurring in cold or hot water pipes.

5.2 Pinhole leak repairs and repair costs

Seventy-seven respondents repaired the leak using a clamp (Table 4). In some cases, a clamp was used initially while the leaking section or all plumbing was replaced for later leaks. One hundred thirty-three respondents repaired the leak by replacing the leaking pipe section. Copper was most often used for repairing leaking sections. Fifty respondents repaired the leak by replumbing the entire house. PEX was most often used for replumbing. Nine respondents applied epoxy coating to their existing plumbing systems.

More than 60% of respondents with leaks spent less than 20 hours dealing with pinhole leaks while more than 30% spent 21 or more hours. Twenty percent spent more than 40 hours dealing with pinhole leaks.

Twenty-nine percent of respondents with leaks reported that the expense of repairing pinhole leaks was less than $100; while 30% reported expenses between $100 and $500; and 37% reported more than $500 in expenses for pinhole leak repairs (Table 5). Seven respondents reported more than $10,000 in costs of repairs.

Repair method	Number of observations	Percent[a]
Clamp over leak	77	7
Replaced leaking pipe section with copper	75	35
Replaced leaking pipe section with CPVC	5	2
Replaced leaking pipe section with PEX	7	3
Replaced leaking pipe section-material not specified	46	22
Applied epoxy coating to all plumbing	9	4
Replumbed with copper	5	2
Replumbed with CPVC	4	2
Replumbed with PEX	32	15
Replumbed-material not specified	9	4
Other	7	1
Don't know	3	1
Total	279	129

[a]Multiple choices per respondent were accepted. Percent = number reported divided by the total number of respondents with leaks (212).

Table 4. Method of leak repair.

Amount	Number of observations	Percent[a]
Less than $100	61	29
$100 to $500	64	30
$501 to $1,000	14	7
$1,001 to $3,000	11	5
$3,001 to $5,000	20	9
$5,001 to $10,000	28	13
$10,001 to $20,000	6	3
More than $20,000	1	0
Do not know	3	1
Missing/not reported	4	2
Total	212	99

[a]Numbers do not sum to 100 due to rounding.

Table 5. Costs of repairing pinhole leaks.

In addition to the expense of repairing leaks, 92% of respondents with leaks reported having to repair property damage caused by leaks. Forty percent of respondents with damage reported less than \$100 of damage, while 49% had over \$100 in damage. Twelve respondents had over \$5,000 in property damage. Thirty-six percent of respondents reporting leaks found the experience of pinhole leaks very stressful, and 46% found it somewhat stressful. Thirteen percent experienced little or no stress.

5.3 Pinhole prevention and water treatment devices

Thirty-five percent of respondents with leaks and 20% of respondents without leaks use some type of pinhole leak prevention strategy (Table 6). The most common strategy among those with leaks is preventive replumbing, which was used by 13% of those with leaks. Water softener / conditioner was the most common strategy used by those without leaks, which was used by 9% of those respondents.

Sixty-seven percent of respondents use some type of water treatment for purposes other than pinhole leak prevention (Table 7). The most common treatment was a refrigerator filter, used by 63%. Thirty-two percent of respondents reported that they purchase drinking water. The most common reasons given for using water treatment devices are to improve taste or smell of drinking water (mentioned by 45% of respondents), and to improve safety of drinking water (mentioned by 33% of respondents).

	Respondents with pinhole leaks[a]		Respondents without pinhole leaks[b]	
	Number	Percent	Number	Percent
Preventive replumbing	28	13	16	2
Preventive epoxy injection	8	4	4	0
Phosphate injection	12	6	26	3
Water softener/water conditioner	11	5	79	9
Copper Knight	5	2	12	1
Other	19	9	64	8
None used	134	63	644	77
Missing/not reported	4	2	29	3
Total	295	139	874	105

[a]Multiple choices per respondent were accepted. Percent = number reported divided by the total number of respondents with leaks (212).
[b]Percent = number reported divided by the total number of respondents without leaks (835).

Table 6. Use of pinhole leak prevention devices.

5.4 Concerns about water safety and quality

Eighty-two percent of respondents were somewhat or very satisfied with home drinking water quality (Figure 2). Only 5% of respondents were not at all satisfied with water quality. Problems with water quality most frequently mentioned were related to taste particularly chlorine. Respondents varied in concern about future pinhole leaks. Forty percent were somewhat or very concerned, while 55% were not very or not at all concerned.

	Number	Percent[a]
Filter for entire home	133	16
Refrigerator filter	523	63
Water softener/water conditioner	66	8
Pitcher or bottle to filter water	136	16
Purchased drinking water	265	32
Filter on faucet or under kitchen sink	117	14
Ultra violet (UV) system	2	0
Other	25	3
None used	249	30
Missing/not reported	19	2
Total	1,535	184

[a]Multiple choices per respondent were accepted. Percent = number reported divided by the total number of respondents (1,047).

Table 7. Use of water treatment for purposes other than corrosion prevention.

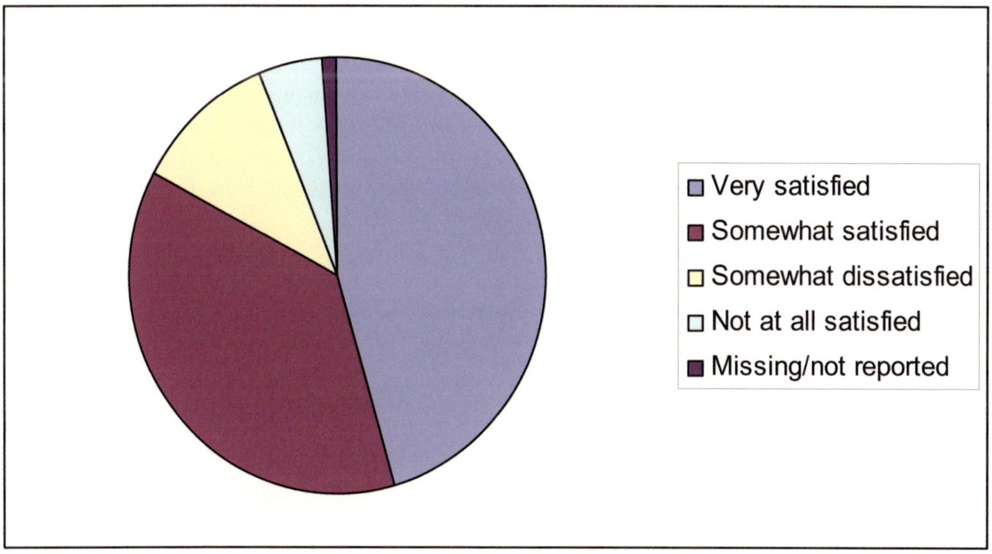

[a]Totals do not sum to 100 due to rounding.

Fig. 2. Satisfaction with home drinking water quality.

6. Household preferences for plumbing material

6.1 Summary of descriptive results

Every respondent to the first Southeastern Community survey was asked to participate in the follow-up survey. Three hundred sixty three respondents agreed to participate, and 245 responded to the follow-up survey. Each respondent evaluated three Conjoint Analysis scenarios describing a set of two plumbing materials (Material A= epoxy coating, Material B = plastic, and Material C = copper that were blinded to avoid survey exposure bias) and answered questions comparing material attributes. Each plumbing material was described by the following attributes: corrosion resistance, taste and odor, health effects, convenience of installation, proven performance on the market, plumbing material cost, and warranty length. Table 1 presents the plumbing material attributes in more detail.

Each respondent was asked to compare a pair of plumbing materials, and evaluate each plumbing material based on a 1-9 preference scale. For example, Material A might be rated as 6, while Material B might be rated as 1. The 1-9 preference scale had a verbal preference assigned to each categorical value. Preference values of 1, 3, 5, 7, and 9 were assigned to 'Not Preferred', 'Moderately Preferred', 'Strongly Preferred', 'Very Strongly Preferred', and 'Extremely Preferred', respectively. Two hundred thirty respondents fully answered all questions, and each viewed three pairs of two plumbing materials resulting in 1,380 preference responses.

As each presented plumbing material had all attributes listed, and there was no randomization of attribute levels across the plumbing materials, the preference score was easily identified with the preferred plumbing material by comparing the attribute levels with the plumbing material descriptions. All preference responses to each plumbing material were then summed, and the plumbing material with the highest number of 'Extremely Preferred' responses and with lowest number of 'Not Preferred' responses was selected as the most preferred plumbing material. Table 8 presents the descriptive statistical summary of preference valuation break down of the 1,380 responses for plumbing materials. Material C (copper) is the least preferred type of plumbing material (211 not preferred responses), while Material A (epoxy coating) is the most preferred material among homeowners (39 extremely preferred responses).

	Preference Response Value				
Plumbing Material	Not Preferred	Moderately Preferred	Strongly Preferred	Very Strongly Preferred	Extremely Preferred
Material A	103	134	99	85	39
Material B	148	151	85	56	20
Material C	211	156	56	31	8
Total	460	441	240	172	67

Table 8. Preference valuation of plumbing materials.

In addition to evaluating three sets of two plumbing material scenarios, each respondent selected the most preferred plumbing material across all three materials displayed at the same time. Table 9 presents that Material A (epoxy coating) is chosen as the preferred plumbing material by more than 50% of respondents. Material C (copper) is the least often chosen as the preferred plumbing material (17.8%). These two separate measures yield the same result of Material A being the most preferred plumbing material across the three alternatives.

Plumbing Material	Frequency	Percent
Material A	116	50.4
Material B	48	20.8
Material C	41	17.8
Neither	4	1.7
Missing	22	9.7
Total	230	100

Table 9. Plumbing material chosen as most preferred.

6.2 Empirical analysis results

6.2.1 Order logit model without socioeconomics variables

For this part of the analysis, the Ordered Logit regression is utilized in the plumbing material estimation of preferences and is estimated at the aggregate response level. The aggregate level analysis implies that average value coefficients are estimated for the participating sample of respondents.

The analysis provides information on the preferences of homeowners for plumbing materials, and the attributes that drive their decision, when making purchasing decision with regards to the type of home plumbing system. Each respondent evaluated a set of two plumbing material portfolios at one time for a total of six portfolios using the valuation metrics 1-9 described earlier. Each of the plumbing materials has a set of attributes described in Table 1. Each material attribute level is employed as the independent variable in the material preference analysis. They are coded as dummy variables taking a value of 1 when that plumbing material characteristic is a part of the product portfolio and zero otherwise. Finally, the socioeconomic characteristics (reported in the first survey) are also included in the Ordered Logit model. These characteristics represent household home value (continuous variables), age of the house (continuous variable), plumbing material type (dummy variable), pinhole leak occurrences in the past (dummy variable), and respondent's previous cost of plumbing material repairs and replacement (continuous variable).[3]

The first step in evaluating the results of the Ordered Logit model is to review the model performance / fitting criteria. The model fitting information indicates the parameters for which the model-fit is calculated. There are four variables that evaluate the goodness of fit: Chi-square statistics[4], p-value[5], log-likelihood value[6], and R-square[7]. The model fitting

[3] Variables for race, education level, and gender were not included in the model, as little variation in these characteristics was observed for the sample of respondents.

[4] Chi-square Test establishes whether or not an observed frequency distribution differs from a theoretical distribution (Aaron, 2005).

[5] P-value is the probability of obtaining a test statistic at least as extreme as the one that was actually observed, assuming that the null hypothesis is true (Aaron, 2005).

[6] Log-likelihood Test compares the fit of two models, one of which (the null model) is a special case of the other (the alternative model) (Aaron, 2005).

[7] R-square represents the proportion of variability in a data set that is accounted for by the statistical model (Aaron, 2005).

information presents that the Chi-square statistic is 114.136 with a p-value of 0.000, and a log-likelihood value of 182.641, which implies the existence of a relationship between the independent variables (plumbing material attributes) and the dependent variable (plumbing material selection) is supported. The goodness-of-fit measure is also employed, and the Nagelkerke's R-square is 0.084, which implies that 8% of variation in the dependent variable is explained by the variation in the independent variables.

In evaluating the Ordered Logit model, threshold represents the response variable in the regression. A different intercept is provided for the different levels of the cumulative logit model. The beta coefficient of the independent variables does not change, and the value of each is subtracted from the intercept. Each threshold level indicates the logit of the odds of being equal to or less than the baseline category when all independent variables are zero (Aaron, 2005). The baseline group is set to 'Extremely Preferred'. The beta estimate represents that a one unit increase in the independent variable increases / decreases the log-odds of being higher than a specific preferred valuation category. Because the beta coefficient is not indexed by each category, a one unit increase affects the log-odds the same regardless of which threshold value is considered (Aaron, 2005).

As represented in Table 10, the regression estimates reveal that when compared to the baseline category ('Extremely Preferred'), the categories 'Moderately Preferred', 'Strongly Preferred', and 'Very Strongly Preferred' have higher threshold estimates. A category 'Not Preferred' has a statistically insignificant negative coefficient estimate. Since the estimate is not statistically significant at the 95% confidence interval, it is not included in comparison analysis between the categories.

The threshold values are also evaluated. These values inform the expected cumulative distribution of categorical preference values for individuals with the independent variables set to zero (Aaron, 2005). This threshold represents a natural tendency for all the responses to all the scenarios presented to respondents when the independent variables are suppressed. When these coefficients are exponentiated, the cumulative odds for each category are obtained (Table 12). By employing the following equation, (odds /(1+odds)), the cumulative probabilities are computed (Aaron, 2005). Table 10 represents the odds ratios and cumulative probabilities (columns 3 and 4 in Table 10). For example, the 'Moderately Preferred" category is 3.7 times more likely to be selected by the respondent compared to the 'Extremely Preferred' category when all independent variables are set to zero.

The independent variable coefficient estimates are statistically significant only for two attribute levels: risk of corrosion variable represented by 'corrosion proof' attribute level and convenience of installation represented by 'no need to tear into the wall and/or floor. Installation takes around 4 days' (Table 10). Other independent variables were considered redundant in the model estimation. The independent variable coefficients represent how the log-odds of these thresholds increase / decrease with one unit of the independent variable. The positive value indicates that one unit of independent variable increases the odds of being in a higher category (Aaron, 2005). For example, the 'corrosion proof' attribute level increases the odds of choosing a higher preference category by 0.654 compared to the independent variable represented by 'some risk of corrosion' attribute level. 'Installation of plumbing material taking about 4 days' increases the odds of choosing a higher preference category by 0.559 compared to 'the installation taking between 7 and 9 days.'

Besides evaluating the directional impact of the independent variables on the preference level of the households, the impact of the statistically significant independent variables on the preference category is evaluated for all three plumbing materials. As the attribute levels describing each of the three hypothetical materials are known, the regression results can be organized by plumbing materials. For example, Material A is described by attribute level called 'corrosion proof' as well as 'installation takes around 4 days'. The coefficient estimates for the statistically significant attribute levels are employed to compute preference valuation categories for each material type. In case of the Material A (epoxy coating) computation of the preference valuation category called 'Moderately Preferred', the following represents the estimate computation: 1.315 - 0.654 - 0.559= 0.102, where 1.315 is the moderately preferred coefficient, 0.654 is 'the corrosion proof' coefficient, and 0.559 is 'the convenience of installation' coefficient; and the odds ratio computation: exp(0.102) = 1.107.

Variable Name	Coefficient Estimate[b]	Standard Error[c]	Wald-Stats[d]	P-Value[d]
Threshold Values (For All Independent Variables Set to Zero)				
Not Preferred	-0.096	0.108	0.790	0.374
Moderately Preferred	1.315	0.115	131.554	0.000
Strongly Preferred	2.289	0.125	333.413	0.000
Very Strongly Preferred	3.742	0.164	521.510	0.000
Independent Variables (Variables that Improve Overall Model Significance)[e]				
1) Corrosion proof	0.654	0.143	20.943	0.000
2) No need to tear into some sections of wall for installation. Installation takes around 4 days.	0.559	0.119	22.016	0.000

[a]The number of observations included in the model is 1086. Independent variables take form of dummy variables with value of one when the characteristic was present in the plumbing material profile and zero otherwise. To avoid a dummy variable trap, one of the attribute levels was excluded from the analysis. The omitted characteristics represent Material C (copper) descriptions.
[b] Coefficient estimates show how much increase in the likelihood of being in a higher category results from a one unit increase in the independent variable.
[c] Standard error represents the variation of the estimate.
d Wald statistics and p-value represent the significance level.
[e] Model Statistics: Log-likelihood value is 182.641 with chi-square of 114.136 and p-value of 0.000; Nagelkerke's R-square is 0.084.

Table 10. Ordered logit regression estimates with categorical answers (dependent variable represents the plumbing material valuation and the independent variables represent the plumbing material attributes (without socioeconomic variables))[a].

When further investigating the Ordered Logit results, the coefficient for each preference category in combination with the coefficients for each independent variable can be expressed as marginal probability estimates to provide a greater insight into the preferred plumbing material (Table 11). Based on the marginal distribution of the probability estimates, Material A has a larger probability estimate for 'Strongly Preferred' to 'Extremely Preferred' category preference. On the other hand, Material C has a higher probability estimates for categories 'Not Preferred' and 'Moderately Preferred'. All three materials have the highest frequency of estimates falling into 'Not Preferred' and 'Moderately Preferred' categories. Based on the overall results, Material A (epoxy coating) is the most preferred material followed by Material B (plastic). Material C (copper) is the least preferred plumbing material.

	Material A	Material B	Material C
	Coefficient Estimates[a]		
Not Preferred	-1.309	-0.750	-0.096
Moderately Preferred	0.102	0.661	1.315
Strongly Preferred	1.076	1.635	2.289
Very Strongly Preferred	2.529	3.088	3.742
Extremely Preferred			
	Odds Ratio Estimates		
Not Preferred	0.270	0.472	0.908
Moderately Preferred	1.107	1.937	3.725
Strongly Preferred	2.933	5.129	9.865
Very Strongly Preferred	12.541	21.933	42.182
Extremely Preferred			
	Marginal Probability Estimates Distribution		
Not Preferred	0.213	0.321	0.476
Moderately Preferred	0.313	0.339	0.312
Strongly Preferred	0.220	0.177	0.120
Very Strongly Preferred	0.180	0.120	0.069
Extremely Preferred	0.074	0.044	0.023

[a]Coefficient estimates are built up from the statistically significant estimates for the attribute levels and threshold values. Coefficients are compared to the base "Extremely Preferred" level.

Table 11. Ordered logit regression results' analysis by plumbing material type (dependent variable represents the plumbing material valuation and the independent variables represent the plumbing material attributes (no socioeconomic variables)).

6.2.2 Order logit model with socioeconomics variables

The second specification of the Ordered Logit model includes the socioeconomic variables alongside of the attributes for plumbing material. As the socioeconomic characteristics do not vary for a given respondent, they should be interacted with the attributes levels of each attribute. As the total number of respondents is rather small (230), there are not enough degrees of freedom to include all interaction variables between the attribute levels and the household characteristics. As a result, the Ordered Logit model was first estimated with socioeconomic variables entering one at a time to measure the impact of household characteristics on the plumbing material preferences. The statistically significant interaction variables were then included in the final model estimation.

When the socioeconomic variables were entered in the Ordered Logit model one at a time, 'corrosion proof' as well as 'installation takes about 4 days' were the two attribute levels appearing statistically significant in many of the model specifications. The coefficient value for corrosion attribute varied from 0.651 to 1.450, and the convenience of installation coefficient varied from 0.554 to 0.754. The only statistically significant interaction effect was observed between attribute level of 'corrosion proof' and respondent's 'previous cost of plumbing materials repairs or replacement' (coefficient estimate = 0.00001; standard error = 0.00000; Wald-statistic[8] = 15.773; p-value = 0.000). This interaction effect was entered into the final model estimation alongside of other plumbing material attributes.

The threshold values, which inform the expected cumulative distribution of categorical preference values for individuals with the independent variables set to zero, are evaluated (Aaron, 2005). Table 12 represents the odds ratios and probabilities. For example, the 'Moderately Preferred" category is 3.67 times more likely to be selected by the respondent than the 'Extremely Preferred' category when all independent variables are set to zero. On the other hand, the 'Not Preferred' category is only 0.86 times as likely to occur compared to the baseline category when no independent variables are considered.

Based on Table 12, the independent variable coefficient estimates are statistically significant only for two attribute levels: 'corrosion proof' and 'installation takes about 4 days'. For example, the 'corrosion proof' variable increases the odds by 1.145 of choosing a higher preference category compared to the variable set at 'some risk of corrosion'. 'Installation of plumbing material taking about 4 days' increases the odds of choosing a higher preference category by 0.575 compared to 'the installation taking between 7 and 9 days'. The only socioeconomic variable entered into the regression is the respondent's previous cost of plumbing repairs and/or material fixing or replacement and is statistically significant when interacted with corrosion proof attribute level. The joint coefficient is 1.197 (1.145+0.0001*$522[9]) and is statistically significant at 5% significance level[10]. This coefficient value further implies that the interaction variable increases the odds by 1.197 of choosing a higher preference category compared to the variable set at 'some risk of corrosion'. This finding can be explained as households, who have accrued cost of plumbing material repairs in the past, value the 'corrosion proof' attribute level more compared to the 'some risk of corrosion' attribute level. Plumbing material with low corrosion risk would imply decrease in the future costs of plumbing material repairs.

As in the previous version of the Ordered Logit model, effects of statistically significant independent variables on the preference category for all three plumbing materials are evaluated. The statistically significant attribute levels were computed together with the thresholds levels by plumbing material into odds ratios and probability values. As attribute levels describing each of the three hypothetical materials are known, the regression results can be organized by plumbing materials. For example, Material A is described by attribute

[8] Wald Test is used to test the true value of the parameter based on the sample estimate (Aaron, 2005).

[9] $522 is the mean cost value of the previous cost spent on plumbing material repairs and replacement.

[10] Cost of Plumbing Material Fixing or Replacement * Corrosion Proof: Wald statistic = 5.684 and p-value = 0.020.

level called 'corrosion proof' and 'installation takes about 4 days'. The coefficient estimates for the statistically significant attribute levels are employed in the material based preference category computation. In case of Material A (epoxy coating) the computation for preference valuation category of 'Moderately Preferred', the following represents the estimate computation: 1.300- 1.145- 0.575 -0.0001*$522 = -0.472; and the odds ratio computation: exp(-0.472) = 0.624 (Table 15).

Variable Name	Coefficient Estimate[b]	Standard Error[c]	Wald-Stats[d]	P-Value[d]
Threshold Values (For All Independent Variables Set to Zero)				
Not Preferred	-0.147	0.089	2.705	0.100
Moderately Preferred	1.300	0.098	176.801	0.000
Strongly Preferred	2.317	0.114	415.544	0.000
Very Strongly Preferred	3.790	0.164	532.389	0.000
Independent Variables for Model Specification with Socioeconomic Variable Interactions[e and f]				
Corrosion Proof	1.145	0.502	5.190	0.023
Need to tear into some sections of wall for installation. Installation takes around 4 days.	0.575	0.134	18.331	0.000
Respondent's previous cost of plumbing repairs and/or replacement * Corrosion Proof	0.0001	0.00006	4.644	0.031

[a] The number of observations included in the model is 1072. Independent variables take form of dummy variables with value of one when the characteristic was present in the plumbing material profile and zero otherwise. To avoid a dummy variable trap, one of the attribute levels was excluded from the analysis. The omitted characteristics represent Material C (copper) descriptions.
[b] Coefficient estimates show how much increase in the likelihood of being in a higher category results from a one unit increase in the independent variable.
[c] Standard error represents the variation of the estimate.
[d] Wald statistics and p-value represent the significance level.
[e] Model Statistics: Log-likelihood value is 1565.522 with chi-square of 119.384 and p-value of 0.000; Nagelkerke's R-square is 0.101.

Table 12. Ordered logit regression estimates with categorical answers (dependent variable represents the plumbing material valuation and the independent variables represent the plumbing material attributes and socioeconomic variables interacted with attribute levels)[a].

As presented in Table 13, Material A has the lowest values of estimates for all preference categories, compared to Materials B and C. Material C has the highest values of preference valuation. Threshold values with smaller absolute values imply smaller differences between preference valuation categories and the base category in the likelihood of that preference category being selected. For example, Material B has a smaller absolute threshold value compared to Material A for the "Not Preferred" category, implying a smaller difference between 'Not Preferred' and 'Extremely Preferred' for Material B (-1.344) compared to Material A (-1.919).

Material C has the highest values of odds ratios for each preference category while Material A has the lowest. The odds ratios that present the likelihood of a preference category being selected are compared to the base category. For example, the category 'Strongly Preferred' is 10.145 times as likely to be selected as the base category for Material C while for Material A it is only 1.724 times as likely. A lower odds ratio for each preference category is more preferred, as it implies that the 'Extremely Preferred' category has a higher chance of being chosen relative to other categories. This finding implies that Material A is a more preferred home plumbing choice for households.

Following further analysis of the marginal distribution probability estimates, Material A has a larger probability estimate for 'Strongly Preferred' to 'Extremely Preferred' category preference. On the other hand, Material C has higher probability estimates for category 'Not Preferred'. Based on these results, Material A (epoxy coating) is again the most preferred material followed by Material B (plastic). Material C (copper) as previously found is the least preferred plumbing material.

	Material A	Material B	Material C
Coefficient Estimates[a]			
Not Preferred	-1.919	-1.344	-0.147
Moderately Preferred	-0.472	0.103	1.300
Strongly Preferred	0.545	1.120	2.317
Very Strongly Preferred	2.018	2.593	3.790
Extremely Preferred			
	Odds Ratio Estimates		
Not Preferred	0.147	0.261	0.863
Moderately Preferred	0.624	1.108	3.669
Strongly Preferred	1.724	3.064	10.145
Very Strongly Preferred	7.522	13.367	44.256
Extremely Preferred			
	Distribution Estimates		
Not Preferred	0.128	0.207	0.463
Moderately Preferred	0.256	0.319	0.323
Strongly Preferred	0.249	0.228	0.124
Very Strongly Preferred	0.250	0.176	0.068
Extremely Preferred	0.117	0.070	0.022

aCoefficient estimates are built up from the statistically significant estimates for the attribute levels and threshold values. Coefficients are compared to the base "Extremely Preferred" level.

Table 13. Ordered logit regression results' analysis by plumbing material type (dependent variable represents the plumbing material valuation and the independent variables represent the plumbing material attributes and the socioeconomic characteristics).

As in the previous model specifications, Material A is the most preferred plumbing material when the CA data is estimated, employing an Ordered Logit Model with and without socioeconomic characteristics. Material C is the least preferred plumbing material. Two plumbing material attributes are important in making the decision on type of pipes to be installed in a house: 'plumbing material installation time' and 'corrosion risk'. The regression coefficients as well as the computed odds ratios and probability estimates differ between the model specification with and without the socioeconomic variables.

For example, for Material A, the odds ratios are lower for all preference categories in the case of model specification with socioeconomic variables, category 'Very Strongly Preferred' has odds ratios ranging from 9.034 to 14.083 for model without socioeconomic variables and 7.522 for model including socioeconomic variables. This finding implies that the socioeconomic variables impact the discrimination level between the plumbing material preference valuations. For example, if a household has experienced previous cost of plumbing repairs and/or replacement, their preference valuation level is lower for a more corrosion prone plumbing material compared to material with an attribute level of 'corrosion proof'.

The marginal distribution of probability estimates (Table 13) has higher values for lower preference categories for Material C in the case of model specification without socioeconomic variables. For example, for Material C, 'Not Preferred' has probability distribution estimate ranging between 0.476 compared to 0.463 (with socioeconomic variables). The marginal distribution estimates for higher preference valuation categories are lower for Material A and B for model without socioeconomic variables. For example, for Material A, 'Extremely Preferred' has a probability distribution estimate ranging from 0.074 (without socioeconomic) compared to 0.117 (with socioeconomic variables). As a result, the inclusion of socioeconomic variables raises the level of preference for Materials A and B, while it decreases the level of preference for Material C.

In conclusion, although the inclusion of socioeconomic variables does not change the final preference ranking of the plumbing materials, it increases the estimated level of preference for Material A (epoxy coating) and Material B (plastic) by increasing the marginal probability distribution of estimates for the higher preference categories (i.e. 'Strongly Preferred'). The increase is the most pronounced in the case of Material A (model with socioeconomic variables) for which the 'Extremely Preferred' category has a probability distribution estimate almost twice as large compared to the model specification without socioeconomic variables (0.117 vs. 0.074). The respondent's previous cost of plumbing material repairs and replacement impacts positively the preference level for plumbing materials described by 'corrosion proof' attribute level. This finding implies that Materials A and B are more highly preferred when socioeconomic factors are taken into consideration. Households experiencing high costs of fixing corrosion related damage in the past are more likely to prefer and choose materials with lower corrosion levels. The decreased corrosion level implies lower future plumbing material failures, and therefore, lower costs associated with repairs of water-related damage.

7. Conclusions and discussion

Due to the fact that homeowners have an important stake in finding plumbing systems appropriate for their households, they should not only rely on expert advice, but also acquire information on plumbing material attributes such as price, health impact, longevity, and corrosion resistance in order to make informed investment decisions about plumbing systems for their homes. Information on consumer preferences for drinking water plumbing attributes can be useful not only to individual households, but also to policymakers, program managers, water utilities, and firms with interests in drinking water infrastructure.

This chapter addressed the issues of household plumbing material decisions. The information was elicited by two surveys of residents residing in a Southeastern Community in the U.S. The first survey elicited information on the prevalence of pinhole leaks and other plumbing material failures, households' experiences with plumbing material failures, the cost of repairs and property damages due to the material failures, and household preferences for corrosion preventive measures. The follow-up survey, sent only to those residents who agreed to participate in future studies related to the plumbing material issues, elicited information on households' preferences for a set of hypothetical plumbing materials.

Overall, the Southeastern Community survey revealed high level of awareness of pinhole leak problem among residents of the community. Twenty percent of the households reported actual pinhole leak incidents. The percent of pinhole leak reports was on par with other hotspot areas of corrosion in the U.S., but above the rate of pinhole leak occurrences in non-hotspots (Scardina et al., 2007). The pinhole leak problem was more prevalent in houses built before the 1990s with copper pipes installed as the plumbing system. This finding is in an agreement with a Maryland Pinhole Leak Survey conducted by Kleczyk and Bosch in 2004.

The total repair expenses due to the pinhole leaks varied between \$100 and \$5,000 with several reports of more than \$5,000 in repairs. Similar results were found by Kleczyk et al. (2006) of selected communities in the East, Southeast, Midwest, and West regions. Over 50% of surveyed respondents spent more than \$100 on repairs with estimates as high as \$12,000. In comparison, in their Maryland Pinhole Leak Survey, Kleczyk and Bosch (2008) found costs from the plumbing material failure repairs as high as \$25,000. Unlike the present survey, however, the study by Kleczyk and Bosch (2008) did not separate the costs associated with pipe failure and property damage. This Southeastern Community survey accounted for this factor, which might have resulted in the differences between the two studies. Furthermore, many households in the Southeastern Community cited using a preventive measure against corrosion, including whole house re-plumbing and installation of water softeners. Over 80% of residents of the Southeastern Community were satisfied with the water quality in their homes.

The follow-up survey data of residents in the Southeastern Community revealed that among three hypothetical plumbing materials (A, B, and C), the households preferred Material A (epoxy coating) followed by Material B (plastic). Material C (copper) was the least preferred material in the set. This result was derived based on each of the respondents' preference evaluation of the different plumbing material groupings. The preference ranking of the materials was the same across both Ordered Logit model specifications (with and without socioeconomics variables). Furthermore, the results were in agreement with the survey baseline method, which ranked Material A as the most preferred and Material C as the least preferred. The baseline ranking of plumbing materials was obtained from households' comparisons of all three plumbing materials at the same time.

The plumbing material attributes that were important in the decision-making process included: 'corrosion risk' and 'time length of plumbing material installation.' In both cases, the attribute level rankings were in agreement with the transitivity assumption of preferences, and the lower corrosion risk attribute level, as well as shorter amount of time required for plumbing material installation was more preferred to the more corrosion risk prone and longer installation period attribute levels.

Only one socioeconomic variable had a statistically significant impact on the chosen plumbing material: 'cost of plumbing material repairs and replacement incurred by the respondent.' This variable was statistically influential when interacted with corrosion attribute levels. Although it did not change the preferences for plumbing materials, the variable skewed the preference valuations favorably towards plumbing materials described by 'corrosion proof' attribute level. This finding implies that the more each household had previously spent on repairs associated with plumbing material failures, the more they preferred a plumbing material with lower corrosion level to avoid future expenditures on drinking water system repairs.

There are several implications for further research that would improve the analysis of preferences for plumbing materials. The information set of plumbing material attributes might not have been the most complete and objective description of the pipe characteristics. Households with copper plumbing materials installed in their houses were more likely to identify Material C as copper (as noted on their questionnaires returned to the researchers), and therefore, might have evaluated it based on their experiences and not based on the comparison with other plumbing materials. This finding, however, is not unexpected, as part of the research question was to examine the impact of previous experiences with plumbing material failures on household decisions for corrosion prevention and plumbing material choices. Furthermore, in his AHP study, Lee (2008) noticed that some of the householders in this community provided a high degree of preference for a specific plumbing material in the survey, but in reality installed other types in their homes (Lee, 2008). As a result, in some cases, there is a mismatch between the stated preferences derived based on the homeowners' survey and the actual behavior exhibited by the households.

The above survey results inform policy makers, utility managers, and home plumbing systems producers on the homeowners' preferences for plumbing materials, and the trade-offs between the risk of corrosion and cost of a leak-free environment based on their experiences with pipe failures in the past. The cost of alternative preventive measures, corrosion risk, and convenience of plumbing material installation drive the decisions of homeowners regarding their plumbing system. As a result, policy makers should take into consideration the implications of new federal and state regulations on the interactions between drinking water and drinking water plumbing. Furthermore, their regulations and standards should accurately test the different types of plumbing materials used in the drinking water infrastructure, as well as their chemical and physical interactions with chemicals used to treat drinking water.

For example, Edwards et al. (2004) suggested that removal of natural organic matter mandated by tighter EPA drinking water standards contributed to the pinhole leak problem in combination with other factors, including faulty installation, since natural organic matter is an inhibitor to the corrosion-inducing chemical reactions. To deal with this problem, Bosch et al. (2006) found that almost 60% of water utilities added corrosion inhibitors, such as phosphate to water treatment. The inhibitors were added to protect water service lines, to comply with the lead and copper rule proposed by EPA, and to give protection to residential customers. Similarly, after adding phosphate to the water treatment process by utility companies who distribute water to the Southeastern Community, the Southeastern Community reported a decrease in the number of pinhole leak reports (Scardina & Edwards, 2007).

Furthermore, the cost associated with employment of different prevention options as well as the convenience of installation has an impact on households' decisions, concerning choosing a plumbing material for their houses. As a result, service providers (i.e. plumbers and material manufactures) should be sensitive to households' financial constraints and convenience of plumbing installation for homeowners. For example, 33% of Southern Community respondents with pinhole leaks spent at least $500 repairing damaged plumbing material, while more than 75% of survey participants with pinhole leaks experienced at least moderate level of stress. In their Maryland study of pinhole leak corrosion, Kleczyk and Bosch (2008) estimated the total cost[11] of fixing damage related to pinhole leaks to range from roughly $1,300 to more than $18,000. As a result, when plumbing services are expensive, the service providers should concentrate on installing plumbing materials that are convenient to install, and present a low failure rate to minimize future financial outlays spent on plumbing material repairs.

Finally, water professionals and policy makers should work on public policy that would address public preferences for drinking water infrastructure. Results of this Southeastern Community analysis can provide information to policy experts and water utility managers who are dealing with extensive corrosion problems in their areas. Information will fill the gaps of knowledge about corrosion occurrences, the financial impact of plumbing material repairs on households, and households' preferences for drinking water infrastructure, as well as the ability of householders to pay for different corrosion prevention options.

8. Acknowledgements

The authors would like to acknowledge the financial support provided by the National Science Foundation under the grant DMI-0329474 and the American Water Works Association Research Foundation under the project #3015. The views expressed in this report are those of the authors, and not of the National Science Foundation nor of the American Water Works Association Research Foundation. In addition, the authors would like to thank James R. Strout, the Book Review Board, as well as InTech Editors for providing comments, and editing earlier versions of this chapter.

[11] Total cost of repairing pinhole leak damage includes the financial and time costs.

9. References

Aaron. G. (2005). *Ordered Logit Model*, Available online at: http://www.uoregon.edu/~arrong/teaching/G4075_Outline/node27.html, Accessed: June 2011.

Agency for Toxic Substances and Disease Registry (ATSDR). (2004). *Toxicological Profile for Copper*, Available Online at: http://www.atsdr.cdc.gov/toxprofiles/tp132.html, Accessed: June 2011.

Bosch, D., Kleczyk, E., Lee, J., & Tanellari, E. (2008). *Southeastern Community Survey Report*, Department of Agricultural and Applied Economics, Virginia Tech, Blacksburg, VA.

Champ, P., Boyle, K., & Brown, T. (2003). *A Primer on Nonmarket Valuation*, Boston: Kluwer Academic Publishers, IBSN 0 792-3649-88.

Dietrich, A., T., Heim, H., Johnson, Y., Zahng, M., Edwards, G. V., Loganathan, et al. (July 2006). *Plumbing Materials: Costs, Impacts on Drinking Water Quality, and Consumer Willingness to Pay*, Proceedings of 2006 NSF Design, Service, and Manufacturing Grantees Conference, St. Louis, Missouri, Available online at http://www.dmigranteeconference.org/paper.htm, Accessed: June 2011.

Dillman, D. A. (1978). *Mail and Telephone Surveys*, New York: John Wiley & Sons, IBSN 0471-3235-43.

Durand, D., & Dietrich, A. (2007). Contributions of Silane Cross-Linked PEX Pipe to Chemical/Solvent Odors in Drinking Water, *Water Science &and Technology 55*(5), pp. 153-160, ISSN 0273-1223.

Edner, P. (2005). *Applied Categorical and Nonnormal Data Analysis: Ordered Logit and Probit Models, Education 231C*, Available online at: http://www.gseis.ucla.edu/courses/ed231c/notes2/ologit.html, Accessed: August 2011.

Edwards, M. (2004). Corrosion Control in Water Distribution Systems, One of the Grand Engineering Challenges for the 21st Century, Edited by Simon Parsons, Richard Stuetz, Bruce Jefferson and Marc Edwards, *Water Science and Technology 49*(2), pp. 1-8, ISSN 0273-1223.

Environmental Protection Agency (EPA) (2006). *Groundwater and Drinking Water Consumer Fact Sheet on Copper*, Available Online at http://www.epa.gov/safewater/contaminants/dw_contamfs/copper.html, Accessed: July 2011.

Farooqi, O., & Lee, J. (2005). *Plumber Telephone Surveys*, Virginia Tech, Blacksburg, VA.

Heim, T., & Dietrich, A. (2007). Sensory Aspects and Water Quality Impacts of Chlorinated and Chloraminated Drinking Water in Contact with HDPE and CPVC Pipe, *Water Research 55*(5), pp. 757 -764, ISSN 0043-1354.

Kleczyk, E., & D. Bosch. (December 2008). Incidence and Costs of Home Plumbing Corrosion, *Journal of American Water Works Association 100*(2), pp. 122-133, ISSN 1551-8833.

Kleczyk, E. J., Tanellari, E., & Bosch, D. J. (November 2006). *Corrosion in Home Drinking Water Infrastructure: Assessment of Causal Factors, Costs, and Willingness to Pay*, 2006 Water Quality Technology Conference, American Water Works Association, Denver, Colorado, Available Online at: http://www.techstreet.com/cgi-bin/detail?product_id=1320028, Accessed: July 2011.

Lee, J. (2008). *Two Issues in Premier Plumbing Contaminants Intrusion at Service Line and Choosing Alternative Plumbing Material*, Doctoral Dissertation, Virginia Polytechnic Institute and State University.

Lee, J., Loganathan, G. V., Bosch, D., Dwyer, S., & Kleczyk, E. (October 2005). *Preference Analysis of Home Plumbing Material*, Virginia Water Resources Research Center, National Water Research Symposium: Balancing water law and science, The Inn at Virginia Tech and Skelton Conference Center, Virginia Tech, Blacksburg, Virginia.

Lee, J., Kleczyk, E., Bosch, D., Tanellari, E., Dwyer, S., & Dietrich, A. (July / August 2009). Case Study: Preference Trade-offs Towards Home Plumbing Attributes and Materials, *Water Resource Planning Management Journal 135*(4), Special Edition in Memory of Dr. G.V. Loganathan, pp. 237-243, ISSN 0733-9496.

Loganathan, G.V. & Lee, J. (2005). Decision Tool for Optimal Replacement of Plumbing Systems, *Civil Engineering and Environmental Systems 22*(4), pp. 189-204, ISSN 1028-6608.

Marshutz, S. (2000). Hooked on Copper. *Reeves Journal*, Available Online at: http://www.reevesjournal.com/CDA/ArticleInformation/features/Features_Index/1,3816,27-820,00.html, Accessed: June 2011.

National Science Foundation (NSF). (2008). *NSF Standard Accepts New Stainless Steel Materials in Drinking Water Applications*, Available Online at: http://www.nsf.org/business/newsroom/press_release.asp?p_id=12241, Accessed: August 2011.

Frustrated by Pinhole Leaks in Their Copper Plumbing, Homeowners Find Relieve with PEX. (July 2007). *Plumbing and Mechanical Magazine 25*(5), pp. 19, ISSN 8750-6041.

PRNews Wire. (2004). Available Online at: http://www.prnewswire.com/cgibin/stories.pl?ACCT=109&andSTORY=/www/story/11-18-2004/0002464315ENDDATE, Accessed: July 2011.

Roberge, P. R. (2000). Searching the Web for Corrosion Intelligence, *Corrosion Reviews 18*(1), pp. 23-40, ISSN 0048-7538.

Scardina, P., Edwards, M., Bosch, D. J., Loganathan, G. V., & Dwyer, S. K. (2007). *Non-Uniform Corrosion in Copper Piping – Assessment*, Final Project Completion Report to American Water Works Association Research Foundation, Blacksburg, Virginia: Virginia Tech.

Scardina, P., & Edwards, M. (2007). *Preliminary Investigation of Copper Pipe Failures*, Report submitted to the Southeastern Community.

Toolbase News (2008). Available Online at: http://www.toolbase.org/pdf/techinv/homerunplumbingsystems_techspec.pdf, Accessed: August 2011.

Train, K. (2003). *Discrete Choice Methods with Simulation.* Cambridge University Press, Available online at: http://elsa.berkeley.edu/books, Accessed: June 2011.

Varian, H. R. (1992). *Microeconomics Analysis.* (3rd ed.), New York: W.W Norton and Company, IBSN 1740-37 18-4 4.

Woodson, R.D. (1999). *Plumber's Standard Handbook.* New York: McGraw-Hill, IBSN 0071-3438-65.

20

Renewable Resources in Corrosion Resistance

Eram Sharmin, Sharif Ahmad and Fahmina Zafar
Department of Chemistry, Jamia Millia Islamia (A Central University), New Delhi, India

1. Introduction

Corrosion of metals or alloys occurs due to chemical or electrochemical reactions with their environment, which often results in drastic deterioration in the properties of metals or materials comprising thereof. Corrosion takes place on a steel surface, due to the development of anodic and cathodic areas, through oxidation and reduction reactions, forming of oxides of metals alloys. There are several corrosion causing agents or "corrodents" such as soot, sulphate salts, chloride ions, temperature, salinity, pH, dissolved gases, humidity, bacteria, sand, gravels, stones, mechanical stresses and also several protection methods employed for corrosion resistance such as the application of alloys, composites, inhibitors, cathodic and anodic protection, protective linings and coatings (Bierwagen, 1996; Ghali et al., 2007; Raja& Sethuraman, 2008; Sorensen et al.,2009). Notwithstanding, corrosion has become a gigantic problem today for every nation. The colossal detrimental impact of corrosion on the economy of a country can be manifested in billions of dollars spent annually to combat or control it.

In the past two decades, research and development efforts in the field have undergone vast changes globally, because of the everyday growing consumer expectations of good quality and performance coupled with lower cost, enormous hikes in the prices of petro-based chemicals out of fear of depleting stocks by the end of twenty first century, serious concerns pertaining to energy consumption and environmental contamination, regulations such as Clean Air Act Ammendments [CAAA, 1990], and above all the "cost of corrosion". These predictions, regulations and innovations have posed constant threats and challenges for anticorrosion industry forcing to change its gears worldwide. The corrosion chemists, researchers and engineers in industry and academics are actively engaged to explore and formulate new strategies to meet the mandatory limits of performance, cost and legislations. The ultimate solution is foreseen through the "excessive utilisation of our naturally available resources" primarily, to cut off the escalating prices of raw materials, to formulate environmentally benign materials, to expedite their post-service degradation, and to add value to a waste material. Consequently "environmentally friendly" or "green" coating technologies (waterborne [WB], powder, high-solid, hyperbranched and radiation-curable) have evolved, with special emphasis being laid on the excessive utilization of naturally available renewable resources thriving on acres of our agricultural lands. These may be formulated as corrosion resistant alloys, corrosion resistant composites, corrosion resistant pigments, corrosion resistant coatings, paints and corrosion inhibitors. Renewable resources provide cheaper and abundant biological feedstocks with numerous advantages, such as cost effectiveness, low toxicity, inherent biodegradability and environment friendliness They yield versatile materials through chemical transformations with plethora of applications, particularly in corrosion resistance against various corrodents [Fig. 1]. (Derksen et al., 1995, 1996; Gandini & Belgacem, 2002; Metzgr, 2001; Weiss, 1997; Ahmad, 2007).

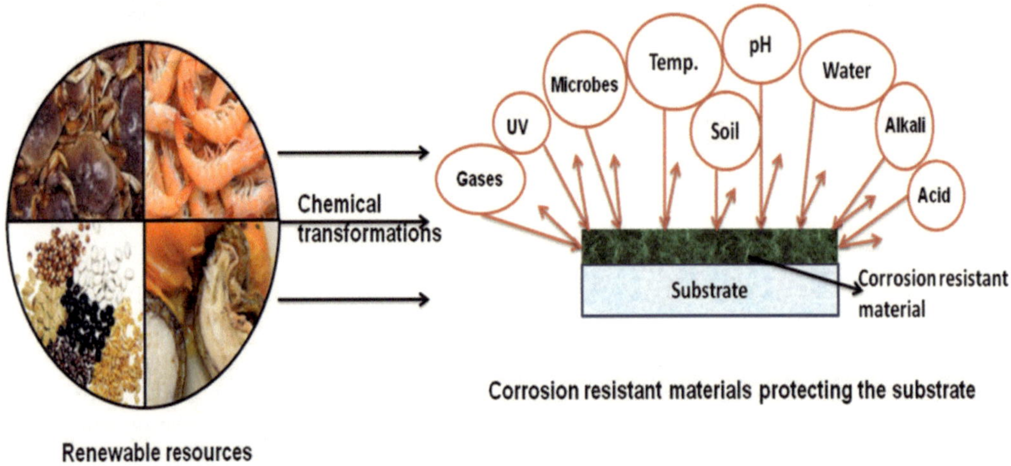

Fig. 1. Renewable resource based materials provide corrosion resistance against various corrodents.

2. Renewable resources in corrosion resistance

Corrosion generally occurs when mild steel comes in contact with oxygen and water. The presence of anodic and cathodic sites on steel surface and their reaction with water and oxygen transforms metal (iron) atom to ions, finally through a series of chemical reactions, hydrated ferric oxide forms (iron) rust. Another anaerobic (without oxygen) corrosion, micro-biological corrosion may occur if conditions favor the growth and multiplication of microbes, i.e., bacteria and fungi (Witte et al., 2006). The preliminary steps to reduce, combat or completely eradicate corrosion require the elimination or suppression of such chemical reactions by the use of corrosion inhibitors, pigments, cathodic protection, coatings and others, providing barrier properties, adhesion between substrate and coatings, corrosion reducing activity and overall an active anticorrosion effect. The effectiveness of coatings as potential anticorrosion agents depends upon their type, the type of substrate, corrodents to which these are exposed and others. For efficient service, coatings should bear very good adhesion to the substrate resulting in low permeability (to oxygen, water) and good "wet" adhesion. The renewable resources or natural biopolymers such as lignin, starch, cellulose, cashewnut shell liquid, rice husk, sucrose, caffeic acid, lactic acid, tannic acid, furan, proteins, glycerol, and vegetable oils contain hydroxyls, aldehydes, ketones, carboxyls, double bonds, ester, ether and other functional groups. These functional groups impart good adhesion and corrosion resistance performance to the substrate. Also, the performance can be further improved by chemical transformations, use of modifiers (inorganic reinforcements, nanomaterials) and other methods.

The proceeding sections provide a brief description of some natural biopolymers and their utilisation in corrosion resistance.

2.1 Cellulose

Cellulose is the largest biopolymer obtained by photosynthesis. It is a crystalline polysaccharide. It is a linear long chain polymer of β(1→4) linked D-glucose units (5,000-10,000), that condense through β(1→4)-glycosidic bonds (Fig. 2). It is mainly obtained from wood pulp and other plants but can also be extracted from algae and bacteria for industrial purposes. Cellulose and their derivativs are used in paper, paperboard, card stock, textiles, cellophane, smokeless gunpowder, pharmaceuticals, biofuels, foods, sponges, cosmetics, reinforced plastics, water-soluble adhesives, binders and coatings.

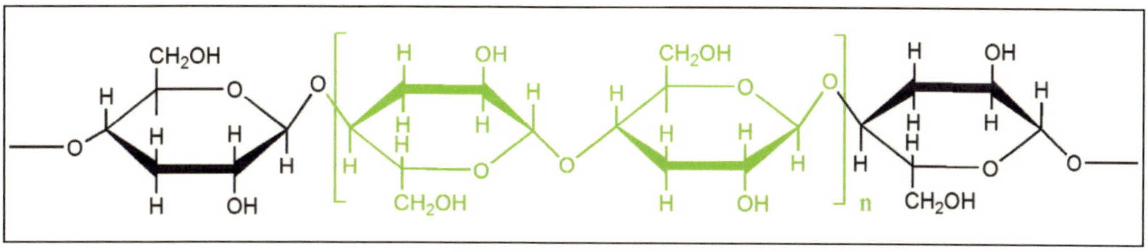

Fig. 2. Structure of cellulose.

Use in corrosion resistance

Cellulose is crystalline in nature. In desirable quantities, it may be used as a modifier rendering toughness in fragile coatings. The primary hydroxyl groups present in the chain may further facilitate adhesion to the substrate. Hydrophoebically modified hydroxyethyl cellulose used in WB coatings and paints provided good gloss, levelling and sag resistance (Kroon 1993). Films obtained from regenerated cellulose (from cotton linter) by coating Castor oil polyurethane/benzyl konjac glucomannan semi-interpenetrating polymer networks were water resistant and biodegradable (Lu et al., 2004). Ethyl cellulose based aqueous dispersions and solvent based films were plasticized with *n*-alkenyl succinic anhydrides -2-octenyl succinic anhydride (OSA) and 2-dodecen-1-ylsuccinic anhydride to overcome the brittleness of cellulose films (Tarvainena et al., 2003). Films obtained showed excellent mechanical properties, low permeability, and good flexibility. Amoxicillin doped cellulose acetate films showed good corrosion resistance on AA2024-T3 substrate (Tamborim et al., 2011). Films doped with 2000ppm of the drug showed good anti-corrosion behavior as observed by Electrochemical Impedance Spectroscopy [EIS] results. These films showed lower current densities up to 3 days of immersion under anodic polarization. Scanning Vibrating Electrode Technique [SVET] results were found to be in close agreement with EIS and polarization results, also informing about the defects in coating. The results also showed a decrease of the electrochemical activity in the doped cellulose acetate films, relative to their undoped counterparts. Liu et al prepared cellulose acetate phthalate free films with diethyl phthalate/triethyl citrate as the plasticizer by spray method under heat-only (50°C for 24 h) and heat-humidity curing (50°C/75% RH for 24 h) conditions (Liu & Williams III, 2002). The latter (despite retaining higher content of plasticizer due to suppressed evaporation) provided increased mechanical strength and decreased water vapor permeability of the films. Triethyl acetate films showed increased % elongation, decreased tensile strength and elastic modulus relative to diethyl phthalate films, however, the latter showed low permeability.

2.2 Lignin

Lignin is the second most common organic polymer. About 50 million tons of lignin is produced worldwide annually as residue in paper production processes. It consists of methoxylated phenyl propane structures. The biosynthesis of complex structure of lignin is thought to involve the polymerization of three primary monomers, monolignols: p-coumaryl, coniferyl, and sinapyl alcohols (Figure 3), which are linked together by different ether and carbon-carbon bonds forming a three-dimensional network. The monolignols are present in the form of p-hydroxylphenol, guaiacyl and syringyl residues in lignin structure. Lignin is non-toxic, inexpensive and abundantly available (Sena-Martins et al.; 2008). It is hydrophoebic, smaller in size and forms stable mixtures (Park et al.; 2008). It is used in dye dispersants, dispersants for crop protection products, to produce low molecular weight chemicals like dimethyl sulphoxide. It is also used as filler in inks, varnishes and paints (Belgacem et al., 2003) and as a dispersing agent in concrete, as binders for wood composites, chelating agents, for treating porous materials, in coatings and paintings (Stewart , 2008; Park et al., 2008; Mulder et al., 2011).

Use in corrosion resistance

Lignin contains hydroxyl, carboxyl, benzyl alcohol, methoxyl, aldehydic and phenolic functional groups. It adsorbs on the metal surface and is capable of forming a barrier between the metal and corrodents (Altwaiq et al., 2011). Extracted alkali lignin as investigated by Altwaiq et al has shown corrosion inhibition behavior in the corrosion of different alloys immersed in HCl solutions. This was investigated by weight loss analysis, surface analysis on the corroded metals by scanning electron microscope (SEM), and micro-beam X-ray fluorescence (μ-XRF), inductively coupled plasma–optical emission spectroscope (ICPOES) and others (Altwaiq et al., 2011). Lignin doped conductive polymers [polyaniline-PANI] are used in corrosion protection. Sulphonated kraft lignin conductive polymers are more dispersible in water and other solvents. Electrochemical analysis revealed that Ligno-PANI is an efficient corrosion inhibitor. A very low loading (1-2%) of the inhibitor brings much (10-20 fold) reduction in corrosion, presumably by the formation of a passive oxide layer (Xu, 2002). Corrosion behavior of Ligno-sulphonate doped PANI coatings on mild steel in neutral saline conditions (salt spray/immersion) was investigated by Sakhri and coworkers by EIS, potentiodynamic measurements [PD] and visual observations. The coatings with highest PANI performed well both in the salt spray and immersion tests (Sakhri et al., 2011).

2.3 Tannic acid [TA]

TA is commercial form of Tannin. It is a polymer of gallic acid molecules and glucose. The pure form of TA is a light yellowish and amorphous powder. It is contained in roots, husks, galls and leaves of plants. It is also found in bark of trees (oak, walnut, pine, mahagony), in tea, nettle, wood, berries and horse chestnuts. TA has astringent, antibacterial, antiviral and antienzymatic properties. TA is used in tanning of leather, staining wood, a mordant for cellulose fibres, dyeing cloth, disinfectant cleansers, pharmaceutical industry, food additives, metal corrosion resistance as rust convertor, slime treatment of petroleum drilling, paper, ink production and oil industry. The structure of TA is shown in Fig. 4.

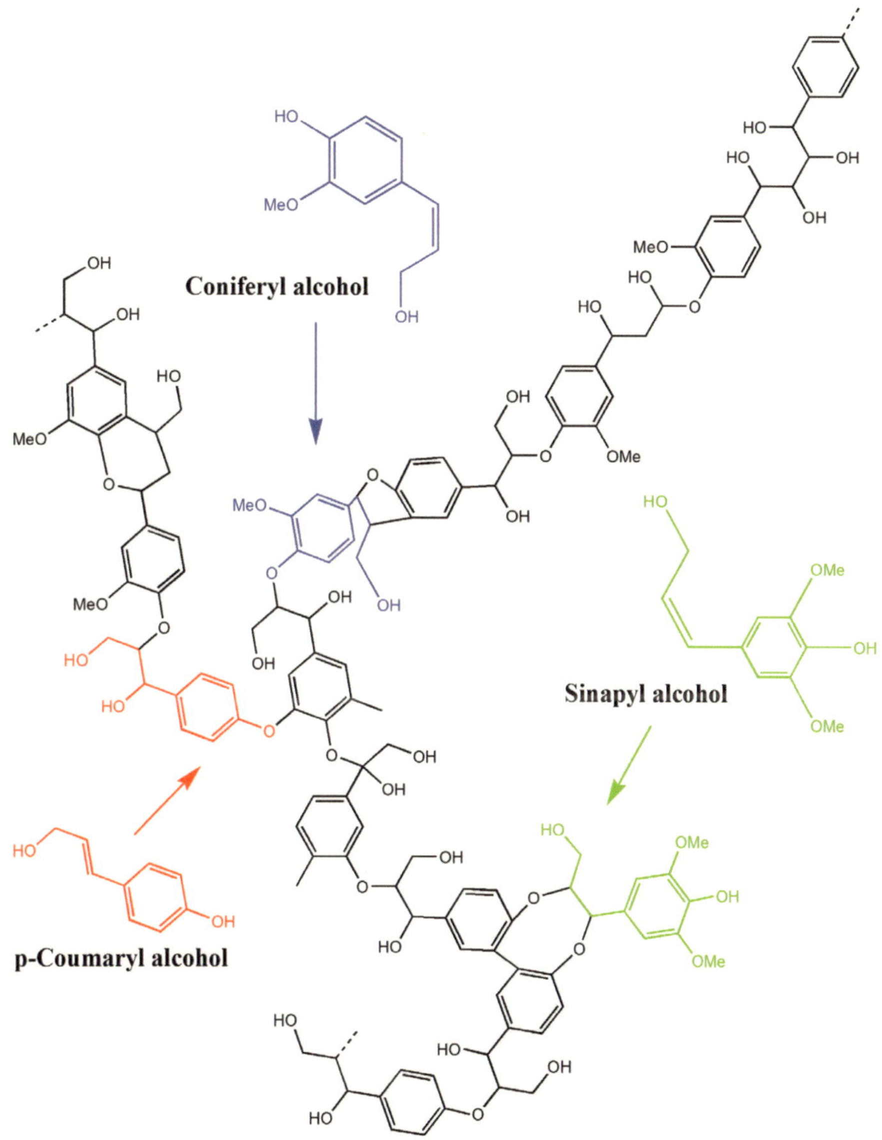

Fig. 3. Structure of lignin.

Use in corrosion resistance

TA has been extensively utilized in anticorrosion methods as investigated by infrared, Mössbauer, Raman spectroscopies, EIS, PD and others (Morcillo et al., 1992; Nasrazadani , 1997; Jaén et al., 2003, 2011; Al-Mayouf, 1999; Ocampo et al., 2004; Galván Jr et al., 1992; Chen et al., 2009). TA is used as conversion coating to prevent corrosion of iron, zinc, copper and their alloys. The (ortho) hydroxyls react with metals forming metal-tannic acid complexes, which protect metal from rusting (Chen et al., 2008). TA based conversion coating can be formed on AZ91D magnesium alloy (Sudagar et al., 2011). Chen et al proposed the formation of organic chromium-free conversion coating on AZ91D magnesium alloy obtained from solution containing TA and ammonium metavanadate. The corrosion resistance performance of these chromate free coatings was compared with the traditional chromate conversion coating. PD revealed that the said coating showed more positive potential and obvious lower corrosion current density relative to traditional chromate conversion coating; salt spray tests also showed the improved anticorrosive behavior of the former (Chen et al., 2008). In another report, mildly rusted steel surface were pretreated with TA based rust converters followed by the application of a Zn rich coating. The rust converters react with iron and rust to form a sparingly soluble iron tannate film on metal surface, which renders low pH adjacent to corroding interface by the diffusion of the unreacted acidic constituents of the rust converter in alkaline concrete solution. The low pH facilitates the formation of passive hydrozincite layer within 50h of exposure to chloride contaminated concrete pore solution relative to 150h for normal zinc coating without rust converter. The mechanism of film formation was investigated by EIS, Potential-time studies, Raman Spectroscopy, SEM, energy dispersive X-ray analysis [EDXA] and X-ray diffraction studies [XRD] (Singh &Yadav, 2008). Methacrylic derivatives of TA [m-digallic acid], toluylene 2,4-diisocyanate [TDI] and 2-hydroxyethyl methacrylate [HEMA] formed UV curable urethane coatings (in molar ratio 1:3:3). The formation occurred by the coupling reaction between TA and TDI followed by HEMA addition (Grassino et al., 1999).

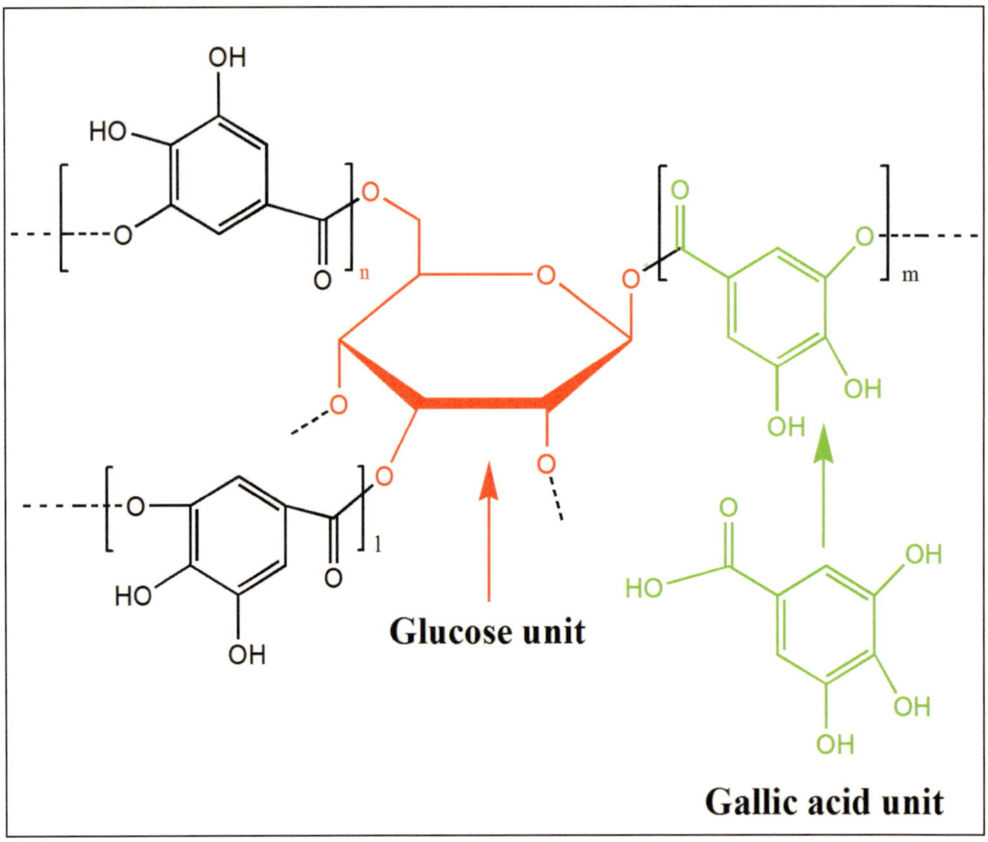

Fig. 4. Structure of tannic acid.

2.4 Chitosan [CHTO]

Chitin and CHTO are polysaccharides. They are chemically similar to cellulose, differing only by the presence or absence of nitrogen. CHTO is deacetylated chitin (degree of deacetylation of chitin ~50%), obtained from the outer shell of crustaceans (crabs, lobsters, krills and shrimps). CHTO primarily consists of β linked 2-amino-2-deoxy-β-D-glucopyranose units. CHTO shows biocompatibility, low toxicity, biodegradability, osteoconductivity and antimicrobial properties (Fig. 5). CHTO is a cationic polyelectrolyte. CHTO forms complexes with metal ions and can gel with polyanions. It contains reactive hydroxyl and amine groups that undergo chemical transformations producing chemical derivatives with plethora of applications. It is used in cosmetics, as preservative, antioxidant, antimicrobial agent and coatings in food, fabrics, drugs, artificial organs and fungicides (Rinaudo, 2006; Bautista-Baños et al., 2006), as metal adsorbants for the removal of metals (mercury, copper, chromium, silver, iron, cadmium) from ground and waste water (Lundvall et al., 2007).

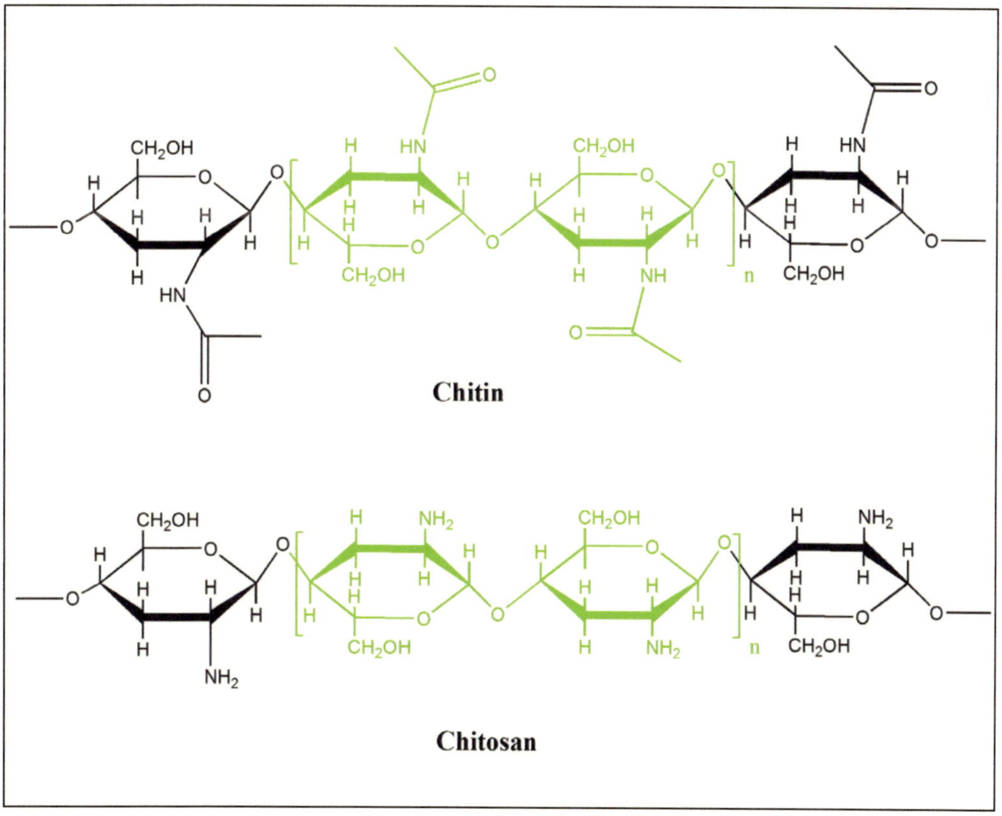

Fig. 5. Structure of chitin and chitosan.

Use in corrosion resistance

CHTO dissolved aqueous solution forms tough and flexible films. CHTO is utilized as anticorrosion material, however, it absorbs moisture from atmosphere, which penetrates the film easily and deteriorates its performance (Lundvall et al., 2007; Sugama & Cook, 2000). As a remedial approach to employ CHTO as an environmentally green water-based coating system for aluminum (Al) substrates, Sugama et al modified CHTO with polyacid electrolyte, poly(itaconic acid) [PI], containing two negatively charged carboxylic acid groups, with CHTO: PI ratio of 100:0, 90:10, 80:20, 70:30, 50:50, 30:70, and 0:100, by weight, applied on 6061-T6 aluminum (Al) sheet by a simple dip-withdrawing method. -COOH and -NH_2 groups of PI and CHTO, respectively, formed (hydrophoebic) secondary amide linkages, which lead to the grafting of PI on CHTO backbone, and at higher temperature crosslinking occurred. Increased "grafts" and "crosslinks" formed coatings that were less susceptible to moisture and prevented the penetration of corrosive electrolyte species, providing good corrosion protection to the substrate. CHTO:PI ratio 80:20 was found to be an ideal composition for efficient corrosion protection (Sugama & Cook, 2000). Sugama et al also modified CHTO with corn-starch derived dextrin and applied on Al-6063. CHTO:dextrin ratio 70/30 provided low moisture resistance and could withstand salt spray test upto 720 h (Sugama & Milian-Jimenez, 1999). CHTO shows high hydrophilicity and poor adhesive strength with Al 2024 T3 alloy. CHTO was modified with epoxy functional silanes [2-(3,4-epoxycyclohexyl)-ethyltrimethoxysilane and (3-Glycidoxypropyl)-trimethoxysilane] as coupling agents and vanadates as corrosion inhibitor (Kumar & Buchheit, 2006). The derivatives of CHTO such as acetylthiourea CHTO, carboxymethyl CHTO are used as efficient corrosion inhibitors as assessed by PD, EIS, SEM, weight loss measurements, conductometric titrations and other studies (Fekry & Mohamed, 2010; Cheng et al., 2007). Hydroxyapatite-CHTO composite coatings on AZ31 Mg alloy by aerosol deposition produce well adherent, corrosion resistant biocompatible coatings (Hahn et al., 2011)

2.5 Starch

As a carbohydrate consisting of a number of glucose units joined together by glycosidic bonds, starch is a low cost, renewable and biodegradable natural polymer. It consists of two types of molecules, amylose (linear) and amylopectin (branched) (Fig. 6). It is the energy store of plants (Sugama & DuVall, 1996). Commercial refined starches are cornstarch, tapioca, wheat and potato starch. Industrial applications include pharmaceutical, papermaking, textile, and in food preparation.

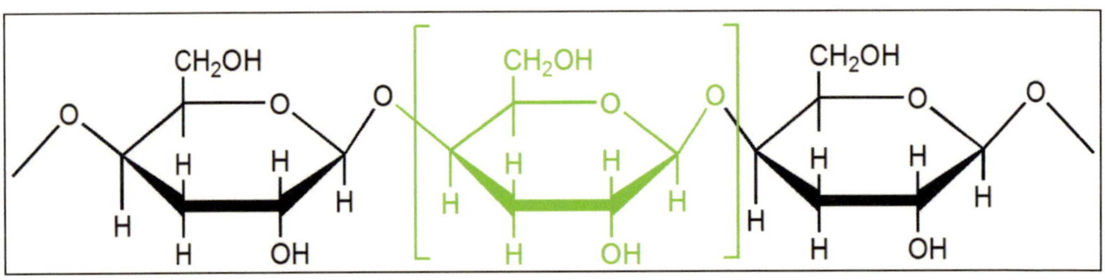

Fig. 6. Structure of starch.

Use in corrosion resistance

Starch is used as a natural corrosion inhibitor. A few reports are available in literature where starch is used to protect metal against corrosion (Sugama & DuVall, 1996). When used at low pH, starch shows low water solubility and poor stability. Thus, for improved performance, certain physical and chemical modifications become necessary. These involve the reactions of their hydroxyl groups with functional groups of the synthetic polymers, such as carboxylic acids, anhydrides, epoxies, urethanes, oxazolines, and others. Another alternative method is via free-radical ring-opening polymerization occurring between their glucose rings and vinyl monomers. Sugama et al carried out the preparation of polyorganosiloxane grafted starch coatings for the protection of aluminium from corrosion (Sugama & DuVall, 1996). The protocol involved the modification of potato starch [PS] with N-[3-(triethoxysilyl)propyl]-4,5,-dihydroimidazole [TSPI]. The constant threat with the use of PS was active bacterial and fungal growth, which caused diminution of its corrosion resistance behavior. TSPI protects the bacterial and fungal growth on PS solution; this was analysed by SEM technique (Sugama & DuVall, 1996). The grafting of organosiloxane occurred by the opening of glycosidic rings. The coating properties were investigated by EIS and salt spray test. PS/TSPI 85/15 and 90/10 ratio-derived coatings displayed good protection of Al against corrosion (salt spray test-288 hours, impedance $>10^5$ Ω cm^2). In another report, Sugama attempted to investigate the effect of cerium (IV) ammonium nitrate modified PS as primer coatings for aluminium substrates (Sugama, 1997).

Bello et al. used modified cassava starch as corrosion inhibitor of carbon steel in an alkaline $200mgL^{-1}$ NaCl solution (chemical composition of tap water) in contact with air at 25^oC. One was cassava starch modified through gelatinization and activation [GAS] and carboxymethylated starch [CMS] with different degrees of substitution [DS]. These were characterized by NMR spectroscopy; estimation of DS was also performed, which was about 0.13±0.03 ($CMS_{0.13}$) and 0.24±0.04 ($CMS_{0.24}$). Electrostatic potential [V(r)] mapping of the repetitive unit of GAS and CMS was based on the model proposed by Politzer and Sjoberg (Bello et al., 2010). Corrosion studies were performed by EIS coupled with a rotating disk electrode with a fixed rotation speed of 1000 rpm. The polarization resistance values followed the order $CMS_{0.13}$ <$CMS_{0.24}$ < GAS. The studies confirmed that starch acts as corrosion inhibitor of carbon steel; the extent of protection against corrosion depended on the amount and type of active groups present [carboxylate (-COO−) and alkoxy (-CO−) groups for CMS, and alkoxy (-CO−) groups for GAS] and also on DS (Bello et al., 2010).

Rosliza and Nik studied the corrosion resistance conferred by tapioca starch [TS] to AA6061 alloy in seawater. The weight loss of AA6061 alloy specimens in seawater diminished with increasing TS concentration as a result of corrosion deposits. PD results revealed that as the concentration of TS increased, corrosion potential [E_{corr}] values shift to more positive value, corrosion current density (i_{corr}) reduced remarkably, the numerical values of both anodic and cathodic Tafel slopes decreased, polarization resistance [R_p] value of AA6061 alloy increased (higher the Rp value, lower the corrosion rate), double layer capacitance value [C_{dl}] decreased, indicating that anodic and cathodic processes are suppressed by TS, that acts as corrosion inhibitor, preferentially reacting with Al^{3+} to form a precipitate of salt or complex on the surface of the aluminum substrate (Rosliza & Nik, 2010). Inhibition efficiency [IE(%)] values obtained from all the measurements viz. gravimetric, PD, linear polarization resistance [LPR] and EIS were in close agreement with each other. IE (%) of TS increased with the corrosion inhibitor concentrations ranging from 200 to 1000 ppm. The protection conferred by TS is attributed to the adsorption on AA6061 alloy surface through all the functional groups present in starch (linear amylose constituted by glucose monomer units joined to one another head to tail forming alpha-1, 4 linkage, and highly branched amylopectin with an alpha-1, 6 linkage every 24–30 glucose monomer units). Other uses of starch include their potential application in edible coatings (Vásconeza et al., 2009; Pagella et al., 2002), coatings for colon-specific drug delivery (Freirea et al., 2009), and in blast cleaning of artificially aged paints (Tangestaniana et al., 2001).

2.6 Plant extracts

Plants naturally synthesize chemical compounds in defence against fungi, insects and herbivorous mammals. Some of these compounds or phytochemicals such as alkaloids, terpenoids, flavonoids, polyphenols and glycosides prove beneficial to humans in unique manner for the treatment of several diseases. These compounds are identical in structure and function to conventional drugs. Extracts from parts of plants such as roots, stems, and leaves also contain such extraordinary phytochemicals that are used as pesticides, antimicrobials, drugs and herbal medicines.

Use in corrosion resistance

Plant extracts are excessively used as corrosion inhibitors. An interesting review in this context is compiled by Raja & Sethuraman, 2008. Plant extracts contain a variety of organic compounds such as alkaloids, flavonoids, tannins, cellulose and polycyclic compounds. The compounds with hetero atoms-N, O, S, P coordinate with (corroding) metal atom or ion consequently forming a protective layer on the metal surface, that prevents corrosion. These serve as cheaper, readily available, renewable and environmentally benign alternatives to costly and hazardous corrosion inhibitors (e.g., chromates). Plant extracts serve as anticorrosion agents to various metals such as mild steel, copper, zinc, tin, nickel, aluminium and its alloys. Literature reveals that there are exhaustive numbers of plant extracts that have shown proven anticorrosion activity as corrosion inhibitors. Examples are *Swertia angustifolia, Accacia conicianna, Embilica officianilis, Terminalia chebula, Terminalia belivia, Sapindus trifolianus, Pongamia glabra, Eucalyptus* leaves, *Annona squamosa, Eugenia jambolans, Azadirachta indica, Accacia Arabica, Vernonia amydalina, Carica papaya, Rosmarinus officinalis, Hisbiscus subdariffa, Opuntia extractd, Mentha pulegium, Occium viridis, Datura* metel, *Ricinus communis, Chelidonium majus, Papaia, Poinciana pulcherrima, Cassia occidentalis and Datura stramonium* seeds, *Papaia, Calotropis procera B, Azydracta indica, Justicia gendarussa, Artemisia pallens, Auforpio turkiale* sap, Black pepper extract, henna extract and several others (Zucchi & Omar, 1985; Dahmani et al., 2010; Ostovaria et al., 2009; Satapathy et al., 2009).

2.7 Vegetable oils [VO]

VO are triglycerides of fatty acids (Fig. 7). They find versatile applications as biofuel, lubricants, adhesives, antimicrobial agents, coatings and paints [Mar et al., 2007; Bruning, 1992. The extensive utilization of VO in several diverse fields is manifested in their rich chemistry-a storehouse of functional groups such as esters, carboxyls, hydroxyls, oxirane, double bonds, active methylenes and others. These functional groups on VO backbone may undergo a host of chemical transformations yielding "green" polymer derivatives, e.g, alkyds, epoxies, polyols, polyurethanes, polyesters, polyesteramides, polyetheramides and others, with versatile applications.

Use in corrosion resistance

VO is the single, largest, well-established, non-polluting, non-toxic, biodegradable family used in coatings and paints, since primeval times particularly in corrosion resistance. Depending on their Iodine value [IV], VO are classified as non-drying, semi-drying and drying, as indicated by their drying index [DI] (DI=linoleic%+(2linolenic%); "drying" VO : IV>130 and DI> 70); "semi-drying" VO: 115<IV<130 and DI 65-75; "non-drying" VO : IV<115; DI< 65). Usually, drying VO are used in coatings and paints. Drying VO are film formers, ie., they have the tendency to form films over the substrate on drying by themselves, without the use of any drier. In drying VO, drying occurs as a natural phenomenon through auto-oxidation initiating from the active methylene groups on VO backbone. However, since these films are not tough enough to meet the desirable performance characteristics, VO are chemically transformed into several derivatives as polyesters, alkyds, polyesteramides, polyetheramides, polyurethanes (Fig. 8) and others, to meet the stringent environmental conditions. These have been further modified through chemical pathways including acrylation, vinylation, metallation, and others, for improvement in their drying, gloss, scratch hardness [SH], impact resistance [IRt] , flexibility [FL], and corrosion resistance of coatings produced therefrom. The presence of hydroxyls, esters, oxiranes, amides, carbonyls, metals, acrylics, carboxyls, urethanes, imparts good adhesion to the substrate due to good electrostatic interactions with the metal substrate.

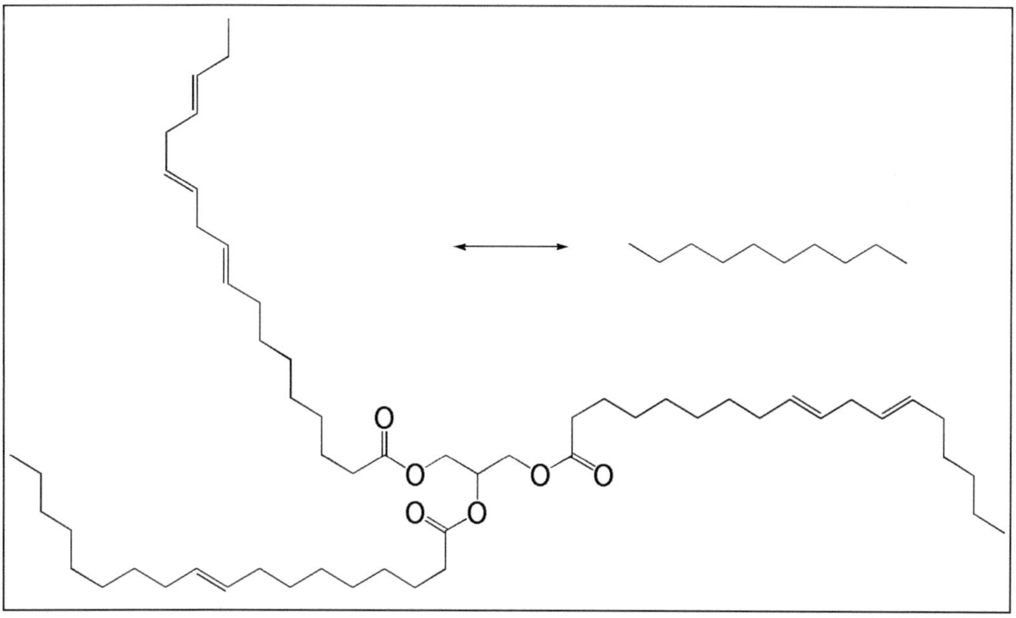

Fig. 7. Chemical structure of VO.

Today, the advancements in knowledge, rise of several innovative technologies, human awareness and concerns related to energy consumption and environmental contamination have brought about manifold changes in the world of VO based coatings and paints. They include VO based low/no solvent coatings, high solids coatings, hyperbranched coatings, WB coatings, UV curable, organic-inorganic hybrids and nanocomposite coatings.

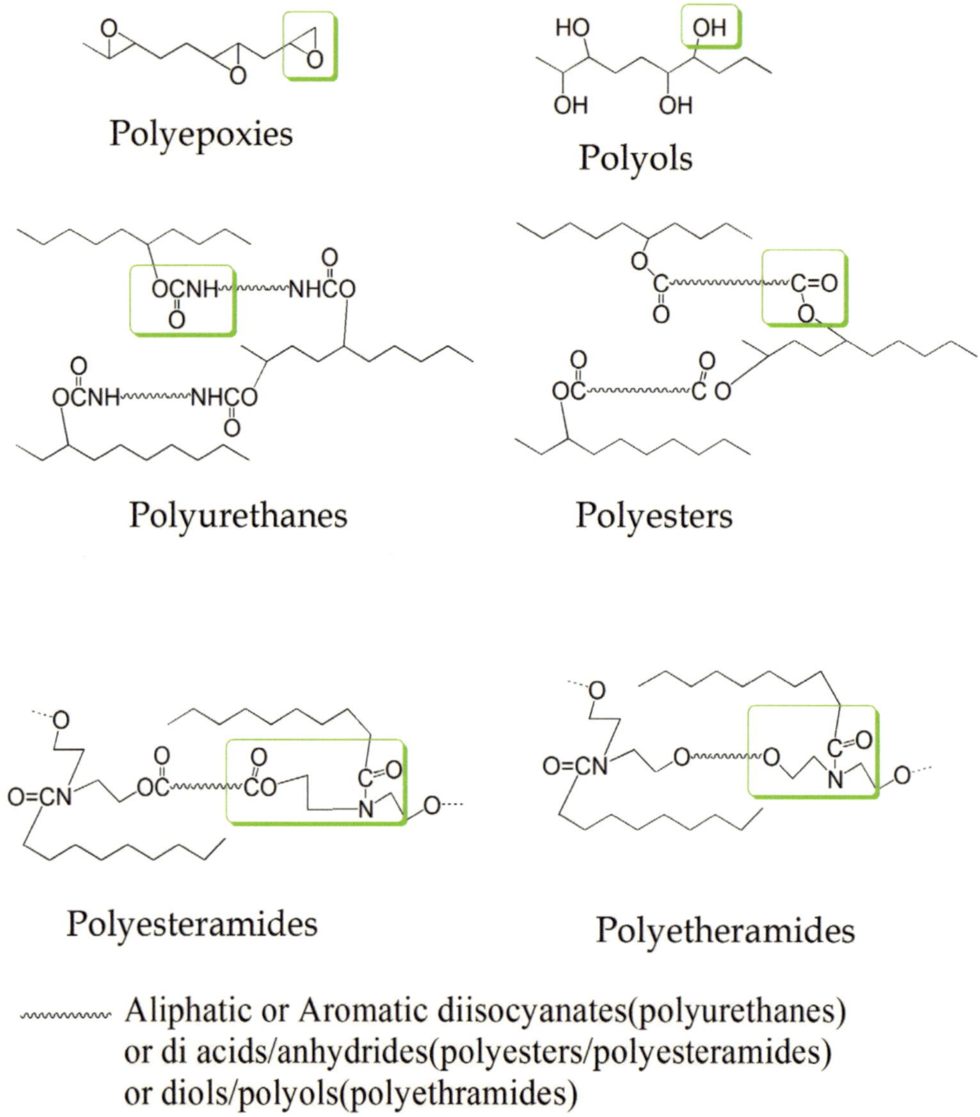

Fig. 8. VO derivatives used in corrosion resistance.

Aigbodion et al prepared WB coatings from rubber seed oil (Aigbodion et al, 2000, 2001, 2003, & 2010). WB polyurethanes with dimer fatty acids showed excellent water and hydrolytic resistance (Liu et al, 2011). Commercially procured acrylated soybean oil modified with acrylated sucrose [ACSU] and hyperbranched acrylates [HYAC], was formulated into UV curable coatings, (Chen, et al., 2011). The addition of HYAC and ACSU improved the adhesion and toughness of coatings, respectively. ACSU acted as reactive flexibilizers in coating formulations. ACSU (in an optimum concentration) modified coatings showed good stability in water, after immersion for seven days, except for slight haziness in smaller portion of the films. Soy alkyd/ PANI conducting coatings showed good SH, IRt, FL and conductivity due to good adhesion between PANI and metal substrate.

While the virgin Soy alkyd coating succumbed to corrosion resistance tests in different corrosive media after 2 h, relatively, Soy alkyd/PANI showed higher performance as monitored for a period of 960 h. The corrosion rate decreased with increased concentration of PANI, being minimum for the highest PANI loading in alkyd. Minimum corrosion rate of 35×10^{-2} mpy in 5% HCl, 32×10^{-2} mpy in 5% NaOH and 30×10^{-2} mpy in 3.5% NaCl was observed for 2.5%-Soyalkyd/PANI (Alam, et al., 2009). Metal containing VO coatings have shown antimicrobial behavior due to the presence of metal, either embedded or incorporated into the matrix. Metal/VO corrosion resistant materials interact with the microbes by adhering to their surface, the long hydrophoebic VO chains engulf the microbes completely cutting off their nutrients, making the cell weak and finally dead.

Mesuea ferrea L. seed oil polyester/clay silver nanocomposite coatings have shown antimicrobial behavior against *Escherichia coli* and *Psuedomonas aeruginosa* (Konwar, et al., 2010). Zafar et al have reported antibacterial activity of Zn containing Linseed polyesteramide coatings (Zafar, et al., 2007, 2007). Sharmin and co-workers recently investigated the coating properties of copper oxide containing poly (ester urethane) metallohybrids from Linseed oil (Sharmin, et al., 2012). Castor polyurethane organo clay composite coatings prepared by Heidarani et al showed good corrosion resistance properties (Heidariani et al., 2010). At 3wt% loading of clay, good corrosion resistance properties could be achieved as determined by PP and EIS. The composite showed i_{corr} (nA/cm^2)=0.139, R_p ($M\Omega cm^2$) polarization resistance= 3819.41, E_{ocp} (mV/Ag|AgCl) (open circuit potential)= −132 after 30 days immersion of samples in 5wt% NaCl. At higher clay loading (>3wt%), the coating material became viscous and the adhesion of the coatings to the substrate deteriorated. The composites prepared through ultrasonication technique did not show any phase separation contrary to their counterparts prepared by mechanical agitation. Zafar et al have for the first time reported the microwave assisted preparation and characterization of Castor oil based zinc containing metallopolyurethane amide coating material. Metallopolyurethaneamide containing 5% metal showed the best performance. The coatings showed good SH (3.5kg), IRt (150lb/inch), FL (1/8in.) and gloss (tested by standard methods and techniques). The coatings were tested by PD in 3.5% HCl, 3.5% NaOH, and 3.5% NaCl solutions. IE% in 3.5% HCl, 3.5% NaOH, and 3.5% NaCl were found as 96.23, 90.81, and 94.50, respectively [Zafar, et al., 2011]. Ahmad et al recently reported the preparation and corrosion resistance performance of Linseed oil based polyurethanefattyamide/ tetraethoxyorthosilane [TEOS-20, 25, 30 phr] based organic-inorganic [PULFAS] prepared at ambient temperature (Ahmad et al., 2012). PD measurements were conducted in HCl (3.5%), NaOH (3.5%), NaCl (5%) and tap water (Cl-ion 63mg/l; conductivity 0.953 mS/A). PULFAS hybrid coatings with 30 phr inorganic content showed the best coating properties, i_{corr} (A/cm^2) 2.65×10^{-8} and IE% 99.77 in 3.5% HCl, i_{corr} (A/cm^2) 1.09×10^{-7}, IE% 99.34 in 3.5% NaOH. Salt spray test of PULFAS coatings was carried out in 3.5% NaCl solution; while the hybrid coatings could withstand the test for 240h, the coatings of virgin polyurethaneamide showed loss in weight and gloss after this time period. Araujo and co-workers investigated the influence of the type of VO on the barrier properties of alkyd paints pigmented with zinc phosphate. They selected Linseed and Soybean oils as modifiers of alkyd paints (Araujo, et al., 2010).

The research work on the use of VO in corrosion resistance is exhaustive. Numerous innovations in the field have occurred in recent years and still more is yet to take place.

2.8 Biofilms

A biofilm consists of a highly organized bacterial community with cells entrapped in an extracellular polymer matrix. Bacteria in biofilms show higher resistance to antibiotics, increased production of exopolysaccharide, morphological changes in cells, different responses to environmental stimuli, and distinct gene expression profile (Zuo, 2007; O'Toole et al., 2000; Videla & Characklis, 1992) (Fig. 9). Biofilm formation on metal surfaces may enhance or hamper corrosion process. The bacterial colonies on metal substrates form anodic (area below thicker colonies, due to more respiration activity and lower oxygen concentration) and cathodic (areas below thinner colonies due to less respiration activity and higher oxygen concentration) areas, resulting in the corrosion of metal surface. The biofilm matrix itself, contrarily, forms a transport barrier, impeding the penetration of corrosive agents (such as oxygen, chloride, and others), decreasing their contact with the metal surface, thus reducing corrosion. Often, the corrosion products themselves form a passive layer that may impede corrosion. The overall process (corrosion or anti-corrosion) depends upon the type of metal and activity of microbes. Some bacteria may become protective or corrosive, depending upon the pH of the medium (Zuo, 2007; O'Toole et al., 2000; Videla & Characklis, 1992; Videla & Herrera, 2005; Lopes et al.; 2006). The mechanism involves the removal of corrodents such as oxygen by aerobic respiration of biofilms, elimination of corrosion causing bacteria by biofilms generated antimicrobials, biofilm secreted corrosion inhibitors form passive layer decreasing contact of metal and corrodents. Such corrosion inhibiting microbes include *Pseudomonas cichorii, Bacillus mycoides, Bacillus licheniformis* and several others. The use of biofilms as anti-corrosion agents requires extensive research to be focussed mainly on interactions between bacteria within the microbial community and interactions between certain bacteria and metal. This requires the collaboration of microbiologists and corrosion chemists for further fruitful results in the field.

2.9 Cashew nut shell liquid (CNSL)

CNSL is obtained as a by-product of the cashew nut industry, mainly containing anacardic acid 80.9%, cardol 10-15%, cardanol, and 2-methyl cardol (Fig. 10). CNSL occurs as a brown viscous fluid in the shell of cashewnut, a plantation product obtained from the cashew tree, *Anacardium oxidentale* (Bhunia, et al., 2000). CNSL is used in the manufacture of industrially important materials such as cement, primers, specialty coatings, paints, varnishes, adhesives, foundry core oils, automotive brake lining industry, laminating and rubber compounding resins, epoxy resins, and in the manufacture of anionic and non-ionic surface active agents. CNSL modified phenolic resins are suitable for many applications and perform improved corrosion and insulation resistance.

Use in corrosion resistance

CNSL has excellent combination of functional groups viz., hydroxyls, double bonds, long aliphatic chain, aromatic ring. It can impart good adhesion to coating material due to its structural attributes. Aggarwal et al prepared epoxy-cardanol resin based paints from epichlorohydrin, bisphenol-A and cardanol (Aggarwal, et al., 2007), in presence of Zn powder, Zn phosphate, micaceous iron oxide and synthetic iron oxide as pigments, some fillers, additives and hardener (aromatic polyamine). The coated panels were subjected to immersion tests in water, 5% NaCl, urea and di-ammonium phosphate for 180 days and humidity cabinet test at 100%RH at 42- 48°C. The coatings showed good SH, adhesion, FL; coatings with micaceous iron oxide showed minimum blistering in immersion and humidity cabinet tests (Aggarwal, et al., 2007). CNSL is also used as a modifier for phenol-formaldehyde [PF] resin. CNSL-PF modified natural rubber has shown improved physico-mechanical performance compared to pure CNSL (Menon, et al., 2002).

Fig. 9. Corrosion resistance by the formation of biofilm.

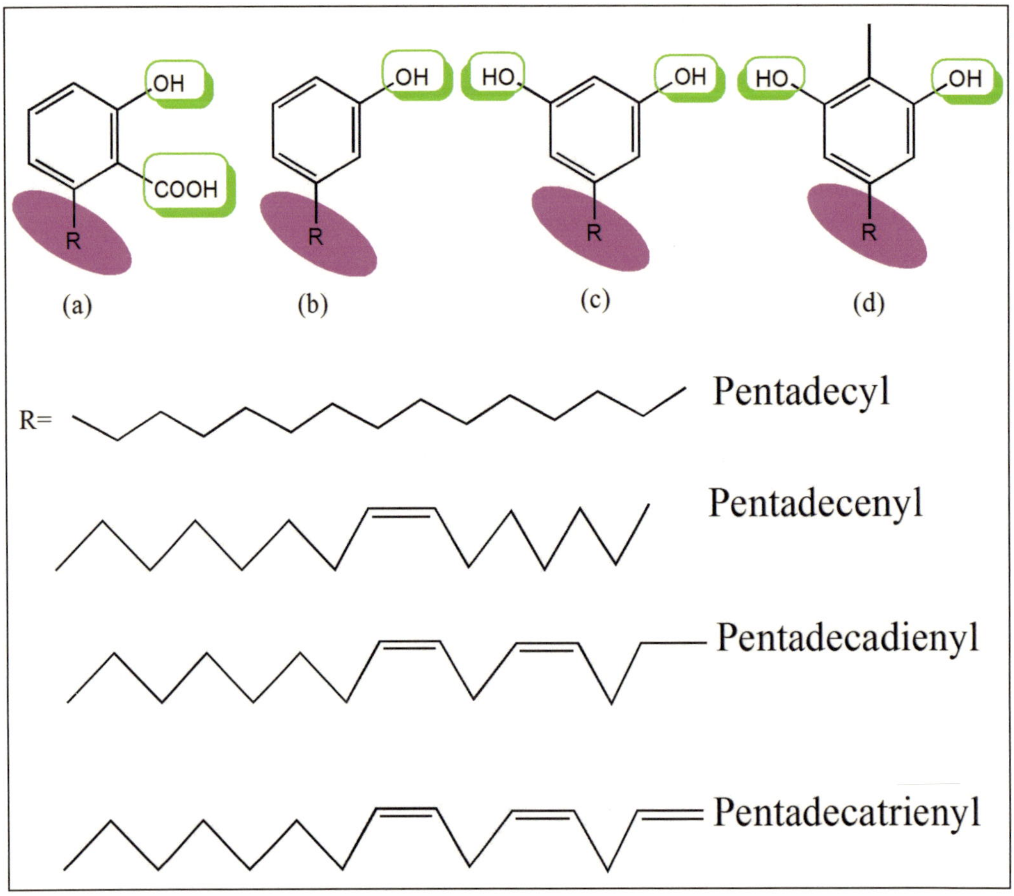

Fig. 10. Chemical structure of the constituents of CNSL, (a) anacardic acid, (b) cardanol, (c) cardol and (d), 2-methyl cardol.

2.10 Others

Other examples include furan, polycaprolactone, glycerol, gums, proteins, pectin, drugs and others, which are also used in corrosion resistance (Hussain et al., 2002; Fabbri et al., 2006; Velayuthama et al., 2009; Umoren, 2008; Umoren et al., 2009; Zuo et al., 2005; Sugama, 1995; Abdallah, 2004; Obot et al., 2009). The role of antibacterial and antifungal drugs like Clotrimazole, Fluconazole, Cefixime, Ampicillin, Ampiclox, Cloxacillin, Tetracycline, Methocarbamol, Orphenadrine, Penicillin G, Azithromycin, and others, in corrosion resistance is basically as corrosion inhibitors. The inhibition mechanism is based mainly on adsorption, significantly influenced by the presence of functional groups -CHO, -N=N, R-OH, steric factors, aromaticity, electron density, molecular weight of inhibitor and others (Abdullah, 2004; Obot, et al., 2009; Naqvi et al., 2011; Eddy et al., 2010). Drugs may often compete with "green" corrosion inhibitors. Similarly, gums (Raphia hookeri gum, gum Arabic) also exert their anticorrosion effect as corrosion inhibitors through the formation of films on metal surface via adsorption and thus, blocking off the corrodents present in the environment (Umoren, 2008).

In an excellent review by Shchukin and Mçhwald(Shchukin & Mçhwald, 2007), they have discussed about the nanoreservoirs containing active materials (corrosion inhibitors) for self-repairing coatings and surfaces. Such an approach can be employed on renewable resources in corrosion. In another review by Nimbalkar and Athawale, they have elaborated the use of VO in WB coatings (Athawale & Nimbalkar, 2011). In another excellent report, use of plant extracts as natural corrosion inhibitors has been briefly described (Raja & Sethuraman, 2008). The target of corrosion engineers and chemists, beyond the boundaries, is to achieve and come to a cost effective, environment friendly, user-friendly and long term solution to corrosion-the metallic cancer. For the present, persistent ongoing research efforts in the direction have shown proven results. The substitution of renewable resources based binders and corrosion inhibitors to conventionally used chemicals will pave way for a fruitful utilisation of our naturally available bioresources. With innovative technologies in hand, green chemistry, nanotechnology and green anticorrosion methods and materials as our tools, we can be fully equipped to combat corrosion and related problems, in near future.

3. Conclusion

Renewable resource based derivatives are cost-effective, abundantly available, biodegradable, environmentally benign alternatives for corrosion resistant coatings, paints and inhibitors. With advancements in knowledge and updated instruments and techniques available, further research in the field may be focussed on the enhanced use of the lesser and highly explored biomaterials for the development of anticorrosion agents in hand with "green" coating technology, for high performance high solids, hyperbranched, waterborne, hybrid and composite coatings that may compete with their petro-based counterparts, both in the terms of cost and performance, in near future. Though we have come a long way, much remains to be done on our palette; we still have a long way to go and explore.

4. Acknowledgements

Dr.Eram Sharmin (Pool Officer) and Dr Fahmina Zafar (Pool Officer) acknowledge CSIR, New Delhi, India for Senior Research Associateships against grant nos. 13(8464-A)/2011-POOL and 13(8385-A)/2010-POOL, respectively. They are also thankful to the Head, Department of Chemistry, Jamia Millia Islamia (A Central University), for providing support to carry out the work.

5. References

Abdallah, M. (2004). Antibacterial Drugs as Corrosion Inhibitors for Corrosion of Aluminium in Hydrochloric Solution. *Corrosion Science,* Vol. 46, No. 8, (August 2004), pp. 1981-1996, ISSN 0010-938X

Aggarwal, L.K.; Thapliyal, P.C. & Karade, S.R. (2007). Anticorrosive properties of the epoxy-cardanol resin based paints. *Progress in Organic Coatings,* Vol. 59, No. 1, (April 2007), pp. 76-80, ISSN 03009440

Ahmad, S.; Zafar, F.; Sharmin, E.; Garg, N. & Kashif, M. (2012). Synthesis and Characterization of Corrosion Protective Polyurethanefattyamide/ Silica Hybrid Coating Material. *Progress in Organic Coatings,* Vol. 73, No. 1, (January 2012), pp.112- 117,ISSN 03009440

Ahmad, S. (2007). Polymer Science, Coatings and Adhesives, Safety Aspects of Coatings in Coatings and Adhesives, National Science Digital Library, NISCAIR, India

Aigbodion A I, Pillai CKS. (2000) Preparation analysis and applications of rubber seed oil and its derivatives in surface coatings. *Progress in Organic Coating,* Vol. 38, No. 3-4, (June 2000), pp. 187-92, ISSN 03009440

Aigbodion AI, Pillai CKS. (2001). Synthesis and molecular weight characterization of rubber seed oil- modified alkyd resins. *Journal of Applied Polymer Science,* Vol. 79, No. 10, (December 2001), pp. 2431–2438, ISSN: 1097-4628

Aigbodion, A.I.; Okiemien, F.E.; Obaza, E.O. & Bakare, I.O. (2003). Utilization of rubber seed oil and its alkyd resin as binders in water borne coatings. *Progress in Organic Coating,* Vol. 46, No.1 (January 2003), pp. 28–31, ISSN 03009440

Aigbodion, A.I.; Okiemien, F.E.; Ikhuoria, E.U.; Bakare, I.O. & Obazaa, E.O. (2003). Rubberseed oil modified with maleic anhydride and fumaric acid and their alkyd resins as binders in water reducible coatings. *Journal of Applied Polymer Science,* Vol. 89, pp. 3256–3259, ISSN 1097-4628

Alam, J.; Riaz, U. & Ahmad, S. (2009). High performance corrosion resistant polyaniline/alkyd ecofriendly coatings, *Current Applied Physics,* Vol. 9, No.1, (January 2009), pp. 80-86, ISSN 1567-1739

Al-Mayouf, A.M. (1999). Inhibitors for Chemical Cleaning of Iron with Tannic Acid. *Desalination,* Vol. 121, No. 2, (12 March 1999), pp. 173-182, ISSN 00119164

Altwaiq, A.; Khouri, S. J.; Al-luaibi, S.; Lehmann, R.; Drücker, H. & Vogt, C. (2011). The Role of Extracted Alkali Lignin as Corrosion Inhibitor. *Journal of Materials and Environmental Science,* Vol. 2, No.3, pp. 259-270, ISSN 2028-2508

Araujo,W.S.; Margarit, I.C.P.; Mattos, O.R.; Fragata, F.L. & de Lima-Neto, P. (2010). Corrosion aspects of alkyd paints modified with linseed and soy oils. *Electrochimical Acta,* Vol. 55 pp. 6204–6211, ISSN 0013-4686

Athawale, V.D. & Nimbalkar, R.V. (2011). Waterborne Coatings Based on Renewable Oil Resources: an Overview. *Journal of American Oil Chemists' Society,* Vol. 88, No. 2, (February 2011), pp. 159-185, ISSN 1558-9331

Bautista-Banos, S.; Hernandez-Lauzardo, A.N.; Velazquez-del Valle, M.G.; Hernandez-Lopez, M.; Ait Barka, E.; Bosquez-Molina, E.; Wilson. C.L. (2006). Chitosan as a Potential Natural Compound to Control Pre and Postharvest Diseases of Horticultural Commodities. *Crop Protection,* Vol. 25, No. 2, (February 2006), pp. 108–118, ISSN 0261-2194

Belgacem,M. N. & Gandini, A. (2008). Materials from Vegetable Oils: Major Sources, Properties and Applications, In Monomers, Polymers and Composites from Renewable Resources, Chap.3, Elsevier, Amsterdam, pp.39-66

Belgacem, M. N.; Blayo, A. & Gandini, A. (2003). Organosolv Lignin as a Filler in Inks, Varnishes and Paints. *Industrial Crops and Products,* Vol. 18, No.2, (September 2003), pp. 145-153, ISSN 09266690

Bello, M.; Ochoa, N.; Balsamo, V.; López-Carrasquero, F.; Coll, S.; Monsalved, A. & Gonzálezd, G. (2010). Modified Cassava Starches as Corrosion Inhibitors of Carbon Steel: An Electrochemical and Morphological Approach. *Carbohydrate Polymers,* Vol. 82, No. 3, (October 2010), pp. 561–568, ISSN 01448617

Bhunia, H.P.; Basakb, A.; Chakia, T.K. & Nandoa, G.B. (2000). Synthesis and characterization of polymers from cashewnut shell liquid: a renewable resource V. Synthesis of copolyester. *European Polymer Journal,* Vol. 36, No. 6, (June 2000), pp. 1157-1165, ISSN 00143057

Bierwagen, G. P. (1996). Reflections on Corrosion Control by Organic Coatings. *Progress in Organic Coatings,* Vol. 28, No.1, (May 1996), pp. 43-48, ISSN 0300-9440

Bruning, H.H. (1992). Utilization of Vegetable Oils in Coatings. *Industrial Crops and Products,* Vol. 1, No. 2-4, (December 1992), pp. 89-99, ISSN 0926-6690

Chen, X.; Li, G.; Lian, J. & Jiang, Q. (2008). Study of the Formation and Growth of Tannic Acid based Conversion Coating on AZ91D Magnesium Alloy. *Surface and Coatings Technology,* Vol. 204, No. 5, (December 2009), pp.736-747, ISSN 0257-8972

Chen, X.; Li, G.; Lian, J. & Jiang, Q. (2008). An Organic Chromium-Free Conversion Coating on AZ91D Magnesium Alloy. *Applied Surface Science,* Vol. 255, No. 5, (December 2008), pp. 2322–2328, ISSN 0169-4332

Cheng, S.; Chen, S.; Liu, T. Chang, X. & Yin. Y. (2007). Carboxymethylchitosan + $Cu2^+$ Mixture as an Inhibitor used for Mild Steel in 1M HCl. *Electrochimica Acta,* Vol: 52, No. 19, (May 2007), pp. 5932–5938, ISSN 00134686

Chen, Z.; Wu, Jennifer, F.; Fernando, S. & Jagodzinski, K. (2011). Soy-based, high biorenewable content UV curable coatings. *Progress in Organic Coatings,* Vol. 71, No. 1, (May 2011), pp. 98–109, ISSN 03009440

Cristina Freire, A.; Fertig, C. C.; Podczeck, F.; Veiga, F.; Sous, J. (2009). Starch -based Coatings for Colon-specific Drug Delivery. Part I: The Influence of Heat Treatment on the Physico-chemical Properties of High Amylose Maize Starches. *European Journal of Pharmaceutics and Biopharmaceutics,* Vol. 72, No. 3, (August 2009), pp. 574-586, ISSN 0191-8141

Dahmani, M.; Et-Touhami, A.; Al-Deyab, S.S.; Hammouti, B.; Bouyanzer, A. (2010). Corrosion Inhibition of C38 Steel in 1 M HCl: A Comparative Study of Black Pepper Extract and Its Isolated Piperine. *International Journal of Electrochemical Sciences,* Vol. 5, pp. 1060 - 1069, ISSN 1452-3981

Derksen, J.T.P.; Cuperus, F.P. & Kolster, P.(1996). Renewable Resources in Coatings Technology: a Review. *Progress in Organic Coatings,* Vol. 27, No. 1-4, (January-April 1996), pp. 45 -53, ISSN 0300-9440

Derksen, J.T.P.; Cuperus, F.P. & Kolster, P. (1995). Paints and Coatings from Renewable Resources. *Industrial Crops and Products,* Vol. 3, No. 4, (May 1995), pp. 225-236, ISSN 0926-6690

Eddy,N.O.; Stoyanov, S. R. and Ebenso, E. (2010). Fluoroquinolones as Corrosion Inhibitors for Mild Steel in Acidic Medium; Experimental and Theoretical Studies. *International Journal of Electrochemical Science,* Vol. 5, No 8, (August 2010), pp. 1127 - 1150, ISSN 1452-3981

Fabbri, P.; Singh, B.; Leterrier, Y.; Månson, J.-A.E.; Messori, M. & Pilati, F. (2006). Cohesive and Adhesive Properties of Polycaprolactone/Silica Hybrid Coatings on Poly(methyl methacrylate) Substrates. *Surface and Coatings Technology, Vol.* 200, No. 24, (November 2005), pp. 6706–6712, ISSN 0257-8972

Fekry, A.M.; Mohamed. R. R. (2010). Acetyl Thiourea Chitosan as an Eco-friendly Inhibitor for Mild Steel in Sulphuric Acid Medium. *Electrochimica Acta,* Vol. 55, No. 6, (February 2010), pp. 1933-1939, ISSN: 0013-4686

Galván Jr, J.C.; Simancas, J.; Morcillo, M.; Bastidas, J.M.; Almeida, E. & Feliua. S. (1992). Effect of Treatment with Tannic, Gallic and Phosphoric Acids on the Electrochemical Behaviour of Rusted Steel. *Electrochimica Acta,* Vol. 37, No. 11, (September 1992), pp. 1983-1985, ISSN 00134686

Gandini, A. & Belgacem, M.N. (2002). Recent contributions to the preparation of polymers derived from renewable resources. *Journal of Polymers and Environment,* Vol. 10, No.3, (July 2002), pp. 105–14, ISSN 1572-8900

Ghali, E.; Sashtri, V.S. and Elboujdaini, M. (2007). Corrosion Prevention and Protection, Practical Solutions, Seiten, Hardcover, John Wiley & Sons, ISBN 047002402X / 0-470-02402-X, United Kingdom

Grassino, S. B.; Strumia, M. C.; Couve, J.; Abadie, M.J.M. (1999). Photoactive Films Obtained from Methacrylo-Urethanes Tannic Acid-based with Potential Usage as Coating Materials: Analytic and Kinetic Studies. *Progress in Organic Coatings,* Vol. 37, No. 1-2, (November 1999), pp. 39-48, ISSN 0300-9440

Hahn, B.-D.; Park, D.-S.; Choi, J.-J.; Ryu, J.; Yoon, W.-H.; Choi, J.-H.; Kim, H.-E. & Kim, S.-G. (2011). Aerosol Deposition of Hydroxyapatite–chitosan Composite Coatings on Biodegradable Magnesium Alloy. *Surface and Coatings Technology, Vol.* 205, No. 8-9, (November 30), 20113112–3118, ISSN 0257-8972

Heidariani, M.; Shishesaz, M.R.; Kassiriha, S.M.; Nematollahi, M. (2010). Characterization of Structure and Corrosion Resistivity of Polyurethane/Organoclay Nanocomposite Coatings Prepared through an Ultrasonication Assisted Process. *Progress in Organic Coatings,* Vol. 68, No. 3, (July 2010), pp. 180-188, ISSN 0300 - 9440

Hintze-Brüning, H. (1992). Utilization of vegetable oils in coatings, *Industrial Crops and Products,* Vol. 1, No. 2-4, (December 1992), pp. 89-99, ISSN 09266690

Hussain, S.; Fawcett, A.H.; Taylor, P. (2002). Use of Polymers from Biomass in Paints. *Progress in Organic Coatings,* Vol. 45, No. 4, (December, 2002), pp. 435–439, ISSN 0300-9440.

Jaén, J. A.; Araúz, E. Y.; Iglesias, J. & Delgado, Y. (2003). Reactivity of Tannic Acid with Common Corrosion Products and Its Influence on the Hydrolysis of Iron in Alkaline Solutions. *Hyperfine Interactions,* Vol. 148-149, No. 1-4, pp. 199-209, ISSN 0304-3843

Jaén, J. A.; De Obaldía, J. & Rodríguez. M. V. (2011). Application of Mössbauer Spectroscopy to the Study of Tannins Inhibition of Iron and Steel Corrosion. *Hyperfine Interactions,* (August 2011), DOI: 10.1007/s10751-011-0337-1Online First.

Konwar, U.; Karak, N. & Mandal, M. (2010). Vegetable oil based highly branched polyester/clay silver nanocomposites as antimicrobial surface coating materials. *Progress in Organic Coatings,* Vol. 68, No. 4, (August 2010), pp. 265–273, ISSN 03009440

Kroon, G. (1993). Associative Behavior of Hydrophobically Modified Hydroxyethyl Celluloses (HMHECs) in Waterborne Coatings. *Progress in Organic Coatings,* Vol. 22, No. 1-4, (May-September 1993), pp. 245-260, ISSN 0300-9440

Kumar,G.; Buchheit, R G. (2005). Development and Characterization of Corrosion Resistant Coatings using Natural Biopolymer Chitosan. *Electrochemical Society Transactions,* Vol. 1, No.9, (October 2005), pp. 101-117, ISSN 1938-6737

Liu, X.; Xu, K.; Liu, H.; Cai, H.; Su, J.; Fu, Z.; Guo, Y. & Chen, M. (2011). Preparation and properties of waterborne polyurethanes with natural dimer fatty acids based polyester polyol as soft segment. Progress *in Organic Coatings,* (July 2011), doi:10.1016/j.porgcoat.2011.07.002 , ISSN 03009440

Liu, J. & Williams III, R. O. (2002). Properties of Heat-Humidity Cured Cellulose Acetate Phthalate Free Films. *European Journal of Pharmaceutical Sciences,*Vol.17, No. 1-2, (October 2002), pp. 31-41, ISSN 0928-0987

Lopes, F.A.; Morin, P.; Oliveira, R. & Melo, L.F. (2006). Interaction of Desulfovibrio Desulfuricans Biofilms with Stainless Steel Surface and its Impact on Bacterial Metabolism. *Journal of Applied Microbiology, Vol.* 101, No.3, pp. 1087–1095, ISSN 1139-6709

Lu, Y.; Zhang, L. & Xiao, P. (2004). Structure, Properties and Biodegradability of Water Resistant Regenerated Cellulose Films Coated with Polyurethane/Benzyl Konjac Glucomannan Semi-IPN Coating. *Polymer Degradation and Stability,* Vol. 86, No.1, (October 2004), pp. 51-57, ISSN 0141-3910

Lundvall, O.; Gulppi, M.; Paez, M.A.; Gonzalez, E.; Zagal, J.H.; Pavez, J. & Thompson, G.E. (2007). Copper Modified Chitosan for Protection of AA-2024. *Surface and Coatings Technology,* Vol. 201, No.12, (March 2007), pp. 5973–5978, ISSN 02578972

Meir, M.A.; Metzger, J.O.; Scubert, U.S. (2007). Plant Oil Renewable Resources as Green Alternatives in Polymer Science. *Chemical Society Review,* Vol. 36, No. 11, pp. 1788–1802, ISSN 0306-0012

Menon, A.R.R.; Aigbodion, A.I.; Pillai, C.K.S.; Mathew, N. M.; Bhagawans, S.S. (2002). Processibility characteristics and phsyco-mechanical properties of natural rubber modified with cashew nut shell liquid and cashew nut shell liquid-formaldehyde resin, *Europeon Polymer Journal,* Vol. 38, No. 1, (January 2002), pp. 163-168, ISSN 0014-3057

Metzger, J.O. (2001). Organic Reactions Without Organic Solvents and Oils and Fats as Renewable Raw Materials for the Chemical Industry. *Chemosphere,* Vol. 43, No. 1, (April 2001), pp. 83-87, ISSN 0045-6535

Morcillo, M.; Feliu, S.; Simancas, J.; Bastidas, J. M.; Galvan, J. C.; Feliu Jr, S.; & Almeida, E. M. (1992). Corrosion of Rusted Steel in Aqueous Solutions of Tannic Acid. *Corrosion,* Vol.48, No.1032, pp. 1-8, ISSN 0010-9312

Mulder, W.J.; Gosselink, R.J.A.; Vingerhoeds, M.H.; Harmsen, P.F.H. & Eastham, D. (2011). Lignin based Controlled Release Coatings. *Industrial Crops and Products,* Vol. 34, No.1, (July 2011), pp. 915-920, ISSN 0926-6690

Naqvi, I.; Saleemi, A. R. & Naveed, S. (2011). Cefixime: A drug as Efficient Corrosion Inhibitor for Mild Steelin Acidic Media. Electrochemical and Thermodynamic Studies. *International Journal of Electrochemical Science,* Vol. 6, No.1, (January 2011), pp. 146 - 161, ISSN 1452-3981

Nasrazadani,S.(1997). The Application of Infrared Spectroscopy to a Study of Phosphoric and Tannic Acids Interactions with Magnetite (Fe_3O_4), Goethite (α-FEOOH) and Lepidocrocite (γ-FeOOH). *Corrosion Science,* Vol. 39, No. 10-11, (October-November 1997), pp. 1845-1859, ISSN 0010-938X

Obot, I.B.; Obi-Egbedi, N.O. & Umoren, S.A. (2009). Antifungal Drugs as Corrosion Inhibitors for Aluminium in 0.1 M HCl. *Corrosion Science,* Vol. 51, No. 8, (August 2009), pp. 1868-1875, ISSN 0010-938X

Ocampo, L.M.; Margarit, I.C.P.; Mattos, O.R.; Cordoba-de-Torresi, S.I. & Fragata, F.L. (2004). Performance of Rust Converter based in Phosphoric and Tannic Acids. *Corrosion Science,* Vol. 46, No. 6, (June 2004), pp.1515-1525, ISSN 0010938x

Ostovari, A.; Hoseinieh, S.M.; Peikari, M.; Shadizadeh, S.R.; Hashemi, S.J. (2009). Corrosion Inhibition of Mild Steel in 1 M HCl Solution by Henna Extract: A Comparative Study of the Inhibition by Henna and its Constituents (Lawsone, Gallic acid, α-D-Glucose and Tannic acid). *Corrosion Science,* Vol. 51, No. 9, (September 2009), pp. 1935-1949, ISSN 0010-938X

O'Toole, G.; Kaplan, H.B. & Kolter, R. (2000). Biofilm Formation as Microbial Development. *Annual Review of Microbiology,* Vol. 54, No. 1, (October 2000), pp. 49–79, ISSN 0066-4227

Pagella, C.; Spigno, G.; De Faveri, D.M.. (2002). Characterization of Starch based Edible Coatings. *Food and Bioproducts Processing,* Vol. 80, No. 3, (September 2002), pp. 193-198, ISSN 0957-0233

Park, Y.; Doherty, W.O.S. & Halley, P. J. (2008). Developing Lignin -based Resin Coatings and Composites. *Industrial Crops and Products,* Vol. 27, No. 2, (March 2008), pp. 163-167, ISSN 0926-6690

Raja, P.B. & Sethuraman M. G. (2008). Natural Products as Corrosion Inhibitor for Metals in Corrosive Media — A review. *Materials Letters,* Vol. 62, No. 1, (month year) pp.113–116, ISSN 0167-577X

Rinaudo, M. (2006). Chitin and Chitosan: Properties and Applications. *Progress in Polymer Science,* Vol. 31, No. 7, (January 2006), pp. 603–632, ISSN 2153-1188

Rosliza,R.; Wan Nik, W.B. (2010). Improvement of Corrosion Resistance of AA6061 Alloy by Tapioca Starch in Seawater. *Current Applied Physics,* Vol. 10, No.1, (January 2010), pp. 221–229, ISSN 15671739

Sakhri, A.; Perrin, F.X.; Benaboura, A.; Aragon, E. & Lamouric, S. (2011). Corrosion Protection of Steel by Sulfo-Doped Polyaniline-Pigmented Coating. *Progress in Organic Coatings,* Vol.72, No. 3, (November 2011), pp.473-479, ISSN 03009440

Satapathy, A.K.; Gunasekaran, G.; Sahoo, S.C.; Amit, K. & Rodrigues, P.V. (2009). Corrosion Inhibition by Justicia Gendarussa Plant Extract in Hydrochloric Acid Solution. *Corrosion Science,* Vol. 51, No. 12, (December 2009), pp. 2848-2856, ISSN 0010-938X

Sena-Martins, G.; Almeida-Vara, E. & Duarte, J.C. (2008). Eco-friendly New Products from Enzymatically Modified Industrial Lignins. *Industrial Crops and Products,* Vol. 27, No. 2, (March 2008), pp. 189-195, ISSN 09266690

Sharmin, E.; Akram, D.; Zafar, F.; Ashraf, S.M. & Ahmad, S. (2012). Plant oil polyol based poly (ester urethane) metallohybrid coatings. *Progress in Organic Coatings,* Vol. 73, No. 1, (January 2012), pp.118– 122, ISSN 03009440

Shchukin, D. G. & Mçhwald, H. (2007). Self-Repairing Coatings Containing Active Nanoreservoirs. *Small,*Vol. 3, No. 6, pp. 926 – 943, ISSN 1613-6829

Singh, D.D.N. & Yadav, S. (2008). Role of Tannic Acid based Rust Converter on Formation of Passive Film on Zinc Rich Coating Exposed in Simulated Concrete Pore Solution. *Surface & Coatings Technology, Vol.* 202, No. , (January 2008), pp.1526–1542, ISSN 0257-8972

Siqueira, G.; Bras, J. & Dufresne, A. (2010). Cellulosic Bionanocomposites: A Review of Preparation. Properties and Applications. *Polymers*, Vol 2, (December 2010), 728-765; ISSN 2073-4360

Sorensen, P.A.; Kiil, S.; Johansen, K.D. & Weinell, C.E. (2009). Anticorrosive Coatings: a Review. *Journal of Coatings Technology Research*, Vol. 6, No.2, (June 2009), pp. 135-176, ISSN 15470091

Stewart, D. (2008). Lignin as a Base Material for Materials Applications: Chemistry, Application and Economics. *Industrial Crops and Products*, Vol. 27, No. 2, (March 2008), pp. 202-207, ISSN 09266690

Sudagar, J.; Jian-she, L.; Xiao-min, Chen.; Peng, L. & Ya-qin, L. (2011). High Corrosion Resistance of Electroless Ni-P with Chromium-Free Conversion Pre-Treatments on AZ91D Magnesium Alloy. *Transaction of Nonferrous Metals Society of China, Vol.* 21, No. 4, (April 2011), pp. 921-928, ISSN 1003-6326

Sugama, T. & Cook, M. (2000). Poly (itaconic acid)-Modified Chitosan Coatings for Mitigating Corrosion of Aluminum Substratesq. *Progress in Organic Coatings*, Vol. 38, No. 2, (May 2000), pp. 79–87, ISSN 0300-9440

Sugama, T. & Milian-Jimenez, S. (1999). Dextrine-modified Chitosan Marine Polymer Coatings. *Journal of Materials Science*, Vol. 34, No. 9,(May 1999), pp. 2003- 2014, ISSN 0022-2461

Sugama, T., & DuVall, J. E. (1996). Polyorganosiloxane-grafted Potato Starch Coatings for Protecting Aluminum from Corrosion. *Thin Solid Films*, Vol. 289, No. 1-2, (November 1996), 39–48, ISSN 0040-6090

Sugama, T. (1997). Oxidized Potato-starch Films as Primer Coatings of Aluminium. *Journal of Materials Science*, Vol. 32, No. 15, (August 1997), pp. 3995- 4003, ISSN 0022-2461

Sugama, T. (1995). Pectin Copolymers with Organosiloxane Grafts as Corrosion-protective Coatings for Aluminium. *Materials Letters*, Vol. 25, No. 5-6, (December 1995), Pages 291-299, ISSN 0167-577X

Tamborim, S.M.; Dias, S.L.P.; Silva, S.N.; Dick, L.F.P. & Azambuja, D.S. (2011). Preparation and Electrochemical Characterization of Amoxicillin-Doped Cellulose Acetate Films for AA2024-T3 Aluminum Alloy Coatings. *Corrosion Science*, Vol. 53, No. 4, (April 2011), pp.1571-1580, ISSN 0010-938X

Tangestanian, P.; Papini, M. & Spelta, J.K. (2001). Starch media blast cleaning of artificially aged paint films. *Wear*, Vol. 248, No. 1-2, (March 2001), PP. 128-139, ISSN 0043-1648

Tarvainena, M.; Sutinen, R.; Peltonen, S.; Mikkonen, H.; Maunusa, J.; Vähä-Heikkiläd, K.; Lehtod, V.-P. & Paronena, P. (2003). Enhanced Film-Forming Properties for Ethyl Cellulose and Starch Acetate using N-Alkenyl Succinic Anhydrides as Novel Plasticizers. *European Journal of Pharmaceutical Sciences*. Vol. 19, No.5, (August 2003), pp. 363-371, ISSN 0928-0987

Umoren, S. A. (2008). Inhibition of Aluminium and Mild Steel Corrosion in Acidic Medium Using Gum Arabic. *Cellulose*, Vol. 15, No. 5, pp. 751–761, ISSN 0969-0239.

Umoren, S. A.; Obot, I. B. & Obi-Egbedi, N. O. (2009). Raphia Hookeri Gum as a Potential Eco-Friendly Inhibitor for Mild Steel in Sulfuric Acid. *Journal of Materials Science*, Vol. 44, No. 1, pp. 274–279, ISSN 0022-2461

Vásconez, M. B.; Flores, S. K.; Campos, C.A.; Alvaradoa, J. & Gerschensonb, L. N. (2009). Antimicrobial Activity and Physical Properties of Chitosan–tapioca Starch based Edible Films and Coatings. *Food Research International*, Vol 42, No. 7, (August 2009), pp 762-769, ISSN 0963-9969

Velayutham, T.S.; Abd Majid, W.H.; Ahmad, A.B.; Kang, G. Y.; Gan, S.N. (2009). Synthesis and Characterization of Polyurethane Coatings Derived from Polyols Synthesized with Glycerol, Phthalic Anhydride and Oleic Acid. *Progress in Organic Coatings, Vol.* 66, No. 4, (December 2009), pp. 367–371, ISSN 03009440

Videla, H.A. & Characklis, W.G. (1992). Biofouling and Microbially Influenced Corrosion. *International Biodeterioration and Biodegradation,* Vol. 29, No.2-3, pp.195–212, ISSN 0964-8305

Videla, H.A. & Herrera, L.K. (2005). Microbiologically Influenced Corrosion: Looking to the Future. *International Microbiology,* Vol. 8, No. 3, pp.169–180, ISSN 1139-6709

Weiss, K.D. (1997). Paint and Coatings: a Mature Industry in Transition. *Progress in Polymer Science,* Vol. 22, No.2, (January 1997), pp. 203-245, ISSN 0079-6700

Witte, F.; Fischer, J.; Nellesen, J.; Crostack, H.-A.; Kaese, V.; Pisch, A.; Beckmann, F. & Windhagen, H. (2006). In vitro and in Vivo Corrosion Measurements of Magnesium Alloys. *Biomaterials,* Vol. 27, No. 7, (March 2006), pp. 1013-1018, ISSN: 0142-9612

Xu, T. (2002). Chemical modification, properties, and usage of lignin, Springer. Biochemistry, 291 pages.

Zafar F.; Mir, M.H.; Kashif, M.; Sharmin, E. & Ahmad, S. (2011). Microwave Assisted Synthesis of Biobased Metallopolyurethaneamide. *Journal of Inorganic and Organometallic Polymers and Materials,* Vol. 21, No.1, (March 2011), pp. 61-68, ISSN 1574-1443

Zafar, F.; Ashraf, S.M.; Ahmad, S. (2007). Cd and Zn-incorporated polyesteramide coating materials from seed oil—A renewable resource. *Progress in Organic Coatings,* Vol. 59, No. 1, (April 2007) pp. 68–75, ISSN 0300-9440

Zafar, F.; Ashraf, S.M.; Ahmad, S. (2007). Studies on zinc-containing linseed oil based polyesteramide. *Reactive and Functional Polymers,* Vol. 67, No. 10, (October 2007), pp. 928–935, ISSN 13815148

Zakzeski, J.; Bruijnincx, P. C. A.; Jongerius, A. L. & Weckhuysen, B. M. (2010). The Catalytic Valorization of Lignin for the Production of Renewable Chemicals. *Chemical Review,* Vol. 110, No. 6, (March 2010), pp. 3552–3599, ISSN 0009-2665

Zucchi, F. & Omar, I. H. (1985). Plant Extracts as Corrosion Inhibitors of Mild Steel in HCl Solutions. *Surface Technology,* Vol. 24, No. 4, (April 1985), pp. 391-399, ISSN 0257-8972

Zuo, R.; Örnek, D. & Wood, T. K. (2005). Aluminum- and Mild Steel-binding Peptides from Phage Display. *Applied Microbiology and Biotechnology,* Vol. 68, No. 4, (September 2005), pp. 505–509, ISSN 0175-7598

Zuo, R. (2007). Biofilms: Strategies for Metal Corrosion Inhibition Employing Microorganisms. *Applied Microbiology and Biotechnology,* Vol. 76, No. 6, (October 2007), pp.1245–1253, ISSN 0175-7598